Discovery of Bioactive Ingredients from Natural Products

Discovery of Bioactive Ingredients from Natural Products

Editors

Simona Fabroni
Krystian Marszałek
Aldo Todaro

MDPI • Basel • Beijing • Wuhan • Barcelona • Belgrade • Manchester • Tokyo • Cluj • Tianjin

Editors

Simona Fabroni
Research Center for Olive, Fruit and Citrus Crops
CREA - Council for Agricultural Research and Economics
Acireale
Italy

Krystian Marszałek
Department of Fruit and Vegetable Product Technology
Prof. Waclaw Dabrowski Institute of Agricultural and Food Biotechnology – State Research Institute (IBPRS- PIB)
Warsaw
Poland

Aldo Todaro
Dipartimento dei Sistemi Agro-Ambientali
University of Palermo
Palermo
Italy

Editorial Office
MDPI
St. Alban-Anlage 66
4052 Basel, Switzerland

This is a reprint of articles from the Special Issue published online in the open access journal *Molecules* (ISSN 1420-3049) (available at: www.mdpi.com/journal/molecules/special_issues/Bioactive_Ingredients).

For citation purposes, cite each article independently as indicated on the article page online and as indicated below:

LastName, A.A.; LastName, B.B.; LastName, C.C. Article Title. *Journal Name* **Year**, *Volume Number*, Page Range.

ISBN 978-3-0365-2106-0 (Hbk)
ISBN 978-3-0365-2105-3 (PDF)

Contents

About the Editors

Simona Fabroni

Simona Fabroni is permanent researcher at CREA—Council for Agricultural Research and Economics. She graduated in Food Science and Technology (2004) at the University of Catania and received her PhD in Food Science and Technology (2009) at the University of Catania (Italy). In 2005, she obtained the Scuola Superiore di Catania (one of the Italian Schools of Excellence) degree in Food Science and Technology. Her areas of interest are focused on: stabilization of food products through new mild technologies aimed at obtaining high-quality productions; physico-chemical and biological aspects of fresh and processed fruits focused on the investigation of the main bioactive constituents of fruit; new isotopic and/or chemical markers for the traceability of organic productions; recovery of potentially high value compounds from fruit processing wastes.

Krystian Marszałek

Krystian Marszałek is an associate professor and head of the Technological Laboratory in the Department of Fruit and Vegetable Product Technology of the Prof. Wacław Dabrowski Institute of Agricultural and Food Biotechnology—State Research Institute. Since 2019, he has been employed in the same position at the Institute of Food Science and Technology at the University of Rzeszow. He graduated in Food Technology and Nutrition (2009) at Warsaw University of Life Sciences and received his PhD in Food Technology and Nutrition (2013) and habilitation (2017) at the same university. His scientific interests cover application of emerging techniques for food processing and preservation such as high pressure processing, high pressure carbon dioxide, high pressure homogenization, microwave heating, ultrasounds etc. He is also focused on green extraction techniques and determination of bioactive compounds in fruit and vegetables as well as determination of bioaccessibility and bioavailability of these compounds in in vitro models.

Aldo Todaro

Aldo Todaro is an associate professor of Food Science and Technology at Università degli Studi di Palermo. He graduated in Food Science and Technology (2001) at the University of Catania and received his PhD in Food Science and Technology (2005) at the same university. His areas of research interest are fruit and vegetable processing; malting and brewing science and technology; design and fabrication of food processing in different factories; chemical and physical food processing engineering; identification and characterization of phenolic compounds and antioxidant capacity in different foods and byproducts; sensory science and rheology; HACCP implementations; food safety engineering and its application in factories. He is a teacher of unit operations in food engineering, food technology, and fruit and vegetable processing.

Preface to "Discovery of Bioactive Ingredients from Natural Products"

The Mediterranean diet is associated worldwide with a lower risk of cancer, heart disease and stroke. This lower risk is universally ascribed to the high intake of vegetables and fruits, thus producing a high ingestion of phytochemicals. At present, there is a great interest in the development of new nutraceuticals of natural origin to be employed within the pharmaceutical field and/or as dietary supplements. For instance, the agri-food industry produces a large amount of wastes which can be conveniently used for the recovery of valuable health-promoting compounds to be used as natural ingredients.

This book is dedicated to original research and review articles that cover the latest findings about new extraction techniques for the recovery of bioactive constituents from fruit and/or vegetables, chemical characterization of phytoextracts with valuable biological activity, in vivo tests and in vitro bioassays to support biological activity of purified compounds, complex extracts and/or combined extracts deriving from different matrices.

Simona Fabroni, Krystian Marszałek, Aldo Todaro
Editors

molecules

MDPI

Article

Evaluation of Antioxidant and Enzyme Inhibition Properties of *Croton hirtus* L'Hér. Extracts Obtained with Different Solvents

Stefano Dall'Acqua [1,*], Kouadio Ibrahime Sinan [2], Stefania Sut [1], Irene Ferrarese [1], Ouattara Katinan Etienne [3], Mohamad Fawzi Mahomoodally [4,*], Devina Lobine [4] and Gokhan Zengin [2]

1 Department of Pharmaceutical and Pharmacological Sciences, University of Padova, Via Marzolo 5, 35131 Padova, Italy; stefania_sut@hotmail.it (S.S.); irene.ferrarese@unipd.it (I.F.)
2 Department of Biology, Science Faculty, Selcuk University, Campus, 42130 Konya, Turkey; sinankouadio@gmail.com (K.I.S.); gokhanzengin@selcuk.edu.tr (G.Z.)
3 Laboratoire de Botanique, UFR Biosciences, Université Félix Houphouët-Boigny, 00225 Abidjan, Côte d'Ivoire; katinan.etienne@gmail.com
4 Department of Health Sciences, Faculty of Medicine and Health Sciences, University of Mauritius, 230 Réduit, Mauritius; devinalobine@gmail.com
* Correspondence: stefano.dallacqua@unipd.it (S.D.); f.mahomoodally@uom.ac.mu (M.F.M.)

Abstract: *Croton hirtus* L'Hér methanol extract was studied by NMR and two different LC-DAD-MS^n using electrospray (ESI) and atmospheric pressure chemical ionization (APCI) sources to obtain a quali-quantitative fingerprint. Forty different phytochemicals were identified, and twenty of them were quantified, whereas the main constituents were dihydro α ionol-*O*-[arabinosil(1-6) glucoside] (133 mg/g), dihydro β ionol-*O*-[arabinosil(1-6) glucoside] (80 mg/g), β-sitosterol (49 mg/g), and isorhamnetin-3-*O*-rutinoside (26 mg/g). *C. hirtus* was extracted with different solvents—namely, water, methanol, dichloromethane, and ethyl acetate—and the extracts were assayed using different in vitro tests. The methanolic extracts presented the highest 1,1-diphenyl-2-picrylhydrazyl (DPPH), 2,2′-azino-bis(3-ethylbenzothiazoline)-6-sulfonic acid (ABTS), and ferric reducing antioxidant power (FRAP) values. All the tested extracts exhibited inhibitory effects on acetylcholinesterase (AChE) and butyrylcholinesterase (BChE), with a higher activity observed for dichloromethane (AChE: 5.03 and BChE: 16.41 mgGALAE/g), while the methanolic extract showed highest impact against tyrosinase (49.83 mgKAE/g). Taken together, these findings suggest *C. hirtus* as a novel source of bioactive phytochemicals with potential for commercial development.

Keywords: *Croton hirtus*; enzyme; flavonoid glycosides; bioactive agents

Citation: Dall'Acqua, S.; Sinan, K.I.; Sut, S.; Ferrarese, I.; Etienne, O.K.; Mahomoodally, M.F.; Lobine, D.; Zengin, G. Evaluation of Antioxidant and Enzyme Inhibition Properties of *Croton hirtus* L'Hér. Extracts Obtained with Different Solvents. *Molecules* **2021**, *26*, 1902. https://doi.org/10.3390/molecules26071902

Academic Editors: Simona Fabroni, Krystian Marszałek and Aldo Todaro

Received: 27 February 2021
Accepted: 25 March 2021
Published: 28 March 2021

1. Introduction

The genus, *Croton* (Euphorbiaceae) comprises of about 1300 species, mostly widespread in tropical and subtropical regions of both hemispheres [1]. Several members of the genus are popular in folk medicine as they are commonly used against a range of illnesses such as inflammatory conditions, pain, diabetes, hypertension, malaria, gastrointestinal disturbances, and ulcers, amongst others [1–3]. *Croton cajucara* Benth., *C. celtidifolius* Baill., *C. eluteria* Bennett, *C. zambesicus* Müll. Arg., and *C. macrostachys* Hochst. ex Rich are some of the *Croton* species used for medicinal purposes. The chemistry of the genus is considerably diverse. Diterpenoids such as phorbol esters, clerodane, neoclerodane, isopimarane, kaurane, and labdane type terpenoids are the principal bioactive components in the genus. Some *Croton* species are also rich in alkaloids such as taspine and julocrotol [1,4].

Croton hirtus L'Hér., also known as *C. glandulosus*, is an erect shrub, and various studies have investigated the chemical composition of its essential oil, leaves, and roots. In particular, Daouda et al. [2] have reported the antimicrobial action of *C. hirtus* essential oil, which was found to contain 93.49% of terpene derivatives, with 15.55% monoterpenes and 77.94% sesquiterpenes. In a study by de Lima, Medeiros, Cunha, da Silva, de Andrade, Neto, Lopes,

Steffen, Araújo, and Reis [3], the presence of spathulenol (26.7%), *E*-caryophyllene (10.0%), bicyclogermacrene (9.5%), α-cadinol (7.7%), and cubenol (7.0%) as main constituents of the essential oils extracted from *C. hirtus* leaves was reported. In 2019, Rosandy, Azman, Khalid, Othaman, Lazim, Choudary, Syah, Latip, Said, and Bakar [4] identified two new diterpenoids from the roots of *C. hirtus*. Preliminary reports suggested the presence of alkaloid, anthraquinone, flavonoid, steroids, and terpenoid in different parts of *C. hirtus* [5,6]. However, as literature shows, information available related to the phytoconstituents of the plant are punctual, and a complete accurate study of C. *hirtus* leaves composition is still missing. Moreover, there is a scarcity of information in literature on pharmacological properties of *C. hirtus*. In a recent paper, Kim et al. [7] reported the anti-inflammatory effects of *C. hirtus* via inhibition of nuclear factor kappa-light-chain-enhancer of activated B cells (NF-κB) signaling pathway.

In this study, detailed phytochemical fingerprinting of *C. hirtus* was obtained using NMR analysis of the methanol extract and LC-DAD-MS^n methods, one using electrospray ionization (ESI) and one using atmospheric pressure chemical ionization (APCI) ion sources.

Once establishing the chemical composition to assess biological activities, plant material was extracted with different solvents—namely, water, methanol, dichloromethane, and ethyl acetate—and the obtained extracts were subjected to different in vitro bioassays. Free radical scavenging, reducing power, metal chelating, and phosphomolybdenum assays were employed for evaluating antioxidant effects of the different extracts, while the enzyme inhibitory activity against targeted enzymes such as cholinesterases, tyrosinase, α-amylase, and α-glucosidase—involved in significant human diseases—were determined. As a general measurement of the content of extracted constituents, the phenolics content of *C. hirtus* were determined by spectrophotometrical assays.

2. Results and Discussion

2.1. Phytochemical Analysis

C. *hirtus* methanol extract was subjected to different NMR experiments as heteronuclear single quantum coherence (HSQC) and heteronuclear multiple bond correlation (HMBC) as well as correlation spectroscopy (COSY). Based on the chemical shifts of ^1H and ^{13}C and the correlations in the 2D spectra, some tentative assignments were obtained and are summarized in Table 1. The presence of simple phenolics such as protocatechuic acid were tentatively established due to the presence of spin system relating to aromatic signals observed in COSY, suggesting the presence of 1,3,4 tri-substituted ring, as well as on the basis of the chemical shift of carbons (obtained from the HSQC) and the long-range correlation with carboxyl function (Table 1; Figure 1). Further signals can be ascribed to hydroxycinnamic acid derivatives due to the olefinic signals at δ 7.59 and 6.45, as well as due to HMBC correlations with aromatic positions. Signals suggesting the presence of flavonoid derivatives can be also observed—namely, the H-6 and H-8 position of flavonols as well as signals ascribable to ring A and B (Table 1; Figure 1). Additionally, different signals support the presence of sugar and glycosidic derivatives. In the 1D and 2D NMR spectrum, some signals are ascribable to megastigmane derivative (Table 1), which is consistent with the literature [8]. In particular, the diagnostic signals are the geminal methyl groups that are partially overlapped with other aliphatic signals in the extract and the sp^2 olefinic protons of the HSQC-DEPT spectrum. Compounds structures were proposed as icariside B and corchionoside C (Figure 2), and an enlargement of the HSQC is reported with some key signals highlighted. Unsaturated proton signals of the α ionol derivative include the one at δ 5.25 (δ_C 126.2) that shows HMBC correlations with carbon signals at δ_C 51.0 (CH in the six membered rings linked to the side chain), at δ_C 36.0 (CH_2), and at δ_C 135.1 (the quaternary carbon signal). From the methyl group at δ 2.05, correlations are observed with the carbon at δ 135.1 and with 126.5. Secondary methyl group at δ 1.14 can be assigned to the methyl group nearby the glycosidic linkage due to the HMBC with carbon at δ 73.1. These data and the signals largely observed in the sugar region supported the MS identification of compound dehydro α ionol-*O*-[arabinosil(1-6) glucoside]. Diagnostic signals for the isomer dehydro β ionol-O-[arabinosil(1-6) glucoside] are the CH sp2 and by the HMBC with the CH at δ 72 and the secondary methyl group.

Table 1. NMR data obtained from 1D and 2D NMR experiments and assignments of constituents identified in *Croton hirtus* extract. COSY, correlation spectroscopy.

Compound No.	Compounds or Class of Compounds Atom Position	δ H	δ c	Correlations in HMBC or COSY
	Protocatechuic acid (HMDB0001856)			
1	2	7.53	115	167.0 (HMBC); 7.48 (COSY)
	5	6.90	115.4	7.48 (COSY)
	6	7.48	124.7	167.0 (HMBC) 7.53; 6.90 (COSY)
	Caffeic acid moieties			
	7	7.59	143.9	167.5; 130.0 (HMBC); 6.45 (COSY)
	8	6.45	115.6	127.4; 7.59 (COSY)
	Aromatic ring protons	7.14–6.80	127.5, 122.3 114.7, 115.6	145.3, 151.0; 127.0, 145; 114 (HMBC)
	Flavonoids			
	H 6-8 of glycosidic flavonols	6.10–6.22	98.5–99.7	165–155 99 101 (HMBC)
	Ring B quercetin	7.40–6-23	115.0, 129, 125, 118	150.0, 145.3, 131.0 (HMBC)
	Ring B kaempferol (HMDB0005801)	7.98	130.5	165.0, 130.5 (HMBC); 6.80 (COSY)
		6.80	115.0	115; 7.98 (COSY)
	Sugar linked to phenolic portions (Flavonol-*O*-glycosides or hydroxycinnamic esters)			
	Anomeric positions	4.85	109.5	
		4.67	99.8	165.0–160 (HMBC with position 7 of flavonol moieties); 3.30–3.40 (COSY)
		4.43-4.50	101.7	133.8 (HMBC with position 3 of flavonol moieties); 3.18 (COSY)
		4.22	104.7	3.18–3.28 (COSY)
		5.42	103	133.5 (HMBC position 3 flavonol)
	Anomeric positions			
	H-1 (rhamnose)	4.60 brs	99.7	3.18 (COSY)
	H-1 (hexose or pentose)	4.50–4.70 d	100.0–101.1	3.16, 3.23 (COSY)
	H-1 (hexose or pentose)	4.50–4.70	100.0–101.1	3.16, 3.23 (COSY)
	H-6 (hexose) free position	3.30–3.50	60.5	
	H-6 (hexose) glycosidic linked	3.30–3.50	64.5	
	CH bearing ester linkage (from sugar residue)	4.95	73.4	165, 104, 71
	CH bearing ester linkage (from sugar residue)	5.08	71.2	165, 89, 63
	Megastigmane (aglycone part)			
	H-2 Icariside B	5.85–5.90	128.7	200, (HMBC)
	CH_2-6 Icariside B	2.58	54.0	200, 72.0 (HMBC)
	CH_3 Icariside B	1.02	24.2	77 (HMBC)
	H-7 Corchionoside	5.75	125.7	
	H-8 Corchionoside	5.86	131.9	
	CH3 Corchioside	1.21	19.6	73.0, 125.7 (HMBC)
	dehydro Ionol derivatives			
	H3	5.30, 5.2	127.5	141, 59, 36 (HMBC)
	CH_3 linked to double bond	2.02	20.5	141, 127
	Geminal methyl groups, secondary methyl group of the butanol chain	0.95–1.04	23.9–24.1	36.0 51.0 72.0
	Fatty acid derivatives			
	terminal methyl groups	0.93	17.5	40.0, 23.0
	CH2	2.02	19.5	24.2, 30.0
	sp2	5.35–5.40	122.5	

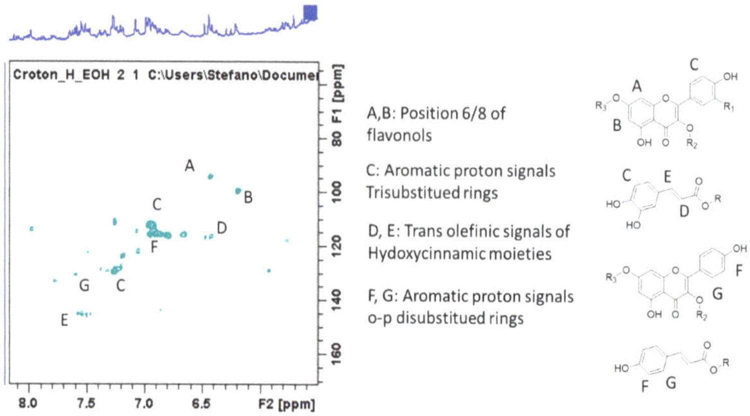

Figure 1. Heteronuclear single quantum coherence- Distortionless enhancement by polarization transfer, (HSCQ-DEPT) portion of the spectrum showing the signals ascribable to phenolic derivatives.

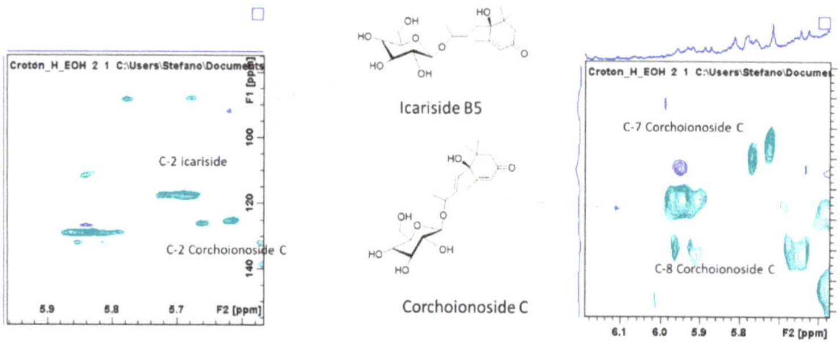

Figure 2. HSQC-DEPT portion of spectrum showing some of the diagnostic signals used to confirm structure identification.

A diffusion order spectroscopy experiment (DOSY) was also performed to observe different behavior of the various extracted phytochemicals present in the mixture. As shown in Figure 3, the spectrum has a group of signals that can be distinguished on the basis of their diffusion coefficients. Ethanol presents its signals largely separated from the other belonging to the plant constituents. The red line indicates the diffusion coefficient of dihydro ionol derivatives, the yellow one shows group of signals that can be ascribed to fatty acid derivatives, while hydroxycinnamic/phenylethanol compounds as well as flavonoid glycosides present larger values of diffusion coefficients.

To have detailed quali-quantitative fingerprint, liquid chromatography coupled with diode array detector and mass spectrometry (LC-DAD-MS^n) methods were developed, and to detect a larger number of the different constituents present in the plant, electrospray (ESI) and atmospheric pressure chemical ionization (APCI) ion sources were used. Phytoconstituents were studied based on MS fragmentation and on the UV spectra (for the phenolic compounds). An exemplificative chromatogram is reported in Figure 4.

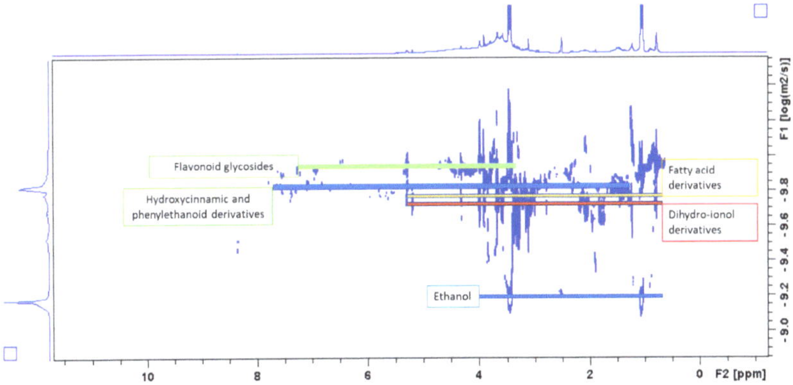

Figure 3. Diffusion order spectroscopy (DOSY) spectrum showing some of the different classes of constituents present in the *C. hirtus* extract.

Table 2. Identification and quantification of compounds in methanolic extract of *Croton hirtus*. * compared with authentic standard.

Compound No.	Retention Time	[M − H]$^-$	Fragments	Formula	Identification and Reference	mg/g
	Flavonoid glycosides					
1	16.5	755	593 300(300→271 255 179 151) (271→243 227)	$C_{33}H_{40}O_{20}$	Quercetin-3-*O*-di(deoxyhexoside)-7-*O*-hexoside	0.61 ± 0.01
2	19.1	755	593 300 (300→271 255 179 151) (271→243 227)	$C_{36}H_{36}O_{18}$	Quercetin-3-deoxyhexoside-hexoside-deoxyhexoside	0.22 ± 0.01
3	18.9	609	301 300 (301→271 255 179 151) (271→243 227)	$C_{27}H_{30}O_{16}$	Quercetin-hexoside-deoxyhexoside	1.59 ± 0.05
4	18.5	739	593 577 447 430 285 257	$C_{33}H_{40}O_{19}$	Kaempferol-7-*O*-hexoside-3-*O*-deoxyhexoside-deoxyhexoside	3.68 ± 0.04
5	19.4	755	609 591 489 300 271 255	$C_{36}H_{36}O_{18}$	Quercetin-3-deoxyhexoside-hexoside-*p*-coumaroyl	3.64 ± 03.04
6	18.5	595	463 300 271 255 (300→271 255) (271→243 227 215 199)	$C_{26}H_{28}O_{16}$	Quercetin-3-apiofuranosyl-glucopyranoside	16.13 ± 0.04
7	19.3	609	301 447 285 255	$C_{27}H_{30}O_{16}$	Rutin*	3.02 ± 0.04
8	20.5	579	447 429 285 (285→255) (255→227 213 211 187)	$C_{25}H_{28}O_{15}$	Kaempferol-3-*O*-hexosyl pentoside	2.52 ± 0.06
9	19.9	463	301 229 179	$C_{21}H_{20}O_{12}$	Quercetin-3-*O*-glucoside*	4.21 ± 0.03
10	23.1	623	315 300 299 271 255 243 (315→300 272 255)	$C_{28}H_{30}O_{16}$	Isorhamnetin-3-*O*-rutinoside*	25.91 ± 0.09
11	21.6	609	315 301 (301→271 255 179 151) (271→243 227)	$C_{27}H_{30}O_{16}$	Isorhamnetin -3-*O*-hexosyl pentoside	4.93 ± 0.03

Table 2. *Cont.*

Compound No.	Retention Time	[M − H]⁻	Fragments	Formula	Identification and Reference	mg/g
	Flavonoid glycosides					
12	22.3	447	301 255	$C_{21}H_{20}O_{11}$	Quercetin-3-*O*-rhamnoside*	1.39 ± 0.04
13	23.5	477	314 285 271	$C_{21}H_{20}O_{11}$	Isorhamnetin-7-*O*-glucoside*	4.46 ± 0.06
14	25.8	447	314 285 271 (314→300 285 271)	$C_{21}H_{20}O_{11}$	Isorhamnetin-7-*O*-rhamnoside*	1.97 ± 0.03
15	21.6	593	447 285	$C_{27}H_{30}O_{15}$	Kaempferol-3-*O*-hexosil-deoxyhexoside	18.00 ± 0.09
16	18.4	329	314 299 271 243 226 199	$C_{17}H_{18}O_7$	Dimethoxy quercetin	4.49 ± 0.02
	Other phenolics					
17	2.3	341	179	$C_{15}H_{18}O_9$	Caffeic acid hexoside	4.36 ± 0.04
18	5.8	315	153	$C_{13}H_{16}O_9$	Protocatechuic acid hexoside	5.66 ± 0.03
19	7.8	401	269 161	$C_{20}H_{18}O_9$	Benzyl alcohol hexose pentose	4.04 ± 0.05
20	11.28	487	337 279 261	$C_{21}H_{28}O_{13}$	Synapoyl pentose-pentose	1.08 ± 0.02
21	17.2	431	261 187 (187→125) (125→97)		Gallic acid benzoic acid derivative	8.84 ± 0.03
22	19.4	769	605 475 315 299	$C_{35}H_{46}O_{19}$	Leonoside A	10.04 ± 0.02
23	20.4	755	623 593 315 297	$C_{34}H_{44}O_{19}$	Forsythoside B	6.31 ± 0.05
24	32.6	797	603 474 456 327 167	$C_{37}H_{50}O_{19}$	Ferruginoside C isomer	7.60 ± 0.05
25	35.5	797	603 474 456 327 167	$C_{37}H_{50}O_{19}$	Ferruginoside C	14.21 ± 0.08
26	33.9	663	517 485 467	$C_{31}H_{36}O_{16}$	Feruloyl-coumaroyl saccharose	15.97 ± 0.08
27	36.4	663	517 485 467	$C_{31}H_{36}O_{16}$	Feruloyl-coumaroyl saccharose	12.21 ± 0.08
28	39.9	663	517 485 467	$C_{31}H_{36}O_{16}$	Feruloyl-coumaroyl saccharose	11.20 ± 0.06
29	41.5	663	517 485 467	$C_{31}H_{36}O_{16}$	Feruloyl-coumaroyl saccharose	13.06 ± 0.09
30	44.2	663	517 485 467	$C_{31}H_{36}O_{16}$	Feruloyl-coumaroyl saccharose	9.06 ± 0.07
	Hydrophylic Terpenoids (positive electrospray (ESI))					
31	13.1	433 [M + HCOOH − H]⁻	387.5 223 205 161 153 (153→138–122)	$C_{19}H_{32}O_8$ + CH_2O_2	Icariside B5	9.15 ± 0.06
32	13.9	431 [M + HCOOH − H]⁻	385.5 223 205 161 153 (153→138–122)	$C_{19}H_{30}O_8$ + CH_2O_2	Corchoionoside C/Roseoside	3.63 ± 0.06
33	22.97	487	355 337 289 279 261 167	$C_{24}H_{40}O_{10}$	dihydro α ionol-*O*-[arabinosil(1-6) glucoside]	132.72 ± 0.11
34	24.07	487	355 337 289 271	$C_{24}H_{40}O_{10}$	dihydro β ionol-*O*-[arabinosil(1-6) glucoside]	79.57 ± 0.11
35	53.4	331	295 277 215 185	$C_{20}H_{28}O_4$	Kongensin D	0.72 ± 0.02

Table 2. *Cont.*

Compound No.	Retention Time	[M − H]⁻	Fragments	Formula	Identification and Reference	mg/g
	Phytosterols and terpenoids		Positive atmospheric pressure chemical ionization (APCI)			
		$[M - H_20 + H]^+$				
36	60.0	397		$C_{29}H_{50}O$	β-sitosterol*	48.60 ± 0.14
37	53.8	383		$C_{28}H_{48}O$	Campesterol*	3.04 ± 0.08
38	51.3	399		$C_{29}H_{52}O$	Stigmastanol*	4.37 ± 0.08

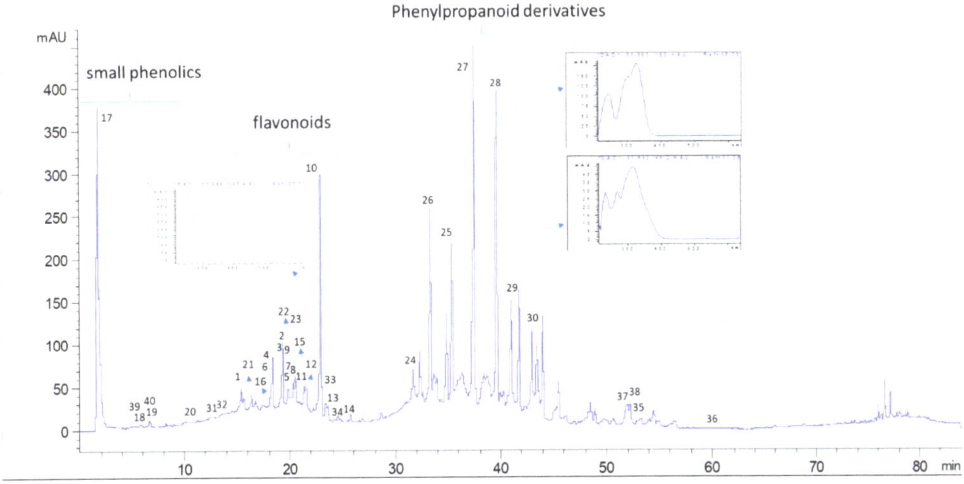

Figure 4. LC-DAD chromatogram (330 nm) showing the peak group assigned to small phenolics, flavonoids, and phenylpropanoid derivatives; exemplificative UV spectra of the peaks are included. Peaks numbers refer to Table 2.

The LC-DAD chromatogram obtained for the methanol extract showed a series of peaks presenting typical UV spectrums ascribable to flavonoids, hydroxycinnamic acids and small phenolics as represented in Figure 1. Mass spectrometric data were acquired both in negative and in positive ion mode using electrospray source, and the study of MS^n fragmentation allowed the identification of several derivatives that were quantified based on the diode array chromatographic traces using reference compounds.

Phenolic compounds were abundant in the plant methanol extract. Some small phenolics were detected in the first minutes of the chromatograms. Peaks were assigned to caffeic acid hexoside **(17)** [9], protocatechuic acid hexoside **(18)** [9], and benzyl alcohol hexose pentose **(19)** [10]. The contents of these simple phenolics in the extract was 19 mg/g of dry extract (Table 2). In the region of 10–13 min in the chromatogram, no significant peaks were detected at wavelength at DAD detector. Nevertheless, two peaks at retention times of 13.1 and 13.8 min were observed in negative mode, suggesting the presence of compounds that lack chromophore (Figure in Supplementary Materials). These peaks present adducts with formic acid $[M + HCOOH - H]^-$ at *m/z* 433 and 431 and also deprotonated molecular ions $[M - H]^-$ at *m/z* 387 and 385, respectively. The fragmentation showed the loss of a hexose unit, followed by the loss of the megastigmane moiety [11,12]. On the basis of the fragmentation schemes, the peak at min 13.1 was assigned to icariside B5 **(31)** (*m/z* 387) [13], while the peak at 13.9 min was assigned to corchoionoside C **(32)** (*m/z* 385). Spectra and proposed structures of main fragments for the two derivatives are summarized in Figures 5 and 6.

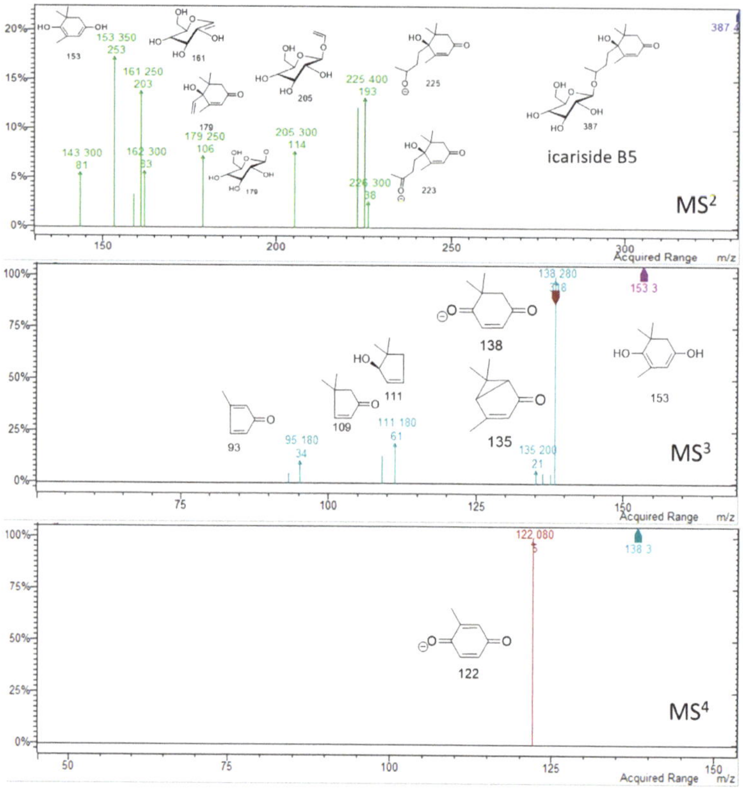

Figure 5. Spectral and fragmentations of icariside B5: MS^2 from parent ion 387 *m/z*, MS^3 from fragment 153 *m/z*, MS^4 from fragment 138 *m/z*.

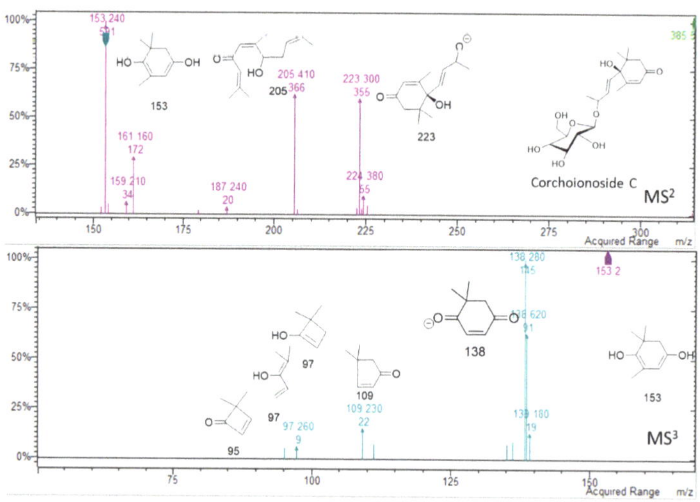

Figure 6. Spectra and fragmentations proposed for corchoionoside C: MS^1 from parent ion 385 *m/z*, MS^2 from fragment 153 *m/z*.

In the region of 16–25 min peaks showing UV spectrum typical of flavonols, characterized by maximum absorption in the range of 350 and 265 nm (Figure 1) [9], were detected. The NMR data showed signals ascribable to flavonoid derivatives as summarized in Table 1, and the main signals related to aromatic portion are indicated in Figure 4. Kaempferol and quercetin as aglycones can be observed due to the MS^n fragmentation spectra [14]. Glycosylation position can be tentatively assigned based on the relative intensity of aglycone fragments [14–16]. The amount of flavonoids and their glycosides **(1–16)** was 96 mg/g of dried extract. Isorhamnetin-3-*O*-rutinoside **(10),** kaempferol-3-*O*-hexosil-deoxyhexoside **(15)**, and quercetin-3-apiofuranosyl-glucopiranoside **(6)** were the most abundant derivatives with 25.91, 18.0, and 16.13 mg/g extract, respectively. LC-MS^n data in negative ion mode, in the region of 23–26 min of the chromatogram, revealed two intense peaks (see Supplementary Material Figure S3) that present $[M-H]^-$ at *m/z* 487. The fragmentation pattern of these two derivatives is reported in Figures 7 and 8 and can be ascribed to two glycosidic derivatives of ionones, based on the comparison with databases, and thus the compounds were identified as dihydro α ionol-*O*-[arabinosil(1-6) glucoside] and the dihydro β ionol-*O*-[arabinosil(1-6) glucoside]. Their amounts are notable in the extract being 132 and 79 mg/g extract, respectively. Furthermore, different peaks observed in the chromatogram region from 32 to 50 min present UV spectra ascribable to hydroxycinnamic acids. The NMR data confirmed the presence of different units of hydroxycinnamic derivatives due to multiple doublet signals ascribed to the double bond of these compounds. Several peaks can be ascribed to phenylethanoid glycosides (Table 1, Figure 4). At retention time 19.4 min, the MS^n fragmentation scheme suggests that the structure can be tentatively assigned to leonoside A (*m/z* 769) **(22),** while at 20.4 min peak presenting $[M-H]^-$ at *m/z* 755 can be annotated as forsythoside B **(23)** [17]. Two derivatives presenting $[M-H]^-$ at *m/z* 797 are annotated as ferrusinogide C **(25)** and its isomer **(24).** Several peaks present adduct with formic acid $[M + HCOOH - H]^-$ at *m/z* 856 and $[M-H]^-$ at *m/z* 810. This compound presents fragmentation MS^3 with a very strong peak at *m/z* 601 and which further forms fragment ions at *m/z* 487, 469, 451, and 337. The latter fragment suggests that they are feruloyl quinic acid derivatives. The UV spectrum confirms the presence of a hydroxycinnamic unit, and the fragmentation scheme indicates the presence of triterpene unit or a big portion of sugar, but due to absence of a similar pattern of fragmentation in the literature, we preferred not to assign any structure to those peaks. As these compounds were not identified, they were not included in the table; however, their isolation is in progress.

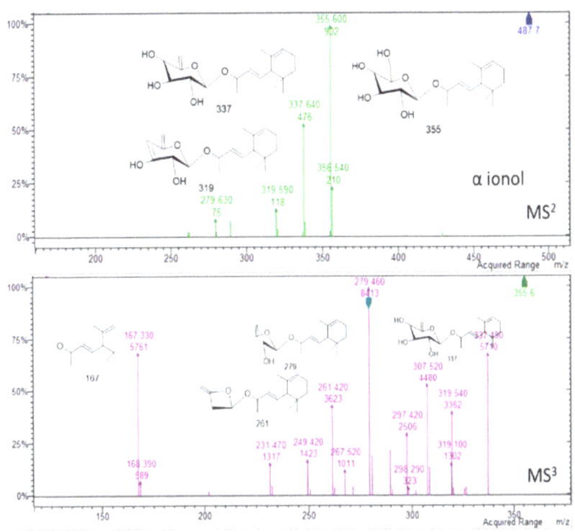

Figure 7. Mass spectral details of dihydro α ionol-*O*-[arabinosil(1-6) glucoside].

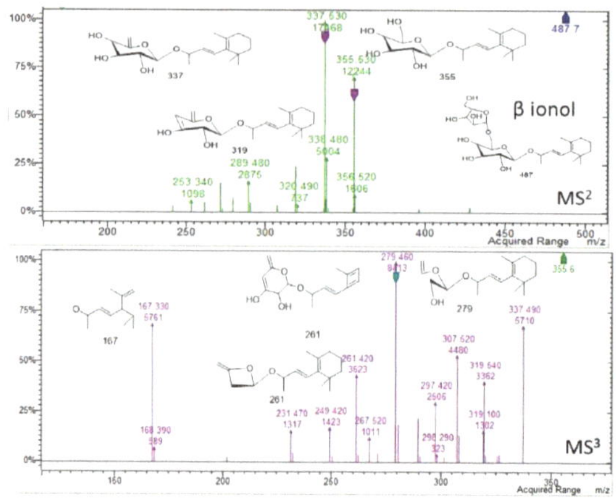

Figure 8. Mass spectral details of dihydro β ionol-*O*-[arabinosil(1-6) glucoside].

Further compounds presenting molecular ion $[M - H]^-$ at *m/z* 663 were assigned to ester of sucrose with ferulic acid and *p*-coumaric acid. Five different isomers **(26–30)** were observed, probably due to the ester linkages at different positions. From MS spectra, it is not possible to assign the different isomers due to lack of specific literature.

In the extract, the overall amount of phenyl propanoid and phenylethanoid derivatives is 100 mg/g. At 53.4 min, kongensin D **(35)**, a diterpenoid derivative, was identified based on *m/z* value at 331. Phytosterols were detected using APCI ion source operating in positive ion mode, and β-sitosterol **(36)** was the most abundant compound. The results of the fingerprinting showed that the extract contains different classes of compounds, such as phytosterols, flavonoid glycosides, hydroxycinnamic derivatives, megastigmane, and other terpenoids, indicating that *C. hirtus* can be considered as a valuable source of bioactive constituents.

Due to the complex nature of constituents detected in the plant, we decided to perform extraction using water and organic solvents (dichloromethane (DCM)), ethyl acetate (EA), and methanol (MeOH)). Different in vitro assays were then performed using the obtained extracts. The results are illustrated in Table 3. A notable amount of total phenolic content was recorded for all the extracts, ranging from 17.96 (MeOH) and 24.24 (DCM) mg GAE/g extract. A significant difference in total flavonoid content using different solvents was observed. Among the solvents used, the MeOH extract (50.16 mg RE/g), followed by EA extract (29.28 mg RE/g extract), resulted in the highest total flavonoid content using the spectrophotometric assay. The variation in the yields can be related to the different solvent polarities and to the different solubility and polarity of the phytoconstituents [18,19]. Such parameters can also be influenced by the ability of the different solvents to penetrate the plant tissue and to solubilize the classes of compounds.

Table 3. Extraction yields and total bioactive components in the tested extracts *.

Extracts	Extraction Yields (%)	Total Phenolic Content (mg GAE/g)	Total Flavonoid Content (mg RE/g)
DCM	2.13	24.24 ± 0.90^{a}	14.37 ± 0.12^{c}
EA	2.34	22.51 ± 0.52^{b}	29.28 ± 1.89^{b}
Infusion	9.95	22.38 ± 0.34^{b}	12.54 ± 0.32^{c}
MeOH	13.18	17.96 ± 0.03^{c}	50.16 ± 2.53^{a}

* Values expressed are means ± S.D. of three parallel measurements. GAE: Gallic acid equivalent; RE: Rutin equivalent. Different letters indicate significant differences in the extracts ($p < 0.05$).

2.2. Antioxidant Ability

Oxidative stress is a phenomenon that arises when reactive oxygen species overwhelms the intrinsic antioxidant defense mechanism. Considerable evidence implicates oxidative stress in the pathophysiology of many diseases and associated complications including diabetes, cardiovascular diseases, and dementia [20,21]. As synthetic antioxidants such as butylated hydroxytoluene (BHT) and butylated hydroxyanisole (BHA) have been reported to exert negative effects on human health, there is a renewed interest to find novel antioxidant compounds with more efficacy and minimal or no side effects. In this context, medicinal plants in particular have been widely studied for discovering potential novel antioxidant molecules [22]. Some of *Croton* species that have been reported for their antioxidant capacity include *C. celtidifolius* Baill. [23], *C. lechleri* Muell.-Arg. [24], *C. zambesicus* Muell.-Arg. [25], and *C. cajucara* Benth. [26].

Therefore, this investigation also endeavored to evaluate the antioxidant potential of the different extracts of *C. hirtus* by employing different in vitro bioassays—namely, free radical scavenging (DPPH and ABTS), reducing power (CUPRAC and FRAP), phosphomolybdenum, and ferrous-ion chelating assays. As illustrated in Table 4, different extracts possessed varied free radical scavenging and reducing capacities. The results of DPPH and ABTS assays indicated that the water extract displayed the highest free radical scavenging ability, with a mean value of 41.08 and 64.84 mg TE/g extract, respectively. For CUPRAC assay, the DCM (88.67 mg TE/g extract) extract, followed by water (78.17 mg TE/g extract) extract, showed highest reducing power. The order of the FRAP assay was water > MeOH > DCM > EA. This indicates that the hydrophilic compounds in the extract are, in general, mostly involved in the antioxidant effect in the assays' conditions. Considering the fingerprinting chemical analysis, these effects can be probably ascribed to hydroxycinnamic and flavonoid derivatives. This observation is in agreement with a previous report that demonstrated that highly polar solvents, such as water and methanol, have a high effectiveness as antioxidants [22]. In the literature, we observed different results for different solvent extracts [27,28]. For example, although polar extracts had the highest levels of phenolic compounds in some studies, apolar extracts exhibited stronger free radical and reducing abilities. This fact can be explained by different reasons. For example, the spectrophotometric assays could have some drawbacks, and not only phenolics but also other classes of compounds could play a role. Furthermore, reaction conditions, organic solvent, aqueous buffer, and pH influence the experimental response of the mixture and compounds, leading to different effects in the antioxidant reactions. Altogether, the utilization of more assays could provide accurate information of the antioxidant capacity of plant extracts.

Table 4. Antioxidant properties of the tested extracts *.

Extracts	DPPH (mgTE/g)	ABTS (mgTE/g)	CUPRAC (mgTE/g)	FRAP (mgTE/g)	Phosphomolybdenum (mmol TE/g)	Chelating Ability (mg EDTAE/g)
DCM	22.78 ± 0.52^{c}	32.32 ± 2.49^{c}	88.67 ± 1.08^{a}	26.35 ± 0.09^{c}	2.70 ± 0.13^{a}	15.18 ± 1.08^{b}
EA	23.66 ± 1.14^{c}	18.59 ± 1.64^{d}	69.04 ± 0.40^{c}	24.15 ± 0.21^{d}	2.33 ± 0.27^{ab}	18.26 ± 0.22^{a}
Infusion	41.08 ± 1.65^{a}	64.84 ± 2.71^{a}	78.17 ± 0.16^{b}	45.67 ± 0.86^{a}	1.46 ± 0.01^{c}	17.94 ± 0.16^{a}
MeOH	30.62 ± 0.54^{b}	42.02 ± 1.11^{b}	62.50 ± 2.30^{d}	30.94 ± 0.35^{b}	1.97 ± 0.10^{b}	13.96 ± 0.10^{b}

* Values expressed are means ± S.D. of three parallel measurements. TE: Trolox equivalent; EDTAE: EDTA equivalent. Different letters indicate significant differences in the extracts ($p < 0.05$).

The phosphomolybdenum assay was employed to estimate the total antioxidant capacity of the *C. hirtus* extracts. The DCM extract was found to show the highest total antioxidant capacity, with a mean value of 2.70 mmol TE/g extract. In the case of ferrous-ion chelating activity, the EA (18.26 mg EDTAE/g extract) and water (17.94 mg EDTAE/g extract) extracts showed the highest activity. In total, despite of its lower total phenolic and flavonoid contents, the water extract showed considerably higher antioxidant capacity as compared with other tested extracts. Such results suggest lipophilic fraction has a role in mediating the antioxidant effects, and some effect can also be ascribed to fatty acid and phytosterols that have been reported as significant antioxidant compounds [29–31].

In the literature, the extracts of some *Croton* species and their isolated compounds exhibited remarkable antioxidant properties. For example, Aderogba et al. [25] isolated three compounds from *C. zambesicus* and one of them displayed good DPPH radical scavenging ability with an IC_{50} value of 200 μM. Similarly, Azevedo et al. [26] reported significant radical scavenging abilities of *C. cajucara* isolated compounds. When combining all data on the genus *Croton*, they could be considered as a significant source of natural antioxidants.

2.3. Enzyme Inhibitory Ability

Due to their essential catalytic role in several physiological processes, enzymes are considered as prime targets for drug design for alleviating diseases. Indeed, the therapy of some important human aliments—namely, hypertension, diabetes, inflammatory and neurodegenerative diseases—includes the use of enzyme inhibitors. For instance, α-amylase and α-glucosidase enzymes are the targeted enzymes in diabetes, while cholinesterases (acetylcholinesterase and butyrylcholinesterase) are the known therapeutic targets in Alzheimer's disease [32,33].

The inhibitory action of the studied *C. hirtus* extracts against cholinesterase, tyrosinase, α-amylase, and α-glucosidase were determined. Table 5 illustrates the inhibitory potential of the different extracts against the studied enzymes, expressed as equivalent values of standards. The tested solvents displayed significantly distinct inhibitory potentials against AChE, with the DCM and EA extracts (5.03 mg GALAE/g extract, 4.84 mg GALAE/g extract) being most potent but with significant effect for methanol (4.05 mg GALAE/g extract). All the extracts showed remarkable anti-BChE effects, with mean values ranging from 16.41 to 15.86 mg GALAE/g extract. This result suggests that different constituents of the plant may exert inhibitory activity on these enzymes. With respect to the composition of the methanol extract, quercetin, kaempferol, and isorhamnetin derivatives detected in the extract have been previously documented as AChE inhibitors [34–36]. In particular, quercetin-3-apiofuranosyl-glucopyranoside, isorhamnetin-3-*O*-rutinoside, and kaempferol-3-*O*-hexosil-deoxyhexoside, which are detected in high level in the methanol extracts, could be potentially the main flavonoid derivatives, attributing to the anti-AChE effects. Rutin was also reported by Szwajgier et al. [37] as a significant cholinesterase agent. Furthermore, as previously reported, the hydroxycinnamic acids, such as caffeic acid hexoside and protocatechuic acid hexoside, detected in the extract could be potent AChE inhibitors [34]. For the lipophilic extracts, it can be postulated that the effects were mediated by the presence of phytosterols, particularly β-sitosterol, which have been reported as AChE and BChE inhibitors [38].

Table 5. Enzyme inhibitory properties of the tested extracts *.

Extracts	AChE Inhibition (mgGALAE/g)	BChE Inhibition (mgGALAE/g)	Tyrosinase Inhibition (mgKAE/g)	Amylase Inhibition (mmolACAE/g)	Glucosidase Inhibition (mmol ACAE/g)
DCM	5.03 ± 0.16 [a]	16.41 ± 1.68 [a]	24.39 ± 0.98 [c]	0.71 ± 0.01 [ab]	1.68 ± 0.14
EA	4.84 ± 0.19 [a]	15.86 ± 0.74 [a]	34.81 ± 2.67 [b]	0.75 ± 0.03 [a]	na
Infusion	1.01 ± 0.13 [c]	14.44 ± 0.34 [a]	na	0.14 ± 0.01 [c]	na
MeOH	4.09 ± 0.02 [b]	16.11 ± 0.25 [a]	49.83 ± 3.94 [a]	0.69 ± 0.02 [b]	na

* Values expressed are means ± S.D. of three parallel measurements. GALAE: Galatamine equivalent; KAE: Kojic acid equivalent; ACAE: Acarbose equivalent. na: not active. Different letters indicate significant differences in the extracts ($p < 0.05$).

All the extracts except the water extract showed considerable tyrosinase inhibition, with values of 49.83, 34.81, and 24.39 mg KAE/g extract for MeOH, EA, and DCM, respectively. Most of the activity can be correlated to hydrophilic phenolic compounds, but as we described for AChE and BChE, a significant role is clearly indicated also by results for the lipophilic constituents. Modest activity against α-amylase activity was exerted by all the extracts (0.14–0.75 nmol ACAE/g extract), whilst only the DCM extract

(1.68 mmol ACAE/g) was effective at inhibiting α glucosidase. Similar to cholinesterases' inhibitory ability, the other enzyme inhibitory abilities could be attributed to the presence of chemical components identified, including rutin, caffeic acid, and protocatechuic acid [39–42]. Literature search on the enzyme inhibitory properties of the genus *Croton* resulted in a few studies on the subject. Aderogba et al. [25] reported on the inhibitory properties of flavonols and indol alkaloids from *C. menyharthii* against acetylcholinesterase, and glucosidase and especially alkaloids exhibited significant acetylcholinesterase abilities. Similarly, Shahwar et al. [43] isolated some alkaloids from *C. sparsiflorus*, and most of them had excellent acetylcholinesterase inhibitory properties. Keerthana et al. [44] reported significant amylase inhibition capacity for *C. bonplandianum* ethanolic leaf extract. The remarkable glucosidase inhibitory properties of *C. thurifer* components were also described by Morocho et al. [45]. With all observations, the genus *Croton* could be considered as a source of great potential for natural enzyme inhibitors to combat global health problems.

2.4. K-Medoids Clustering

Following the initial screening of biological activities of the prepared extracts by univariate analysis, *k*-medoids cluster analysis was performed with aim of assessing the effects of the solvents used for extraction. The *k*-medoids model is a robust clustering approach for identifying and partitioning a dataset into an optimal number of homogeneous clusters. This approach allowed us to visually evaluate the clustering tendency of the samples from the very beginning. As observed in Figure 9, there is a low similarity among the subjects representing each solvent (Figure 9A). This points out a high intra-class similarity and low inter-class similarity and confirms that there is a cluster structure in the *C. hirtus* biological activities dataset. Based on Figure 9B, it can be observed the optimal number of clusters is 4. Figure 9C depicts the *k*-medoids scatter plot, and it can be observed that the samples were grouped depending on the solvents used and distributed differently in the hyperspace along the two dimensions. Therefore, this statistical approach demonstrated that the biological activities of *C. hirtus* were influenced by the extraction solvents. Notably, the cluster validation was found to be excellent, being 0.69 for the silhouette coefficient, thus confirming the goodness of the *k*-medoids model (Figure 9D).

Many researchers have studied and evaluated the effect of different types of solvents on biological activities of various *Croton* species [1,23–25]. These studies indicated that the selectivity of the solvent is crucial not only for obtaining optimum yield of one or more desired molecules but also for pharmacological effects of the resulting molecules. In fact, secondary metabolites possess different degrees of polarity; thereby, the extractive solvent should be chosen carefully to ensure a good dissolution of secondary metabolites intended to be studied. For example, aqueous solvent with a dielectric constant of 80 dissolves polar molecules more rapidly than non-polar solvent hexane due to its low dielectric constant of 1.89.

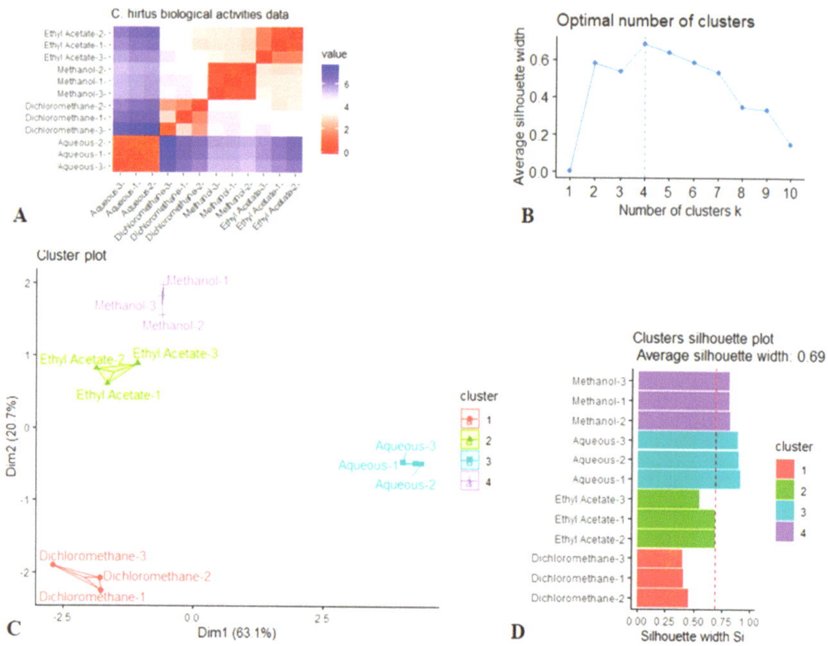

Figure 9. A *k*-medoids clustering analysis of biological activities of *C. hirtus*. (**A**) Distance matrix showing the dissimilarity between each pair of subjects (Red color: high similarity, Blue color: low similarity). (**B**) Number of clusters. (**C**) Scatter plot. (**D**) Silhouette plot displaying the average distance between clusters.

3. Materials and Methods

3.1. Preparation of Extracts

Plant samples were collected from Lolobo (Yamoussoukro) of Ivory Coast in 2019 (during summer season) and identified by Ouattara Katinan Etienne (Botanist at Université Félix Houphouet Boigny, Ivory Coast). The plant samples were deposited in the Department of Biology at Selcuk University. The roots were removed and the aerial parts (whole plant with flowers, stems, and foliar tissue) were used. The plant materials were dried in shade for about 10 days and then grounded using a laboratory mill.

In this study, different extraction procedures were employed. As starting point, methanol was used as solvent. At a later stage, solvents with different polarities (dichloromethane, ethyl acetate, water) were also used. Maceration technique was used for dichloromethane, ethyl acetate, and methanol. For this purpose, 5 g of the samples were macerated with respective solvents (individually) for 24 h at room temperature (24 °C ± 1) on a magnetic stirrer, and solvents were changed twice. The mixtures then were filtered and concentrated using a rotary-evaporator. With respect to the water extract, 5 g of the plant material was infused in 100 mL boiled water. Then, the water was dried by using freeze-drying. All the extracts were stored + 4 °C until further studies.

3.2. Spectrophotometric Assays for Total Phenolic and Flavonoids

Total phenolic and flavonoid contents of *C. hirtus* extracts were assayed by colorimetric methods according to Zengin and Aktumsek [46]. Gallic acid (GAE) and rutin (RE) were chosen as standards for phenols and flavonoids, respectively.

3.3. Fingerprinting of Phytoconstituents by LC-DAD-ESI-MS^n, LC-DAD-APCI-MS^n

Dried methanol extract (100 mg) was dissolved in 15 mL of methanol and then subjected to sonication for 15 min at room temperature. The sample was centrifuged at

13,000 rpm, and the supernatant was removed and used for analysis. For the mass spectrometry detection, we used electrospray (ESI) source for phenolics, atmospheric pressure chemical ionization (APCI) ion source for lipophilic compounds as phytosterols and triterpene. Apparatus used was Agilent 1260 chromatograph (Santa Clara, CA, USA) equipped with 1260 diode array detector (DAD) and Varian MS-500 ion trap mass spectrometer equipped with ESI or APCI ion source (Varian inc, Palo Alto, CA, USA). A "T" splitter was installed immediately after chromatographic column allowing the separation of the eluate, half to DAD and half to Agilent/Varian MS-500 ion trap mass spectrometer. UV-Vis spectra were acquired in the range of 190–400 nm.

For the phenolic analysis, synergy RP polar (Phenomenex, Bologna, Italy) was used as stationary phase (3.0 × 150 m^2; 4 micron) and as eluents, water 1% formic acid (A) and methanol (B) were used. Elution gradient started with 95% of A and went to 10% of A in 50 min, then the amount of B was increased to 100% in 55 min. The flow rate was 400 μL/min. The MS spectra were acquired in negative ion mode in 50–2000 *m/z* range using an ESI source; drying gas temperature started at 295 °C and in 40 min decreased to 250; drying gas pressure was 25 psi; nebulizer was 30 psi; needle was set at 4500 V, RF 85%, capillary 80 V. Quantification of compounds was obtained using DAD. Chlorogenic acid solutions (of concentration ranging from 100 to 1 μg/mL) were used to create calibration curve for the quantification of the different hydroxycinnamic acid derivatives at 330 nm. Calibration curve was Y = 43.325x + 0.236 (R^2 = 0.9989). For the quantification of flavonols, quercetin and kaempferol solutions (range 100–1 μg/mL) were used at 350 nm, and calibration curves were Y = 25.3x + 0.55 (R^2 = 0.9998) and Y = 20.3x + 0.91 (R^2 = 0.9991), respectively. Rutin was used for quantification of quercetin glycosides; isorhamnetin-7-*O*-glucoside and kaempferol-3-*O*-glucoside were used for quantification of isorhamnetin and kaempferol derivatives. Solutions were prepared in the range of 120–0.61 μg/mL, and chromatograms were recorded at 350 nm; calibration curves were Y = 31.62x + 1.021 (R^2 = 0.9993), Y = 33.11x + 0.841 (R^2 = 0.9991), and Y = 23.42x + 1.042 (R^2 = 0.9989) for rutin, isorhamnetin-7-*O*-glucoside, and kaempferol-3-*O*-glucoside, respectively. Benzoic acid was used for protocatechuic acid and small phenolics using solutions (of concentration ranging from 130–1.30 μg/mL) at 280 nm and calibration curve was Y = 81.1x + 0.98 (R^2 = 0.9996). For terpenoids, a mixture of dihydro α ionol-*O*-[arabinosil(1-6) glucoside] and dihydro β ionol-*O*-[arabinosil(1-6) glucoside was purified using an Agilent 1260 chromatograph equipped with analytical fraction collector. As stationary phase, an Agilent XDB C18 3.0 × 150 m^2 was used, and isocratic elution was used (70% water 0.1% formic acid, 30% methanol) at flow rate of 4 mL/min. Fraction containing compounds were pooled and dried under nitrogen flow and redissolved in deuterated methanol. The amount of α ionol-*O*-[arabinosil(1-6) glucoside] and dihydro β ionol-*O*-[arabinosil(1-6) glucoside] in the mixture was determined using quantitative NMR with a previously published procedure that used caffeine as internal standard [47]. Solvent was then removed and a solution in methanol was prepared and diluted preparing solutions in the range of 300–1.0 μg/mL. Calibration curve was Y = 115.1x + 10.321 (R^2 = 0.9989) using MS detector in negative ion mode.

3.4. HPLC-(APCI)-MS Analysis of Phytosterols

For the phytosterol constituents, an Agilent Zorbax Eclipse XDB-C18 column (Agilent, Santa Clara, CA, USA) (3.0 mm × 150 mm, 3.5 μm) was used as stationary phase. All analytical details are described in our previous publications [48]. The compounds were identified based on the comparison with the literature and reference compounds, when available. As standards, known solutions of β-sitosterol (of concentration ranging from 176 to 1.76 μg/mL) and stigmasterol (of concentration ranging from 185.6 to 1.9 μg/mL) were used.

Compound's quantification was obtained via calibration curve derived from β-sitosterol and stigmasterol solutions. Calibration curves were Y = 0.63x + 0.2705 (R^2 = 0.99611) for β-sitosterol and Y = 2.1153x − 15.216 (R^2 = 0.9949) for stigmasterol.

3.5. NMR Analysis

1D and 2D NMR spectra were obtained on a Bruker Avance III 400 Ultrashield spectrometer with 400 MHz magnet. NMR spectra were acquired in MeOD-d4 (Sigma-Aldrich) with TMS as an internal standard. Duran® 4.95 mm NMR tubes (Duran Group, Mainz, Germany) were used. Chemical shifts are expressed in δ values in ppm. ^{1}H-NMR and HSQC-DEPT, HMBC, and COSY experiments were acquired using standard Bruker sequences measuring p1 and d1 for each acquired sample. Samples of dried extract were initially dissolved in 100 mg in MeOD-d_4 then centrifuged, and the liquid was used for NMR measurements. DOSY experiments were performed using the ledbpgp2s sequence.

3.6. Determination of Antioxidant and Enzyme Inhibitory Effects

To detect antioxidant properties, we used several chemical assays including different mechanisms—namely, radical scavenging, reducing power and metal chelating. Trolox (TE) and ethylenediaminetetraacetic acid (EDTA) were used as standard antioxidant compounds. The results are expressed as equivalents of the aforementioned compounds. To detect inhibitory effects on enzymes, we used colorimetric enzyme inhibition assays, and these assays included tyrosinase, α-glucosidase, α-amylase, and cholinesterases. Standard inhibitors (galantamine for cholinesterases, kojic acid for tyrosinase, and acarbose for α-glucosidase and α-amylase) were used as positive controls [49]. The experimental details are given in Supplementary Materials.

3.7. Statistical Analysis

A one-way analysis of the variance (ANOVA) was performed to investigate significant differences ($p < 0.05$, Turkey's post hoc test) for each assay done. Thereafter, the dataset was submitted to *k*-medoids cluster analysis after the distance matrix was visualized for measuring the (dis)similarity between the observations. Euclidean distance measure was used for this purpose. To estimate the optimal number of clusters, the average silhouette method was used. A high average silhouette width indicated a good clustering solution. Finally, the goodness of the *k*-medoids clustering was evaluated by estimating the silhouette coefficient (S_i). A value of S_i close to 1 confirmed a good clustering solution. The statistical analysis was performed by using R 3.5.1 software (R Core Team, Vienna, Austria).

4. Conclusions

The present study offers a new perspective in the potential use of *C. hirtus* as a source of bioactive constituents. Fingerprinting showed the different plant constituents and suggested the possibility of extracting significant amounts of different classes of compounds—terpenoids, phytosterols, phenolics—from plant material using methanol as solvent. Notable antioxidant and enzyme inhibitory effects against cholinesterases (AChE and BChE) and tyrosinase were displayed by the extracts. Among the tested solvents, the polar solvents (water and MeOH) were the best solvents for extracting bioactive compounds with antioxidant properties from *C. hirtus*, whilst the MeOH, DCM, and EA were suitable for obtaining biomolecules with enzyme inhibitory potentials, suggesting that multiple constituents in the extracts being soluble in different solvents exerted inhibitory activity on the same enzyme. This suggests a rationale for the use of the overall mixture, which can offer the opportunity to act on the same target compounds with different physicochemical properties. Based on the present findings, *C. hirtus* can be considered as a valuable source of pharmacologically active agents and used for designing phytomedicines. Nonetheless, further investigations are essential to elucidate the mode of action of the antioxidant and the enzyme inhibition activity, toxicity, safety, and bioavailability of the extracts.

Supplementary Materials: Supplementary materials are available online, S1: Croton methanol extract H-NMR spectrum; Croton methanol extract HSQC-NMR spectrum enlargement of aromatic portion; Croton methanol extract HMBC spectrum; Croton methanol extract DOSY spectrum; S2: LC-MS in negative ion mode showing the two peaks and relative mass spectra ascribable to Icariside

B5 ([M + HCOOH − H]- ion at m/z 433) and Corchionoside([M + HCOOH − H]- ion at m/z 431); S3: LC-MS in negative ion mode showing the two peaks and relative mass spectra ascribable to dihydro α ionol-*O*-[arabinosil(1-6) glucoside] and the dihydro β ionol-*O*-[arabinosil(1-6) glucoside]. Ions are detected as [M + HCOOH − H]-and also as [M − H]- as visible in figure.

Author Contributions: Conceptualization, S.D., K.I.S., S.S., and G.Z.; methodology, S.D., S.S., I.F., and G.Z.; software, K.I.S. and I.F.; validation, S.D., M.F.M., and G.Z.; formal analysis, G.Z.; investigation, K.I.S. and D.L.; resources, K.I.S. and O.K.E.; data curation, S.D., M.F.M., and G.Z.; writing—original draft preparation, S.D., M.F.M., and D.L.; writing—review and editing, G.Z.; visualization, K.I.S., I.F., and S.S.; supervision, S.S.; project administration, G.Z.; funding acquisition, S.S. All authors have read and agreed to the published version of the manuscript.

Funding: This research received no external funding.

Institutional Review Board Statement: Not applicable.

Informed Consent Statement: Not applicable.

Data Availability Statement: Data is not available from the authors.

Conflicts of Interest: The authors declare no conflict of interest.

Sample Availability: Samples of the compounds are not available from the authors.

References

1. Salatino, A.; Salatino, M.L.F.; Negri, G. Traditional uses, chemistry and pharmacology of Croton species (Euphorbiaceae). *J. Braz. Chem. Soc.* **2007**, *18*, 11–33. [CrossRef]
2. Daouda, T.; Prevost, K.; Gustave, B.; Joseph, D.A.; Nathalie, G.; Raphaël, O.; Rubens, D.; Claude, C.J.; Mireille, D.; Felix, T. Terpenes, Antibacterial and Modulatory Antibiotic Activity of Essential Oils from Croton hirtus L' Hér. (Euphorbiaceae) from Ivory Coast. *J. Essent. Oil Bear. Plants* **2014**, *17*, 607–616. [CrossRef]
3. De Lima, S.G.; Medeiros, L.B.P.; Cunha, C.N.L.C.; da Silva, D.; de Andrade, N.C.; Neto, J.M.M.; Lopes, J.A.D.; Steffen, R.A.; Araújo, B.Q.; Reis, F.d.A.M. Chemical composition of essential oils of Croton hirtus L'Her from Piauí (Brazil). *J. Essent. Oil Res.* **2012**, *24*, 371–376. [CrossRef]
4. Rosandy, A.R.; Azman, A.A.; Khalid, R.; Othaman, R.; Lazim, A.M.; Choudary, I.M.; Syah, Y.M.; Latip, J.; Said, I.M.; Bakar, M.A. Isolation of New Rotundone from the Roots of Croton Hirtus (Euphorbiaceae). *Malays. J. Anal. Sci.* **2019**, *23*, 677–681.
5. Subin, M.; Reghu, N. Phytochemical screening and antibacterial properties of Croton hirtus L'Her. plant against some important pathogenic bacteria. *Nat. Environ. Pollut. Technol.* **2012**, *11*, 59–64.
6. Ezeabara, C.A.; Okonkwo, E. Comparison of phytochemical and proximate components of leaf, stem and root of Croton hirtus L'Herit and Croton lobatus Linn. *J. Med. Health Res.* **2016**, *24*, 33.
7. Kim, M.J.; Kim, J.G.; Sydara, K.M.; Lee, S.W.; Jung, S.K. Croton hirtus L'Hér Extract Prevents Inflammation in RAW264. 7 Macrophages Via Inhibition of NF-κB Signaling Pathway. *J. Microbiol. Biotechnol.* **2020**, *30*, 490–496. [CrossRef]
8. Morreel, K.; Saeys, Y.; Dima, O.; Lu, F.; Van de Peer, Y.; Vanholme, R.; Ralph, J.; Vanholme, B.; Boerjan, W. Systematic Structural Characterization of Metabolites in Arabidopsis via Candidate Substrate-Product Pair Networks. *Plant Cell* **2014**, *26*, 929. [CrossRef] [PubMed]
9. Chen, H.-J.; Inbaraj, B.S.; Chen, B.-H. Determination of phenolic acids and flavonoids in Taraxacum formosanum Kitam by liquid chromatography-tandem mass spectrometry coupled with a post-column derivatization technique. *Int. J. Mol. Sci.* **2012**, *13*, 260–285. [CrossRef] [PubMed]
10. Bystrom, L.M.; Lewis, B.A.; Brown, D.L.; Rodriguez, E.; Obendorf, R.L. Characterisation of phenolics by LC–UV/Vis, LC–MS/MS and sugars by GC in Melicoccus bijugatus Jacq.'Montgomery' fruits. In *Food Chem.*; 2008; Volume 111, pp. 1017–1024.
11. Kamel, M.S.; Mohamed, K.M.; Hassanean, H.A.; Ohtani, K.; Kasai, R.; Yamasaki, K. Iridoid and megastigmane glycosides from *Phlomis aurea*. *Phytochemistry* **2000**, *55*, 353–357. [CrossRef]
12. Kawakami, S.; Matsunami, K.; Otsuka, H.; Shinzato, T.; Takeda, Y. Crotonionosides A–G: Megastigmane glycosides from leaves of Croton cascarilloides Räuschel. *Phytochemistry* **2011**, *72*, 147–153. [CrossRef] [PubMed]
13. Anh, N.Q.; Yen, T.T.; Hang, N.T.; Anh, D.H.; Viet, P.H.; Van Doan, V.; Van Kiem, P. 1H-Indole-3-acetonitrile glycoside, phenolic and other compounds from *Stixis suaveolens*. *Vietnam J. Chem.* **2019**, *57*, 558–561. [CrossRef]
14. Fabre, N.; Rustan, I.; de Hoffmann, E.; Quetin-Leclercq, J. Determination of flavone, flavonol, and flavanone aglycones by negative ion liquid chromatography electrospray ion trap mass spectrometry. *J. Am. Soc. Mass Spectrom.* **2001**, *12*, 707–715. [CrossRef]
15. Kachlicki, P.; Piasecka, A.; Stobiecki, M.; Marczak, Ł. Structural characterization of flavonoid glycoconjugates and their derivatives with mass spectrometric techniques. *Molecules* **2016**, *21*, 1494. [CrossRef]
16. Yang, W.Z.; Qiao, X.; Bo, T.; Wang, Q.; Guo, D.A.; Ye, M. Low energy induced homolytic fragmentation of flavonol 3-O-glycosides by negative electrospray ionization tandem mass spectrometry. *Rapid Commun. Mass Spectrom.* **2014**, *28*, 385–395. [CrossRef]

17. Georgiev, M.I.; Ali, K.; Alipieva, K.; Verpoorte, R.; Choi, Y.H. Metabolic differentiations and classification of Verbascum species by NMR-based metabolomics. *Phytochemistry* **2011**, *72*, 2045–2051. [CrossRef] [PubMed]
18. Abarca-Vargas, R.; Peña Malacara, C.F.; Petricevich, V.L. Characterization of Chemical Compounds with Antioxidant and Cytotoxic Activities in Bougainvillea x buttiana Holttum and Standl, (var. Rose) Extracts. *Antioxidants* **2016**, *5*, 45. [CrossRef] [PubMed]
19. Rocchetti, G.; Pagnossa, J.P.; Blasi, F.; Cossignani, L.; Hilsdorf Piccoli, R.; Zengin, G.; Montesano, D.; Cocconcelli, P.S.; Lucini, L. Phenolic profiling and in vitro bioactivity of Moringa oleifera leaves as affected by different extraction solvents. *Food Res. Int.* **2020**, *127*, 108712. [CrossRef]
20. Pizzino, G.; Irrera, N.; Cucinotta, M.; Pallio, G.; Mannino, F.; Arcoraci, V.; Squadrito, F.; Altavilla, D.; Bitto, A. Oxidative stress: Harms and benefits for human health. *Oxid. Med. Cell. Longev.* **2017**, *2017*. [CrossRef]
21. Tan, B.L.; Norhaizan, M.E.; Liew, W.-P.-P.; Sulaiman Rahman, H. Antioxidant and Oxidative Stress: A Mutual Interplay in Age-Related Diseases. *Front. Pharmacol.* **2018**, *9*, 1162. [CrossRef]
22. Altemimi, A.; Lakhssassi, N.; Baharlouei, A.; Watson, D.G.; Lightfoot, D.A. Phytochemicals: Extraction, Isolation, and Identification of Bioactive Compounds from Plant Extracts. *Plants* **2017**, *6*, 42. [CrossRef] [PubMed]
23. Nardi, G.M.; Felippi, R.; DalBó, S.; Siqueira-Junior, J.M.; Arruda, D.C.; Delle Monache, F.; Timbola, A.K.; Pizzolatti, M.G.; Ckless, K.; Ribeiro-do-Valle, R.M. Anti-inflammatory and antioxidant effects of *Croton celtidifolius* bark. *Phytomedicine* **2003**, *10*, 176–184. [CrossRef]
24. Lopes, M.I.L.e.; Saffi, J.; Echeverrigaray, S.; Henriques, J.A.P.; Salvador, M. Mutagenic and antioxidant activities of *Croton lechleri* sap in biological systems. *J. Ethnopharmacol.* **2004**, *95*, 437–445. [CrossRef] [PubMed]
25. Aderogba, M.A.; McGaw, L.J.; Bezabih, M.; Abegaz, B.M. Isolation and characterisation of novel antioxidant constituents of *Croton zambesicus* leaf extract. *Nat. Prod. Res.* **2011**, *25*, 1224–1233. [CrossRef]
26. Azevedo, M.M.B.; Chaves, F.C.M.; Almeida, C.A.; Bizzo, H.R.; Duarte, R.S.; Campos-Takaki, G.M.; Alviano, C.S.; Alviano, D.S. Antioxidant and Antimicrobial Activities of 7-Hydroxy-calamenene-Rich Essential Oils from *Croton cajucara* Benth. *Molecules* **2013**, *18*, 1128–1137. [CrossRef]
27. Bouasla, I.; Hamel, T.; Barour, C.; Bouasla, A.; Hachouf, M.; Bouguerra, O.M.; Messarah, M. Evaluation of solvent influence on phytochemical content and antioxidant activities of two Algerian endemic taxa: *Stachys marrubiifolia* Viv. and *Lamium flexuosum* Ten. (*Lamiaceae*). *Eur. J. Integr. Med.* **2021**, *42*, 101267. [CrossRef]
28. Amalraj, S.; Mariyammal, V.; Murugan, R.; Gurav, S.S.; Krupa, J.; Ayyanar, M. Comparative evaluation on chemical composition, in vitro antioxidant, antidiabetic and antibacterial activities of various solvent extracts of *Dregea volubilis* leaves. *S. Afr. J. Bot.* **2021**, *138*, 115–123. [CrossRef]
29. Yoshida, Y.; Niki, E. Antioxidant effects of phytosterol and its components. *J. Nutr. Sci. Vitaminol.* **2003**, *49*, 277–280. [CrossRef] [PubMed]
30. Hsu, C.-C.; Kuo, H.-C.; Huang, K.-E. The Effects of Phytosterols Extracted from Diascorea alata on the Antioxidant Activity, Plasma Lipids, and Hematological Profiles in Taiwanese Menopausal Women. *Nutrients* **2017**, *9*, 1320. [CrossRef] [PubMed]
31. Richard, D.; Kefi, K.; Barbe, U.; Bausero, P.; Visioli, F. Polyunsaturated fatty acids as antioxidants. *Pharmacol. Res.* **2008**, *57*, 451–455. [CrossRef] [PubMed]
32. Goncalves, S.; Romano, A. *Inhibitory Properties of Phenolic Compounds Against Enzymes Linked with Human Diseases*; IntechOpen: London, UK, 2017.
33. Ramsay, R.R.; Tipton, K.F. Assessment of enzyme inhibition: A review with examples from the development of monoamine oxidase and cholinesterase inhibitory drugs. *Molecules* **2017**, *22*, 1192. [CrossRef]
34. Vinholes, J.; Silva, B.M.; Silva, L.R. Hydroxycinnamic acids (HCAS): Structure, biological properties and health effects. *Adv. Med. Biol.* **2015**, *88*, 105–130.
35. Szwajgier, D.; Borowiec, K. Phenolic acids from malt are efficient acetylcholinesterase and butyrylcholinesterase inhibitors. *J. Inst. Brew.* **2012**, *118*, 40–48. [CrossRef]
36. Khan, H.; Marya; Amin, S.; Kamal, M.A.; Patel, S. Flavonoids as acetylcholinesterase inhibitors: Current therapeutic standing and future prospects. *Biomed. Pharmacother.* **2018**, *101*, 860–870. [CrossRef] [PubMed]
37. Szwajgier, D.; Borowiec, K.; Zapp, J. Activity-guided isolation of cholinesterase inhibitors quercetin, rutin and kaempferol from Prunus persica fruit. *Z. Nat. C* **2020**, *75*, 87–96. [CrossRef]
38. Ali, M.; Muhammad, S.; Shah, M.R.; Khan, A.; Rashid, U.; Farooq, U.; Ullah, F.; Sadiq, A.; Ayaz, M.; Ali, M. Neurologically potent molecules from Crataegus oxyacantha; isolation, anticholinesterase inhibition, and molecular docking. *Front. Pharmacol.* **2017**, *8*, 327. [CrossRef] [PubMed]
39. Oboh, G.; Ademosun, A.O.; Ayeni, P.O.; Omojokun, O.S.; Bello, F. Comparative effect of quercetin and rutin on α-amylase, α-glucosidase, and some pro-oxidant-induced lipid peroxidation in rat pancreas. *Comp. Clin. Path.* **2015**, *24*, 1103–1110. [CrossRef]
40. Taira, J.; Tsuchida, E.; Uehara, M.; Ohhama, N.; Ohmine, W.; Ogi, T. The leaf extract of Mallotus japonicus and its major active constituent, rutin, suppressed on melanin production in murine B16F1 melanoma. *Asian Pac. J. Trop. Biomed.* **2015**, *5*, 819–823. [CrossRef]
41. Park, K.-Y.; Kim, J. Synthesis and biological evaluation of the anti-melanogenesis effect of coumaric and caffeic acid-conjugated peptides in human melanocytes. *Front. Pharmacol.* **2020**, *11*, 922. [CrossRef]

42. Truong, X.T.; Park, S.-H.; Lee, Y.-G.; Jeong, H.Y.; Moon, J.-H.; Jeon, T.-I. Protocatechuic acid from pear inhibits melanogenesis in melanoma cells. *Int. J. Mol. Sci.* **2017**, *18*, 1809. [CrossRef]
43. Shahwar, D.; Ahmad, N.; Yasmeen, A.; Khan, M.A.; Ullah, S.; Atta-ur, R. Bioactive constituents from *Croton sparsiflorus* Morong. *Nat. Prod. Res.* **2015**, *29*, 274–276. [CrossRef]
44. Keerthana, G.; Kalaivani, M.; Sumathy, A. In-vitro alpha amylase inhibitory and anti-oxidant activities of ethanolic leaf extract of *Croton bonplandianum*. *Asian J. Pharm. Clin. Res.* **2013**, *6*, 32–36.
45. Morocho, V.; Sarango, D.; Cruz-Erazo, C.; Cumbicus, N.; Cartuche, L.; Suárez, A.I. Chemical Constituents of *Croton thurifer* Kunth as α-Glucosidase Inhibitors. *Chemistry* **2020**, *19*, 20. [CrossRef]
46. Zengin, G.; Aktumsek, A. Investigation of antioxidant potentials of solvent extracts from different anatomical parts of *Asphodeline anatolica* E. Tuzlaci: An endemic plant to Turkey. *Afr. J. Tradit. Complement. Altern. Med.* **2014**, *11*, 481–488. [CrossRef] [PubMed]
47. Sut, S.; Pavela, R.; Kolarčik, V.; Lupidi, G.; Maggi, F.; Dall'Acqua, S.; Benelli, G. Isobutyrylshikonin and isovalerylshikonin from the roots of *Onosma visianii* inhibit larval growth of the tobacco cutworm *Spodoptera littoralis*. *Ind. Crops Prod.* **2017**, *109*, 266–273. [CrossRef]
48. Sharan Shrestha, S.; Sut, S.; Ferrarese, I.; Barbon Di Marco, S.; Zengin, G.; De Franco, M.; Pant, D.R.; Mahomoodally, M.F.; Ferri, N.; Biancorosso, N. Himalayan Nettle *Girardinia diversifolia* as a Candidate Ingredient for Pharmaceutical and Nutraceutical Applications—Phytochemical Analysis and In Vitro Bioassays. *Molecules* **2020**, *25*, 1563. [CrossRef]
49. Grochowski, D.M.; Uysal, S.; Aktumsek, A.; Granica, S.; Zengin, G.; Ceylan, R.; Locatelli, M.; Tomczyk, M. In vitro enzyme inhibitory properties, antioxidant activities, and phytochemical profile of *Potentilla thuringiaca*. *Phytochem. Lett.* **2017**, *20*, 365–372. [CrossRef]

molecules

Article

Valorization of Apple Peels through the Study of the Effects on the Amyloid Aggregation Process of κ-Casein

Valeria Guarrasi [1,*,†], Giacoma Cinzia Rappa [1,†], Maria Assunta Costa [1], Fabio Librizzi [1], Marco Raimondo [1], Vita Di Stefano [2], Maria Antonietta Germanà [3] and Silvia Vilasi [1]

1 Istituto di Biofisica, Consiglio Nazionale delle Ricerche, Via Ugo La Malfa 153, 90146 Palermo, Italy; cinziarp@hotmail.it (G.C.R.); mariaassunta.costa@cnr.it (M.A.C.); fabio.librizzi@cnr.it (F.L.); raimondo.marco@gmail.com (M.R.); silvia.vilasi@cnr.it (S.V.)
2 Dipartimento Scienze e Tecnologie Biologiche Chimiche e Farmaceutiche, Università degli Studi di Palermo, Via Archirafi 32, 90123 Palermo, Italy; vita.distefano@unipa.it
3 Dipartimento di Scienze Agrarie, Alimentari e Forestali, Università degli Studi di Palermo, Viale delle Scienze Ed. 4, 90128 Palermo, Italy; mariaantonietta.germana@unipa.it
* Correspondence: valeria.guarrasi@ibf.cnr.it; Tel.: +39-0916809356
† These authors contributed equally to this work.

Abstract: Waste valorization represents one of the main social challenges when promoting a circular economy and environmental sustainability. Here, we evaluated the effect of the polyphenols extracted from apple peels, normally disposed of as waste, on the amyloid aggregation process of κ-casein from bovine milk, a well-used amyloidogenic model system. The effect of the apple peel extract on protein aggregation was examined using a thioflavin T fluorescence assay, Congo red binding assay, circular dichroism, light scattering, and atomic force microscopy. We found that the phenolic extract from the peel of apples of the cultivar "Fuji", cultivated in Sicily (Caltavuturo, Italy), inhibited κ-casein fibril formation in a dose-dependent way. In particular, we found that the extract significantly reduced the protein aggregation rate and inhibited the secondary structure reorganization that accompanies κ-casein amyloid formation. Protein-aggregated species resulting from the incubation of κ-casein in the presence of polyphenols under amyloid aggregation conditions were reduced in number and different in morphology.

Keywords: polyphenolic extract; fruit waste; κ-casein amyloid aggregation

Citation: Guarrasi, V.; Rappa, G.C.; Costa, M.A.; Librizzi, F.; Raimondo, M.; Di Stefano, V.; Germanà, M.A.; Vilasi, S. Valorization of Apple Peels through the Study of the Effects on the Amyloid Aggregation Process of κ-Casein. *Molecules* **2021**, *26*, 2371. https://doi.org/10.3390/molecules26082371

Academic Editors: Simona Fabroni, Krystian Marszałek and Aldo Todaro

Received: 6 March 2021
Accepted: 15 April 2021
Published: 19 April 2021

1. Introduction

The recycling and utilization of agricultural wastes is considered to be an important step in environmental sustainability and agricultural development as well as in promoting a circular economy [1]. Large amounts of residue with high levels of organic matter cause serious environmental pollution if discarded or disposed of improperly [2].

In this regard, an important opportunity is provided by the valorization and recycling of peels from apples, which are a rich source of bioactive compounds, like polyphenols [3]. In fact, similar to the whole fruit, the apple peel polyphenol content includes abundant amounts of flavan-3-ols/procyanidines, phenolic acids, flavonols, dihydrochalcones, and anthocyanins [4], known for their beneficial effects in human health [5]. There is evidence that apple polyphenols are strong antioxidants [6] and have, among other benefits, anti-tumor [7], anti-allergy [8], and life-extending properties [9].

Extensive research demonstrated how the polyphenols present in apples are highly promising as potential therapeutics for disorders resulting from protein amyloid aggregation, the process at the basis of severe neurodegenerative diseases, like Parkinson's, Alzheimer's (AD), and Huntington's disease [10–13]. These pathologies are characterized by the conversion of a normally soluble protein into an insoluble aggregated form, with further deposition onto a cross-β fibrillary structure [14–17]. A crucial feature of amyloidosis is the complex polymorphism of the species formed whose ultrastructure and morphology are strictly correlated to their specific toxic potentiality [18,19]. The specific low-molecular-weight oligomers that form during the intermediate steps of the process are thought to play a crucial role in interacting with cell membranes and producing reactive oxygen species (ROS), thus causing cellular function impairment that ultimately leads to the death of the involved cells [20–22]. Thereby, the polyphenols known for their marked antioxidant properties are very effective at fighting the oxidative stress associated with amyloid toxicity [23–25].

An increasing number of studies have demonstrated how several specific polyphenols present in apples, including quercetin, procyanidins, (−)-epigallocatechin-3-gallate (EGCG), (+)-catechin and (−)-epicatechin, ferulic acid, and vanillin, possess the proper structural features to directly interfere with the proteins involved in the amyloid aggregation so as to inhibit their pathological assembly [26–33]. In this regard, a protein that has recently been used as a model system for characterizing the direct interference of polyphenols with proteins susceptible to amyloid formation is κ-casein from bovine milk, which is able to form amyloid fibrillar structures in vitro under physiological conditions (37 °C and neutral pH) [28,34,35].

This is the smallest of the caseins and is responsible for the steric stability of the casein micelles [36]. In native environments, κ-casein forms multimeric colloidal systems through several types of interactions, such as disulfide bonds, electrostatic, hydrogen bonds, and hydrophobic interactions [37–40]. Researchers observed that κ-casein, both in its native and reduced-carboxymethylated form, is able to form amyloid fibrils in vitro if incubated at 37 °C [41–44]. The aggregation in vivo is prevented by the activity of the other two caseins, β- and αs1-casein, which are also naturally present in bovine milk [43,44] and known for their chaperone-like activity [45–49]. When the chaperone-system fails, κ-casein forms corpora amylacea in the bovine mammary glands and in milk [45]. Differently from a typical nucleation-polymerization scheme, the rate-limiting step involved in κ-casein amyloid fibrillogenesis is not an initial nuclei formation but the dissociation of amyloidogenic monomers from the initial multimeric state [41,42,50]. This amyloid precursor has a native β-strand conformation with the proper structural features leading to fibrillar aggregation and it has the same biophysical features of the amyloid core formed by Aβ peptide in Alzheimer's disease [41,42,50]. Similar to the Aβ peptide, κ-casein is the archetype of an amyloidogenic protein belonging to the class of Intrinsically Disordered Proteins (IDPs) [51]. More importantly, native κ-casein forms micelle-like complexes and studies demonstrated that aggregates with micellar properties could play a role in amyloid formation of several proteins [52–54]. Studying the effect of some selected compounds on amyloid aggregation of proteins forming micelles could help to individuate inhibitors with a specific role towards such micelle-like conformations also gaining insight into their mechanism of action. For all these reasons, bovine milk protein κ-casein was used to study the effect of specific polyphenols, many of which present in apples, on the amyloid fibril formation mechanism [28,34,35].

In this study, κ-casein was used as a protein model to investigate the effect of polyphenolic extracts from apple peels in the amyloid aggregation processes. The apple cultivar was Fuji grown in Sicily. While several studies concerned the effect on amyloids of individual polyphenols found in apples, only an experiment, carried out by thioflavin T (ThT) assay, focused on the ability of the whole polyphenolic extract obtained from apples to inhibit an amyloid aggregation process [31]. However, the molecular mechanism underlying this inhibitory action is far from understood. This is a critical point to evaluate the therapeutic potentiality of fruit and vegetable total phenolic extracts, already tested for important conditions regarding human health [55], in the field of amyloid diseases.

To the best of our knowledge, no study has been performed on extracts from a waste product, like apple peel, in the amyloid aggregation process. These investigations, aimed at agricultural and food industry waste valorization, offer important incentives in the recovery and reutilization of valuable materials, thus sustaining the eco-system health and circular economy strategies. In fact, the waste from apples used for commercial products, like jam, juice, and jelly, constitute about 30% of the original fruits. This large amount of organic residue seriously contributes to increasing air pollution [2].

Here, we investigated the influence of the apple peel polyphenol extracts on the amyloid assembly process of κ-casein by incubating them with the protein before thermally inducing fibrillization. We used light scattering (LS) and thioflavin T (ThT) fluorescence assay to monitor the effect of the extracts on the κ-casein aggregation, and fibril formation kinetics. Using Congo red binding assay, circular dichroism (CD), and atomic force microscopy (AFM), we monitored the structural and morphological features of the κ-casein species formed in the presence of the apple peel polyphenols (APP).

Our results show that the APP extract inhibited, in a dose-dependent manner, κ-casein amyloid fibril formation, and the structural reorganization that occurs during the protein assembly was inhibited in the presence of APP. The aggregates resulting from κ-casein incubation with APP under amyloid formation conditions were different in number and morphology.

2. Results

2.1. *Total Phenolic Content (TPC), Antioxidant Properties, and Phenolic Characterization of Extracts from Apples Peels*

As a prior step to the experiments, we conducted a quantitative determination of the apple peel extract phenolic compounds using the Folin–Ciocalteu method. The results are reported in Table 1 where the TPC is expressed as catechin equivalents. For comparison, in the same table, TPC values from the same apple variety other tissues (flesh and whole fruit) are also reported. The results show that the peel had a significantly higher TPC value over both the flesh and whole fruit.

Table 1. The peel, flesh, and whole Fuji apple total phenolic content evaluated by Folin–Ciocalteu and the antioxidant properties assessed by ORAC assay. ORAC = oxygen radical absorbance capacity; TE = Trolox equivalents; FW = 100 g fresh weight; and CTE = catechin equivalents.

Tissue	ORAC Value (μmol TE/g FW)	TPC (mg CTE/100 g of FW)
Peel	118 ± 5	315 ± 20
Flesh	13 ± 5	42 ± 5
Whole fruit	22 ± 7	54 ± 8

We also evaluated the antioxidant capacity of the three tissues (peel, whole, and flesh) by using an oxygen radical absorbance capacity (ORAC) assay (Table 1). The ORAC method uses a biologically relevant free radical source (peroxyl radical), which is the most prevalent free radical in human biology [56]. Therefore, ORAC results can be correlated to in vivo studies on the antioxidant activity shown by apple polyphenols [57–61].

As shown in the table, the antioxidant activity of the polyphenolic extract from apple peel was significantly more marked when compared to the other tissues of the fruit.

The polyphenolic profile of Fuji apple peel extract was studied using ultra-high-performance liquid chromatography with heated electrospray ionization coupled to mass spectrometry (UHPLC–HESI–MS) techniques, and seven phenolic compounds were identified (Supplementary Materials, Figure S1). In good agreement with the literature data [5], the qualitative profile showed vanillin, gentisic acid, sinapinic acid, ferulic acid, epicatechin, quercetin, and procyanidin A2.

2.2. *Ability of the APP Extract to Influence κ-Casein Fibrillogenesis and Disaggregate κ-casein Preformed Fibrils*

The effect of polyphenolic extracts from the apple peel on the kinetics of κ-casein aggregation was analyzed by Thioflavin T (ThT) fluorescence assay. ThT is a fluorescent dye that is known to reliably detect the formation of amyloid fibrils due to the significant increase in its emission intensity upon binding to the linear array of β-strands [62,63].

We analyzed the effect of APP at the concentrations of 5, 10, 20, and 40 μg/mL, after testing their toxicity on NHI-3T3 cells. This is an easy-to-handle cell line that immortalized spontaneously [64] and, for this reason, it is considered more representative of normal cells than a tumor cell line. Hence, NIH/3T3 is one of the most frequently used lines in material/cell interaction research, and it has been previously used for studying the toxic effect induced on cells by amyloid insults [22,65,66].

The above-mentioned concentrations were not cytotoxic after 72 h of treatment as reported in Figure S2 (Supplementary Materials). Figure 1 shows the ThT signal during the fibrillization kinetics of 50 μM κ-casein incubated at 37 °C in the absence and in the presence of different APP concentrations. In the absence of polyphenols, ThT from κ-casein exhibited an increase in fluorescence emission, which is indicative of amyloid formation. The absence of a lag phase in the kinetics profile is due to an aggregation mechanism in which the rate-limiting step is not the initial nuclei aggregation but the dissolution of native protein multimers with subsequent exposition to the solvent of buried amyloidogenic regions [41,42].

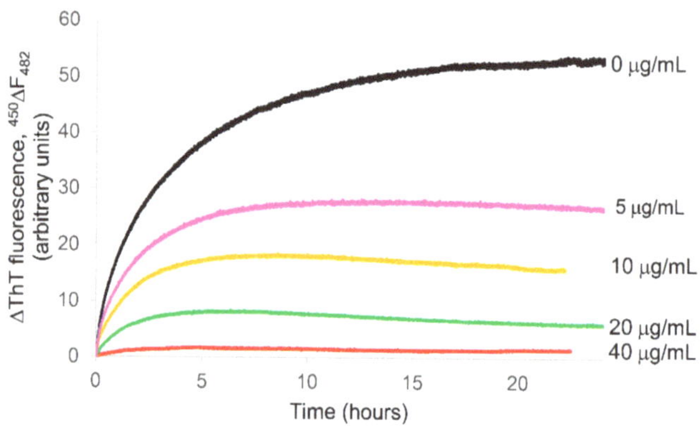

Figure 1. In situ real-time thioflavin T fluorescence assay for monitoring the aggregation kinetics of 50 μM (~1 mg/mL) κ-casein incubated at 37 °C in the absence (black) and in the presence of different APP concentrations: 5 μg/mL (pink), 10 μg/mL (yellow), 20 μg/mL (green), and 40 μg/mL (red).

In the presence of APP, both the aggregation rate and the fluorescence plateau, corresponding to the total amount of fibrils, were reduced in a concentration dependent manner (Figure 1).

As the next step, we verified that, in addition to inhibiting protein aggregation, APP were also able to disaggregate preformed κ-casein amyloid structures. To do this, we added APP at 40 μg/mL to the 50 μM κ-casein sample taken at the end of the aggregation kinetics (after about 24 h at 37 °C) and followed the dye fluorescence for the next 8 h at 37 °C. After an initial phase taken for the system to achieve the thermal equilibration, the fluorescence did not vary in time (Figure 2), thus revealing that, while APP were able to inhibit protein aggregation, they could not disrupt already formed amyloid assemblies.

This experiment was also useful in order to exclude the occurrence of potential interference between the dye and APP. Caution is recommended in the use of thioflavin T assay when probing anti-amyloidogenic compounds that, even if not spectroscopically active at ThT wavelengths, like the APP under analysis (data not shown), can interfere with the dye fluorescence causing a bias affecting the experimental data interpretation [35,67]. Our experiment showed that APP did not cause ThT fluorescence reduction by interfering with the dye in a time of about 8 h, thus excluding the occurring of fluorescence quenching, in this time, by the introduction of phenols.

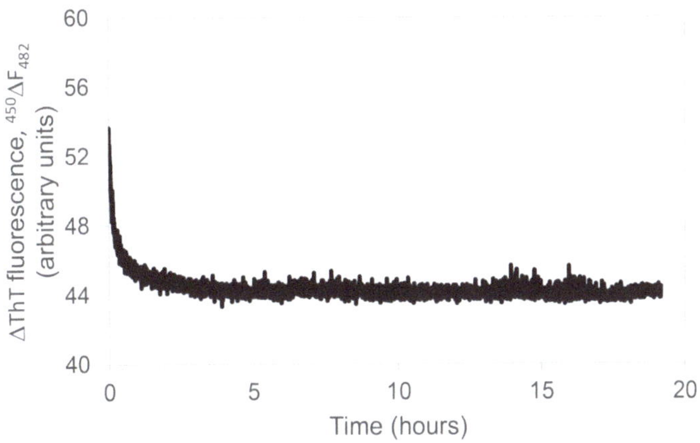

Figure 2. Fluorescence of Thioflavin T as a function of time after the addition of 40 μg/mL APP to 50 μM (~1 mg/mL) κ-casein species formed after 24 h incubation at 37 °C.

2.3. *β-Sheet-Amyloid Formation Assessed by the Congo Red Binding Assay*

We used other experimental tools to confirm the inhibitory APP effect found by the ThT assay. First, we used the absorptive dye, Congo red (CR), which has been demonstrated as a viable alternative to ThT in the case of potential interference with exogenous compounds precisely tested with κ-casein [35]. CR allows the detection of the presence of amyloid β-pleated sheets, because, when bound to β-sheet-rich amyloid fibrils, the CR molecules adopt a specific orientation with their long axis lying parallel to the fibril axis. This causes a molecular CR torsional restriction that induces a characteristic increase in absorption and a red shift in the absorption maximum from 490 to 540 nm [68]. To further investigate inside the β-sheet-amyloid formation in κ-casein aggregates and evaluate the effect of APP on κ-casein assembly process, we added 20 μM CR to aliquots of 50 μM fresh native κ-casein and the protein after 24 h of incubation at 37 °C. The experiment was performed on samples incubated both in the presence and absence of APP at 40 μg/mL. After the dye addition, the CR absorption band was recorded immediately. In Figure 3a, the spectra of CR alone, CR in the presence of fresh native κ-casein, and CR in the presence of κ-casein after 24 h of incubation at 37 °C are presented. In the presence of fresh native κ-casein, the CR spectrum showed, in comparison to free dye, a slight shift from 500 to 510 nm likely due to the dye binding to protein initial micelles. However, a more marked change occurred when CR was incubated with the κ-casein preformed structure formed after 24 h at 37 °C. A shoulder at 540 nm appeared, indicative of CR binding to β-sheet-rich amyloid fibrils [68]. In the presence of polyphenols, no significant change in the absorbance maximum wavelength was observed between the spectra of CR added to the native κ-casein + APP and the same sample taken after 24 h incubation at 37 °C (Figure 3b). This further confirms the APP inhibitory action toward amyloid formation.

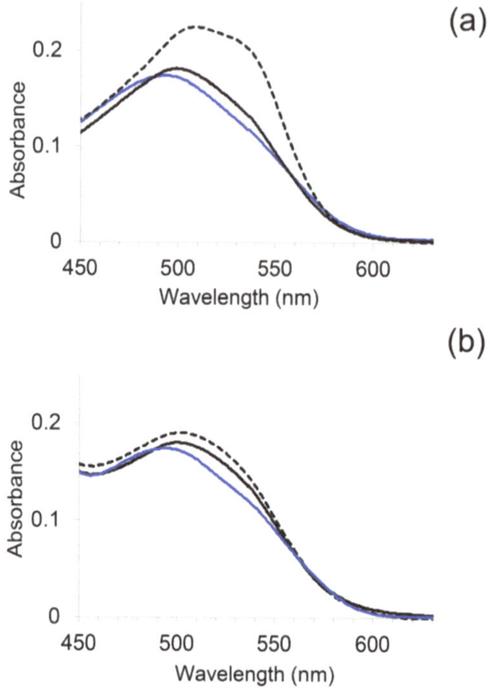

Figure 3. (**a**) Absorption spectra of 20 μM Congo red (blue), Congo red with 50 μM fresh native κ-casein (solid black line), and with 50 μM κ-casein incubated for 24 h at 37 °C (dashed black line). (**b**) Absorption spectra of 20 μM Congo red + 40 μg/mL APP (blue), CR with 50 μM κ-casein + 40 μg/mL APP before (solid black line), and after incubation 24 h at 37 °C (dashed lack line).

2.4. Apple Peel Polyphenols Influence on κ-Casein Secondary Structure

We used circular dichroism (CD) to monitor the secondary structure variation of κ-casein incubated under amyloid aggregation conditions in the absence and in the presence of different APP concentrations. In Figure 4a, the κ-casein spectra at the beginning of the aggregation process, which starts with thermal incubation at 37 °C (t_0), and after about 24 h (t_{end}), corresponding to the plateau in protein fibrillogenesis kinetics (Figure 1), are reported. We found that, similar to what was reported by Farrell et al. [40], the protein during amyloid formation underwent a conformational transition with two spectral changes, one at 200 nm, and the other, more marked, at around 220 nm. As shown by electron microscopy experiments, these spectral changes are associated, by Farrell et al., with the formation of structures with fibrillar morphology. The fibrillization of κ-casein starts from the dissociation from the oligomeric state, which is considered the rate limiting step in the protein fibril formation [41,42]. The dissociated species, characterized by a significant percentage of β-native structure, would be the amyloidogenic precursor of fibrillar assembly through a β-sheet stacking that would not dramatically increase the total β-structure content but would be responsible for variations in the distribution of secondary β-strand and turn elements, thus determining the shape changes in the CD spectrum. These variations were not detectable when the protein was incubated with extracts from apple peel at 40 μg/mL polyphenolic content (Figure 4b), which was the most effective in preventing ThT fluorescence increase. The two spectra, at the beginning (t_{start}) and after 24 h (t_{end}) of incubation, almost overlap. At the APP concentrations' intermediate range (5, 10, and 20 μg/mL), the higher the polyphenol content (Supplementary Materials, Figure S3), the more evident the changes in the spectra between that recorded at initial time t_{start} and after 24 h (t_{end}) of incubation. To better visualize these results, we reported, in Figure 4c, the difference squared between the κ-casein spectra recorded at t_{start} and t_{end} in the wavelength region around 222 nm. As the figure shows, the extract from apple peel was able, in a concentration-dependent manner, to reduce the spectra changes sensitive to the conformational transitions that accompany the κ-casein amyloid formation.

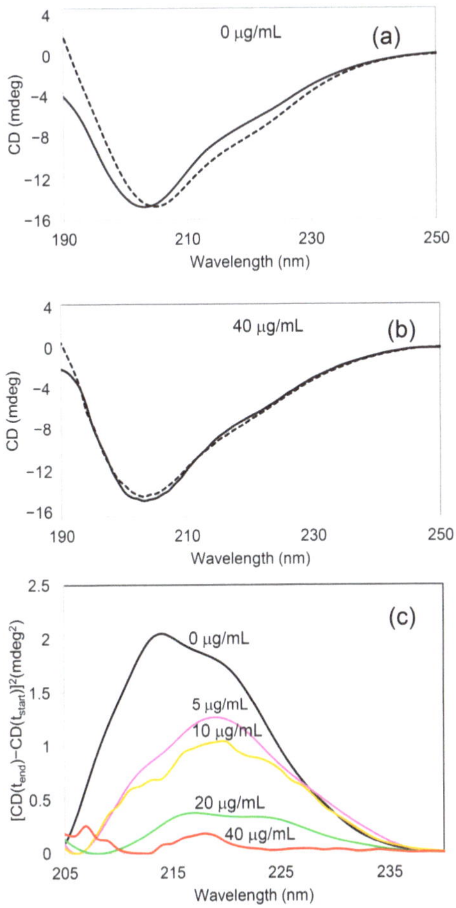

Figure 4. Circular dichroism (CD) spectra of 50 μM κ-casein at the beginning (t_{start}) of the amyloid aggregation process (solid black) and after 24 h (t_{end}) of incubation at 37 °C (dashed black) in the absence (**a**) and presence of 40 μg/mL of APP (**b**). (**c**) Squared difference between the spectrum at t_{start} and t_{end} in the absence (black) or the presence of different APP concentrations: 5 μg/mL (pink), 10 μg/mL (yellow), 20 μg/mL (green), and 40 μg/mL (red).

2.5. *Apple Peel Polyphenols Influence on κ-Casein Aggregation Monitored by Light Scattering*

To further evaluate the κ-casein aggregation process, we carried out static and dynamic light scattering measurements of κ-casein incubated at 37 °C in the absence and in the presence of APP at the concentration of 40 μg/mL that, as revealed by ThT assay and CD measurements, was the most effective concentration among the tested ones in inhibiting amyloid formation. The light-scattering intensity depends on the molecular mass of a species in solution and, therefore, is a suitable technique to monitor the aggregation process. κ-casein is initially in an oligomeric micelle-like state, and this is reflected in the light-scattering intensity value that is detected at the beginning of the process.

When κ-casein was incubated at 37 °C, an increase in the intensity scattered from the particles in solution occurred in time. The static scattered intensity, in particular, switched from 540 kcps to 700 kcps, from the beginning to the end of the process, with an increase of 30% indicating the occurrence of an aggregation process (Figure 5a). By the cumulant analysis of the intensity autocorrelation function generated by LS in dynamic modality, a correspondent increase in the average hydrodynamic diameter of species in solution from 100 to 110 nm was also found (Figure 5b).

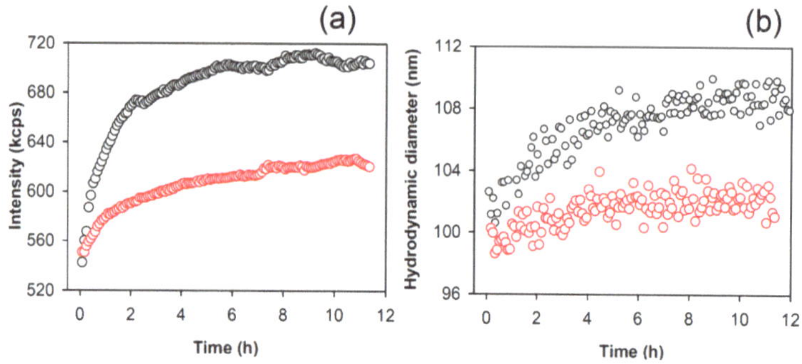

Figure 5. (**a**) The time course of the light scattered intensity (measured as kilo count per seconds (kcps), referring to the average number of photons per second arriving at the detector) from κ-casein (50 μM = 1 mg/mL) incubated at 37 °C in the absence (black circles) and presence of 40 μg/mL of apple peel polyphenols (red circles). (**b**) Time-course of the hydrodynamic diameters of κ-casein (50 μM = 1 mg/mL) incubated in the absence (black circles) and presence of 40 μg/mL of apple polyphenols (red circles) obtained from the cumulant analysis of the intensity autocorrelation function.

When κ-casein was incubated with 40 μg/mL of APP extract, the increase in the scattered intensity was significantly reduced (Figure 5a). It switched from 540 to 610 kcps with an increase of only 13% compared to the κ-casein alone condition. A slight variation in the average hydrodynamic diameter of species in solution was observed as it remained at about 100 nm (Figure 5b).

2.6. Morphology of κ-Casein Aggregates Incubated with Apple Peel Polyphenols

To analyze the morphology of the κ-casein species formed at the end of kinetics in the presence or absence of 40 μg/mL APP, AFM imaging was applied. When κ-casein was incubated alone, fibrillar rod-shaped structures with average height of 5.13 nm and length of 100 nm (Figure 6a) were observed. According to the other techniques, when κ-casein was incubated in the presence of APP extract (40 μg/mL) (Figure 6b), the number of fibers and amyloid aggregates was found to be significantly reduced (Figure 6c).

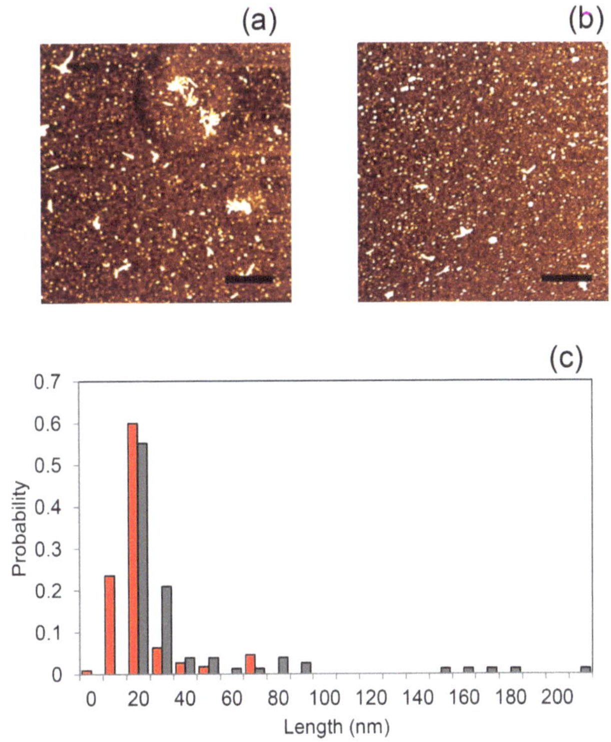

Figure 6. The morphology of κ-casein species formed in the absence and presence of polyphenols (**a**) κ-casein aggregates (50 μM) formed at the end of the aggregation kinetics (12 h). (**b**) κ-casein aggregates (50 μM) formed after 12 h in the presence of 40 μg/mL apple polyphenols. Scale bar = 400 nm and z-range = 5.3 nm. (**c**) Distributions of the lengths for the sample populations in the absence (grey) and presence of polyphenols (red).

2.7. Apple Peel Polyphenols Effect on Aβ$_{1–42}$ Peptide Aggregation Monitored by Light Scattering

To add value to the results thus far obtained, we concluded our study by obtaining information on APP action on the amyloid aggregation process of a protein that, differently from κ-casein, is involved in a severe and common disease for human health. Hence, we tested the APP effect on the Aβ$_{1–42}$ peptide involved in Alzheimer' Disease by using the free-dye light scattering technique. It is known that heating from 0 to 37 °C induces, in the Aβ$_{1–42}$ peptide, a conversion from coil to beta-strand structures typical of amyloid formation [69]. As shown in Figure 7a, when the Aβ$_{1–42}$ peptide is incubated at 37 °C in the presence of APP, the increase found in the light scattered for the Aβ$_{1–42}$ peptide alone, and indicative of the aggregation process, was strongly reduced.

However, as shown by the analysis of the hydrodynamic radii, also in this case, in the presence of APP, Aβ$_{1–42}$ forms aggregated species, although lower in size with respect to the control (Figure 7b). Further studies should aim to better understand the structural and pathological features of the Aβ$_{1–42}$ peptide species resulting from the APP; however, this is out of the objectives of the present study. We just want to highlight the importance of considering the synergic action of phenols from natural sources in the amyloid aggregation mechanisms at the basis of dangerous conditions for human health.

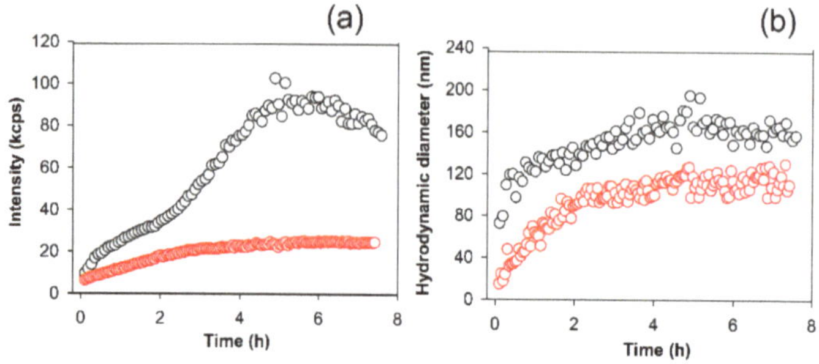

Figure 7. (**a**) Time-course of the light scattered intensity (measured as kilo count per seconds (kcps), referring to the average number of photons per second arriving at the detector) from 50 μM Aβ_{1-42} incubated at 37 °C in the absence (black circles) and presence of 40 μg/mL of apple peel polyphenols (red circles). (**b**) Time-course of the hydrodynamic diameters of Aβ_{1-42} incubated in the absence (black circles) and presence of 40 μg/mL of apple polyphenols (red circles) obtained from the cumulant analysis of the intensity autocorrelation function.

3. Discussion

The aim of our study was to gain insight into the ability of polyphenolic extracts to prevent the amyloid aggregation of the κ-casein, used as amyloidogenic protein model. The extracts were from apple peels, which are normally disposed of as waste in agriculture and the food industry. Several studies reported the ability of specific single polyphenols to inhibit both oxidative stress and protein aggregation associated with neurodegenerative diseases. However, few studies focused on the direct action on amyloid aggregation pathways exerted by the whole polyphenolic extract from fruit tissues generally considered as waste.

Research has shown that apples are a great source of bioactive compounds [5] with marked antioxidant activity and antiproliferative activity [5–7], and we demonstrated that the peels from the Fuji apples under study presented higher phenolic contents and more marked antioxidant activity with respect to the flesh and whole fruit (Table 1). The polyphenolic profile of Fuji apple peel extracts was studied using UHPLC–HESI–MS techniques, and seven phenolic compounds were identified (Supplementary Materials, Figure S1). In agreement with the literature data, a qualitative profile found vanillin, gentisic acid, sinapinic acid, ferulic acid, epicatechin, quercetin, and procyanidin A2, many of which have been revealed to be effective in preventing protein amyloid aggregation when singly tested [26–33], but they have never been tested in a natural phenolic pool.

To monitor the APP effect on amyloid formation mechanisms, we used κ-casein from bovine milk, a suitable model protein to study the aggregation processes in vitro. First, κ-casein forms fibrils at pH and temperature conditions (pH 7.0 and 37 °C) that well reproduce the physiological cell environment [41,42]. In addition, according to the Ecroyd model, a κ-casein monomer contributes more than one β-strand to the fibril structure, in a similar manner to that shown for the amyloid Aβ peptide involved in AD [70]. Similar to the peptides and proteins involved in several amyloid diseases (e.g., Alzheimer's and Parkinson's disease), κ-casein adopts an essentially unstructured conformation in its normal biological state (i.e., it belongs to the class of Intrinsically Disordered Proteins (IDPs)) [43,51]. More importantly, κ-casein forms micelles, and several studies hypothesized the crucial role of potential intermediate micellar conformations in amyloid fibrillogenesis [52–54].

Here, we show that APP at concentrations not toxic for cells (Figure S2) significantly inhibited κ-casein amyloid fibril formation in a concentration-dependent manner. Both the rate and the plateau of the protein fibrillogenesis kinetics profile were reduced in the presence of APP, and the effect was more marked at increasing concentrations of incubated polyphenols (Figure 1). At the highest phenol concentration tested (40 μg/mL), the thioflavin T increase appeared to be totally inhibited. The marked increase in ThT fluorescence results from the selective immobilization in the conformational ensemble of the dye. When free, the benzylamine and benzathiole rings of ThT can rotate with no constraints about their shared carbon–carbon bond. This rotation rapidly quenches the excited states generated by photon excitation causing low fluorescence emission for free ThT. Instead, in the presence of amyloid fibrils, the rings are sterically immobilized, and this preserves the excited state, resulting in a high fluorescence quantum yield. Therefore, ThT can be considered as a "molecular rotor" [71]. To exclude the potentially occurring interactions of polyphenols-ThT [35,67], we verified that fluorescence of the dye was not influenced by APP. In order to do this, we added 40 µg/mL APP to a sample of preformed κ-casein fibrils and ThT and verified that the dye fluorescence, enhanced by the presence of fibrillar structures, did not change in time (Figure 2), thus excluding the occurring of fluorescence quenching by the introduction of phenols.

Furthermore, we also used alternative dyes and dye-free tools to confirm the APP inhibitory properties. First, we confirmed the β-sheet formation in κ-casein incubated at 37 °C by using Congo red (CR) whose spectral shift assay is an alternative to ThT fluorescence for quantifying amyloid fibril formation [35]. The spectral shift present when CR is added to κ-casein after 24 h at 37 °C is absent when the dye is added to the protein incubated, under the same conditions, with APP (Figure 3). Then, we used circular dichroism to evaluate the APP ability to prevent the secondary structure variation that accompanies κ-casein amyloid formation. In agreement with thioflavin T experiments, this secondary structure variation was lower for higher APP concentrations, and the κ-casein spectrum registered after 24 h of incubation at 37 °C in the presence of APP at 40 µg/mL did not present significant changes in respect to the κ-casein at the initial state (Figure 4).

When κ-casein was incubated with APP at 40 µg/mL, the light scattered intensity variation from the sample was significantly reduced when compared to the control sample (Figure 5a) with very little variation in the average hydrodynamic diameter of species in solution observed over time (Figure 5b). The presence of aggregated species formed from κ-casein in the presence of 40 µg/mL APP, even if lower in size, morphologically different, and reduced in number with respect to the control, was also observed by AFM (Figure 6).

This is a very critical point because, as well established, amyloid oligomers smaller in size than the mature fibrillary structures are greatly more toxic than fibrils, due to a greater number of open active ends and their ability to diffuse in tissues [20–22]. However, the κ-casein aggregates formed in the presence of APP do not possess the amyloid cross-β spine structure that, in restricting the thioflavin rings rotation, determines a high fluorescence quantum yield. CD experiments further prove that these aggregates are prevented from forming the β-sheet stacking, which is responsible for the secondary structure content variation observed during κ-casein amyloid formation.

According to the model proposed by Ecroyd et al. [41,42], the κ-casein dissociation from the oligomeric state is the rate-limiting step in κ-casein fibril formation. The dissociated species, characterized by a significant percentage of β-native structure, would be an amyloidogenic precursor of fibrillar assembly through β-sheet stacking that would not dramatically increase the total β-structure content but would be responsible for variations in the distribution of secondary β-strands and turn elements. We hypothesize that this reorganization, leading to β-sheet packing occurring inside the first oligomers, is prevented by the interference with polyphenols.

Research reported that polyphenols are capable of directly interacting with the protein amyloidogenic core through specific structural constraints and aromatic interactions between the phenolic compounds and aromatic residues in the amyloidogenic sequence. The packing in κ-casein amyloid formation would involve residues from position to 20 to 68 that contain not-stranded, highly hydrophobic β-sheets (from 20 to 25, 29 to 34, 39 to 45, and 49 to 55) connected by γ or β-turns [50]. These segments belong to a tyrosine-rich region (Figure 8), which, therefore, is highly susceptible to aromatic interactions with APP that we believe to occur before potential further packing leading to the amyloid core formation.

```
          10         20         30         40         50
QEQNQEQPIR CEKDERFFSD KIAKYIPIQY VLSRYPSYGL NYYQQKPVAL

          60         70         80         90        100
INNQFLPYPY YAKPAAVRSP AQILQWQVLS NTVPAKSCQA QPTTMARHPH

         110        120        130        140        150
PHLSFMAIPP KKNQDKTEIP TINTIASGEP TSTPTTEAVE STVATLEDSP

         160       169
EVIESPPEIN TVQVTSTAV
```

Figure 8. The amino acid sequence (signal sequence-deprived) of κ-casein from bovine milk (UniProtKB-P02668). The region involved in protein packing during amyloid formation is in light yellow. Residues belonging to native β-sheets are in green.

The aromatic interactions could be at the basis of the preventive action exerted by tannin and (−)-epigallocatechin-3-Gallate (EGCG) against κ-casein aggregation and, in general, for polyphenol inhibitors in amyloid aggregation through direct interactions with the proteins involved [28,34,35].

Our studies also revealed that APP did not show disrupter activity on κ-casein fibrils. This could be indicative of fibrillary stiffness or most likely, since only few protein segments are involved in the formation of the cross-β-sheet fibrillar structure, the remainder of the molecule could prevent the APP from reaching the fibril core, thus, impeding the formation of the preferential noncovalent interactions responsible for the dissolution of the fibrils.

4. Materials and Methods

4.1. Materials

κ-casein (UniProtKB-P02668) from bovine milk and all the reagents and chemicals were purchased from Sigma-Aldrich (St. Louis, MO, USA) except when mentioned specifically.

4.2. Plant Materials

Fuji apples were harvested at commercial maturity from "Scannale s.r.l.", an organic farm located in Caltavuturo (37°49′ N and 850 m a.s.l.) (Sicily, Italy). Fruits were harvested at optimum maturity assessed by the starch iodine test (starch index value of 6; [72]). The apples were washed and stored at 4 °C.

Weight, longitudinal (LD) and transverse (TD) diameters, and flesh firmness of apples were detected. The fruits were individually weighted with a precision balance. LD and TD were measured with a digital caliper. The flesh firmness was measured on half of fruits with a penetrometer with a cylindrical 8 mm head (EFFEGI texture analyzer). Chemical parameters such as titratable acidity (TA), pH, and soluble solid content (SSC) were also detected. TA was performed by titrating an aliquot of 5 mL of filtered apple juice with a solution of NaOH 0.1 N and expressed in percentage of malic acid. The SSC were measured with digital refractometer (Atago) and reported as °Brix. Ripe Index (RI) was also calculated as the ratio between SSC, expressed in °Brix, and TA, and expressed in percentage (%). Physical parameters data: weight = (158.76 ± 0.45) g; LD = (60.10 ± 0.29) mm; TD = (66.70 ± 0.17) mm; firmness = (4.46 ± 0.45) Kg/cm^2. Chemical parameters data: pH = 3.9 ± 0.1; TA = (0.30 ± 0.03) % malic ac.; SSC = 18.52 ± 0.10 °Brix; RI= 61.73 ± 0.83. All parameters were obtained by a mean of five measures. Thirty Fuji apples free of defects were selected and peeled (~1 mm thickness) then the apple peels were pooled and frozen in liquid nitrogen and stored at −80 °C until analysis.

4.3. APP Extraction, Phenolic Content Determination and Antioxidant Properties Evaluation

4.3.1. Polyphenol Extraction

Polyphenols were extracted by a peel pool of at least thirty Fuji apples according to the method described by Ceymann et al. [73], with some modifications. Frozen peels were ground to fine powder by using a grinder and then the polyphenolic extraction was carried out. Briefly, an aliquot of the powder (2.50 g) was mixed with 50 mL of methanol containing 1% formic acid (*v*/*v*), homogenized for 15 min, and centrifuged at 10,000 rpm for 10 min. The supernatant was collected and filtered through 0.45 µm nylon filters. The polyphenols' extraction was repeated three times. The same protocol was used to extract polyphenols from whole and flesh Fuji apple and compare the results with those obtained for the peel. For toxicity and k-casein experiments, the extraction solvent was removed by rotatory evaporator at 30 °C and the concentrated peel polyphenolic extract was dissolved in water, aliquoted, and kept at −20°C until analysis, avoiding direct contact with light and oxygen.

4.3.2. Total Polyphenol Content (TPC) by Folin–Ciocalteu

The TPC analysis was performed following the polyphenols extraction by the Folin–Ciocalteu method in Ceymann et al. [73] with slight modifications. Two microliters of the Folin–Ciocalteu reagent and 2 mL of distilled water were automatically pipetted to 0.2 mL of the methanolic extract. After 1 min, 0.8 mL sodium carbonate solution (20%) and 0.8 mL distilled water were added, thoroughly mixed, and incubated for 30 min at 37 °C. Absorption was measured automatically at 750 nm. Total polyphenolic content was calculated with the external standard catechin calibration curve and expressed as mg catechin equivalents (CTE) per 100 g fresh weight (FW). The test was carried out in triplicate.

4.3.3. Identification of Polyphenols by UHPLC–HESI-MS

Identification of polyphenols in Fuji apple peel extract was based on a reported procedure [74,75]. Phenolic compounds were identified by ultra-high-performance liquid chromatography, heated electrospray coupled with high-resolution mass spectrometry (UHPLC-HESI-MS) analysis using a quadrupole Orbitrap mass spectrometer (Thermo Fisher Scientific, Bremen, Germany). UHPLC analysis was performed using a Dionex Ultimate 3000 System (Dionex Softron GmbH, Germering, Germany) equipped with an auto sampler controlled by Chromeleon 7.2 Software (Thermo Fisher Scientific, Bremen, Germany). The column was a Phenomenex Luna C18(2) 50 × 1 mm, packed with core–shell particles of 2.5 µm. The flow rate was set at 50 µL min^{-1} at 20 °C and the total chromatographic analysis time was 52 min. The eluent A was water with 0.1% formic acid (*v*/*v*) pH 3.2, and eluent B was acetonitrile with 0.1% formic acid (*v*/*v*). The elution gradient program was 0–5 min 10% B; 5–45 min linear increase to 99% B, 45–50 min 10% B coming back to the initial conditions until full stabilization. The MS detection was conducted in two acquisition modes: full scan (negative ion mode) and targeted selected ion monitoring. For targeted selected ion monitoring analyses, a mass inclusion list containing exact masses of target phenolic compounds was built and applied.

4.3.4. Oxygen Radical Absorbance Capacity (ORAC) Assay

The method reported by Cao et al. [76] was slightly modified and applied. The reaction mixture was prepared in a 96-well black microplate as follows: 160 µL of 0.04 µM Fluorescein in 0.075 M Na-K phosphate buffer pH 7.0, 20 µL of appropriately diluted extract or 20 µL of 100 µM Trolox used as control standard, or 20 µL of extraction solvent as blank. Each mixture was kept 10 min at 37 °C in the dark, and the reaction was started with the addition of 20 µL of 40 mM 2,2′-azobis (2-methylpropionamidine) dihydrochloride (AAPH). The fluorescence decay was measured at 37 °C every 1 min at 485 nm excitation and 538 nm emission, using a Thermo Scientific Fluoroskan Ascent F2 Microplate. The ORAC value refers to the net area under the curve (AUC) of fluorescein decay in the presence of ginger extract or Trolox, subtracted of the blank area. The activity of the sample was expressed as μmol of Trolox Equivalents (TE)/g of FW, with the following equation (Equation (1)):

$$\text{ORAC value } (\mu\text{mol TE/g FW}) = k \times a \times h \times [(S_{sample} - S_{blank})/(S_{Trolox} - S_{blank})] \quad (1)$$

where k is the extract dilution; a is the ratio between the volume (liters) of the extract and grams of sample used for the extraction; h is the final concentration of Trolox expressed as μmol/L; and S is the area under the curve of fluorescein in the presence of sample, Trolox, or blank.

4.4. Cell Culture and Cytotoxicity Assay

Mouse embryonic fibroblast cell line NIH-3T3 (Sigma-Aldrich) was cultured in Dulbecco's Modified Eagle Medium (DMEM)-high glucose supplemented with 100 U/mL penicillin, 100 μg/mL streptomycin, and 10% bovine calf serum at 37 °C in 5% CO_2 humidified atmosphere. NIH-3T3 cells were seeded into 96-well plates at a density of 104 cells per well in 100 μL of growth medium, cultured for 24 h, and then treated Fuji apple peel polyphenolic extract at final concentrations of 80, 40, 20, 10, or 5 μg/mL. All treatments were performed for 24 and 48 h of incubation in triplicate. Cell viability was analyzed by the CellTiter 96® AQueous One Solution Cell Proliferation Assay (MTS assay, Promega, Milano, Italy) following the manufacturer's instruction. In brief, after cell treatments, 20 μL of the MTS solution was added to each well and incubated with cells for 3 h at 37 °C, 5% CO_2. The absorbance was read at 490 nm with the Bio-Rad iMarktm Microplate Reader (Promega, Milano, Italy). Cell viability was quantified as the percentage of viable cells using untreated cells as a control. All the experiments were repeated three times.

4.5. κ-Casein Amyloid Formation

A fresh stock solution of κ-casein in 50 mM phosphate buffer pH 7.4 was continuously stirred for 24 h and filtered through 0.22 μm filters before using. As verified by dynamic light scattering, this procedure assures sample homogeneity but not reduction to monomeric species, which cannot be obtained due to the presence, in the κ-casein sample, of stable self-associating oligomeric micelle-like species [40]. The protein concentration was determined by recording the absorbance spectrum after filtering the sample by using an extinction coefficient at 280 nm of 0.95 $mg^{-1} \cdot mL \cdot cm^{-1}$ [39] and considering the purity of the sample (70%) provided by Sigma Aldrich. κ-casein fibril formation is highly temperature-dependent [44]. To generate amyloid fibrils, all the experiments were performed incubating protein solutions at 50 μM without shaking at the mammalian physiological temperature of 37 °C.

4.6. Atomic Force Microscopy (AFM) Measurements

AFM measurements were performed by using a Nanowizard III (JPK Instruments, Berlin, Germany) mounted on an Axio Observer D1 (Carl Zeiss, Berlin, Germany) or on an Eclipse Ti (Nikon, Tokyo, Japan) inverted optical microscope. Aliquots of protein solutions were deposited onto freshly cleaved mica surfaces (Agar Scientific, Assing S.P.A., Monterotondo, Roma, Italy) and incubated for up to 20 min before rinsing with deionized water and drying under a low-pressure nitrogen flow. Imaging of the protein was carried out in intermittent contact mode in air by using NCHR silicon cantilever (Nanoworld, Neuchâtel, Switzerland) with nominal spring constant ranging from 21 to 78 N/m and typical resonance frequency ranging from 250 to 390 kHz.

4.7. Thioflavin T (ThT) Spectrofluorometric Measurements

ThT fluorescence emission was monitored by using a JASCO FP-6500 spectrometer (JASCO Corporation, Tokyo, Japan). The excitation and emission wavelengths were 450 and 485 nm, respectively, with a slit width of 3 nm. ThT concentration was 12 μM. κ-casein (50 μM) was dissolved with the dye before being placed at 37 °C in the thermostated cell compartment (10 mm). It is known that, upon binding to fibrils, ThT displays a dramatic shift of the excitation maximum (from 385 nm to 450 nm) and of the emission maximum (from 445 nm to 482 nm) [77]. Due to the binding of native κ-casein with Thioflavin T [40], data were presented as ΔF in function of time, where $\Delta F = F - F_0$ is the difference between the fluorescence reading value and F_0 the fluorescence value registered once samples reached thermal equilibrium at 37 °C, after about 20 min.

4.8. Congo Red Binding Assay

The binding of Congo red was assessed by using absorption spectroscopy (Spectrophotometer Shimadzu UV2401PC, Tokyo, Japan). A 4 mM stock solution of Congo red (Sigma Aldrich) was prepared in 50 mM phosphate buffer pH 7.4. Congo red solution (20 μM) was mixed with κ-casein species (50 μM) taken at the beginning and after 24 h of protein incubation at 37 °C in the absence and in the presence of 40 μg/mL APP.

4.9. $A\beta_{1\text{–}42}$ Peptide Amyloid Formation

The synthetic peptide $A\beta_{1\text{–}42}$ (AnaSpec, Inc., Fremont, CA, USA) was solubilized in NaOH. 5 mM, pH 10, and lyophilized according to Fezoui et al. protocol [78]. The lyophilized peptide was then dissolved in 50 mM phosphate buffer pH 7.4 and filtered with two filters in series of 0.20 μm (Whatman, Merck KGaA, Darmstadt, Germany) and 0.02 μm (Millex-LG, Merck KGaA, Darmstadt, Germany) respectively, in order to eliminate large aggregates. The sample preparation was operated in asepsis using a cold room at 4°C. Aβ concentration was determined by tyrosine absorption at 276 nm using an extinction coefficient of 1390 $cm^{-1}\cdot M^{-1}$. The aggregation kinetics of 50 μM $A\beta_{1\text{–}42}$ was monitored by light scattering after incubating the sample at controlled temperature (37 °C) [69,79]. The samples containing 50 μM Aβ and APP were obtained by appropriate aseptic mixing of the protein solutions and placed in closed cuvettes in a cold room at 4 °C, before incubation at higher temperatures. The aggregation kinetics was followed, as for the control, at a controlled temperature (37 °C) and under controlled stirring (200 rpm) for 8 h.

4.10. Static and Dynamic Light Scattering

The aggregation of proteins in the presence or absence of polyphenols was investigated by static and dynamic light scattering [80,81]. The samples were placed into a dust-free quartz cell without further filtering and kept at 37 °C in the thermostatic cell compartment of a Brookhaven Instruments BI200-SM goniometer (Brookhaven Instruments, Holtsville, NY, USA). The temperature was controlled within 0.1 °C using a thermostatic recirculating bath. The light scattered intensity and its autocorrelation function were measured at θ = 90° by using a Brookhaven BI-9000 correlator (Brookhaven Instruments, Holtsville, NY, USA) and a 50 mW He-Ne laser tuned at a wavelength λ = 632.8 nm. Static light scattering data were corrected for the background scattering of the solvent and normalized by using toluene as a calibration liquid. The intensity I(q) is measured as kilo count per seconds kcps, referred to as average number of photons per second arriving at the detector.

Due to their Brownian motion, particles moving in solution give rise to fluctuations in the intensity of the scattered light. In a light scattering experiment carried out in dynamic modality, the autocorrelator measures the homodyne intensity–intensity correlation function that, for a Gaussian distribution of the intensity profile of the scattered light, is related to the electric field correlation function (Equation (2)):

$$g^{(2)}(q,t) = \left[A + Bg^{(1)}(q,t)\right]^2 \tag{2}$$

where A and B are the experimental baseline and the optical constant, respectively. For polydisperse particles, $g^{(1)}(q,t)$ is given by Equation (3):

$$g^{(1)}(q,t) = \int_0^{\infty} G(\Gamma)exp(-\Gamma t)d\Gamma \tag{3}$$

Here, $G(\Gamma)$ is the normalized number distribution function for the decay constant $\Gamma = q^2 D_T$, where $q = (4\pi n/\lambda)\sin(\theta/2)$ is the scattering vector defining the spatial resolution with n and D_T being the solvent refractive index and the translational diffusion coefficient, respectively. The hydrodynamic diameter D_H is calculated from D_T through the Stokes–Einstein relationship (Equation (4)):

$$D_T = \frac{k_B T}{3\pi\eta D_H} \tag{4}$$

where k_B is the Boltzmann constant, T is the absolute temperature, and η is the solvent viscosity. D_H was obtained by the intensity autocorrelation functions by means of the cumulant method [82].

4.11. Circular Dichroism

CD spectroscopic measurements were carried out at 20 °C by using a JASCO J-810 spectrometer (JASCO Corporation, Tokyo, Japan) equipped with a temperature control unit. A quartz cell with a path length of 0.2 mm was used for FAR-UV (190–250 nm) measurements. Samples of 50 µM κ-casein in the presence and in the absence of different APP concentrations were incubated at 37 °C. Aliquots of 100 µL were removed from each sample at the beginning and after 24 h of thermal incubation and quenched at 4 °C before CD experiments. Each CD spectrum was obtained by averaging over eight scans and subtracting the blank solvent contribution.

5. Conclusions

In this work, we demonstrated that the whole pool of polyphenols extracted from the peels of Fuji apple—a widespread apple variety—was effective in inhibiting the structural conversion accompanying the aggregation of κ-casein, a protein from bovine milk that forms amyloid fibrils under physiological conditions (neutral pH and 37 °C). Although more insights into the toxic properties of the protein species resulting from incubation with APP is required, we believe that our study lays the groundwork for considering a by-product of the apple food industry as a beneficial source for human health.

In fact, while the action of some individual polyphenols from apples on κ-casein and other amyloid proteins has been extensively proven, to our knowledge, the effect exerted by the pool of polyphenols extracted from apple peels (which, although rich in bioactive compounds, are normally considered as waste) has never been analyzed. Considering some common features between the Aβ peptides involved in Alzheimer's Disease and κ-casein amyloid structures, including the cross-beta spine of the inner core [70] and the potential role of micellar conformations [52,53], we also verified that APP inhibit the aggregation of the $A\beta_{1-42}$ peptide involved in Alzheimer's Disease.

Therefore, our study could provide the basis for further research on the pathology of the protein species resulted from the APP inhibitory action thus favoring the valorization of agriculture and food industry waste products for recycling and circular economy purposes. The Fuji cultivar was introduced in Sicily only recently [83], and new knowledge on its functional properties could incentive this cultivation and help to preserve apple biodiversity.

More generally, our study further confirms the importance of designing new therapeutic effective strategies in the neurodegeneration field based on natural compounds and of investigating functional food consumption for prevention and health purposes.

Supplementary Materials: The following are available online at. Figure S1: UHPLC–HESI-MS targeted single ion monitoring chromatogram obtained from Fuji apple peel phenolic extract, Figure S2: Effects of Fuji apple peel phenolic extract on NIH-3T3 cell viability. The cells were incubated for 24, 48, and 72 h with the appropriate concentrations of phenolic extracts. At the end of incubation, cell viability was measured by MTS assay as described in Materials and Methods. Values are means ± SD of cell viability calculated from at least three separate experiments, Figure S3: CD spectra of 50 µM κ-casein at the beginning of the amyloid aggregation process (black) and after 24 h of incubation at 37 °C (dashed black) in the presence of 5 μg/mL of APP (a); 10 μg/mL of APP (b); 20 μg/mL of APP (c).

Author Contributions: Conceptualization, V.G., M.A.C., F.L., M.A.G., and S.V.; methodology, V.G., G.C.R., M.A.C., F.L., V.D.S., and S.V.; investigation, G.C.R., M.R., and V.D.S.; data curation, G.C.R., M.A.C., F.L., and S.V.; writing—original draft preparation, V.G., G.C.R., M.A.C., and S.V. writing—review and editing, V.G., G.C.R., M.A.C., F.L., M.R., V.D.S., M.A.G., and S.V. All authors have read and agreed to the published version of the manuscript.

Funding: This work was supported by Italian grant FIRB "Future in research" RBFR12SIPT MIND: "Multidisciplinary Investigations for the development of Neuroprotective Drugs", by the Sicily Region grants CAFIS (PO FESR 2007–2014) and "Rafforzare l'occupabilità nel sistema R&S e la nascita di spin off di ricerca in Sicilia" (P.O. FSE 2014/2020). This research was also financially supported by the Ph.D. Program Mediterranean Fruit Crops, Palermo University, Italy.

Data Availability Statement: Data are contained within the article and Supplementary Materials.

Acknowledgments: We express our sincere thanks to Donatella Bulone and Pier Luigi San Biagio for their critical suggestions. We are grateful to Rita Carrotta and Maria Rosalia Mangione for their scientific advice during the experiments and to Fabrizio Giambertone, Roberto Megna and Alessia Provenzano for their help and technical assistance.

Conflicts of Interest: The authors declare no conflict of interest. The funders had no role in the design of the study; in the collection, analyses, or interpretation of data; in the writing of the manuscript, or in the decision to publish the results.

Sample Availability: Samples of APP are available from the authors.

Abbreviations

APP	Apple Peel Polyphenols
CD	Circular Dichroism
LS	Light Scattering
ThT	Thioflavin T
TE	Trolox Equivalents
CTE	Catechin Equivalents
FW	Fresh Weight
AFM	Atomic Force Microscopy
FAR-UV	Far-UltraViolet
ORAC	Oxygen Radical Antioxidant Capacity
TPC	Total Polyphenol Content
CR	Congo Red

References

1. Diacono, M.; Persiani, A.; Testani, E.; Montemurro, F.; Ciaccia, C. Recycling agricultural wastes and by-products in organic farming: Biofertilizer production, yield performance and carbon footprint analysis. *Sustainability* **2019**, *11*, 3824. [CrossRef]
2. Gulsunoglu, Z.; Purves, R.; Karbancioglu-Guler, F.; Kilic-Akyilmaz, M. Enhancement of phenolic antioxidants in industrial apple waste by fermentation with *Aspergillus* spp. *Biocatal. Agric. Biotechnol.* **2020**, *25*, 101562. [CrossRef]
3. Wolfe, K.L.; Liu, R.H. Apple peels as a value-added food ingredient. *J. Agric. Food Chem.* **2003**, *51*, 1676–1683. [CrossRef]
4. Wandjou, J.G.N.; Lancioni, L.; Barbalace, M.C.; Hrelia, S.; Papa, F.; Sagratini, G.; Vittori, S.; Dall'Acqua, S.; Caprioli, G.; Beghelli, D.; et al. Comprehensive characterization of phytochemicals and biological activities of the Italian ancient apple 'Mela Rosa dei Monti Sibillini'. *Food Res. Int.* **2020**, *137*, 109422. [CrossRef]
5. Lončarić, A.; Matanović, K.; Ferrer, P.; Kovač, T.; Šarkanj, B.; Skendrović Babojelić, M.; Lores, M. Peel of traditional apple varieties as a great source of bioactive compounds: Extraction by micro-matrix solid-phase dispersion. *Foods* **2020**, *9*, 80. [CrossRef]
6. Wolfe, K.; Wu, X.; Liu, R.H. Antioxidant activity of apple peels. *J. Agric. Food Chem.* **2003**, *51*, 609–614. [CrossRef] [PubMed]
7. Miura, T.; Chiba, M.; Kasai, K.; Nozaka, H.; Nakamura, T.; Shoji, T.; Kanda, T.; Ohtake, Y.; Sato, T. Apple procyanidins induce tumor cell apoptosis through mitochondrial pathway activation of caspase-3. *Carcinogenesis* **2008**, *29*, 585–593. [CrossRef] [PubMed]

8. Akiyama, H.; Sato, Y.; Watanabe, T.; Nagaoka, M.H.; Yoshioka, Y.; Shoji, T.; Kanda, T.; Yamada, K.; Totsuka, M.; Teshima, R.; et al. Dietary unripe apple polyphenol inhibits the development of food allergies in murine models. *FEBS Lett.* **2005**, *579*, 4485–4491. [CrossRef] [PubMed]
9. Sunagawa, T.; Shimizu, T.; Kanda, T.; Tagashira, M.; Sami, M.; Shirasawa, T. Procyanidins from apples (*Malus pumila* Mill.) extend the lifespan of Caenorhabditis elegans. *Planta Med.* **2011**, *77*, 122–127. [CrossRef] [PubMed]
10. Zhao, S.; Zhang, L.; Yang, C.; Li, Z.; Rong, S. Procyanidins and Alzheimer's disease. *Mol. Neurobiol.* **2019**, *56*, 5556–5567. [CrossRef] [PubMed]
11. Khan, H.; Ullah, H.; Aschner, M.; Cheang, W.S.; Akkol, E.K. Neuroprotective effects of quercetin in Alzheimer's disease. *Biomolecules* **2020**, *10*, 59. [CrossRef]
12. Boyina, H.K.; Geethakhrishnan, S.L.; Panuganti, S.; Gangarapu, K.; Devarakonda, K.P.; Bakshi, V.; Guggilla, S.R. In Silico and In Vivo Studies on Quercetin as Potential Anti-Parkinson Agent. In *GeNeDis 2018: Advances in Experimental Medicine and Biology*; Vlamos, P., Ed.; Springer: Cham, Switerlands, 2020; Volume 1195, pp. 1–11.
13. Phan, H.T.; Samarat, K.; Takamura, Y.; Azo-Oussou, A.F.; Nakazono, Y.; Vestergaard, M.D.C. Polyphenols modulate alzheimer's amyloid beta aggregation in a structure-dependent manner. *Nutrients* **2019**, *11*, 756. [CrossRef]
14. Carrell, R.W.; Lomas, D.A. Conformational disease. *Lancet* **1997**, *350*, 134–138. [CrossRef]
15. Makin, O.S.; Serpell, L.C. Examining the structure of the mature amyloid fibril. *Biochem. Soc. Trans.* **2002**, *30*, 521–525. [CrossRef] [PubMed]
16. Serpell, L.C.; Sunde, M.; Benson, M.D.; Tennent, G.A.; Pepys, M.B.; Fraser, P.E. The protofilament substructure of amyloid fibrils. *J. Mol. Biol.* **2000**, *300*, 1033–1039. [CrossRef] [PubMed]
17. Dobson, C.M. Protein folding and misfolding. *Nature* **2003**, *426*, 884–890. [CrossRef] [PubMed]
18. Koike, H.; Katsuno, M. Ultrastructure in Transthyretin Amyloidosis: From Pathophysiology to Therapeutic Insights. *Biomedicines* **2019**, *7*, 11. [CrossRef]
19. Koike, H.; Nishi, R.; Ikeda, S.; Kawagashira, Y.; Iijima, M.; Sakurai, T.; Shimohata, T.; Katsuno, M.; Sobue, G.J. The morphology of amyloid fibrils and their impact on tissue damage in hereditary transthyretin amyloidosis: An ultrastructural study. *J. Neurol. Sci.* **2018**, *394*, 99–106. [CrossRef]
20. Canale, C.; Seghezza, S.; Vilasi, S.; Carrotta, R.; Bulone, D.; Diaspro, A.; San Biagio, P.L.; Dante, S. Different effects of Alzheimer's peptide Aβ (1–40) oligomers and fibrils on supported lipid membranes. *Biophys. Chem.* **2013**, *182*, 23–29. [CrossRef]
21. Walsh, D.M.; Klyubin, I.; Fadeeva, J.V.; Rowan, M.J.; Selkoe, D.J. Amyloid-β oligomers: Their production, toxicity and therapeutic inhibition. *Biochem. Soc. Trans.* **2002**, *30*, 552–557. [CrossRef]
22. Vilasi, A.; Vilasi, S.; Romano, R.; Acernese, F.; Barone, F.; Balestrieri, M.L.; Maritato, R.; Irace, G.; Sirangelo, I. Unraveling amyloid toxicity pathway in NIH3T3 cells by a combined proteomic and 1 H-NMR metabonomic approach. *J. Cell. Physiol.* **2013**, *228*, 1359–1367. [CrossRef]
23. Choi, D.Y.; Lee, Y.J.; Hong, J.T.; Lee, H.J. Antioxidant properties of natural polyphenols and their therapeutic potentials for Alzheimer's disease. *Brain Res. Bull.* **2012**, *87*, 144–153. [CrossRef]
24. Ngoungoure, V.L.N.; Schluesener, J.; Moundipa, P.F.; Schluesener, H. Natural polyphenols binding to amyloid: A broad class of compounds to treat different human amyloid diseases. *Mol. Nutr. Food Res.* **2015**, *59*, 8–20. [CrossRef]
25. Stefani, M.; Rigacci, S. Beneficial properties of natural phenols: Highlight on protection against pathological conditions associated with amyloid aggregation. *Biofactors* **2014**, *40*, 482–493. [CrossRef]
26. Francioso, A.; Punzi, P.; Boffi, A.; Lori, C.; Martire, S.; Giordano, C.; D'Erme, M.; Mosca, L. β-sheet interfering molecules acting against β-amyloid aggregation and fibrillogenesis. *Bioorg. Med. Chem.* **2015**, *23*, 1671–1683. [CrossRef]
27. Porat, Y.; Abramowitz, A.; Gazit, E. Inhibition of amyloid fibril formation by polyphenols: Structural similarity and aromatic interactions as a common inhibition mechanism. *Chem. Biol. Drug Des.* **2006**, *67*, 27–37. [CrossRef]
28. Hudson, S.A.; Ecroyd, H.; Dehle, F.C.; Musgrave, I.F.; Carver, J.A. (−)-Epigallocatechin-3-gallate (EGCG) maintains κ-casein in its pre-fibrillar state without redirecting its aggregation pathway. *J. Mol. Biol.* **2009**, *392*, 689–700. [CrossRef]
29. Wang, J.B.; Wang, Y.M.; Zeng, C.M. Quercetin inhibits amyloid fibrillation of bovine insulin and destabilizes preformed fibrils. *Biochem. Biophys. Res. Commun.* **2011**, *415*, 675–679. [CrossRef]
30. Ono, K.; Yoshiike, Y.; Takashima, A.; Hasegawa, K.; Naiki, H.; Yamada, M. Potent anti-amyloidogenic and fibril-destabilizing effects of polyphenols in vitro: Implications for the prevention and therapeutics of Alzheimer's disease. *J. Neurochem.* **2003**, *87*, 172–181. [CrossRef]
31. Toda, T.; Sunagawa, T.; Kanda, T.; Tagashira, M.; Shirasawa, T.; Shimizu, T. Apple Procyanidins Suppress Amyloid-Protein Aggregation. *Biochem. Res. Int.* **2011**, *2011*, 784698. [CrossRef]
32. Cui, L.; Zhang, Y.; Cao, H.; Wang, Y.; Teng, T.; Ma, G.; Li, Y.; Li, K.; Zhang, Y. Ferulic acid inhibits the transition of amyloid-β42 monomers to oligomers but accelerates the transition from oligomers to fibrils. *J. Alzheimers Dis.* **2013**, *37*, 19–28. [CrossRef]
33. Iannuzzi, C.; Borriello, M.; Irace, G.; Cammarota, M.; Di Maro, A.; Sirangelo, I. Vanillin Affects Amyloid Aggregation and Non-Enzymatic Glycation in Human Insulin. *Sci. Rep.* **2017**, *7*, 15086. [CrossRef]
34. Ma, W.; Tribet, C.; Guyot, S.; Zanchi, D. Tannin-controlled micelles and fibrils of k-casein. *Chem. Phys.* **2019**, *151*, 245103. [CrossRef]
35. Hudson, S.A.; Ecroyd, H.; Kee, T.W.; Carver, J.A. The thioflavin T fluorescence assay for amyloid fibril detection can be biased by the presence of exogenous compounds. *FEBS J.* **2009**, *276*, 5960–5972. [CrossRef] [PubMed]

36. Holt, C.; Sawyer, L. Caseins as rheomorphic proteins: Interpretation of primary and secondary structures of the αS1-, β-and κ-caseins. *J. Chem. Soc. Faraday Trans.* **1993**, *89*, 2683–2692. [CrossRef]
37. Thurn, A.; Burchard, W.; Niki, R. Structure of casein micelles I. Small angle neutron scattering and light scattering from β-and χ-casein. *Colloid Polym. Sci* **1987**, *265*, 653–666. [CrossRef]
38. Vreeman, H.J.; Brinkhuis, J.A.; Van der Spek, C.A. Some association properties of bovine SH-k-casein. *Biophys. Chem.* **1981**, *14*, 185–193. [CrossRef]
39. McSweeney, P.L.; Fox, P.F. *Advanced Dairy Chemistry: Volume 1A: Proteins: Basic Aspects*; Springer Science & Business Media: Berlin/Heidelberg, Germany, 2013.
40. Farrell, H.M.; Cooke, P.H.; Wickham, E.D.; Piotrowski, E.G.; Hoagland, P.D. Environmental influences on bovine κ-casein: Reduction and conversion to fibrillar (amyloid) structures. *J. Protein Chem.* **2003**, *22*, 259–273. [CrossRef]
41. Ecroyd, H.; Koudelka, T.; Thorn, D.C.; Williams, D.M.; Devlin, G.; Hoffmann, P.; Carver, J.A. Dissociation from the oligomeric state is the rate-limiting step in fibril formation by κ-casein. *J. Biol. Chem.* **2008**, *283*, 9012–9022. [CrossRef]
42. Ecroyd, H.; Thorn, D.C.; Liu, Y.; Carver, J.A. The dissociated form of κ-casein is the precursor to its amyloid fibril formation. *Biochem. J.* **2010**, *429*, 251–260. [CrossRef]
43. Thorn, D.C.; Ecroyd, H.; Carver, J.A. The two-faced nature of milk casein proteins: Amyloid fibril formation and chaperone-like activity. *Aust. J. Dairy Technol.* **2009**, *64*, 34.
44. Leonil, J.; Henry, G.; Jouanneau, D.; Delage, M.M.; Forge, V.; Putaux, J.L. Kinetics of fibril formation of bovine κ-casein indicate a conformational rearrangement as a critical step in the process. *J. Mol. Biol.* **2008**, *381*, 1267–1280. [CrossRef] [PubMed]
45. Thorn, D.C.; Meehan, S.; Sunde, M.; Rekas, A.; Gras, S.L.; MacPhee, C.E.; Dobson, C.M.; Wilson, M.R.; Carver, J.A. Amyloid fibril formation by bovine milk κ-casein and its inhibition by the molecular chaperones αS-and β-casein. *Biochemistry* **2005**, *44*, 17027–17036. [CrossRef]
46. Carrotta, R.; Canale, C.; Diaspro, A.; Trapani, A.; San Biagio, P.L.; Bulone, D. Inhibiting effect of α(s1)-casein on Aβ(1–40) fibrillogenesis. *Biochim. Biophys. Acta Gen. Subj.* **2012**, *1820*, 124–132. [CrossRef]
47. Carrotta, R.; Vilasi, S.; Librizzi, F.; Martorana, V.; Bulone, D.; San Biagio, P.L. α-Casein inhibition mechanism in concanavalin A aggregation process. *J. Phys. Chem. B* **2012**, *116*, 14700–14707. [CrossRef]
48. Librizzi, F.; Carrotta, R.; Spigolon, D.; Bulone, D.; San Biagio, P.L. α-Casein inhibits insulin amyloid formation by preventing the onset of secondary nucleation processes. *J. Phys. Chem. Lett.* **2014**, *5*, 3043–3048. [CrossRef]
49. Feng, B.Y.; Toyama, B.H.; Wille, H.; Colby, D.W.; Collins, S.R.; May, B.C.; Prusiner, S.B.; Weissman, J.; Shoichet, B.K. Small-molecule aggregates inhibit amyloid polymerization. *Nat. Chem. Biol.* **2008**, *4*, 197–199. [CrossRef]
50. Kumosinski, T.F.; Brown, E.M.; Farrell, H.M., Jr. Three-dimensional molecular modeling of bovine caseins: A refined, energy-minimized k-casein structure. *J. Dairy Sci.* **1993**, *76*, 2507–2520. [CrossRef]
51. Jain, N.; Bhattacharya, M.; Mukhopadhyay, S. Chain collapse of an amyloidogenic intrinsically disordered protein. *Biophys. J.* **2011**, *101*, 1720–1729. [CrossRef]
52. Sabaté, R.; Estelrich, J.J. Evidence of the existence of micelles in the fibrillogenesis of beta-amyloid peptide. *Phys. Chem. B* **2005**, *109*, 11027–11032. [CrossRef]
53. Morel, B.; Hingant, E.; Fontes, P.; Alvarez-Martinez, M.T.; Arnaud, J.D.; Liautard, J.P.; Pujo-Menjouet, L.; Hingant, E.; Fontes, P.; Alvarez-Martinez, M.T.; et al. Dynamic micellar oligomers of amyloid beta peptides play a crucial role in their aggregation mechanisms. *Phys. Chem. Chem. Phys.* **2018**, *20*, 20597–20614. [CrossRef]
54. Hingant, E.; Fontes, P.; Alvarez-Martinez, M.T.; Arnaud, J.D.; Liautard, J.P.; Pujo-Menjouet, L. A micellar on-pathway intermediate step explains the kinetics of prion amyloid formation. *PLoS Comput. Biol.* **2014**, *10*, e1003735. [CrossRef]
55. Liu, R.H. Health benefits of fruit and vegetables are from additive and synergistic combinations of phytochemicals. *Am. J. Clin. Nutr.* **2003**, *78*, 517S–520S. [CrossRef]
56. Prior, R.L. Oxygen radical absorbance capacity (ORAC): New horizons in relating dietary antioxidants/bioactives and health benefits. *J. Funct. Foods* **2015**, *18*, 797–810. [CrossRef]
57. Jairath, G.; Chatli, M.K.; Biswas, K. Comparative Study on In vitro and In vivo Evaluation of Antioxidant Potential of Apple Peel Extract and Aloe Vera Gel. *J. Food Process. Preserv.* **2016**, *40*, 607–614. [CrossRef]
58. Bitalebi, S.; Nikoo, M.; Rahmanifarah, K.; Noori, F.; Gavlighi, H.A. Effect of apple peel extract as natural antioxidant on lipid and protein oxidation of rainbow trout (*Oncorhynchus mykiss*) mince. *Int. Aquat. Res.* **2019**, *11*, 135–146. [CrossRef]
59. He, R.R.; Wang, M.; Wang, C.Z.; Chen, B.T.; Lu, C.N.; Yao, X.S.; Chen, J.X.; Kurihara, H.J. Protective effect of apple polyphenols against stress-provoked influenza viral infection in restraint mice. *Agric. Food Chem.* **2011**, *59*, 3730–3737. [CrossRef]
60. Xu, Z.R.; Li, J.Y.; Dong, X.W.; Tan, Z.J.; Wu, W.Z.; Xie, Q.M.; Yang, Y.M. Apple Polyphenols Decrease Atherosclerosis and Hepatic Steatosis in ApoE-/- Mice through the ROS/MAPK/NF-κB Pathway. *Nutrients* **2015**, *7*, 7085–7105. [CrossRef]
61. Ogino, Y.; Osada, K.; Nakamura, S.; Ohta, Y.; Kanda, T.; Sugano, M. Absorption of dietary cholesterol oxidation products and their downstream metabolic effects are reduced by dietary apple polyphenols. *Lipids* **2007**, *42*, 151–161. [CrossRef]
62. Levine, H., III. Thioflavine T interaction with synthetic Alzheimer's disease β-amyloid peptides: Detection of amyloid aggregation in solution. *Protein Sci.* **1993**, *2*, 404–410. [CrossRef]
63. Hawe, A.; Sutter, M.; Jiskoot, W. Extrinsic fluorescent dyes as tools for protein characterization. *Pharm. Res.* **2008**, *25*, 1487–1499. [CrossRef]

64. Leibiger, C.; Kosyakova, N.; Mkrtchyan, H.; Glei, M.; Trifonov, V.; Liehr, T. First molecular cytogenetic high resolution characterization of the NIH 3T3 cell line by murine multicolor banding. *J. Histochem. Cytochem.* **2013**, *61*, 306–312. [CrossRef] [PubMed]
65. Sirangelo, I.; Iannuzzi, C.; Vilasi, S.; Irace, G.; Giuberti, G.; Misso, G.; D'Alessandro, A.; Abbruzzese, A.; Caraglia, M. W7FW14F apomyoglobin amyloid aggregates-mediated apoptosis is due to oxidative stress and AKT inactivation caused by Ras and Rac. *J. Cell. Physiol.* **2009**, *221*, 412–423. [CrossRef] [PubMed]
66. Iannuzzi, C.; Vilasi, S.; Portaccio, M.; Irace, G.; Sirangelo, I. Heme binding inhibits the fibrillization of amyloidogenic apomyoglobin and determines lack of aggregate cytotoxicity. *Protein Sci.* **2007**, *16*, 507–516. [CrossRef]
67. Coelho-Cerqueira, E.; Pinheiro, A.S.; Follmer, C. Pitfalls associated with the use of Thioflavin-T to monitor anti-fibrillogenic activity. *Bioorg. Med. Chem. Lett.* **2014**, *24*, 3194–3198. [CrossRef]
68. Frid, P.; Anisimov, S.V.; Popovic, N. Congo red and protein aggregation in neurodegenerative diseases. *Brain Res. Rev.* **2007**, *53*, 135–160.
69. Vilasi, S.; Carrotta, R.; Ricci, C.; Rappa, G.C.; Librizzi, F.; Martorana, V.; Ortore, M.G.; Mangione, M.R. Inhibition of Aβ_{1-42} Fibrillation by Chaperonins: Human Hsp60 Is a Stronger Inhibitor than Its Bacterial Homologue GroEL. *ACS Chem. Neurosci.* **2019**, *10*, 3565–3574. [CrossRef]
70. Petkova, A.T.; Ishii, Y.; Balbach, J.J.; Antzutkin, O.N.; Leapman, R.D.; Delaglio, F.; Tycko, R. A structural model for Alzheimer's β-amyloid fibrils based on experimental constraints from solid state NMR. *Proc. Natl. Acad. Sci. USA* **2002**, *99*, 16742–16747. [CrossRef]
71. Stsiapura, V.I.; Maskevich, A.A.; Kuzmitsky, V.A.; Turoverov, K.K.; Kuznetsova, I.M. Computational study of thioflavin T torsional relaxation in the excited state. *J. Phys. Chem. A* **2007**, *111*, 4829–4835. [CrossRef]
72. Chu, G.C.L.; Wilson, K. Evaluating Maturity of Empire, Idared and Spartan Apples. Factsheet—Ontario. Ministry of Agriculture, Food and Rural Affairs. Order No. 00-027. March 2000. Available online: http://www.omafra.gov.on.ca/english/crops/facts/00-027.htm (accessed on 13 February 2021).
73. Ceymann, M.; Arrigoni, E.; Scharer, H.; Bozzi Nising, A.; Hurrell, R.F. Identification of apples rich in health-promoting flavan-3-ols and phenolic acids by measuring the polyphenol profile. *J. Food Compos.* **2012**, *26*, 128–135. [CrossRef]
74. Di Stefano, V.; Pitonzo, R.; Novara, M.E.; Bongiorno, D.; Indelicato, S.; Gentile, C.; Avellone, G.; Bognanni, R.; Scandurra, S.; Melilli, M.G. Antioxidant activity and phenolic composition in pomegranate (*Punica granatum* L.) genotypes from south Italy by UHPLC-Orbitrap-MS approach. *J. Sci. Food Agric.* **2019**, *99*, 1038–1045. [CrossRef] [PubMed]
75. Di Stefano, V.; Melilli, M.G. Effect of storage on quality parameters and phenolic content of Italian extra-virgin olive oils. *Nat. Prod. Res.* **2020**, *34*, 78–86. [CrossRef] [PubMed]
76. Cao, G.; Alessio, H.M.; Cutler, R.G. Oxygen-radical absorbance capacity assay for antioxidants. *Free Radic. Biol. Med.* **1993**, *14*, 303–311. [CrossRef]
77. Higuchi, K.; Hosowaka, M.; Takeda, T. Fluorometric determination of amyloid fibrils in vitro using the fluorescent dye, thioflavin T1. *Anal. Biochem.* **1989**, *177*, 244–249.
78. Fezoui, Y.; Hartley, D.M.; Harper, J.D.; Khurana, R.; Walsh, D.M.; Condron, M.M.; Selkoe, D.J.; Lansbury, P.T., Jr.; Fink, A.L.; Teplow, D.B. An improved method of preparing the amyloid beta-protein for fibrillogenesis and neurotoxicity experiments. *Amyloid* **2000**, *7*, 166–178. [CrossRef]
79. El-Agnaf, O.M.; Mahil, D.S.; Patel, B.P.; Austen, B.M. Oligomerization and toxicity of beta-amyloid-42 implicated in Alzheimer's disease. *Biochem. Biophys. Res. Commun.* **2000**, *273*, 273–1007. [CrossRef]
80. Pusey, P.N. Introduction to scattering experiments. In *Neutrons, X-rays, and Light: Scattering Methods Applied to Soft Condensed Matter*; Zemb, T., Lindner, P., Eds.; Elsevier: Amsterdam, The Netherlands, 2002.
81. Berne, B.J.; Pecora, R. *Dynamic Light Scattering*; John Wiley: New York, NY, USA, 1976.
82. Stepanek, P. The method and some applications. In *Dynamic Light Scattering*; Brown, W., Ed.; Clarendon Press: Oxford, UK, 1993.
83. Francaviglia, D.; Farina, V.; Avellone, G.; Lo Bianco, R. Fruit yield and quality responses of apple cvars Gala and Fuji to partial rootzone drying under Mediterranean conditions. *J. Agric. Sci.* **2013**, *151*, 556–569. [CrossRef]

molecules

MDPI

Review

Sustainable Use of Bioactive Compounds from *Solanum tuberosum* and *Brassicaceae* Wastes and By-Products for Crop Protection—A Review

Daniela Pacifico [1,*], Chiara Lanzanova [1,†], Eleonora Pagnotta [1,†], Laura Bassolino [1], Anna Maria Mastrangelo [1], Daniela Marone [1], Roberto Matteo [1], Roberto Lo Scalzo [2] and Carlotta Balconi [1]

1 CREA Council for Agricultural Research and Economics—Research Centre for Cereal and Industrial Crops, 00198 Rome, Italy; chiara.lanzanova@crea.gov.it (C.L.); eleonora.pagnotta@crea.gov.it (E.P.); laura.bassolino@crea.gov.it (L.B.); annamaria.mastrangelo@crea.gov.it (A.M.M.); daniela.marone@crea.gov.it (D.M.); roberto.matteo@crea.gov.it (R.M.); carlotta.balconi@crea.gov.it (C.B.)

2 CREA Council for Agricultural Research and Economics—Research Centre for Engineering and Agro-Food Processing, 00198 Rome, Italy; roberto.loscalzo@crea.gov.it

* Correspondence: daniela.pacifico@crea.gov.it; Tel.: +39-0516316811

† These authors contributed equally.

Citation: Pacifico, D.; Lanzanova, C.; Pagnotta, E.; Bassolino, L.; Mastrangelo, A.M.; Marone, D.; Matteo, R.; Lo Scalzo, R.; Balconi, C. Sustainable Use of Bioactive Compounds from *Solanum tuberosum* and *Brassicaceae* Wastes and By-Products for Crop Protection—A Review. *Molecules* **2021**, *26*, 2174. https://doi.org/10.3390/molecules26082174

Academic Editors: Simona Fabroni, Krystian Marszałek and Aldo Todaro

Received: 12 March 2021
Accepted: 1 April 2021
Published: 9 April 2021

Abstract: Defatted seed meals of oleaginous Brassicaceae, such as *Eruca sativa*, and potato peel are excellent plant matrices to recover potentially useful biomolecules from industrial processes in a circular strategy perspective aiming at crop protection. These biomolecules, mainly glycoalkaloids and phenols for potato and glucosinolates for Brassicaceae, have been proven to be effective against microbes, fungi, nematodes, insects, and even parasitic plants. Their role in plant protection is overviewed, together with the molecular basis of their synthesis in plant, and the description of their mechanisms of action. Possible genetic and biotechnological strategies are presented to increase their content in plants. Genetic mapping and identification of closely linked molecular markers are useful to identify the loci/genes responsible for their accumulation and transfer them to elite cultivars in breeding programs. Biotechnological approaches can be used to modify their allelic sequence and enhance the accumulation of the bioactive compounds. How the global challenges, such as reducing agri-food waste and increasing sustainability and food safety, could be addressed through bioprotector applications are discussed here.

Keywords: sustainable food industry; waste residues; potato peel; brassica defatted meals; antifungal activity; mycotoxigenic fungi control; cereal protection; circular economy

1. Introduction

Food loss and waste are a complex problem involving various drivers along the food supply chain. The worldwide intensification of agro-industrial processing and the current linear food economy based on converting natural resources into waste lasted so long, creating large quantities of residues, and their management poses economical, hygienic, and environmental problems [1]. The future state of global natural resources, especially in the frame of the current climatic change, forces the world towards a turnaround in order to achieve significant outcomes in waste reduction, though keeping high yield production levels and improving accessibility and quality of the resources. Among the possible solutions, international organizations implement the idea of sustainable development, which satisfies the requirements of the circular economy in the view of a new agro-industrial system (UN 2030 Agenda for Sustainable Development, the UN Convention on Biodiversity Strategic Plan for 2020, the EU Green Deal "Farm to Fork strategy" for a fair, healthy, and environmentally friendly food system and the upcoming Horizon Europe 2021-2027 Pillar II-Cluster "Food, bioeconomy, natural resources, agriculture and environment").

The challenge for the circular economy to implement the sustainability of a modern agricultural management is facing the increase in human population, the request for high crop yield, food production, and safety. To achieve these goals, the main issue to be solved is to guarantee a sustainable and circular way to control crop pests. The most common tool used for crop protection consists of synthetic pesticides, which offer a practical and rapid solution to plant pathogen disease outbreaks. However, numerous problems arise from the wide-scale use of chemical pesticides: Their negative impact on environment [2], several fruit and vegetable industrial processing residues [3], and the development of fungal resistance to their active substances [4]. Therefore, the use of chemicals has long been viewed as a strategic management to control crop diseases, and nowadays, EU farmers have to manage plant protection without many pesticides because of the obligation of their gradual dismissal [5]. The sprout inhibitor Chlorpropham used to protect stored potatoes has been recently banned (January 2020), as was Chlorothalonil, a fungicide for wheat and barley protection, banned from May 2020, and Mancozeb, approved in the EU in 2006 for very wide use, especially for controlling potato blight, and under increasing scrutiny from the EU Commission and from member states. With the aim to highlight alternative tools to chemical products, researchers started to be strongly focused on innovative strategies such as identifying plant molecules useful to develop natural biocides [6].

Botanical extracts can be effective in controlling a wide range of plant pathogens providing an innovative and sustainable solution to be applied in organic and conventional agriculture. Studies on extracts from garlic, carnation, cinnamon, and thyme showed, for example, their effectiveness on *Fusarium solani* and *Rizochotonia solani* in faba beans [7]. Similarly, lupine seeds were successfully treated with extracts of *Nerium oleander*, *Eugenia jambolana*, and *Citrullus colocynthis* [8]. Clove, garden quinine, Brazilian pepper, anthi mandhaari, black cumin, white cedar, neem, and rice extracts were used in the control of leaf rust in wheat [9,10]. It is unclear if these products can improve plant vigor and response to other fungal pathogens, especially under organic farming conditions, in which plants are more susceptible to the pathogen-induced damage. These extracts could act by enhancing the host defense mechanisms [11], either by increasing the activity of peroxidase, the accumulation of phenolic compounds [12], or through inhibition of some antioxidant enzymes and catalases, thereby leading to the production of elevated amounts of H_2O_2 [13].

In view of the increasing interest in the concept of circular economy, using molecules recovered from other productive activities, rooted in the territory with no further input of new matter, which will reduce the impact of pathogens, pests, biotic stresses on cereal crop quality and yield, is of strategic importance to guarantee food for new generations with respect to the future state of global natural resources. Minimizing the 'yield gap' and increasing yield stability under different stress conditions is the worldwide trend being emphasized. However, despite the increasing interest around the concept of circular economy, there are currently few systematic studies on extracts from wastes of *Solanaceae* and *Brassicaceae* that can explain their effects on plant pathogens due to a defined concentration of bioactive molecules [14].

Potato is the fourth major food crop in the world, with a production around 388 million tons [15], of which <50% is consumed fresh [14]. The large consumption of processed potato products (French fries, potato crisp, and frosted potatoes) causes a huge quantity of residues every year as Potato Peel Waste (PPW), a by-product of the large-scale peeling from manufacturing industries [16]. This amount accounts for 15 to 40% of the initial product's mass depending on the peeling methods [17,18] and ranges between 70,000 and 140,000 tons worldwide [19], posing a pressing challenge of how to use it [16] in order to avoid considerable environmental and economic concerns, such as pollution derived from organic waste decomposition. It has been a long time that the whole tuber can be converted into energy, as a source for biogas, biofuel, and biomanure [20], but this destination of use does not consider the high valuable utilization of peel. The presence of active compounds such as phenols and glycoalkaloids should favor the area of re-use or recovery, i.e., of green extracts suitable as nutraceutical and drugs [21] and as additives for improving the shelf life of fresh-cut fruits [22].

Concerning *Brassicaceae*, a green biorefinery approach was successfully applied in recent years in particular to oleaginous *Brassicaceae* with the aim to create novel agriculture inputs such as fertilizers, amendments, biostimulants, and biopesticides for the control of pest and pathogens both at the open-field level and in the post-harvest phase [23]. Vegetable oils, in particular those from *Brassica* oilseeds, are well known for their potential applications in green chemistry, including their use as hydraulic fluids, biolubricants, and cosmetics [24]. A fair biorefinery approach, devoted to the production of oil as the main product, should be closely related, in the frame of circular economy, to the exploitation of one or more co-products with the aim of minimizing or avoiding waste production [25]. Biomolecules produced from *Brassica* oilseed co-products, which can be applied in industry and as sustainable tools in agriculture, represent an urgent option considering the health and environment problems caused by conventional chemical application. The perspective of fractioning the *Brassica* co-products permits the separation of high added molecules, allowing the preparation of a pool of bio-based materials useful for several applications, which is a fundamental approach for innovative biorefineries [23].

This review aims to provide a comprehensive perspective of the scientific literature focused on the biological role of secondary plant metabolites derived from potato and *Brassicaceae* wastes such as alkaloids, glycoalkaloids, terpenoids, organic acids, glucosinolates (GSLs), and their hydrolysis products. These biocompounds are promising sources of plant-protecting tools and could be used as bio-weapons against pests, for example reducing the attractiveness of plants to different insects or reducing fungal growth development [26–29].

2. Biocompounds in Potato Peel

Potato germplasm is characterized by a huge variability in composition and concentration of secondary metabolites that play a role in increasing plant ability to cope with environmental challenges, due to their reported biocide activity on insects, bacteria, and fungi [27,30,31]. Their distribution within the tuber is not uniform: Most of them are concentrated in the peel, made of periderm tissue, whose cell layers contain corky cell walls, which confer protection from phytopathogens, especially during tuber growth and storage. Thus, considering that potato peel is constantly exposed to biotic stresses, it is not surprising that it is a precious source of bioactive compounds, mainly phenolics and alkaloids, which have an enormous potential to deliver new bioprotectors.

Potato Glycoalkaloids

Steroidal glycoalkaloids (SGAs), a class of nitrogen-containing steroid glyosides, are the most important alkaloids in the *Solanaceae* family [32]. In potato germoplasm, there are more than 80 different molecules [33–35], but in cultivated potato, α-chaconine (β-D-Glucopyranoside, (3β)-solanid-5-en-3-yl O-6-deoxy-α-L-mannopyranosyl-(1-2)-O- (6-deoxy-α-L-mannopyranosyl-(1-4)), and α-solanine (solanid-5-en-3β-yl α-L-rhamnopyranosyl -(1→2)-[β-D-glucopyranosyl- (1→3)]-β-D-galactopyranoside) account for up to 95% of the total tuber SGAs [36,37]. SGA structures share a common aglycone, a six-ring steroid skeleton (solanidine), to which a branched triose is attached. The triose is a sugar moiety attached to the 3-position of the first ring and a nitrogen atom in the sixth ring end of the molecule: In the case of α-chaconine it consists of β-chacotriose (bis-α-L-rhamnopyranosyl-β-D-glucopyranose), while for α-solanine, it is a β-solatriose (α-L-rhamnopyranosyl-β-D-glucopyranosyl-β-galactopyranose (Figure 1B) [36]. The SGA biosinthetic pathway has not yet been completely determined, even though it putatively derives from isopropanoid pathway. Cholesterol has been identified as metabolic precursor: It is cyclized into solanidine, a steroidal alkaloid (SA) that is subsequently glycosylated to α-solanine and α-chaconine [38]. A similar SGA, tomatine, has been found in tomato and eggplants, with small but significant structural differences. It has a branched chain composed of lycotetraose, a tetrasaccharide constituted by two glucose units, xylose and galactose (22S,25S)-5α-spirosolan-3β-yl-β-D-glucopyranosyl-(1→2)-[β-D-xylopyranosyl-(1→3)]-β-D-glucopyranosyl-(1→4)-β-D-galactopyranoside [36].

The distribution of SGAs in plant is not uniform, including leaves, tubers, roots, and sprouts, especially in green parts, flowers, and fruits [36]; their accumulation is affected by many different factors, including developmental and environmental conditions such as high temperature [39], light exposure [40], light quality [41], and wounding [42], but it also depends on genotype [43]. SGA content in the peel was often reported to be high, ranging between 0.83 and 352.6 mg/100 gr DW (dry weight) in eight potato cultivars [44] and resulting in 50 mg/100 gr DW in *cv* Bionica [45]. Two interesting studies indicated that this variability could be correlated with flesh color: White-flesh potatoes showed higher content of SGAs in peel than blue-flesh, red-flesh, and yellow-flesh potatoes [46]. White-flesh potatoes can reach up to 3526 mg Kg^{-1}, blue-flesh potatoes 245 mg Kg^{-1}, red-flesh potatoes 1264 mg Kg^{-1}, and yellow-flesh potatoes 425 mg Kg^{-1} FW (fresh weight) [47].

SGAs are considered toxic to humans: Daily oral doses from 3 to 6 mg GAs/kg body weight can even be lethal [48]. The widely accepted safety limit in tubers is 20 mg/100 g FW, even though a maximum level has not been officially established at the EU level [43]. A recent opinion from the German Federal Institute for Risk Assessment (BfR) suggests lowering the safety limit in potatoes to less than 10 mg/100 g FW [49]. However, most edible mature tubers contain low amounts of SGAs in their flesh [50], as a consequence of their selective reduction by breeding due to potato domestication for human consumption. Commonly commercialized potatoes rarely exceed the recommended SGAs level, with some exceptions such as *cv.* Lenape and *cv.* Bonum that were consequently withdrawn from the US and Swedish markets [51,52].

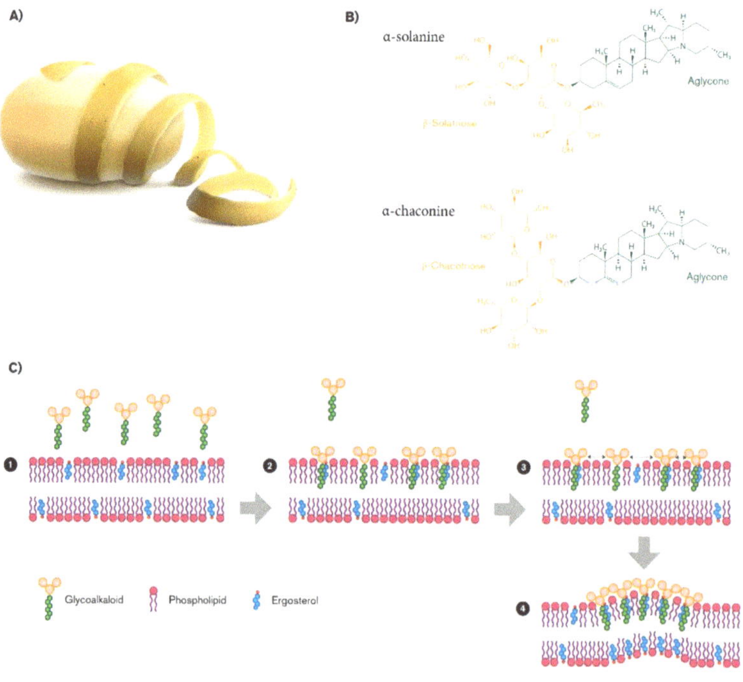

Figure 1. Biovalorization of glycoalkaloids α-chaconine and α-solanine from potato peel waste (**A**). Their chemical structures (**B**), and their role in fungal membrane disruption (**C**).

Potato Phenols

Thousands of phenolic compounds have been isolated in the plant kingdom with a role of either oxidative stress protectors or as pest control agents, and have been classified in several subgroups based on their structure: Phenolic acids, flavonoids, tannins, coumarins, lignans, quinones, stilbens, and curcuminoids. Their structure contains hydroxylated aromatic rings, with the hydroxy group being attached directly to the phenyl, substituted phenyl, or different aryl group. Phenolic compounds are synthesized via the shikimic acid and phenylpropanoid pathways. Potato peel is a well-established source of phenolic acids and flavonoids: Their content varies, respectively, among 1.02–2.92 g/100 g and 0.51–0.96 g/100 g DW [21] and significantly decreases toward the skin close sections (cortex) and flesh, as confirmed either by direct measurement, such as HPLC analysis [45], and indirect measurement, such as the radical scavenging activity [53]. The most abundant phenolic compound is chlorogenic acid, CGA (5-O-caffeoylquinic acid; 5-CQA), together with its isomers (3,4-diCQA, 3,5-diCQA, and 4,5-diCQA) [36]. CGA was also truly called 3-caffeoylquinic acid before 1976, when, according to new IUPAC rules, the CA structure changed in 5-caffeoylquinic acid and the 3-caffeoylquinic acid isomer was referred to as neochlorogenic acid [54]. CGA resulted in nearly 2115 $\mu g\ g^{-1}$ DW in skin, decreasing in adjacent cortex to 276 $\mu g\ g^{-1}$ DW (*cv.* Bionica) [45]. Conversely, hydroxycinnamic and hydroxybenzoic acids are present only in trace, with the exception of caffeic acid ((2*E*)-3-(3,4-dihydroxyphenyl) prop-2-enoic acid; CA). CA and its derivatives, and catechin (C_6-C_3-C_6), belonging, respectively, to the class of hydroxycinnamics (C_6-C_3) and flavonoids, specifically flavan-3-ols, resulting in the lowest concentration in the tuber.

CA possesses relevant antioxidant activity both in vitro and in vivo, higher than that observed for CGA [55] and it can be accumulated in plants, mainly in conjugated forms due to esterification by quinic acid (1*S*,3*R*,4*S*,5*R*)-1,3,4,5-tetrahydroxycyclohexane -1-carboxylic acid), in two configurational isomers, such as *trans* or *cis*. In biological systems, the most relevant form is the *trans* isomer, because of its best stability at subacid pH in the plant microenvironment [56], even though its photoexcitation for the absorption in the UVA region (400-315 nm) can lead from *trans* to *cis* structure [57]. Its ortho-diphenolic moiety confers important antioxidant features to CA: Lowering the OH bond dissociation enthalpy, thereby increasing the rate of H-atom transfer to peroxyl [58]. Moreover, the *ortho*-diphenolic system has a relatively higher oxidative reaction rate, with the negative secondary effect of the tuber turning brown when cut or damaged, with the essential contribution of molecular O_2 and of specific enzymatic activities. The molecular mechanism of the browning phenotype is not yet completely elucidated [59].

Catechin (2R,3S)-2-(3,4-dihydroxyphenyl)-3,4-dihydro-2H-chromene-3,5,7-triol) is an ortho-diphenolic belonging to the flavan-3-ol family. Differently from other phenol compounds, catechin is found in free form. Peculiarly, it can be condensed in oligomers or in higher molecular weight polymers, commonly named "condensed tannic compounds", present in black tea and red wine [60]. Its chemical structure confers interesting antioxidant properties, higher than those of other common biological antioxidants, such as glutathione and ascorbate [61]. Although its level in peel is lower compared to CA, catechin might be investigated as a potential bioprotector due to its antimicrobial and antifungal activities [62].

2.1. *Potato Eco-Friendly Plant Bioprotector Activity Against Biotic Stresses*

2.1.1. SGA Activity Against Phytophages

Chowansky reported an exhaustive study on SGA activity on insects, such as Heliothis virescens, Manduca sexta, Spodoptera frugiperda, Podisus maculiventris, Schizaphis graminum, Leptinotersa decemlineata, Ceratitis capitata, Empoasca fubae, Mizus persicae, Galleria mallonella, Tribolium custeneum, and Zophobas atratus [27]. The hypothesis about the mechanisms by which they act involve molecular, cellular, and organismal levels. As are most pesticides, SGAs are also inhibitors of the acetylcholinesterase and butyrylcholinesterase enzymes, which catalyze the hydrolysis of acetylcholine at the synapse in the nervous system [63]. These features are the main issues determining that they should not exceed limit level in human consumption and, at the same time, are the reasons why many studies on their biocide activity were performed. Additionally, the bitter/burning taste that SGAs confer lead to insect-feeding deterrence: The pre-ingestive antifeedant effects could be due to the taste sensations they impart, especially related to α-chaconine, while their putative synergism in antifeedant activity has not been shown yet [64,65]. Among the large number of studies on SGAs activity, most of them focused on the correlation between genotype-resistance and SGA content in vivo and on in vitro dose–response testing of pure metabolites against pests. The resistance of five potato genotypes to Phhorimaea operculella (Lepidoptera: Gelechiidae), the causative agent of potato tuber moth (PTM), was correlated to peel SGAs and peel phenolic compound contents, suggesting a putative key role of α-chaconine and CA in potato PTM resistance [45]. Data on sublethal or lethal toxicity of pure metabolites from a dose–response test are also available. On the whole, the activity of α-chaconine was frequently higher than α-solanine one by a factor of 3–10 times [64,66,67], with only a few exceptions [68], but their synergistic effects remain questionable, and the dose effect of pure SGAs solutions compared to their mixtures have been scarcely studied. Smith exhaustively discussed their activity on snail *Helix aspersa* L. [64] already known for its high susceptibility to SGAs [69]. In pure form, both singly α-solanine and singly α-chaconine showed an increasing feeding deterrence at higher concentrations. A combination of both synergistically increased the feeding inhibition. Additionally, in order to understand if peel could contain other substances that may interfere with SGAs, the authors compared the activity of the raw potato peel extracts with the equivalent concentration of SGAs solutions in mixtures. The peel extracts of cv. Majestic and cv. Sharpe's Express did not show significantly different inhibition activity with respect to the equivalent SGA mixture. On the other hand, peel extracts of cv. Homeguard inhibited feeding more effectively than solutions of SGAs suggesting their presence in the peel of additional inhibitory biocompounds. The most interesting aspect is that cv Homeguard also showed lower SGAs content in the peel among the three evaluated genotypes. In 2011, Nenaah carried out a similar study on adults of the red flour beetle Tribolium castaneum Herbst and on the rice weevil Sitophilus oryzae L. [70]. The SGA activity increased in a dose-depending manner, and once more, raw peel extract resulted in being more toxic than pure molecules. Thus, a lack of correlation between the bioactivity of raw peel extract and the dose equivalence of the most active pure compounds, determined through the peel profiling composition, was observed. Antagonistic, additive, or synergistic interaction against the pest could be an explanation, such as concurrent or competitive binding of two or more compounds to the receptor site of the pest.

Similarly, Friedman [68] showed a high SGA activity in a dose–response test against human and animal pathogenic *Trichomonads* strains. Raw potato peel extracts have different antiprotozoal inhibitory effects depending on both genotype and strain, but, as observed in inhibition of complex organisms [64], peel of the *cv*. Russet, despite the lowest content of SGAs, showed the highest inhibitory effect, suggesting the presence of additional compounds able to influence SGA bioactivity modulating their final effect. Therefore, the various substances present in extracts may reciprocally increase [64] or decrease the toxic effects of SGAs [68].

Based on the synergistic effect of the two predominant SGAs, Gee and co-workers suggested that SGA activity against pests is more likely mediated through effects on the membranes of chemoreceptor cells rather than via signal transduction by neurotransmitter enzyme [71]. Indeed, SGAs are known to alter membrane potential/trans membrane ion transport and to synergize in this action. In vitro studies of SGAs indicate that they may have cytotoxic effects that often compromise cell membrane integrity, through the disruption of phosphatidylcholine/cholesterol liposomes [36,72], causing negative effect on intestinal permeability and altered metabolism [36,73]. This mechanism is also confirmed in frog embryo cells [74], rabbit erythrocytes, beet cells, and protoplast of *Penicillium notatum* [75], where α-chaconine has been observed as the most active compound.

2.1.2. SGA Antifungal Activity

SGAs are inhibitory to a wide range of fungi. Fewel and Roddick have reported in vitro activity against four different fungi, not all potato specific (*Alternaria brassicicola, Phoma medicaginis Ascobolus crenulatus,* and *Rhizoctonia solani*) [66], and afterward, Cipollini and Levey have been proven against additional 10 different strains [63]. A synergism between α-chaconine and α-solanine in inhibiting the percentage of radial growth of *A. brassicicola* and *P. medicaginis* has been observed. Indeed, a significantly higher activity of inhibition of them was reported, upon coadministration of mixed compounds rather than pure ones [66]. The ratios producing their maximal synergism and maximal inhibition are of the order of those naturally occurring in *Solanum* species; however, an amount of them as little as 10-20% in a mixture in vitro was sufficient to activate an important synergism. Similarly, SGAs from tomato (solasonine and solamargine) showed the capacity to reduce the spore germination of *A. brassicola* and *P. medicaginis* either alone or in combination 1:1 [67]. Additionally, SGAs from potato were not responsible for the resistance to the oomycete *Phytoftora infestans*, the causative agent of the late blight disease responsible for major damage to potato crops [76]. *P. infestans* was also tested with SGAs from tomato and aubergine and the most interesting finding was that the last one appeared to exercise a greater toxic effect on the oomycete than compounds from potato, implying some level of host adaptation [77].

2.1.3. Interaction Between SGAs and Fungal Membrane

A possible explanation of the great variation in SGA resistance among the fungi species is the composition of the sterol patterns present in the fungal membrane, closely related to the fungi taxonomic classification [78]. The SGA biological activity depends on their membrane disruption by which they compromise the structural and functional integrity of cells and tissues of the organisms [75]. Indeed, SGAs consist of an aglycone (solanidine) with a carbohydrate side chain thought to be important for the interaction with sterols of the fungal lipidic bilayer (Figure 1B). This property has been confirmed by solasonine and solamargine, using synthetic membrane vesicles, liposomes [79], and protoplasts [75]. The leakage of SGAs with membrane sterols suggests that the sterol pattern of fungal membranes influences SGAs activity on fungi. In liposomes, SGAs interact more efficiently with membrane sterols with planar ring structures and a three b-OH: β-sitosterol and fucosterol made the bilayer much more susceptible to SGAs than cholesterol and ergosterol, present in fungal membrane [80,81]. Recently, Sánchez Maldonado confirmed the membrane-disruptive effect of glycoalkaloids: α-chaconine, but not α-solanine, disrupted phosphatidylcholine/cholesterol liposomes, and confirmed that the fungal resistance to α-chaconine is due to the sterol pattern, particularly related to the relative presence of some specific unsaponifiable lipids [82]. α-chaconine resulted in being invariably more damaging than α-solanine, and the synergism in sterol binding significantly enhances the membrane-disruptive activity of SGA mixtures, even though this effect has only been demonstrated in vitro [82].

The proposed mechanism of activity involves insertion of the aglycone in the bilayer in an aglycone/sterol ratio of 1:1 (step1 and step 2), followed by sugar–sugar interactions between the sugar moieties of glycoalkaloids that start the formation of an irreversible complex (step 3; Figure 1C). As a result, a rigid sterol-glycoalkaloid matrix is formed (step 4), which disturbs membrane function and causes lysis of the cell (Figure 1C) [80]. This hypothesis is also supported by the loss of activity after cleaving monosaccharides from the glycosidic moiety of glycoalkaloids in liposomes containing sterols [80].

The sterol profile of fungi, a chemotaxonomic tool [78,83,84], seems to relate to their resistance to SGAs due to their effective interface with fungal membrane. The same mechanism could be also the explanation as to why a pH dependency of SGAs on fungal growth inhibition was observed: their effect was generally lower at pH 6 than at pH 7. At lower pH, this behavior could be hampered by a decreased alkaloid solubility in a lipophilic environment, due to an increased protonation of the steroidal N-moiety. Indeed, it has been shown that the successful invasion of tomato fruits by fungal pathogens was related to the ability of the fungus to decrease the pH and hence the activity of tomatine [66,85]. This fact could be the consequence of fungal detoxification capacity through pH change [86]. However, at pH 6, a valuable SGA synergism was observed in all fungi tested whose magnitude decreased at pH 7. A possible explanation is the masking effect due to the higher activity of individual glycoalkaloids at higher pH.

Moreover, fungal resistance to SGAs activity can also depend on hydrolytic enzyme secretion [87]. Extracts derived from virulent fungi contain enzymes with SGA degrading activity. In that way, some phytopathogenic fungi can overcome SGA toxicity by their enzymatic deglycosylation: The removal of the trisaccharide from SGAs would lead to no toxic steroidal alkaloid (SA) solanidine. Only a few exceptions are reported. In a recent study, solanidine resulted in being more active than glycosylated forms to inhibit the growth of *P. infestans* [77]. Very recently, a bacterial gene cluster (isolated from *Arthrobacter* sp. S41) involved in the complete deglycosylation of SGAs has been characterized, as suitable for a potential application in the bioconversion of feed proteins to food ones, useful for human nutrition [88].

2.1.4. Role of Phenols in Plant Protection

The role of phenols in defense against predators and diseases in plants seems to be well documented. For example, in potato skin, CGA and CA contents are strongly related to PTM larval mortality [45]. However, dose–response testing of pure phenols from potato (CA, CGA, and quercetin) showed only mild inhibition activity against *Thricomonas* [68]. Moreover, CA alone showed only weak antifungal activity when compared to other phenolics and alkyl esters tested, showing the lowest minimum inhibitory concentration but also the highest protection factor [89,90]. It would be interesting to study, in depth, whether the phenols characterized by limited inhibitory effects, when tested alone, could instead show the ability to synergize with SGAs. Indeed, the effect of SGAs could be strongly enhanced by phenols. Accordingly, CA increases the α-chaconine activity against fungi 1000 times, decreasing its minimum inhibitory concentration and the physiological mechanism causing a synergistic effect with SGAs; a change in membrane fluidity could be the reason [81]. The possibility of an added effect given by catechins and their derivatives, also well known as modifiers of membrane fluidity, could be a useful tool to enhance this biological process [91], but up to now, no studies on it have been reported. In Table 1, the literature about SGA and phenolic activities is summarized.

2.2. *Recovery of Eco-Friendly Bioprotectors from PPW*

In the framework of an eco-friendly, integrated, and circular agro-economy, the demand for reducing waste through the re-use of the residues from agro-industry, during the last years, has been relevant. For this reason, the release of the environmental pressure from potato peel, considered as one of the most important agro-wastes [16], and the high PP surplus value due to rich biocompound content [92], confer to PPW an enormous potential to deliver new products for crop protection [21]. Highlighting PPW recovery potential benefits and considering the connected limits will have a huge impact on production-integrated systems through the development of new highly sustainable products and tools suitable for low-impact agriculture and organic farming systems.

To date, information concerning a bio-formulate based on SGAs and phenols to be used in organic defense or in Integrated Pest Management (IPM) is not available. On a dry weight basis, PP represent 5 to 9% of the whole potato and often contains 50% or more of the total SGAs in the whole tuber [44]. The fact that environmental agro-technological factors, together with the genotype, may greatly influence chemical composition is well known; for example, SGAs and phenols are higher in peels from organic tubers, with the only exception being the red genotype [93]. Thus, the determination of SGAs and phenolic content in the peel of the most-used potato-processing cultivars grown in different agro-technical regimes (conventional and organic tubers) would provide useful information to manage the levels of secondary metabolites. In addition, post-harvest technology of potato storage has been historically aimed at lowering the levels of SGAs in edible tuber as soon as possible, but factors such as light, temperature, humidity, wounding, and processing conditions may elevate SGAs content to toxic level [94].

Table 1. Potato metabolites activity against biotic stresses.

Metabolite/Extract: Glycoalkaloids	Target: Molluscs	Effect	References
α-chaconine; α-solanine	*Helix aspersa* L.	Feeding deterrence	[64]
	Target: Protoza		
α-chaconine; α-solanine	*Trichomonas vaginalis*	Growth inhibition	[68]
Peel extracts	*Salmonella typhimurium; Eschrichia coli*	Mutagenic activity	[30]
Peel extracts	*S.aureus (Gram +); P.aeruginosa (Gram -)*	Growth inhibition	[31]
	Target: Insects		
α-chaconine	*Phorimaea operculella*	Strongly related with PTM larval mortality	[45]
α-chaconine; α-solanine	*Tribolium castaneum, Sitophilus oryzae* L.	Toxic against adults	[70]
α-chaconine; α-solanine	*Trogoderma granarium*	Antifeedant activity	[25]
α-solanine	*Galleria mallonella* L.	Increased mortality of larvae; pupae and adults, decreased fertility and fecundity	[27]
α-solanine	*Myzus persicae*	Antifeedant activity, decreased fecundity, increase mortality of pupae	[27]
α-chaconine; α-solanine	*Zophabas atratus*	Decrease heart activity in pupae and adults	[27]
vegetable waste extract	*Culex quinquefasciatus, Nopheles stephensi*	Larvicidal activity	[27]
leaf extract	*Zophobas atratus, Tenebrio molitor*	In vivo cardioinhibitory activity in pupae and adults, increased mortality	[27]
leaf extract	*Leptinotarsa decemlineata, Spadoptera exigua*	Increased mortality of larvae, pupae and adults, disturbance in fertility and fecondity	[27]
	Target: Amphibia		
α-chaconine; α-solanine	*Frog embryos*	Impact on membrane	[74]
	Target: Fungi		
α-chaconine; α-solanine	*Ascobus crenulatus, Alternaria brassicicola, Phoma medicaginis, Rhizoctonia solani*	Effect on the fungal growth	[66]
α-chaconine; α-solanine	*Ascobus crenulatus, Alternaria brassicicola, Phoma medicaginis, Rhizoctonia solani*	Spore germination inhibition Fungal growth inhibition	[67]
α-chaconine; α-solanine	*Alternaria alternata, Pyrenophora teres f. teres, Pyrenophora tritici-repentis*	Effect on the fungal growth	[82]
solanidine	*Phythophtora infestans*	Inhibition of mycelial growth	[77]
Metabolites/Extract: Phenols	**Target: Protozoa**	**Effect**	**References**
CA; CGA	*Trichomonas vaginalis*	Growth inhibition	[68]
	Target: Insects		
CA; CGA	*Phorimaea operculella*	Strongly related with potato PTM larval mortality	[45]

A crucial and limiting issue is the optimization of the industrial multi-step process in the frame of a green strategy necessary for sustainable PPW exploitation [14]. The first critical point to be addressed is how to dry the starting fresh sample to obtain a stable and extractable matrix. A non-conventional, effective, and cheap approach could be solar drying [95]. However, it can induce a variable retention of phytochemicals in comparison with conventional techniques. Indeed, a forced-air oven, moved by electric power, still leads to a well-dried product in a short amount of time [96], even though it has high energy consumption. The following step, the extraction of bio-compounds from PPW, is the most critical, to a great extent depending on the choice of solvent to be used in the process. Actually, the best procedure for phytochemical extraction employs protocols including the use of alcohols, usually ethanol, and is characterized by long extraction times, a large quantity of solvent, and high temperature. The utilization of supercritical systems of fluids, although very expensive and not eco-friendly, resulted in being very effective. The "green" extraction alternative protocol should employ the use of water, but often this solvent is not effective, especially because of the low yields [97]. Ultrasound-assisted extraction (UAE) is a promising and innovative technique, well described by Bankeblia [14] and Hossain [98]. Additionally, the characterization and quantification of active principles in recovered by-products is carried out at the lab-scale by using complex chromatographic methods, such as HPLC (high-performance liquid chromatography) and HPTLC (high-performance thin-layer chromatography). For this reason, a future perspective to improve the sustainable bio-compound extraction procedure from PPW could be based on the development of user-friendly and non-destructive monitoring methods, already applied in the internal quality evaluation process of some fruits [99], by means of devices easily used by non-specialized personnel. To date, these non-destructive approaches are not validated and do not fulfil economic and environmental criteria, requiring high costs and energy input machines. In conclusion, at the industrial scale, the recycle technology to process PPW could reach a large potential market, resulting in being more attractive for consumers. However, the challenging task to be addressed is to overcome specific technological limits in order to maximize the yield reducing costs, energy, and solvent consumption.

2.3. Molecular Approaches to Modulate Eco-Friendly Bioprotector Production in Potato Peel

Addressing the genetic basis that influences SGAs content and composition is a key challenge to exploit PPW as bioprotector. Genome-wide association mapping studies and marker-assisted selection to find QTLs (quantitative trait locus) and markers involved in the trait have been reported [100], even if the expression pattern of SGA biosynthesis genes to be employed in breeding programs should be better understood. The biosynthesis of SGAs has been highlighted to be via the mevalonate/isoprenoid pathway (Figure 2) [101,102]. The key enzymes involved in the mevalonate biosynthetic pathway, from the top to the bottom, are, respectively, HMRG1 (3-hydroxy-3-methylglutaryl coenzyme A reductase 1) and PSS1 or SQS (Squalene synthase), that act at the pre-cycloartenol part and for which a coordinated regulation is known [103]. The *PVS1* (*Vetispiradiene sequiterpene*) gene facilitates the branching pathway that converts the farnesyl-PP into sequiterpenoid. The genes leading to solanidine from sterol precursors, including cholesterol, are Δ^{24}-*reductase*, *STM1* also named *SMT1* (*Sterol C24-methyltransferase* type 1), and *CH* (*Cholestrole hydroxylase*). Finally, SGT1 (*Solanidine galactosyltransferase*), SGT2 (*Solanidine glucosyltransferase*), and SGT3 (*Glycosterol rhamnosyltransferase*) represent key enzymes in the biosynthesis of solanine and chaconine. The role of most of these enzyme encoding genes has been studied in transgenic potato lines. Ginzberg proposed a feedback regulation cycle in transgenic plants where over-expression of *HMG1* resulted in increased transcript levels of endogenous *PSS1*, and, consequently, the over-expression of *PSS1* caused a reduction of *HMG1* transcript level [104]. Several studies on the characterization of the *STM1* gene have been reported. Over-expression of *STM1* led to an increase of sterol level, due to an increased flow of sterol precursors into the 24-alkylated pathway, and to a reduced cholesterol and glycoalkaloid content in leaves and tubers, suggesting that cholesterol is a precursor for the glycoalkaloid biosynthesis [105]. Factors such as wounding, light, and microbial pathogens exposure have been demonstrated to be regulators of SGA synthesis, via modification of STM1 activity. In particular, down-regulation of *STM1* acts as a fungal elicitor, affecting sterol synthesis by down-regulation of the *PSS1* gene [106,107]. A study reporting the down-regulation of the Δ^{24}-*reductase* in transgenic lines resulted in low levels of cholesterol and glycoalkaloids, demonstrating the involvement of this gene in the SGA pathway [108]. Moreover, another key enzyme involved in the biosynthesis of cholesterol and related SGAs, the *Sterol Side Chain Reductase 2* (SSR2) that acts between C-24 alkylsterols and cholesterol steps, has been identified and characterized in potato by genome editing [109]; in fact, loss of function of this gene showed highly reduced levels of cholesterol and SGAs in plants. Tubers genetically modified to reduce the glycoalkaloid content, via down-regulation of the three genes *SGT1*, *SGT2,* and *SGT3* acting at the end of the SGA pathway, resulted in significant changes in specific glycolalkaloids compared to wild type [110]. Moreover, a compensatory effect has been described, when the *SGT1* gene was down-regulated, α-solanine accumulation was inhibited whereas α-chaconine content increased; conversely, when SGT2 was down-regulated, α-chaconine was lowered compared to α-solanine, which increased. On the other hand, down-regulation of *SGT3* lowered both above-mentioned glycoalkaloids to varying degrees.

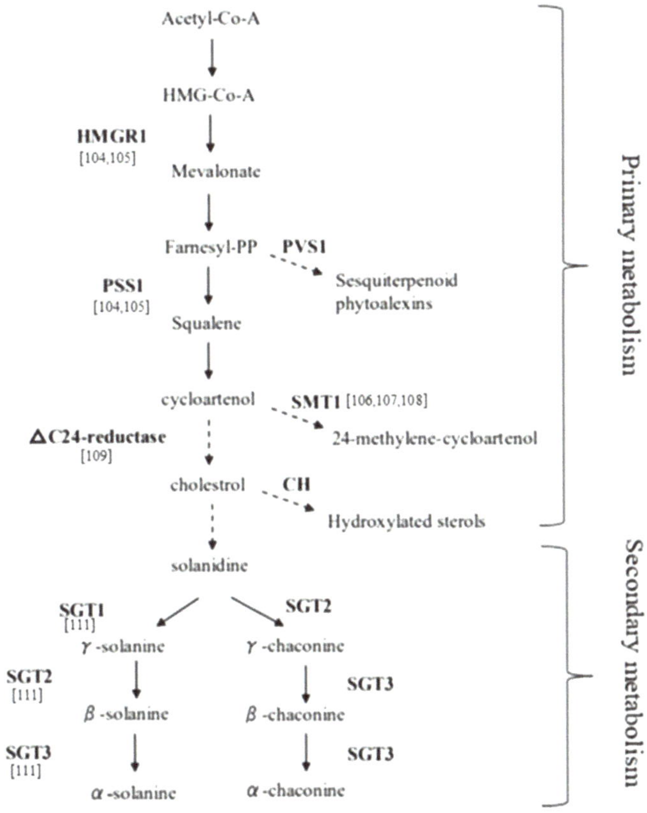

Figure 2. Glycoalkaloid biosynthetic pathway in potato (modified from Khan et al.) [102].

Natural variation in candidate SGA-related biosynthetic genes and whole-genome genotyping represent a good strategy to determine the relationship between genes and SGA accumulation. A research focusing on five candidate genes related to the primary (*3-hydroxy-3-methylglutaryl coenzyme A reductase* 1 and 2, *HMG1* and *HMG2*; *2.3-squalene epoxidase*, *SQE*) and the secondary metabolism (*SGT1* and *SGT2*) has highlighted a higher level of polymorphism in introns than exons and in genes of the secondary metabolism compared to the primary [111]. In addition, haplotypes of informative single -nucleotide polymorphisms (SNPs) in these candidate genes able to discriminate among high, intermediate, and low levels of SGAs in wild potatoes have been identified, therefore useful as functional markers to test segregating populations or association mapping panel. Similarly, Hardigan used a panel of 67 genotypes to capture the genome variation in cultivated and wild potatoes, by analyzing the sequence and the structural variants in the form of SNPs and copy number variation (CNV), against the reference genome of *S. tuberosum* Group Phureja [112]. Signatures of selection have been observed in *squalene synthase (SQS)* and in *glycoalkaloid metabolism 9 (GAME9)*, an *APETALA2/Ethylene* response factor, being candidates for landraces and cultivars. Moreover, *GAME9* was located on chromosome 1, where a highly significant QTL was reported, explaining a major proportion of the SGA content in potato tubers [113]. The region spanning 230 kilobases pair included many transcription factors together with *GAME9*. Transcriptomic and proteomic analyses are important tools to provide insight into the regulation of SGA biosynthesis. Understanding regulatory players of response to environmental stress conditions can be useful in breeding and quality assessment in order to predict the levels of SGAs expression. Tubers of two potato cultivars, characterized by low (cv. Atlantic) and high SGAs content (cv. Haryoung), showed a very different transcript accumulation of *HMG1* and *PSS1* genes, under drought stress conditions, being doubled in the cultivar Haryoung [114]. The mRNA level of these two genes could be used as selection markers for breeding potatoes with low SGAs level. In addition, the abundance of transcripts of *SGT* genes was also detected, indicating their direct involvement in glycoalkaloid accumulation. A transcript profiling on two different cultivars during glycoalkaloid-inducing treatments (wounding and light exposure) has also been carried out and only a small number of differentially expressed genes, covering important steps of the entire SGAs biosynthetic pathway, was found to be associated with increased SGAs levels [115]. In particular, four genes (*MVD, Mevalonate diphosphate decarboxylase*; *FPS2, Farnesyl diphosphate synthase 2*; *SMO1-like, Sterol C4-methyl oxidase 1-like*; *DWF1-like, Sterol* Δ^{24}*-reductase-like)* were found to be strongly induced upon wounding, whereas *MVD* and *FPS2* resulted in being not up-regulated under light exposure. The differences in the expression of these genes underlies the existence of cultivar variations in basal SGAs levels. A relation between SGAs content and the expression of *GAME*, *SGT1*, and *SGT3* genes was also reported in potato tubers by Mariot et al. [116]. A detailed analysis of *GAME*, *SGT1*, and *SGT3* promoter regions highlighted *cis*-elements related to the response of potato plants to biotic and abiotic stresses, confirming that unpredictable variations in SGA levels could be related to these stressors. Indeed, diverse studies also reported that higher levels of SGAs can impart strong resistance against pests [36,117]. In particular, Zhang et al. investigated the gene expression profiles of SGAs, induced by light exposure, in potato tubers under biotic stress [118]. A strong correlation between the stress response and SGA accumulation has been found, with both disease resistance and SGA biosynthesis genes resulting in being up-regulated. A study aimed to elucidate the defense response activated by BABA (β-aminobutyric acid), known to induce resistance in a wide range of plants against several types of pathogens, highlighted the down-regulation of genes involved in sterol biosynthesis and up-regulation of sesquiterpene phytoalexin biosynthesis enzymes, whereas a high level of pathogenesis proteins was accumulated [119,120]. Another example of correlation between SGA biosynthesis and plant defense against biotic stresses has been reported in two lines of *Solanum tuberosum* in which the *Glycoalkaloid metabolism 4 (GAME4)* enzyme encoding gene involved in the conversion of cholesterol to SGA aglycones was silenced by RNA interference [121]. When exposed to insect pest CPB (Colorado Potato Beetle), a gregarious defoliator for the *Solanaceae* plant family, inoculated with *Verticillium dahliae* pathogen, *GAME4* RNAi lines showed changes in metabolite profile, including increased levels of phytoecdysteroids, thus affecting the growth of the insect pest and, in one of the two lines, also affecting the colonization by the pathogen. Taken as a whole, these results demonstrate that in potato, targeted modifications of secondary metabolic SGA pathways can affect plant disease resistance.

In conclusion, in order to increase extraction yield from PPW and contribute to its full potential of economic value both as a by-product and as. a crop bioprotector from biotic stresses, the development of genotypes with a high level of SGAs in peel, via breeding or New Breeding Technologies (NBTs), should be a future perspective.

3. Biocompounds in *Brassicaceae*

Plants of the *Brassicaceae* family are characterized by a very effective and specific chemical defense system based on the production of a group of secondary anionic metabolites called glucosinolates (GSLs) [26]. GSLs consist of a S-β-D-glucopyrano unit anomerically connected to an O-sulfated (Z)- thiohydroximate function, which represents the invariant backbone of the molecule, which is linked to a variable side chain (or R-Group). More than 130 GSL structures have been discovered and validated in the whole Brassicales order of plants to date [121]. They are most abundant in species belonging to the *Brassicaceae* family such as mustards, broccoli, cauliflower, cabbage, and the model plant *Arabidopsis thaliana* [122,123].

3.1. Glucosinolates, Myrosinases, and Hydrolysis Products

GSLs are not toxic per se but become biologically active upon hydrolysis by myrosinases and associated proteins to form GSL hydrolysis products (GHPs). In plants, GSLs and myrosinases are compartmentalized in different tissues or different parts of the same cell and get together only after tissue disruption leading to the formation of several hydrolysis products characterized by different physicochemical properties and biological activities [6]. The myrosinase-catalyzed reaction starts with the cleavage of the thioglucosidic linkage, resulting in the release of a D-glucose and an unstable thiohydroximate-O-sulfate, known as aglucone, that undergoes "Lossen-like" rearrangements and degradation to afford a wide range of hydrolysis products [121]. The outcomes of the unstable aglucone have been shown to depend on (i) the structure of the GSL side chain, (ii) the presence of supplementary proteins known as specifier proteins, and/or (iii) the physiochemical reaction condition (Figure 3). Several myrosinases have been characterized from more than 20 species of Brassicales, insects, and many bacteria residing in the human intestine. Plant myrosinases are reported to be generally activated by ascorbic acid, while in insects and bacteria, myrosinases seem not to be influenced or inhibited from it. This highly regulated defense system has been continuously updated during the plant and herbivore evolution paths. Plants evolve to defend themselves against insects by producing many different chemicals that are toxic to herbivores and other pests, and the latter ones evolve to defend themselves through detoxification pathways [124,125]. As a consequence, some plants have gained the ability to produce more than one type of chemical defense. In the *Brassicaceae* family, wallflowers of genus *Erysimum*, for example, produce two types of toxic chemicals: Hydrolysis products of GSLs, and cardenolides, which are otherwise found only in distantly related plants such as foxglove and milkweed. The combination of these two chemical defense compounds within the same plant may explain the evolutionary success of this genus within the last 2 million years [125].

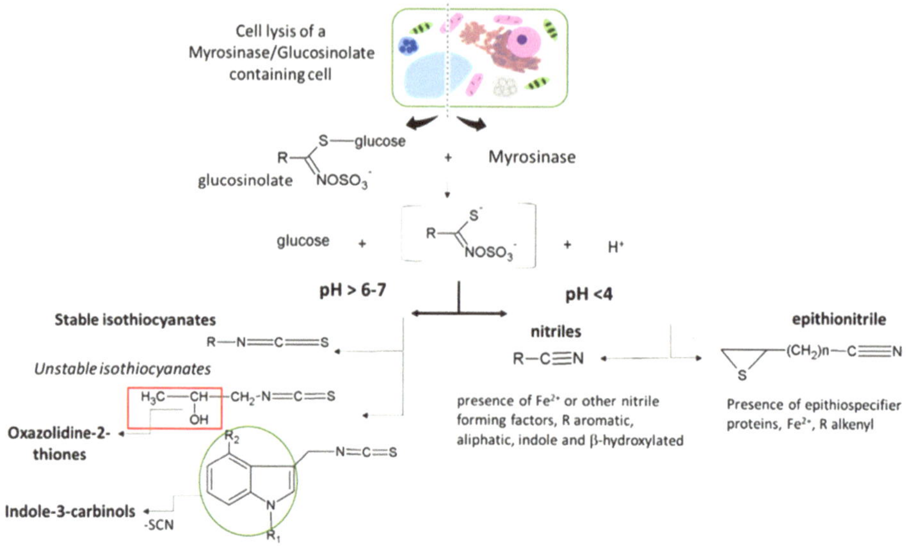

Figure 3. Glucosinolate hydrolysis products reactivity according to the structure of the glucosinolate side chain (R), the presence of supplementary specifier proteins, and/or the physiochemical reaction condition.

3.2. Brassicaceae Eco-Friendly Plant Bioprotector Activity against Biotic Stresses

A role of GLs in constitutive or inducible defenses against microbial pathogens and insect herbivores has been reported [126], as they act as signaling molecules, and may initiate pathways such as stomatal closure, apoptosis, and callose accumulation [127,128].

The action of these compounds against biotic stresses suggests that external application of GSLs could promote plant resistance as an alternative to chemical pesticides. The most effective GSLs have been searched by testing them in vitro against different pathogens, in order to proceed to develop new formulations for field treatments. Strong effects of GSLs have been found against many biotic agents, from bacteria to herbivores, and they are summarized in Figure 4.

3.2.1. GSL Antimicrobial Activity

A clear effect on certain bacteria has been shown for different specific GSL compounds. Indeed, when the antibacterial activities of 4 isothiocyanates (ITCs) (3-butenyl, 4-phentenyl, 2-phenylethyl, and benzyl isothiocyanate) were investigated in vitro against four Gram-positive bacteria (*Bacillus cereus*, *Bacillus subtilis*, *Listeria monocytogenes*, and *Staphylococcus aureus*) and seven Gram-negative bacteria (*Aeromonas hydrophila*, *Pseudomonas aeruginosa*, *Salmonella choleaesuis*, *Salmonella enterica*, *Serratia marcescens*, *Shigella sonnei*, and *Vibrio parahaemolyticus*), benzyl and 2-phenylethyl isothiocyanate (2-PEITC) showed higher activity against most of the pathogenic bacteria than 3-butenyl and 4-pentenyl isothiocyanate, and were more effective against Gram-positive bacteria than against Gram-negative ones [129]. For these properties, GSL-myrosinase system has also been employed to prevent bacterial and fungal (see below) spoilage and to improve shelf life of food product in advanced packaging systems [130].

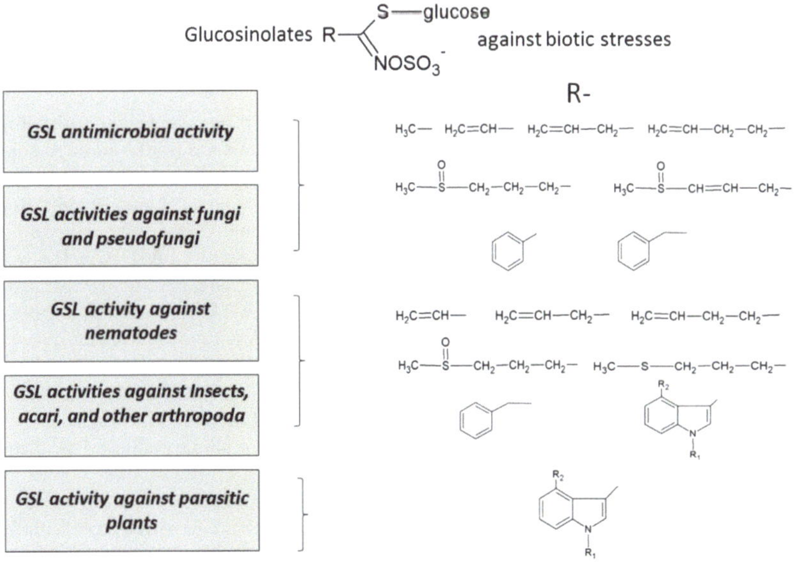

Figure 4. Main activities of glucosinolates and their hydrolysis products against biotic stresses in plant protection. The side chains (R) of glucosinolates, studied to date for each effect, have also been reported.

3.2.2. GSL Antifungal Activity

Other molecules have been shown to be characterized by a wide spectrum of activities, as in the case of some *B. carinata* GSL-derived ITCs, which were significantly effective in vitro control of four assayed pathogens seedborne fungi tested by Pane et al. [131] (*Alternaria dauci, Alternaria radicina, Colletotrichum lindemuthianum,* and *Ascochyta rabiei*), confirming their high antimicrobial activity [132]. Recent experimental evidence showed that ITCs were also promising for treatments against black spot, one of the most important diseases of pear fruit during storage caused by *Alternaria alternata.* The development of black spot rot on the pear fruit inoculated with *A. alternata* was significantly decreased by 2-PEITC fumigation [133]. The authors proposed that the antifungal effect of 2-PEITC against *A. alternata* might be mediated by a reduction in toxin content and breakdown of cell membrane integrity. A similar effect on fungal toxins has been observed for prop-2-enyl or allyl isothiocyanate (AITC)s in the treatment of post-harvest products such as in maize. Nazareth et al. stated that AITC caused a transcriptional alteration of genes involved in aflatoxin B1 and other processes key for normal fungal growth and development [134,135]. The authors reported prevention of the growth and development of *A. niger, A. parasiticus,* and *Fusarium verticillioides,* as well as the reduction of mycotoxins (aflatoxins and fumonisins) by GHPs application. Additional studies showed that gaseous AITC, benzyl, and phenyl isothiocyanate inhibited the growth of different fungal species and the production of mycotoxins in in vitro studies and food products [136–138]. Similar effects, starting from flours of oriental and yellow mustard, were obtained for wheat tortillas as preservatives against aflatoxins B1, B2, G1, and G2 [139]. Numerous data reported in the scientific literature encourage the use of GSLs to develop new formulations for the sustainable protection of crops from abiotic stresses, and a similar strategy can be imagined also for other cereals crops.

3.2.3. GSL Activity Against Nematodes and Insects

Eleven GSLs and their degradation products were evaluated for the biocidal activity of on second-stage juveniles of the root-knot nematode *M. incognita* in vitro [140]. None of the intact GSLs showed any biological effect. Following myrosinase addition, GHPs (essentially ITCs) resulted in highly different biocidal activities. Similarly, Bhushan et al. showed the insecticidal potential of 3-isothiocyanato-1-propene or AITC on the growth and development of a polyphagous pest, *Spodoptera litura* (Fab). AITC resulted in 100% larval mortality when applied at the highest concentration (3125 ppm) [141]. Larval period, pupal period, total development period, and pupal weight were also influenced by AITC. Wireworms are widely distributed throughout the world and are important pests of a wide range of crops including small and large grain cereals. *Brassica carinata* defatted seed meals (DSMs) used on maize plant at early development stage in pots and in open field experiments showed levels of wireworm control and prevention of plant damage comparable to the conventional insecticide (Regent®, BASF Italia [142]. In these trials, the efficacy of AITC, the main GHPs in *B. carinata* was tested; in literature several applications especially for maize, are reported. Indeed, AITC was recently proposed as a potential tool to be employed in control strategies against maize weevils, to overcome resistance to phosphine and other conventional insecticides [143,144]. Other Brassicales, such as *Eruca sativa* and *Brassica rapa* seed powders, showed allelopathic activities due to the presence of different GSLs, as selective bioherbicides for controlling *Cyperus rotundus*, improving both growth parameters and carbohydrate contents in maize cultivation [145].

As for the interaction with the herbivores, a high level of indole GSLs had a positive action against pea aphids (*Acyrthosiphon pisum*) but not against peach aphids (*Myzus persicae*) [146]; the last mentioned study showed that GSLs may have both direct and indirect effects on dodder-feeding herbivores.

3.2.4. GSL Activity Against Parasitic Plants

The effect of GSLs is not limited to microorganisms and animals. Recently, an interesting study by Smith et al. investigated the effect of GSLs on a parasitic plant and on the herbivores that grow on it [146]. Transgenic *Arabidopsis* lines with elevated indole GSLs or without them, as control, were developed and allowed to interact with the parasitic dodder vines (*Convolvulaceae*; *Cuscuta gronovii*). Parasitic plants acquire diverse secondary metabolites from their hosts, including defense compounds that target insect herbivores; in this case, concentrations of aliphatic and indole GSLs were higher in parasite tissues than those observed in corresponding host tissues. Dodder growth was enhanced on plants without indole GSLs and inhibited on plants with elevated indole GSLs compared to wild-type hosts. Therefore, an additional defensive role of indole GSLs against parasitic plants can be argued.

3.2.5. GSLs Role in Biofumigation Crop Protection Management

Biofumigation is a sustainable agronomic practice for pest management based on the release of volatile hydrolysis products of GSLs. It involves the sustainable and circular use of renewable and biodegradable plant materials, determines a significant reduction of CO_2 emissions, returns organic matter to the soil, and is less toxic to the general soil environment than synthetic pesticides. It can be achieved through several methods. One of the most beneficial practices for soil health is the green manure technique, often coupled with appropriated crop rotations, which consists in growing selected *Brassicaceae* as a cover crop and tilling them into the soil where they break releasing by-products of GSLs. Biofumigation tissues or bio-based products may also be applied as industrially formulated DSM or concentrated plant essential oils or extracts [147–151].

The agronomic technique of biofumigation is based on the idea that *Brassicaceae* tissues may sustain the molecular defense system of a profitable crop providing it with peculiar hydrolysis products of GSLs. In fact, at least since 2005, when the process of phasing out from agriculture of methyl bromide occurred in all the major countries [152], several non-synthetic chemical alternatives have been explored in order to substitute this widespread fumigant. Among these low-impact alternatives, the biofumigation techniques have established and widespread in the last 30 years. The first applications were plants, selected for rusticity, biomass yield, and concentration of specific GSLs in epigeal and/or hypogean tissues, used as intercrops or biofumigant green manures [153,154]. Afterwards, beside plants, seed- and plant-derived materials were optimized to release allyl-isothiocyanate. In fact, *Brassica* plants translocate GSLs in the seeds at a high concentration, becoming a starting material for biofumigant bio-products. After seed defatting, the residual meal contains a high level of GSLs. The obtained meal, or pellet, can be formulated by a patented procedure able to modulate ITCs release [155,156]. *Brassica* crops or seed meal amendments incorporated into soil may have the potential to control soil-borne plant pathogens, by changing soil pH, microbial populations, and enhancing enzymatic activities concurrently with the release of ITCs [157]. These processes may trigger a new microbial balance manifested as soil suppressiveness, that is "soil's ability to delay pathogen infection and disease progress in a susceptible host, even in the presence of virulent pathogens". In fact, even when less effective, different *Brassica* treatments—both green manure and seed meal—either improve or maintain soil microbial activity and fertility compared to the chemical treatments [158]. Generally, meals and pellets are distributed on dry soil, incorporated at 20–30 cm depth at a rate of 250/300 g m^{-2} and activated by light irradiation. In this way, the meal releases ITCs directly into the soil limiting significantly active compound losses in the air by volatilization. Afterward, liquid formulations based on a *Brassica* oil in water emulsion with the addition of a reduced amount of biofumigant meals were conceived. In this case, GSL hydrolysis is activated by the water in the emulsion, and, according to its hydrophobicity, the released ITCs are solubilized in the oil fraction of the emulsion. Liquid emulsion, once distributed, forms an oil microfilm on the plant organs, determining a potential repellent effect and a physical suffocating action on some pests and pathogens, meanwhile improving the biofumigant effect of ITCs [159].

In Table 2, the experimental evidence mainly reported in the last 10 years of literature about the molecular system of *Brassicaceae* involving GHPs released by the GSL/myrosinase system, including a clearly or tentatively defined GSL concentration in experimental trials and the plant pathogens involved in the studies, is summarized. Table 2 [160–177] shows the growing and keen interest for biofumigation applications, particularly as regards plant protection against nematodes, fungi, pseudofungi, and some arthropoda. Nevertheless, soil organic matter addition through plant-based products should be carefully evaluated. As an example, even if in vitro assays, containment of *Fusarium* spp. through AITC was established [124], and the control effect on *F. graminearum*, a severe wheat pathogen, was demonstrated in the field condition [178]. Several doubts are still raised on *Brassicas* applied as green manure, given their host status for some *Fusarium* species, for example some *formae speciales* of *Fusarium oxysporum* known for being pathogenic on *Brassica* crops that become a vehicle for increasing infection and some non-reproducible results [179]. Despite their effectiveness, Brassicales-based products' application in commodity crops is still underestimated. This is principally due to the economic costs, which are rather high to date. On the other hand, as already mentioned, future EU strategy will lead to a progressive reduction in chemicals applied in agriculture. This process already begun years ago, since many of the high-impact chemical products used in European agriculture are being phased out (Directive 2009/128/EC) [5], and non-chemical alternatives are strongly needed as suggested by Regulation (CE) no.1907/2006 (REACH) [180]. Furthermore, the secondary effects of such products have been underestimated. In fact, besides their biological effects, most *Brassica* DSMs are characterized by a valuable level of Nitrogen [181] with a good C/N ratio, making them suitable soil amendments. Again, in the near future, bio-product environmental impacts in terms of CO_2 sequestration or, more in general, their LCA (life-cycle assessment) should be considered. Besides the suitable C/N ratio, a *Brassica*-based formulated product could consistently reduce CO_2 emissions [25], while still being effective and less harmful to the soil food web functioning [120,162] and to the beneficial soil invertebrate [177], thus reducing agricultural environmental impact [156].

Table 2. Biofumigation in vitro, in pots, and in open-field trials published in the last 10 years with defined glucosinolate concentrations.

Glucosinolate/isothiocyanate	Target: Nematodes	Effect	References
Allyl GSL from leaf flour of *Brassica macrocarpa* Guss.	*Meloidogyne* spp. root-knot nematodes, target crop: tomato in greenhouse	*B. macrocarpa* leaf flour inserted in the soil at a GSL dose of 300 µmol m^{-2} showed similar effect to Vydate 5G® on root disease index, root weight, and marketable yield	[160]
GSL from the DSMs of 13 *Brassicaceae* species: *Barbarea verna*, *B. carinata*, *B. nigra*, *B. rapa B. tournefortii*, *B. oleracea* var. *acephala Crambe abyssinica*, *Eruca sativa*, *Lepidium densiflorum*, *Lepidium sativum*, *Raphanus sativus*, *Rapistrum rugosum*, and *Sinapis alba*.	*Meloidogyne incognita* in pots in glasshouse-controlled conditions, with *Solanumlycopersicum L. cv. UC82* as host plant.	Among the tested DSMs, the best results for all inoculations were achieved by *Eruca sativa*, *Barbarea verna* (300 µmol m^{-2} GSLs) and *Brassica nigra* (370 µmol m^{-2} GSLs), whereas the other species gave either alternate results or results not different from untreated or sunflower DSM controls. All the DSMs, including sunflower, determined a clear positive effect on tomato vigour.	[149]
Allyl GSL (98% of total GSLs) from *B. juncea*; 4-(Methylsulfinyl)butyl GSL (72 % of total GSLs) from *R. sativus*; and 4-pentenyl GSL (38% of total GSLs) from *E. sativa*	cyst nematode *Globodera pallida*, target crop: potato	10 kg ha^{-1} *E. sativa*, 8 kg ha^{-1} *B. juncea*, and 20 kg ha^{-1} *R. sativus* were sown, cultivated, and incorporated in soil, in open field trials, both in summer and winter. The incorporation of green materials was done at complete flowering for summer trials or one week prior the potato planting for winter trials. A positive linear regression of GSL concentration (mol m^{-2}) in incorporated biomasses and *G. pallida* mortality was determined for all brassicaceous species cultivated during summer; only *R. sativus* and *E. sativa* demonstrated a significant relationship between GSL concentration and *G. pallida* mortality.	[161]
GSL from *Brassica carinata* DSM	*Meloidogyne incognita* root-knot nematodes, target crop: tomato in greenhouse.	A pot trial was conducted on tomato plants grown in a soil naturally infested with *M. incognita*, amended with *B. carinata* DSM (3 t ha^{-1} or 40 mmol m^{-2} GSL), and finally compared to a soil fumigated with Vapam (Sodium methyldithiocarbamate) and to an untreated control. Both *B. carinata* DSM and Vapam treatments were effective in protecting tomato plants against *M. incognita* but they exhibited different effects on soil biota. In general, nematode populations strongly responded to *B. carinata* DSM amendments both in terms of abundance and structure. Although the free-living nematode structure was negatively influenced by the two treatments, *B. carinata* DMS proved to be the best compromise between efficiency to control *M. incognita* and environmental impact.	[162]

Table 2. *Cont.*

Glucosinolate/isothiocyanate	Target: Nematodes	Effect	References
GSL from four cultivars: three mustards (*Brassica juncea* 'Caliente 61', 'Caliente 199', and 'Pacific Gold') and one broccoli (*Brassica oleracea* var. *botrytis* 'Arcadia')	2-year open field study of biofumigant of the four Brassicales in a chile pepper *Capsicum annuum* 'AZ-20', rotation system in southern New Mexico.	Broccoli produced lower biomass and lower GSL concentrations than the mustard treatments but may be a valuable crop for growers with nematode issues because *Meloidogyne incognita* populations decreased in its presence. Based on high biomass production and high GSL concentration, *B. juncea* 'Caliente 199' showed the most potential as a biofumigant crop for southern New Mexico	[163]
Leaf flour of dry plants of *B. juncea*, *E. sativa*, *R. sativus* and *B. macrocarpa*, characterized for sinigrin content.	*Meloidogyne* spp. root-knot nematodes, target crop: tomato in greenhouse	Leaf flours were distributed before planting (60 and 90 g m^{-2}), with the mean dose corresponding to allyl GSL content in the commercial formulate (Nemathorin) applied at 3 g m^{-2}. Disease index detected on the tomato roots at the end of the growing cycle resulted in all thesis lower than the control and Nemathorin, whereas it was lower with 60 g m^{-2} *E. sativa* and 90 g m^{-2} *R. sativus*, in comparison to 90 g m^{-2} *B. juncea*.	[164]
Glucosinolate/isothiocyanate	**Target: Fungi**	**Effect**	**References**
Pure AITC, and macerated plant tissues from 18 different cultivars amongst *Raphanus sativus*, *Sinapis alba*, *Brassica carinata*, *Brassica juncea*	*Fusarium gramineaum* and *Fusarium poae* in vitro and pot experiment	*B. carinata* and *B. junceae* (AITC containing tissues) performed a better mycelial growth reduction in Petri dish. *Fusarium poae* resulted more tolerant to AITC than *F. graminearum*. In general, all pots added with *Brassicaceae* plant material presented a reduced fungal infection, but only *B. juncea* plant material alleviated *F. graminearum* negative effect on maize growth.	[165]
Pure Allyl isothiocanate	maize grains contaminated with *Aspergillus flavus* in glass jars	maize grains contaminated with A. flavus in glass jars of 1 L and treated with 0.125, 0.25, 0.5, 1 and 5 µL of AITC. After 7 days of storage, the mycelial growth was significantly reduced in doses higher than 0.125 µL/L of AITC. All doses of AITC significantly reduced the fungal growth and Aflatoxin B1 production in maize after 30 d, regardless of moisture content.	[135]
GSL derived from bio-based experimental formulations con-taining either *B. carinata* oil 1.5% and *B. carinata* DSM 3 g L^{-1} (270 µmol GSL L^{-1}), or 2% *B. carinata* oil and *B. carinata* DSM 4.5 g L^{-1} (405 µmol GSL L^{-1}).	*Podosphorea xanthii* control on melon in open field	The field trials carried out over two years demonstrated the efficacy of the two bio-based experimental formulations based on *B. carinata* biomasses. In particular, the formulation with the highest concentration of oil and DSM gave results statistically not different from those of penconazole (Topas).	[166]

Table 2. *Cont.*

Glucosinolate/isothiocyanate	Target: Fungi	Effect	References
AITC from six *B. juncea* (L.) Czern cultivars, two *B. rapa* cultivars and one *B. oleracea*, as macerated of frozen plant tissues	*Rhizoctonia solani* AG1-1A, in vitro assays. Target plant: *Oryza sativa*	3 g of macerated frozen tissues, amended or not with 3 different soils, were confined to the lid of an upside-down Petri dish, containing potato dextran agar (PDA) medium with a disc of agar inoculated with *R. solani* AG1-1A. The dishes were incubated for 72 h at 25 °C. All six *B. juncea* cultivars consistently inhibited mycelium growth (90% inhibition) in all soils tested; *E. sativa* which was not considered as a brassicaceous, achieved 60% *R. solani* AG1-1A inhibition if amended into soils with the lowest levels of pH, organic matter, proteins, calcium and magnesium.	[167]
GSLs from above-ground parts of *Brassica juncea* (L.) Czern. & Coss, 'Negro Caballo', *Eruca sativa* Miller, and *Sinapis alba* L., 'Asta'	Arbuscular mycorrhizal fungi (AMF) colonization, and *Fragaria x ananassa* Duch. var. *'Marmolada'* strawberry yield	91 g of each biofumigant plant per kg of soil were added to AMF inoculated or not inoculated soil, corresponding to 78.9 mg GSL per kg of soil to the *B. juncea* treatment, 75.2 mg GSL per kg of soil to the *E. sativa* treatment and 28.6 mg GSL per kg of soil to the *S. alba* treatment. The soil treatments with biofumigant plants revealed moderate inhibitory effects on strawberry plant AMF colonization, whereas they increased the plant growth and fruit production, especially for the *B. juncea* and *S. alba* treatments. Effects of solarization were also investigated.	[168]
GSLs from *Brassica juncea*, *Raphanus sativus*, and *Sinapis alba*	*Verticillium dahlia* in vitro and in soils	Commercial standards of methyl ITC, propenyl AITC, 4-(methylsulfinyl)but-3-enyl4- methyl s u l f i n yl-3-butenyl ITC, benzyl ITC and 2-phenylethyl ITC were tested in vitro against V. dahliain sand at ITC concentrations of 1, 5, 25, 125, 625 nmol g^{-1} sand. Furthermore, the effect of propenyl ITC with a dose of 150 nmol g^{-1} on *V. dahliae* in natural infested soil samples from 22 sites with different crop rotation history and infestation levels was tested. All ITCs tested suppressed microsclerotia of *V. dahlia* in vitro in sterile sand, and the ITCs containing an aromatic moiety were considerably more toxic than the aliphatic ITCs. In natural soils the ITC toxicities seem negatively correlated to organic carbon content in the soils. In experiments with biomass incorporation in soil, *B. juncea* reduced the infection significatively (69-80% efficacy), while *S. alba* and *R. sativus* gave mortalities between 9-37%. Overall, the study demonstrates that brassicaceous green manures are hardly able to release ITCs at levels necessary for an adequate suppression of *V. dahlia* microsclerotia in natural soils and because organic matter can reduce the availability of ITCs and their effect. The authors conclude that more promising is the incorporation of high GSL-containing seed meal formulations, which should generate more effective ITC concentrations.	[169]

Table 2. *Cont.*

Glucosinolate/isothiocyanate	Target: Fungi	Effect	References
GSL-derived AITC released from *B. carinata* DSM	*Botrytis cinerea*, in vitro and in vivo with strawberries as plant host.	In in vitro trial AITC had a fungistatic effect against the pathogen. In in vitro trials two varieties of organic grown strawberries, infected with *B. cinerea* were exposed for 4 h in an atmosphere enriched either with synthetic AITC or ITC derived from DSM (0.1 mg L^{-1}). The AITC treatment (pure or GSL-derived ITC) reduced the decay caused by the pathogen significantly different from the untreated fruit. Residue analysis performed on fruit at the end of storage showed values lower than 1 mg kg^{-1}. Total phenolic content and antioxidant capacity estimated in treated and untreated strawberries showed no significant difference between control and AITC treated fruit.	[170]
Glucosinolate/isothiocyanate	**Target: Pseudofungi**	**Effect**	**References**
GSLs from DSMs of *Brassica napus*, *Brassica carinata* and *Brassica juncea* genotypes	*Phytophthora cinnamomi*, in vitro and *in planta* on *Lupinus luteus*	DSMs with high levels of allyl GSL inhibited mycelial growth and effectively inhibited the viability of chlamydospores in treated soils. Roots symptoms were less when plants grew in soils biofumigated with *B. carinata* and *B. juncea* DSMs with highest allyl GSL contents in comparison with plants in control soils. In particular *B. juncea* DSM (3 ÷ 30 μmol allyl GSL per gram of soil) had the largest effect on decreasing root necrosis by *P. cinnamomi* in *Lupinus*.	[171]
GSL from *Brassica carinata* pellets (Biofence)	*Phytophthora nicotianae* in vitro, in vivo with pepper plant as host.	Sensitivity of the vegetative structures of *P. nicotianae* to *Brassica carinata* pellets (Biofence) was evaluated in vitro at different doses and temperatures. The effectiveness of the pellets varied depending on the dose. The highest dose of pellets tested (24 mg) was fungitoxic to mycelium regardless of temperature for all the isolates. Moreover, biofumigation was effective in suppressing chlamydospores germination when the pellets were incorporated into the soil (1.5 and 3 g L−1 of soil) under different temperature regimes. In bioassays with pepper plants, both rates of *B. carinata* pellets (1.5 and 3 g L−1 of soil) reduced populations of *P. nicotianae* totally controlled the disease after a 4-week biofumigation treatment.	[172]

Table 2. *Cont.*

Glucosinolate/isothiocyanate	Target: Pseudofungi	Effect	References
GSLs from above-ground parts of *B. napus*, *B. carinata* and *B. juncea* genotypes at different phenological stages	*Phytophthora cinnamomi*, in vitro and *in planta* on *Lupinus luteus*	Genotypes of Brassica with high levels of allyl GSL inhibited mycelial growth, decreased sporangial production, and effectively inhibited the viability of chlamydospores in soil, but only *B. carinata* (10 g/75 mL soil) decreased disease symptoms in *L. luteus* roots.	[173]
GSL from *B. carinata* pellets (Biofence)	*Phytophthora cinnamomi* in vitro and in vivo on *Quercus cerris*	Maximum inhibition of vegetative or reproductive structure in vitro occurred at 15 °C and decreased as temperature increased. In vivo assays confirmed efficacy of pellets (3 g L^{-1}) in reducing the pathogen, but a total inhibition was not reached even if at high doses in comparison to the maximum dose tested in vitro assays (0.4 g L^{-1}).	[174]
Glucosinolate/isothiocyanate	**Target: Insects, acari, and other arthropoda**	**Effect**	**References**
GSL from *Brassica juncea* granulated seed meal (Kosmalski Herbs & Spices)	*Melolontha melolontha* grubs	In dose–response experiments the mortality of the grubs at each instar was significantly dependent on the GSL concentration applied with the granulate. The mortality reached 100% in the smallest grubs at 320 µmol L^{-1}, whereas at the same GSL concentration 95% of the bigger grubs (4.5÷7 mm) died. In field tests the mortality was 67.4%.	[175]
GSL derived from bio-based experimental formulations con-taining either *B. carinata* oil 1.5% and *B. carinata* DSM 3 g L^{-1} (270 µmol GSL L^{-1}), or 2% *B. carinata* oil and *B. carinata* DSM 4.5 g L^{-1} (405 µmol GSL L^{-1}).	red spider mite *Tetranychus urticae* on eggplant in open field trials.	The 2-year results indicated that the application of both formulations have a clear effect in containing mites, statistically different from the untreated control. Moreover, the ability of pest control of the formulation with the higher concentrations of oil and DSM was not different from the commercial chemical acaricide (fenazaquin),	[176]
GSL from leaf material of purple sprouting broccoli 'Santee', Savoy cabbage 'Wintessa', and the wild *B. oleracea* accession Winspit	*Folsomia candida* (springtail), *Eisenia andrei* (earthworm) and the soil bacterial community.	Biofumigation experiments were performed using the springtail *Folsomia candida* and the earthworm *Eisenia andrei*, each representing a functional soil invertebrate group with important effects on soil processes. Biofumigation was performed using freeze-dried leaves of the three different *B. oleracea* genotypes: One percent of freeze-dried leaf material relative to total soil (that is about 200 µmol kg^{-1} soil in GSL for Winspit accession; 125 µmol kg^{-1} soil in GSL for Santee accession, and 10 µmol kg^{-1} soil in GSL for Wintessa accession) was used for biofumigation. After 28 days, Winspit (but-3-enyl GSL as dominant GSL) was the genotype displaying highest toxicity to soil invertebrates. Earthworm survival was not affected by the *B. oleracea* plant material, and overall, the bacterial community was quite resilient to biofumigation.	[177]

Table 2. *Cont.*

Glucosinolate/isothiocyanate	Target: Insects, acari, and other arthropoda	Effect	References
GSLs derived by chopped fresh plants from *Brassica juncea*, sel. ISCI 99 and biofumigant meals derived from defatted seeds of *Brassica carinata* sel. ISCI 7	Wireworm populations (*Agriotes brevis* Candeze, *Agriotes sordidus* Illiger, and *Agriotes ustulatus* Schäller) was evaluated under both pot assays (on maize and lettuce) and field conditions on maize and potato.	In pot assays a clear rate effect was demonstrated, with sufficient seed meal to supply approximately 160 moles of GSL L^{-1} of soil resulting in significant wireworm mortality. At field level the protection of maize and potato crops comparable to that provided by Regent®.	[142]

3.3. Molecular Approaches to Enhance GSL Content in Brassicales

GHPs from aliphatic, indolic, or aromatic are associated with diverse phytochemical activities spanning from beneficial effects for agriculture (pest management and biofumigation) to nutraceutical properties when brassicaceous vegetables are part of human diet. Understanding GSL properties has led to the identification of molecular target for both classical breeding and biotechnological approaches to specifically enhance the content of GSLs and thus address a specific trait of interest. GSL profiles vary significantly among *Brassicaceae*, and GSL synthesis is a complex quantitative trait; the molecular basis of these variations is largely unknown. The history of GSL research can be seen as a sort of "opposite sides of the same coin" since the identification of targets for agricultural improvement moved in parallel with findings on GSL phytochemical properties. Indeed, in the first instance, the main target of GSL research was to reduce their content in a specific tissue or in the whole plant, thus reducing the goitrogenic effect and use Brassicales-derived seed cake as animal feed [182]. To achieve this goal, researchers exploited germplasm biodiversity, first finding the *B. napus* L. variant with low erucic acid (the 0-variant) and later on the Bronowski variety (the 00-variant) with both low erucic acid and GSLs. The latter variety is nowadays used as genetic background for the *B. napus* cultivars employed in agriculture [183]. Conversely, once the GSL beneficial effects for both human and plants were discovered, the research goal shifted from lowering the content of GSLs to increasing it. In particular, certain GHPs like sulforaphane or 4-(Methylsulfinyl)butyl ITC, which is the isothiocyanate derivative of 4-(methylsulfinyl)butyl GSL found in broccoli, have been identified as potent cancer-preventing agents and for this reason, increasing their content has been the focus of many studies in the past decade.

Wild relatives are valuable sources of desirable quality traits that can be subjected to introgression via breeding into elite cultivars, developing new and improved crop varieties. The *Brassicaceae* family encompasses many important crops [184] and includes the model species *Arabidopsis thaliana*. Wild relatives of the *Brassica oleracea* species have been shown to be characterized by high levels of aliphatic GSLs whose analysis revealed candidate taxa for broccoli breeding programs aiming at specifically increasing the concentration of 4-(methylsulfinyl)butyl GSL. An example is represented by members of *Brassica villosa-rupestris* with a non-functional *GSL-ALK* allele and thus can be considered as the progenitors of cultivated broccoli [185]. The beneficial effect of cruciferous-derived food products was the main target of classical breeding programs aiming at increasing the concentration of glucoraphanin, an aliphatic thiofunctionalized GSL. Indeed, an ITC-enriched broccoli variety, called Beneforté and nowadays commercialized in the UK, was obtained through introgression of three QTLs from the wild relative *B. villosa* to the commercial variety *Brassica oleracea* var. *Italica* via marker-assisted selection (MAS) [186]. Interestingly, this variety is commercialized with a health claim addressing the beneficial effects of cruciferous-derived food products. An additional example of a successful breeding program is provided by Li and co-workers, who obtained segregating populations of *Brassica oleracea* L. by three crosses: Broccoli × cauliflower, collard × broccoli, and collard × cauliflower [187]. These segregating populations showed a high content of 4-(methylsulfinyl)butyl GSLglucoraphanin and a low amount of anti-nutritional GSLs, such as (R)-2-hydroxybut-3-enyl GSL progoitrin, thus selectively modifying the aliphatic GSL composition. Furthermore, these recombinant inbred lines allowed to better elucidate the functional role of *A. thaliana* ortholog genes, *GSL-ELONG, GSL-PRO, GSL-OH,* and *GSL-ALK,* which are involved, respectively, inside chain elongation, hydroxylation, and secondary modification. Indeed, non-functional *GSL-OK* or *GSL-ALK* alleles were responsible for the observed biochemical GSL profile. Based on this study, Liu et al. [188] used an RNAi strategy to knockdown the expression of *GSL-ALK* gene thus obtaining *B. napus* lines with a strong reduction of (R)-2-hydroxybut-3-enyl GSL progoitrin and an increase of 4-(methylsulfinyl)butyl GSL glucoraphanin. Based on the assumption that amino acids like methionine and phenylalanine boost the synthesis of aliphatic and aromatic

GSLs, respectively, strategies aiming at increasing their availability have been successfully addressed [189].

3.3.1. Biotechnology Approaches to Enhance GSL Content

Genetic engineering technologies have been applied to modulate GSL profiles via heterologous expression of CYP79 enzymes. Indeed, Brader and co-workers expressed the *CYP79* transgenes into *A. thaliana* plants [190]. Transgenic plants expressing the CYP79D2 showed high amounts of isopropyl and methylpropyl GSLs and resistance to *Erwinia carotovora*, a bacterial pathogen. *A. thaliana* plants over expressing CYP79A1/A2 were less susceptible to *P. syringe* but more susceptible to *A. brassicicola* suggesting that different GSLs act differently as pathogens' defense, and this may be due to their specific modes of action. Wentzell et al. demonstrated that the *AOP2* enzyme encoding gene is the major regulator of aliphatic GSL biosynthesis and accumulation; furthermore, its overexpression resulted in upregulation of the entire biosynthetic pathway [191,192]. The recent release of *Eruca sativa* genome [193] will represent a valuable resource to identify the gene target for genetic engineering strategies in this species. Elicitation, through diverse approaches both physical and chemical, is an experimental procedure well established to modulate the content of secondary metabolites in plant tissues/organs and examples about GSLs were also recorded. Due to the fact that methionine-derived GSLs are sulfur-rich metabolites, sulfur fertilization affects their concentration leading to an increase in the range of 25–50% depending on (*i*) the plant species, (*ii*) amount of sulfur applied, and (*iii*) type of treatment. The higher concentration of GSLs is related to upregulation of the entire biosynthetic pathway; conversely, sulfur-deficient plants showed a down regulation of GSL biosynthetic genes [194].

Taking advantage of knowledge of the biosynthetic pathway leading to these secondary metabolites, mainly derived from research on *A. thaliana*, several strategies have been applied to engineer enzymes encoding genes in the first instance, and only upon the discovery as master regulator was the *MYB28* transcription factor was addressed too [195,196]. Traka and co-workers [196] found that the high-glucoraphanin4-(methylsulfinyl)butyl GSL F1 broccoli hybrids developed via an introgression breeding program with *B. villosa* showed a high induction of *MYB28* allele from the wild *B. villosa* suggesting its role in regulating the expression of glucoraphanin4-(methylsulfinyl)butyl GSL biosynthetic pathway components.

The use of plant tissues, hairy roots cultivation, and cell cultures, even in the presence of elicitors, may represent an efficient plant platform for the production of active metabolites [197] in a plant molecular farming scenario. Despite several attempts, the use of cell suspension culture for GSL production did not lead to successful yields compared to *in planta* content highlighting that it is not a suitable strategy for large scale production [189]. Tissues/organ and liquid cells culture mainly from *A. thaliana* var. Col0 were exploited by using both metabolic engineering and elicitation (e.g., hormones supply) approaches. The hairy root system has been widely used in aromatic plant species to specifically target and scale-up the production of phytochemically relevant metabolites. Kastell et al. [198] used this system to express the transgenes encoding for entry point enzymes of the. aliphatic pathway, *CYP79F1* and *CYP79F2* genes; however, the overall GSL rate was very low compared to leaf tissue. Studies combining elicitor treatments have also been conducted; acetylsalicylic acid in combination with phenylalanine and cysteine amino acids resulted in being the best elicitors among tested hormones, inducing an increase of benzyl GSLs [199]. Table 3 shows the biotechnology approaches to enhance GSL content mentioned herein.

Table 3. Biotechnological approaches to enhance glucosinolate (GSL) content.

Brassicaceae species	Biotechnological approach to enhance GSL content	Results	References
Brassica oleracea var. Italica	Introgression of three QTLs from *B. villosa* to commercial variety via classical breeding via MAS	Broccoli variety, Beneforté, with high level of ITC	[186]
Brassica oleracea var. *Italica* (broccoli) x *Brassica oleracea* var. *botrytis* (cauliflower) *Brassica oleracea var. Italica x Brassica oleracea var. Lacinato* *Brassica oleracea var. Lacinato* (collard) x *Brassica oleracea var. botrytis*	Classical breeding via MAS	RILs with high content of of 4-(methylsulfinyl)butyl and low (*R*)-2-hydroxybut-3-enyl GSL	[187]
Brassica napus	iRNA to target the GSL-ALK gene	*B. napus* iRNA lines with high content of of 4-(methylsulfinyl)butyl and low (*R*)-2-hydroxybut-3-enyl GSL	[188]
A. thaliana	Overexpression of *CYP79D2 and CYP79A1/A2* genes	p35S: *CYP79D2* plants with enhanced isopropyl and methylpropyl GSLs and resistant to *Erwinia carotovora*p35S: *CYP79A1/A2* plants were less susceptible to *P. syringe* but more susceptible to *A. brassiciola*	[190]
A. thaliana	Overexpression of *AOP2 gene*	Upregulation of aliphatic alkenyl GSL biosynthetic pathway	[192]
A. thaliana	Overexpression of *CYP79F1 and CYP79F2* genes	Hairy roots overexpressing the transgenes showed an increased aliphatic GSL rate compared to wild type roots but lower than leaf extract	[198]
B. rapa x B. oleracea	Conventional breeding via MAS	Cabbage lines with increased GSL content	[200]
B. oleracea	Kncock-out of *MYB28* gene via CRISPR/Cas9 mediated editing	downregulation of aliphatic GSL biosynthetic genes and reduction in methionine-derived GSL content in *myb28* mutant broccoli plants	[201]

3.3.2. GSL Molecular Markers and Gene Mapping

Based on *A. thaliana* GSL-related gene sequence information, a positional cloning strategy was applied to identify orthologs in *B. oleracea* genome. The *B. oleracea* genes *BoGSL-ELONG* and *BoGSL-PRO* together with *BoGSL-ALK* and *BoGSL-OH* were mapped on a high-density *B. oleracea* linkage map [202], which may be part of a QTL region for GSLs [203]. QTLs were identified in *B. rapa* too by means of comparative genomics and synteny with *A. thaliana* [204,205]. Biosynthetic genes identified in *B. rapa* and QTLs were reported [206]. A recombinant inbred line (RIL) population was developed from crossing Chinese Cabbage *(Brassica rapa* L. *chinensis*) and yellow sarson (*Brassica rapa* L. subsp. *trilocularis*) and used to fine map seed GSL trait in *Brassica rapa* L. [207]. Taking advantage of a previously reported ultra-dense genetic map of *B. rapa*, molecular markers associated with GSLs gene and SSR were found, and a QTL analysis was performed. The authors reported the identification of a major QTL for 4C and 5C GSL, which colocalized with *GSL-ELONG* locus *SCAR* marker *BrMAM1-1* in *B. rapa* seeds, thus identifying a candidate gene, *Br- GSL-ELONG*, for 5C side chain elongation of aliphatic GSL in this species. Recently, Zhang et al. [208] identified a major QTL controlling aliphatic GSL accumulation in *B. rapa* leaves. The QTL, which encompasses three tandem *MAM* genes and two *MYB* genes, was detected in two BC2DH populations [208]. Interestingly, the authors reported that the *BrMAM-3* gene was correlated with the accumulation of aliphatic GSLs in *B. rapa* leaves. Furthermore, a naturally occurring insertion within exon 1 of *BrMAM-3* causing a loss of function, mutation was associated with the low GSL content in *B. rapa* accessions, thereby this gene is a good candidate to manipulate aliphatic GSL in *B. rapa* via metabolic engineering and classical breeding approaches. QTLs and associated molecular markers for GSL were also discovered for amphidiploid *Brassicaceae* species like *B. napus* [209,210] and *B. juncea* [211,212]. Based on QTLs analysis, five loci on A2, A9, C2, C7, C9 *B. napus* chromosomes were associated with GSLs in seeds [212,213] while five other QTLs explain 30–45% of the total aliphatic GSL variation in *B. juncea* [214]. Zou and co-workers [215] identified five QTLs associated with GSL variance in roots of two F2 populations of *Raphanus sativus*, thus opening new breeding perspectives to enhance GSLs in a different genus. It is well established that the GSL profile is greatly variable between species, within species, and individual cultivars, thus varietal selection and development of breeding lines with uniform GSL profiles will contribute to fix plant breeding materials. Furthermore, another source for GSLs accumulation traits in breeding programs is represented by Gene bank accessions including wild genotypes, which are underutilized. Phenotyping these large germplasm collections will again contribute to identifying individual pre-breeding materials useful for GSL content improvement. Despite that GSL content and profile in *Brassicales* is influenced by several factors related to both environment and cultivation conditions, a wide genetic variation does exist.

GSL content has a quantitative genetic inheritance, regulated by complex genetic factors and affected by environment [204]. Genetic and genomic resources also have to be developed for *Brassicaceae* species not deeply studied, like rocket, radish Chinese cabbage, and *B. carinata*; there is an urgent need to identify species-specific SNPs associated with GSLs and GHPs as well as molecular markers to assist selection for highly beneficial GSLs. A successful example of MAS for GSL increase was provided by cabbage lines derived from a cross between *B. rapa* and *B. oleracea* [216]. Furthermore, a genetic linkage map and QTL analysis revealed seed quality traits like erucic acid content and GSL in *Sinapis alba* L., providing useful tools for breeding strategy mediated by molecular markers associated with GSLs components [216]. Recently, a CRISPR/Cas9-mediated editing of *MYB28* gene in *Brassica oleracea* has been achieved by Neequaye et al. [201], and although the paper has not been subjected to peer review process so far, this result shed light on the opportunity that new biotechnological approaches could be applied to modify the content of GSLs *in planta*.

4. Perspectives

Pollution problems in the environment and the toxic effects of synthetic chemicals on non-target organisms have prompted investigations on exploiting pesticides of plant origin. Natural plant products and their analogues resulted in being an important source of new agricultural chemicals [217,218] used in control of insect pests [219], plant diseases [9,10,220–224], and as a bird repellent [225–227]. Several studies have shown the importance of natural chemicals as a possible source of non-phytotoxic, systemic, and easily biodegradable alternative pesticides [228,229].

Plants constitutively accumulate compounds that may be involved in several different physiological roles, such as defense against pathogens and/or involved in regulatory and developmental processes [230–233]

Such antimicrobial bioactive peptides or compounds could be either toxic or inhibitory against pathogens, acting against bacteria and/or fungi and/or pests and may be expressed in the plant constitutively or induced following infection [234,235]. In their long association with pathogens, plants evolved an intricate and elaborate array of defensive tools acting in constitutive and/or inducible defense strategies, including pathogenesis-related proteins, antifungal proteins, and secondary metabolites [232,236,237]. The availability of plant-derived compounds with antifungal activities sufficient to make them feasible for agronomic use in disease control still remains inadequate to increasing environmental requirements.

In this perspective, the use of bioactive molecules derived from *Brassicaceae* defatted seed meals and from *Solanaceae* waste (peel potato recovery) for cereal crop protection could be a successful strategy with the main goal to find out new formulations of eco-friendly products with fungicidal/fungistatic/insecticidal properties. In particular, the circular utilization of Brassicaceous and Solanaceous industry waste derived by-products for disease biocontrol of (*Zea mays* L.) and wheat (*Triticum aestivum* L.) could meet the criteria of benign environmental profiles and low toxicity to humans and wildlife, with specific attention to the sustainability of cultivation and the quality of products.

In fact, mycotoxigenic fungal diseases strongly limit crop yield (up to 20% annual cereal loss) [238,239], quality, and safety, especially in maize and wheat, two of the most important sources of food all over the world with a global production of 1100×10^6 and 129×10^6 tons, respectively (FAOSTAT 2019). Fungi of the genus *Fusarium* and *Aspergillus* are widely distributed maize pathogens, causing diseases to occur on seedlings, roots, stalks, ears, and kernels, affecting grain quality through the production of mycotoxins, including fumonisins produced by *Fusarium verticillioides* [240] and aflatoxin B1 produced by *Aspergillus flavus* [241]. As described for maize, *Fusarium*, specifically *Fusarium graminearum*, also threatens wheat grain yield and safety due to the production of mycotoxins (deoxynivalenol-DON, zearalenone-ZEA). Mycotoxins are secondary metabolites produced by fungi, which may be toxic or have other debilitating effects on living organisms [242,243]. Bioactive molecules extracted from *Brassicaceae* defatted seed meals and from *Solanaceae* waste (peel potato recovery) should be preliminary evaluated in in vitro bio-assays for their efficacy against *A. flavus*, *F. verticillioides*, *F. graminearum* growth, and mycotoxin production [229,244] with the aim to highlight the most promising formulation to be used in field trials. Maize field protection could be planned by the application of bioactive extracts to maize-developing ear at the flowering stage with the aim to control the development of mycotoxigenic fungal infections and the subsequent accumulation of mycotoxins in the grain. Mycotoxin contamination in grains is a global threat to both safety of human food and animal feed, [241,245,246]. New regulations for the allowable mycotoxin limits in food and feed have been put in place in many countries. The binding European Union regulations on toxin contamination for human consumption and recommendations for animal feeding [247,248] have forced renewed efforts in finding reliable tools and solutions for control of mycotoxin contamination.

Furthermore, infection by fungal pathogens on maize can be favored by the attack of insects such as *Diabrotica virgifera virgifera* or *Ostrinia nubilalis* [248,249] that, both at larval and adult stage, can create vehicle inputs of the fungal inoculum. Interestingly, in this perspective, the utilization of bioactive extracts from *Brassicaceae* and *Solanaceae* could also be tested as possible bioprotector for maize pest treatment. An additional wheat disease to be mentioned and that could be addressed by bioactive extracts is rust infection, which may cause up to 50% yield losses, mainly due to a reduction in biomass, harvest index, and kernels per square meter [250]. The recent emergence of new widely virulent and aggressive strains of rusts (particularly stripe and stem rust) threatens wheat production worldwide, especially under the trend of higher temperature and humidity observed in the frame of the current climate changes. The bioactive molecules from *Brassicaceae* and *Solanum tuberosum* wastes could represent an innovative tool to be applied also in organic farming systems, to address the reported rust wheat disease, and for reducing phytosanitary emergencies caused by Fusarium Head Blight disease.

5. Conclusions

The need for effective, suitable, and alternative crop protection tools is particularly urgent considering that recent results estimate, at a global level, the losses for wheat, rice, and maize to medium values of 21.5%, 30.0%, and 22.5%, respectively, also highlighting that the highest losses are associated with food-deficit regions with fast-growing populations, and frequently with emerging or re-emerging plant pests and diseases [251]. This trend and the consequent evolution have accelerated over the past 70 years, favoring phenomena such as the rapid adaptation of populations of pests and pathogens to used synthetic pesticides and chemical protection products.

Biodegradable pesticides derived from Potato and *Brassicaceae* wastes could have a large potential market especially in a view of an integrated, environmentally friendly, and circular agro-economy. Based on the reported comprehensive results and evidence derived from scientific literature, some relevant considerations could be highlighted in the perspective of their optimal and innovative use.

In this perspective, interesting suggestions and intriguing opportunities could take advantage from the valorization of the combined effect of SGAs and phenols. The future use of potato peel could depend on the fact that major crop protection could be achieved with mixtures of compounds at levels that individually are almost or completely inactive [71]. In the frame of a green strategy for sustainable PPW exploitation, it would be necessary to investigate the most suitable genotype to obtain an innovative by-product of agro-industrial processing, which can be applied as a plant pathogen biocontrol agent in crop cultivation. Additionally, further investigations are required to optimize the recycle technology to process potato peel waste.

The great value of bioactive molecules present in Brassicaceae, a number between 88 and 137 different GSLs, with reference only to the MYR/GSL system, could be more exploited than it is currently being studied, in the perspective of increasing development of innovative plant bioprotectors; no more than 10% of available molecules have been tested to date. Nevertheless, taken together, all the *Brassicaceae* effects reported in recent years show the great potential of *Brassica* bio-based products both in field applications and in food formulation or storage.

There is an urgent need to accelerate the development and feasibility of genetic, genomic resources, and biotechnological tools (*i*) in potato for the development of genotypes with an effective compositions of biocompounds in peel and (*ii*) in Brassicaceae not deeply studied until now like *E. sativa* and *B. carinata*, with the aim to boost research to increase the content of beneficial compounds in these species, leading, meanwhile, to sustainable and circular industrial-agronomic practices.

Author Contributions: Writing—original draft preparation, D.P., C.L., E.P., L.B., A.M.M., D.M., C.B., R.M. and R.L.S.; writing—rivew and editing, D.P. and C.B.; supervision, D.P. and C.B. All authors have contributed to the manuscript. All authors have read and agreed to the published version of the manuscript.

Funding: This research was funded by CARIPLO FOUNDATION, in the frame of SUSINCER Project (Bioactive Compounds from Brassicaceae and Solanaceae waste for cereals crop protection; Project code. 2019-2538).

Institutional Review Board Statement: Not applicable.

Informed Consent Statement: Not applicable.

Data Availability Statement: Data sharing is not applicable to this article. No new data were created or analyzed in this study.

Conflicts of Interest: The authors declare no conflict of interest.

Abbreviations

AITC	allyl isotiocyanate
CA	3,4-Dihydroxyphenyl)prop-2-enoic acid
CQA	5-O-caffeoylquinic acid
CNV	copy number variation
CPB	Colorado Potato Beetle
DSMs	defatted seed meals
DW	dry weight

FW	fresh weight
GHPs	GSL hydrolysis products
GSLs	glucosinolates
HPLC	High Performance Liquid Chromatography
HPTLC	High Performance Thin Layer Chromatography
IPM	integrated pest management
ITCs	isotiocyanates
MAS	marker assisted selection
PPW	Potato Peel Waste
PTM	Potato Tuber Moth
QTL	quantitative trait locus
RIL	recombinant inbred line
SGAs	Steroidal glycoalkaloids
SA	steroidal alkaloid
SNPs	single nucleotide polymorphisms
UAE	Ultrasound-assisted extraction

References

1. Ishangulyyev, R.; Kim, S.; Lee, S.H. Understanding Food Loss and Waste-Why Are We Losing and Wasting Food? *Foods* **2019**, *8*, 297. [CrossRef]
2. Köhler, H.R.; Triebskorn, R. Wildlife Ecotoxicology of Pesticides: Can We Track Effects to the Population Level and Beyond? *Science* **2013**, 759–765. [CrossRef] [PubMed]
3. Keikotlhaile, B.M.; Spanoghe, P.; Steurbaut, W. Effects of food processing on pesticide residues in fruits and vegetables: A meta-analysis approach. *Food Chem. Toxicol.* **2010**, *48*, 1–6. [CrossRef] [PubMed]
4. Yang, L.N.; He, M.H.; Ouyang, H.B.; Zhu, W.; Pan, Z.C.; Sui, Q.J.; Shang, L.P.; Zhan, J. Cross-resistance of the pathogenic fungus *Alternaria alternata* to fungicides with different modes of action. *BMC Microbiol.* **2019**, *19*, 205. [CrossRef]
5. Directive 2009/128/EC of the European Parliament and of the Council of 21 October 2009, 2009. Establishing a Framework for Community Action to Achieve the Sustainable Use of Pesticides. Off. J. Eur. Union. Available online: https://www.europarl.europa.eu/RegData/etudes/STUD/2018/627113/EPRS_STU(2018)627113_EN.pdf (accessed on 20 February 2021).
6. Poveda, J.; Eugui, D.; Velasco, P. Natural control of plant pathogens through glucosinolates: An effective strategy against fungi and oomycetes. *Phytochem. Rev.* **2020**, *19*, 1045–1059. [CrossRef]
7. El-Mougy, N.S.; Abdel-Kader, M.M. Antifungal effect of powdered spices and their extracts on growth and activity of some fungi in relation to damping-off disease control. *J. Plant. Prot. Res.* **2007**, *47*, 267–277.
8. Abdel-Monaim, M.F.; Abo-Elyousr, K.A.M.; Morsy, K.M. Effectiveness of plant extracts on suppression of damping-off and wilt diseases of lupine (Lupinus termis Forsik). *Crop. Prot.* **2011**, *30*, 185–191. [CrossRef]
9. Ahmed, S.S.; Abd El-Aziz, G.H.; Abou-Zeid, M.A.; Fahmy, A.H. Environmental impact of the use of some eco-friendly natural fungicides to resist rust disease in wheat. *CATRINA* **2019**, *18*, 87–95.
10. Draz, I.S.; Elkhwaga, A.A.; Elzaawely, A.A.; El-Zahaby, H.M.; Anter Ismai, A.W. Application of plant extracts as inducers to challenge leaf rust of wheat. *Egypt J. Biol Pest. Control.* **2019**, *29*, 6. [CrossRef]
11. Geetha, H.M.; Shetty, H.S. Induction of resistance in pearl millet against mildew disease caused by Sclerospora graminicola using benzothiadiazole, calcium chloride and hydrogen peroxide—A comparative evaluation. *Crop. Protection* **2012**, *21*, 601–610. [CrossRef]
12. Hassan, M.E.M.; Abd El-Rahman, S.S.; El-Abbasi, I.H.; Mikhail, M.S. Change in peroxidase activity due to resistance induced against faba bean chocolate spot disease. *Egypt J. Phytopathol.* **2007**, *35*, 35–48.
13. Radwan, D.E.; Lu, G.; Fayez, K.A.; Mahmoud, S.Y. Protective action of salicylic acid against been yellow mosaic virus infection in *Vicia faba* leaves. *J. Plant. Physiol.* **2008**, *165*, 845–857. [CrossRef] [PubMed]
14. Benkeblia, N. Potato Glycoalkaloids: Occurrence, biological activities and extraction for biovalorisation—A review. *Int. J. Food Sci. Technol.* **2020**, *55*, 2305–2313. [CrossRef]
15. FAO International Year of Potato. Available online: http://www.fao.org/potato-2008/en/potato/cultivation.html (accessed on 20 February 2020).
16. Akyol, H.; Riciputi, Y.; Capanoglu, E.; Caboni, M.F.; Verardolnt, V. Phenolic Compounds in the Potato and Its Byproducts: An Overview. *Int. J. Mol. Sci.* **2016**, *17*, 835. [CrossRef] [PubMed]
17. Sepelev, I.; Galoburda, R. Industrial potato peel waste application in food production: A Review. *Res. Rural Dev. Int. Sci. Conf. Proc. (Latvia)* **2015**, *1*, 130–136.
18. Samotyja, U. Potato Peel as a Sustainable Resource of Natural Antioxidants for the Food Industry. *Potato Res.* **2019**, *62*, 435–451. [CrossRef]
19. Di, W. Recycle Technology for Potato Peel Waste Processing: A Review. *Procedia Env. Sci.* **2016**, *31*, 103–107.
20. Arapoglou, D.; Varzakas, T.; Vlyssides, A.; Israilides, C. Ethanol production from potato peel waste (PPW). *Waste Manage.* **2010**, *30*, 1898–1902. [CrossRef]

21. Calcio Gaudino, E.; Colletti, A.; Grillo, G.; Tabasso, S.; Cravotto, G. Emerging Processing Technologies for the Recovery of Valuable Bioactive Compounds from Potato Peels. *Foods* **2020**, *9*, 1598. [CrossRef] [PubMed]
22. Venturi, F.; Bartolini, S.; Sanmartin, C.; Orlando, M.; Taglieri, I.; Macaluso, M.; Lucchesini, M.; Trivellini, A.; Zinnai, A.; Mensuali, A. Potato Peels as a Source of Novel Green Extracts Suitable as Antioxidant Additives for Fresh-Cut Fruits. *Appl. Sci.* **2019**, *9*, 2431. [CrossRef]
23. Lazzeri, L.; Riva, G.; D'Avino, L.; Pedretti, E.F. Short introduction to the VALSO and EXTRAVALORE project activities. *Ind. Crops Prod.* **2015**, *75*, 1–7. [CrossRef]
24. Fanigliulo, R.; Pochi, D.; Bondioli, P.; Grilli, R.; Fornaciari, L.; Folegatti, L.; Malaguti, L.; Matteo, R.; Ugolini, L.; Lazzeri, L. Semi-refined Crambe abyssinica (Hochst. EX R.E.Fr.) oil as a biobased hydraulic fluid for agricultural applications. *Biomass Conv. Bioref.* **2021**. [CrossRef]
25. D'Avino, L.; Dainelli, R.; Lazzeri, L.; Spugnoli, P. The role of co-products in biorefinery sustainability: Energy allocation versus substitution method in rapeseed and carinata biodiesel chains. *J. Clean. Prod.* **2015**, *94*, 108–115. [CrossRef]
26. Maina, S.; Misinzo, G.; Bakari, G.; Kim, H. Human, animal and plant health benefits of glucosinolates and strategies for enhanced bioactivity: A systematic review. *Molecules* **2020**, *25*, 3682. [CrossRef]
27. Chowa'nski, S.; Adamski, Z.; Marciniak, P.; Rosia'nski, G.; Büyükgüzel, E.; Büyükgüzel, K.; Falabella, P.; Scrano, L.; Ventrella, E.; Lelario, F.; et al. Review of Bioinsecticidal Activity of *Solanaceae* Alkaloids. *Toxin* **2016**, *8*, 1–28.
28. Bennett, R.N.; Wallsgrove, R.M. Secondary metabolites in plant defence mechanisms. *New Phytologist* **1994**, *127*, 617–633. [CrossRef]
29. Pusztahelyi, T.; Holb, U.; Pocsi, I. Secondary metabolites in fungal plant interaction. *Front. Plant. Sci.* **2015**, *6*, 1–23.
30. De Sotillo, D.R.; Hadley, M.; Wolf-Hall, C. Potato peel extract a nonmutagenic antioxidant with potential antimicrobial activity. *J. Food Sci.* **1998**, *63*, 907–910. [CrossRef]
31. Amanpour, R.; Abbasi-Maleki, S.; Neyriz-Naghadehi, M.; Asadi-Samani, M. Antibacterial effects of *Solanum tuberosum* peel ethanol extract in vitro. *J. HerbMed. Pharmacol.* **2015**, 4.
32. Friedman, M.; McDonald, G.M.; Filadelfi-Keszi, M. Potato Glycoalkaloids: Chemistry, Analysis, Safety, and Plant Physiology. *CRC Crit. Rev. Plant. Sci.* **1997**, *16*, 55–132. [CrossRef]
33. Kozukue, N.; Yoon, K.S.; Byun, G.I.; Misoo, S.; Levin, C.E.; Friedman, M. Distribution of glycoalkaloids in potato tubers of 59 accessions of two wild and five cultivated *Solanum* species. *J. Agric. Food Chem.* **2008**, *56*, 1920–1928. [CrossRef] [PubMed]
34. Manrique-Carpintero, N.C.; Tokuhisa, J.G.; Ginzberg, I.; Holliday, J.A.; Veilleux, R.E. Sequence diversity in coding regions of candidate genes in the glycoalkaloid biosynthetic pathway of wild potato species. *G3 Genes Genomes Genet.* **2013**, *3*, 1467–1479. [CrossRef] [PubMed]
35. Mweetwa, A.M.; Hunter, D.; Poe, R.; Harich, K.C.; Ginzberg, I.; Veilleux, R.E.; Tokuhisa, J.G. Steroidal glycoalkaloids in *Solanum chacoense*. *Phytochemistry* **2012**, *75*, 32–40. [CrossRef]
36. Friedman, M. Potato Glycoalkaloids and Metabolites: Roles in the Plant and in the Diet. *J. Agric. Food Chem.* **2006**, *54*, 8655–8681. [CrossRef]
37. Distl, M.; Wink, M. Identification and Quantification of Steroidal Alkaloids from Wild Tuber-Bearing Solanum Species by HPLC and LC-ESI-MS. *Potato Res.* **2009**, *52*, 79–104. [CrossRef]
38. Itkin, M.; Heinig, U.; Tzfadia, O.; Bhide, A.J.; Shinde, B.; Cardenas, P.D.; Bocozeba, S.E.; Unger, T.; Malitsky, S.; Finkers, R.; et al. Biosynthesis of antinutritional alkaloids in solanaceous crops is mediated by clustered genes. *Science* **2013**, *341*, 175–179. [CrossRef]
39. Lafta, A.M.; Lorenzen, J.H. Influence of High Temperature and Reduced Irradiance on Glycoalkaloid Levels in Potato Leaves. *J. Am. Soc. Hort. Sci.* **2000**, *25*, 563–566.
40. Dale, M.F.B.; Griffiths, D.W.; Bain, H.; Todd, D. Glycoalkaloid increase in *Solanum tuberosum* on exposure to light. *Ann. Appl. Biol.* **1993**, *123*, 411–418. [CrossRef]
41. Mekapogu, M.; Sohn, H.-B.; Kim, S.-J.; Lee, Y.-Y.; Park, H.-M.; Jin, Y.-I.; Hong, S.-Y.; Suh, J.-T.; Kweon, K.; Jeong, J.-C.; et al. Effect of Light Quality on the Expression of Glycoalkaloid Biosynthetic Genes Contributing to Steroidal Glycoalkaloid Accumulation in Potato. *Am. J. Potato Res.* **2016**, *93*, 264–277.
42. Choi, D.; Bostock, R.M.; Avdiushko, S.; Hildebrand, O.F. Lipid-derived signals that discriminate wound- and pathogenresponsive isoprenoid pathways in plants: Methyl jasmonate and fungal elicitor arachidonic acid induce different 3-hydroxy3-methylglutaryl-coenzyme A reductase genes and antimicrobial isoprenoids in *Solanum tuberosum* L. *Proc. Natl. Acad. Sci. USA* **1994**, *91*, 2329–2333.
43. Zarins, R.; Kruma, Z. Glycoalkaloids in potatoes: A review. *Foodbalt* **2017**, 7–11.
44. Friedman, M.; Roitman, J.N.; Kozukue, N. Glycoalkaloid and calystegine contents of eight potato cultivars. *J. Agric. Food Chem.* **2003**, *51*, 2964–2973. [CrossRef]
45. Pacifico, D.; Musmeci, S.; Sanchez del Pulgar, J.; Onofri, C.; Parisi, B.; Sasso, R.; Mandolino, G.; Lombardi-Boccia, G. Caffeic acid and α-chaconine influence the resistance of potato tuber to Phthorimaea operculella (Lepidoptera: Gelechiidae). *Am. Potato J.* **2019**, *96*, 403–413.
46. Friedman, M. Analysis of biologically active compounds in potatoes (*Solanum tuberosum* L.), tomatoes (*Lycopersicon esculentum* L.), and jimson weed (*Datura stramonium* L.) seeds. *J. Chromatogr. A* **2005**, *1054*, 143–155. [CrossRef] [PubMed]
47. Rytel, E.; Czopek, A.T.; Aniolowska, M.; Hamouz, K. The influence of dehydrated potatoes processing on the glycoalkaloids content in coloured-fleshed potato. *Food Chem.* **2013**, *141*, 2495–2500. [CrossRef] [PubMed]

48. OECD. Consensus document on compositional considerations for new varieties of potatoes: Key food and feed nutrients, anti-nutrients and toxicants. In *OECD Environmental Health and Safety Publications, Safety of Novel Foods and Feeds No. 4*; OECD: Paris, France.
49. BfR Opinion No 010/2018 of 23 April 2018. Table Potatoes Should Contain Low Levels of Glycoalkaloids (Solanine). Available online: http://www.bfr.bund.de/cm/349/table-potatoes-should-contain-low-levels-of-glycoalkaloids-solanine.pdf (accessed on 20 February 2021).
50. Uluwaduge, D.I. Glycoalkaloids, bitter tasting toxicants in potatoes: A review. *Int. J. Food Sci. Nutr.* **2018**, *3*, 188–193.
51. Hellenäs, K.E.; Branzell, C.; Johnsson, H.; Slanina, P. High levels of glycoalkaloids in the established Swedish potato variety Magnum Bonum. *J. Sci. Food Agric.* **1995**, *68*, 249–255. [CrossRef]
52. Knuthsen, P.; Jensen, U.; Schmidt, B.; Larsen, I.K. Glycoalkaloids in potatoes: Content of glycoalkaloids in potatoes for consumption. *J. Food Compos. Anal.* **2009**, *22*, 577–581. [CrossRef]
53. Wu, Z.G.; Xu, H.Y.; Ma, Q.; Cao, Y.; Ma, J.N.; Ma, C.M. Isolation, identification and quantification of unsaturated fatty acids, amides, phenolic compounds and glycoalkaloids from potato peel. *Food Chem.* **2012**, *135*, 2425–2429. [CrossRef]
54. Kremr, D.; Bajer, T.; Bajerová, P.; Surmová, S.; Ventura, K. Unremitting problems with chlorogenic acid Nomenclature: A review. *Quím. Nova* **2016**, *39*, 530–533.
55. Sato, Y.; Itagaki, S.; Kurokawa, T.; Ogura, J.; Kobayashi, M.; Hirano, T.; Sugawara, M.; Iseki, K. In vitro and in vivo antioxidant properties of chlorogenic acid and caffeic acid. *Int. J. Pharm.* **2011**, *403*, 136–138. [CrossRef]
56. Clifford, M.N. Chlorogenic acids and other cinnamates—nature, occurrence, dietary burden, absorption and metabolism. *J. Sci. Food Agric.* **2000**, *80*, 1033–1043. [CrossRef]
57. Horbury, D.; Baker, L.A.; Quan, W.D.; Greenough, S.E.; Stavros, W.G. Photodynamics of potent antioxidants: Ferulic and caffeic acids. *Phys. Chem. Chem. Phys.* **2016**, *18*, 17691–17697. [CrossRef]
58. Lucarini, M.; Pedulli, G.F. Bond dissociation enthalpy of α-tocopherol and other phenolic antioxidants. *J. Org. Chem.* **1994**, *59*, 5063–5070. [CrossRef]
59. Le Bourvellec, C.; Renard, C.M.G.C. Interactions between Polyphenols and Macromolecules: Quantification Methods and Mechanisms. *Crit. Rev. Food Sci. Nutr.* **2012**, *52*, 213–248. [CrossRef]
60. Santos-Buelga, C.; Scalbert, A. Proanthocyanidins and tannin-like compounds—nature, occurrence, dietary intake and effects on nutrition and health. *J. Sci. Food Agric.* **2000**, *80*, 1094–1117. [CrossRef]
61. Grzesik, M.; Naparło, K.; Bartosz, G.; Sadowska-Bartosz, I. Antioxidant properties of catechins: Comparison with other antioxidants. *Food Chem* **2018**, *241*, 480–492. [CrossRef] [PubMed]
62. Veluri, R.; Weir, T.L.; Bais, H.P.; Stermitz, F.R.; Vivanco, J.M. Phytotoxic and Antimicrobial Activities of Catechin Derivatives. *J. Agric. Food Chem.* **2004**, *52*, 1077–1082. [CrossRef] [PubMed]
63. Cipollini, M.L.; Levey, D.J. Antifungal activity of Solanum fruit glycoalkaloids: Implications for frugivoryand seed dispersal. *Ecology* **1997**, *78*, 799–809. [CrossRef]
64. Smith, D.B.; Roddick, J.G.; Jones, J.L. Synergism between the potato glycoalkaloids alpha-chaconine and alpha-solanine in inhibition of snail feeding. *Phytochemistry* **2001**, *57*, 229–234. [CrossRef]
65. Zitnak, A.; Filadelfi, M.A. Estimation of taste thresholds of three potato glycoalkaloids. *Can. Inst. Food Technol. J.* **1985**, *18*, 337–339. [CrossRef]
66. Fewell, A.M.; Roddick, J.G. Interactive antifungal activity of the glycoalkaloids α-solanine and α-chaconine. *Phytochemistry* **1993**, *33*, 323–328. [CrossRef]
67. Fewell, A.M.; Roddick, J.G.; Weissenberg, M.A. Interactions between the glycoalkaloids solasonine and solamargine in relation to inhibition of fungal growth. *Phytochemistry* **1994**, *37*, 1007–1011. [CrossRef]
68. Friedman, M.; Huang, V.; Quiambao, Q.; Noritake, S.; Liu, J.; Kwon, O.; Chintalapati, S.; Young, J.; Levin, C.E.; Tam, C.; et al. Potato Peels and Their Bioactive Glycoalkaloids and Phenolic Compounds Inhibit the Growth of Pathogenic Trichomonads. *J. Agric. Food Chem.* **2018**, *66*, 7942–7947. [CrossRef] [PubMed]
69. Marston, A.; Hostettmann, K. Plant molluscicides. *Phytochemistry* **1984**, *24*, 639–652. [CrossRef]
70. Nenaah, G. Individual and synergistic toxicity of Solanaceous glycoalkaloids against two coleopteran stored-product insects. *J. Pest. Sci.* **2011**, *84*, 77–86. [CrossRef]
71. Gee, J.M.; Worley, G.M.; Jonhson, I.T.; Price, K.R.; Rutten, A.A.J.J.L.; Houben, G.F.; Penninks, A.H. Effects of saponins and glycoalkaloids on the permeability and visibility of mammalian intestinal cells and on the integrity of tissue preparation in vitro. *Toxicol. Vitr.* **1996**, *10*, 117–128. [CrossRef]
72. Ruprich, J.; Rehurkova, I.; Boon, P.E.; Svensson, K.; Moussavian, S.; Van der Voet, H.; Bosgra, S.; Van Klaveren, J.D.; Busk, L. Probabilistic modeling of exposure doses and implications for health risk characterization: Glycoalkaloids from potatoes. *Food Chem Toxicol.* **2009**, *47*, 2899–2905. [CrossRef]
73. Patel, B.; Schutte, R.; Sporns, P.; Doyle, J.; Jewel, L.; Fedorak, R.N. Potato glycoalkaloids adversely affect intestinal permeability and aggravate inflammatory bowel disease. *Inflamm. Bowel Dis.* **2002**, *8*, 340–346. [CrossRef]
74. Blankemeyer, J.Y.; Stringer, B.K.; Rayburn, J.R.; Bantle, J.A.; Friedman, M. Effect of potato glycoalkaloids, alpha-chaconine and alpha-solanine on membrane potential of frog embryos. *J. Agric. Food Chem.* **1992**, *40*, 2022–2025. [CrossRef]
75. Roddick, J.G.; Rijnenberg, A.L.; Weissenberg, M. Membrane-disrupting properties of the steroidal glycoalkaloids solasonine and solamargine. *Phytochemistry* **1990**, *29*, 1513–1518. [CrossRef]

76. Sarquis, J.I.; Coria, N.A.; Aguilar, I.; Rivera, A. Glycoalkaloid content in *Solanum* species and hybrids from a breeding program for resistance to late blight (*Phytophthora infestans*). *Am. J. Potato Res.* **2000**, *77*, 295–302. [CrossRef]
77. Dahlin, P.; Müller, M.C.; Ekengren, S.; McKee, L.S.; Bulone, V. The Impact of Steroidal Glycoalkaloids on the Physiology of Phytophthora infestans, the Causative Agent of Potato Late Blight. *Mol. Plant. Microbe Interact.* **2017**, *30*, 531–542. [CrossRef]
78. Weete, J.W.; Abril, M.; Blackwell, M. Phylogenic distribution of fungal sterols. *PLoS ONE* **2010**, *5*, e10899. [CrossRef] [PubMed]
79. Roddick, J.G.; Rijnenberg, A.L.; Weissenberg, M. Alterations to the permeability of liposome membranes by the solasodine-based glycoalkaloids solasonine and solamargine. *Phytochemistry* **1992**, *31*, 1951–1954. [CrossRef]
80. Keukens, E.A.; de Vrije, T.; van den Boom, C.; de Waard, P.; Plasman, H.H.; Thiel, F.; Chupin, V.; Jongen, W.M.; de Kruijff, B. Molecular basis of glycoalkaloid induced membrane disruption. *Biochim. Biophys. Acta.* **1995**, *13*, 216–228. [CrossRef]
81. Roddick, J.G.; Rijnenberg, A.L.; Osman, S.F. Synergistic interaction between potato glycoalkaloids α-solanine and α-chaconine in relation to destabilization of cell membranes Ecological. *Implic. J. Chem. Ecol.* **1988**, *14*, 889–902. [CrossRef]
82. Sánchez-Maldonado, A.F.; Schieber, A.; Gänzle, M.G. Antifungal activity of secondary plant metabolites from potatoes (*Solanum tuberosum* L.): Glycoalkaloids and phenolic acids show synergistic effects. *J. Appl. Microbiol.* **2016**, *120*, 955–965. [CrossRef]
83. Müller, M.M.; Kantola, R.; Kitunen, V. Combining sterol and fatty acid profiles for the characterization of fungi. *Mycol. Res.* **1994**, *98*, 593–603.
84. Mejanelle, L.; Lopez, J.F.; Gunde-Cimerman, N.; Grimalt, J.O. Sterols of melanized fungi from hypersaline environments. *Org. Geochem.* **2000**, *31*, 1031–1040.
85. Kesten, C.; Gámez-Arjona, F.M.; Menna, A.; Scholl, S.; Dora, s.; Huerta, A.I.; Huang, H.Y.; Tintor, N.; Kinoshita, T.; Rep, M.; et al. Pathogen-induced pH changes regulate the growth-defense balance in plants. *EMBO J.* **2019**, *38*, e101822. [CrossRef]
86. Ford, J.E.; McCance, D.J.; Drysdale, R.B. The detoxification of α-tomatine by *Fusarium oxysporum* f. sp. *Lycopersici Phytochemistry* **1977**, *16*, 545–546. [CrossRef]
87. Oda, Y.; Saito, K.; Ohara-Takada, A.; Mori, M. Hydrolysis of the potato glycoalkaloids alpha-chaconine by filamentous fungi. *J. Biosci. Bioeng.* **2002**, *94*, 321–325. [CrossRef]
88. Hennessy, R.C.; Nielsen, S.D.; Greve-Poulsen, M.; Larsen, L.B.; Sørensen, O.B.; Stougaard, P. Discovery of a Bacterial Gene Cluster for Deglycosylation of Toxic Potato Steroidal Glycoalkaloids α-Chaconine and α-Solanine. *J. Agric. Food Chem.* **2020**, *68*, 1390–1396. [CrossRef] [PubMed]
89. Andrews, J.M. Determination of minimum inhibitory concentrations. *J. Antimicrob. Chemother.* **2001**, *48*, 5–16. [CrossRef] [PubMed]
90. Merkl, R.; Hrádková, I.; Filip, V.; Šmidrkal, J. Antimicrobial and antioxidant properties of phenolic acids alkyl esters. *Czech. J. Food Sci.* **2010**, *28*, 275–279. [CrossRef]
91. Tsuchiya, H. Effects of green tea catechins on membrane fluidity. *Pharmacology* **1999**, *59*, 34–44. [CrossRef] [PubMed]
92. Deußer, H.; Guignard, C.; Hoffmann, L.; Evers, D. Polyphenol and glycoalkaloid contents in potato cultivars grown in Luxembourg. *Food Chem.* **2012**, *135*, 2814–2824.
93. Friedman, M.; Kozukue, N.; Kim, H.J.; Choi, S.H.; Mizuno, M. Glycoalkaloid, phenolic, and flavonoid content and antioxidative activities of conventional nonorganic and organic potato peel powders from commercial gold, red, and Russet potatoes. *J. Food Compos. Anal.* **2017**, *62*, 69–75. [CrossRef]
94. Wang, Y.; Naber, M.R.; Crosby, T.W. Effects of Wound-Healing Management on Potato Post-Harvest Storability. *Agronomy* **2020**, *10*, 512.
95. Janjai, S.; Bala, B.K. Solar Drying Technology. *Food Eng. Rev.* **2012**, *4*, 16–54. [CrossRef]
96. Paolo, D.; Bianchi, G.; Morelli, C.F.; Speranza, G.; Campanelli, G.; Kidmose, U.; Lo Scalzo, R. Impact of drying techniques, seasonal variation and organic growing on flavor compounds profiles in two Italian tomato varieties. *Food Chem.* **2019**, *298*, 125062. [CrossRef]
97. Pandey, A.; Tripathi, S. Concept of standardization, extraction and pre phytochemical screening strategies for herbal drug. *J. Pharmacogn. Phytochem* **2014**, *2*, 115–119.
98. Hossain, M.B.; Tiwari, B.K.; Gangopadhyay, N.; O'Donnell, C.P.; Brunton, N.P.; Rai, D.K. Ultrasonic extraction of steroidal alkaloids from potato peel waste. *Ultrason. Sonochem.* **2014**, *21*, 1470–1476. [CrossRef] [PubMed]
99. Jie, D.; Wei, X. Review on the recent progress of non-destructive detection technology for internal quality of watermelon. *Comput. Electron. Agric.* **2018**, *151*, 156–164. [CrossRef]
100. Sørensen, K.K.; Kirk, H.G.; Olsson, K.; Labouriau, R.; Christiansen, J. A major QTL and an SSR marker associated with glycoalkaloid content in potato tubers from *Solanum tuberosum* × *S. sparsipilum* located on chromosome I. *Theor. Appl. Genet.* **2008**, *117*, 1–9. [CrossRef] [PubMed]
101. Krits, P.; Fogelman, E.; Ginzberg, I. Potato steroidal glycoalkaloid levels and the expression of key isoprenoid metabolic genes. *Planta* **2007**, *227*, 143–150. [CrossRef]
102. Khan, M.S.; Munir, I.; Khan, I. The potential of unintended effects in potato glycoalkaloids. *Afr. J. Biotechnol.* **2013**, *12*, 754–766.
103. Cui, T.; Bai, J.; Zhang, J.; Zhang, J.; Wang, D. Transcriptional expression of seven key genes involved in steroidal glycoalkaloid biosynthesis in potato microtubers, N.Z.J. *Crop. Hortic. Sci.* **2014**, *42*, 118–126. [CrossRef]

104. Ginzberg, I.; Thippeswamy, M.; Fogelman, E.; Demirel, U.; Mweetwa, A.M.; Tokuhisa, J.; Veilleux, R.E. Induction of potato steroidal glycoalkaloid biosynthetic pathway by overexpression of cDNA encoding primary metabolism HMG-CoA reductase and squalene synthase. *Planta* **2012**, *235*, 1341–1353. [CrossRef] [PubMed]
105. Arnqvist, L.; Dutta, P.C.; Jonsson, L.; Sitbon, F. Reduction of cholesterol and glycoalkaloid levels in transgenic potato plants by overexpression of a type 1 sterol methyltransferase cDNA. *Plant. Physiol.* **2003**, *131*, 1792–1799. [CrossRef] [PubMed]
106. Zook, M.N.; Kuc, J.A. Induction of sesquiterpene cyclase and suppression of squalene synthetase activity in elicitor treated or fungal infected potato tuber tissue. *Physiol. Mol. Plant. Pathol.* **1991**, *39*, 377–390. [CrossRef]
107. Shi, J.; Gonzales, R.A.; Madan, J.; Bhattacharyya, K. Identification and Characterization of an S-Adenosyl-L-methionine: D24-Sterol-C-methyltransferase cDNA from Soybean. *J. Biol. Chem.* **1996**, *271*, 9384–9389. [CrossRef] [PubMed]
108. Nurun, N. *Regulation of Sterol and Glycoalkaloid Biosynthesis in Potato (*Solanum tuberosum *L.)—Identification of Key Genes and Enzymatic Steps*; Acta Universitatis Agriculturae Sueciae: Uppsala, Sweden, 2011; ISSN 1652-6880.
109. Sawai, S.; Ohyama, K.; Yasumoto, S.; Seki, H.; Sakuma, T.; Yamamoto, T.; Takebayashi, Y.; Kojima, M.; Sakakibara, H.; Aoki, T.; et al. Sterol side chain reductase 2 is a key enzyme in the biosynthesis of cholesterol, the common precursor of toxic steroidal glycoalkaloids in potato. *Plant. Cell* **2014**, *26*, 3763–3774.
110. Shepherd, L.; Hackett, C.; Alexander, C.; McNicol, J.; Sungurtas, J.; Stewart, D.; McCue, K.; Belknap, W.; Davies, H. Modifying glycoalkaloid content in transgenic potato—Metabolome impacts. *Food Chem.* **2015**, *187*, 10–1016. [CrossRef] [PubMed]
111. Manrique-Carpintero, N.C.; Tokuhisa, J.G.; Ginzberg, I.; Veilleux, R.E. Allelic variation in genes contributing to glycoalkaloid biosynthesis in a diploid interspecific population of potato. *Theor. Appl. Genet.* **2014**, *127*, 391–405. [CrossRef] [PubMed]
112. Hardigan, M.; Laimbeer, P.; Newton, L.; Crisovan, E.; Hamilton, J.; Vaillancourt, B.; Wiegert-Rininger, K.; Wood, J.; Douches, D.; Farré, E.; et al. Genome diversity of tuber-bearing *Solanum* uncovers complex evolutionary history and targets of domestication in the cultivated potato. *Proc. Natl. Acad. Sci.* **2017**, *114*, 14380. [CrossRef] [PubMed]
113. Cárdenas, P.D.; Sonawane, P.; Pollier, J.; Bossche, R.V.; Dewangan, V.; Weithorn, E.; Tal, L.; Meir, S.; Rogachev, I.; Malitsky, S.; et al. GAME9 regulates the biosynthesis of steroidal alkaloids and upstream isoprenoids in the plant mevalonate pathway. *Nat. Commun.* **2016**, *7*, 10654. [CrossRef]
114. Manjulatha, M.; Hwang-Bae, S.; Yul-Ho, K.; Su-Jeong, K.; Kwang-Soo, C.; Oh-Keun, K.; Yong-Ik, J.; Su-Young, H.; Jeong-Hwan, N.; Jong-Taek, S.; et al. Comparative Expression of Key Genes Involved in Steroidal Glycoalkaloid Biosynthesis in Tubers of Two Potato Cultivars, Atlantic and Haryoung. *Plant. Breed. Biotechnol.* **2014**, *2*, 257–267.
115. Nahar, N.; Westerberg, E.; Arif, U.; Huchelmann, A.; Guasca, A.O.; Beste, L.; Dalman, K.; Dutta, P.c.; Jonsson, L.; Sitbon, F. Transcript profiling of two potato cultivars during glycoalkaloid-inducing treatments shows differential expression of genes in sterol and glycoalkaloid metabolism. *Sci. Rep.* **2017**, *7*, 43268. [CrossRef] [PubMed]
116. Mariot, R.F.; de Oliveira, L.A.; Voorhuijzen, M.M.; Staats, M.; Hutten, R.C.B.; van Dijk, J.P.; Kok, E.J.; Frazzon, J. Characterization and Transcriptional Profile of Genes Involved in Glycoalkaloid Biosynthesis in New Varieties of *Solanum tuberosum* L. *J. Agric. Food Chem.* **2016**, *64*, 988–996.
117. Sanford, L.L.; Deahl, K.L.; Sinden, S.L.; Ladd, T.L. Glycoalkaloid contents in tubers from Solanum tuberosum populations selected for potato leafhopper resistance. *Am. Potato J.* **1992**, *69*, 693–703. [CrossRef]
118. Zhang, W.; Zuo, C.; Chen, Z.; Kang, Y.; Qin, S. RNA Sequencing Reveals That Both Abiotic and Biotic Stress-Responsive Genes Are Induced during Expression of Steroidal Glycoalkaloid in Potato Tuber Subjected to Light Exposure. *Genes* **2019**, *10*, 920. [CrossRef]
119. Bengtsson, T.; Weighill, D.; Proux-Wéra, E.; Levander, F.; Resjö, S.; Burra, D.D.; Moushib, L.I.; Hedley, P.E.; Liljeroth, E.; Jacobson, D.; et al. Proteomics and transcriptomics of the BABA-induced resistance response in potato using a novel functional annotation approach. *BMC Genom.* **2014**, *15*, 315. [CrossRef] [PubMed]
120. Paudel, J.R.; Davidson, C.; Song, J.; Maxim, I.; Aharoni, A.; Tai, H.H. Pathogen and pest responses are altered due to RNAi-mediated knockdown of Lycoalkaloid Metabolism4 in *Solanum tuberosum*. *Mol. Plant. Microbe Interact.* **2017**, *30*, 876–885. [CrossRef]
121. Blažević, I.; Montaut, S.; Burčul, F.; Olsen, C.E.; Burow, M.; Rollin, P.; Agerbirk, N. Glucosinolate structural diversity, identification, chemical synthesis and metabolism in plants. *Phytochemistry* **2020**, *169*. [CrossRef]
122. Halkier, B.A.; Gershenzon, J. Biology and biochemistry of glucosinolates. *Annu. Rev. Plant. Biol.* **2006**, *57*, 303–333. [CrossRef]
123. Buxdorf, K.; Yaffe, H.; Barda, O.; Levy, M. The effects of glucosinolates and their breakdown products on necrotrophic fungi. *PLoS ONE* **2013**, *8*, e70771. [CrossRef]
124. Galletti, S.; Sala, E.; Leoni, O.; Cinti, S.; Cerato, C. *Aspergillus flavus* transformation of glucosinolates to nitriles by an arylsulfatase and a β-thio-glucosidase. *Soil Biol. Biochem.* **2008**, *40*, 2170–2173.
125. Züst, T.; Strickler, S.R.; Powell, A.F.; Mabry, M.E.; An, H.; Mirzaei, M.; York, T.; Holland, C.K.; Kumar, P.; Erb, M.; et al. Independent evolution of ancestral and novel defenses in a genus of toxic plants (*Erysimum*, Brassicaceae). *eLife* **2020**, *9*, e51712. [CrossRef] [PubMed]
126. Barco, B.; Clay, N.K. Evolution of glucosinolate diversity via whole-genome duplications, gene rearrangements, and substrate promiscuity. *Annu. Rev. Plant. Biol.* **2019**, *70*, 585–604. [CrossRef]
127. Chhajed, S.; Misra, B.B.; Tello, N.; Chen, S. Chemodiversity of the glucosinolate-myrosinase system at the single cell type resolution. *Front. Plant. Sci.* **2019**, *10*, 618. [CrossRef]

128. Zhang, K.X.; Hao, H.Y.; Li, J. Recent research advances in the mustard oil glycoside-black mustard enzyme defense system. *J. Plant. Physiol.* **2017**, *12*, 2069–2077.
129. Jang, M.; Hong, E.; Kim, G.H. Evaluation of antibacterial activity of 3-butenyl, 4-pentenyl, 2-phenylethyl, and benzyl isothiocyanate in *Brassica* vegetables. *J. Food Sci.* **2010**, *75*, 7. [CrossRef] [PubMed]
130. Melrose, J. The Glucosinolates: A sulphur glucoside family of mustard anti-tumour and antimicrobial phytochemicals of potential therapeutic application. *Biomedicines* **2019**, *7*, 62. [CrossRef] [PubMed]
131. Pane, C.; Villecco, D.; Roscigno, G.; De Falco, E.; Zaccardelli, M. Screening of plant-derived antifungal substances useful for the control of seedborne pathogens. *Arch. Phytopathology Plant. Protect.* **2013**, *46*, 1533–1539. [CrossRef]
132. Tierens, K.F.J.; Thomma, B.P.; Brouwer, M.; Schmidt, J.; Kistner, K.; Porzel, A.; Broekaert, W.F. Study of the role of antimicrobial glucosinolate-derived isothiocyanates in resistance of *Arabidopsis* to microbial pathogens. *Plant. Physiol.* **2001**, *125*, 1688–1699. [CrossRef] [PubMed]
133. Zhang, M.; Li, Y.; Bi, Y.; Wang, T.; Dong, Y.; Yang, Q.; Zhang, T. 2-Phenylethyl isothiocyanate exerts antifungal activity against Alternaria alternata by affecting membrane integrity and mycotoxin production. *Toxins* **2020**, *12*, 124. [CrossRef] [PubMed]
134. Nazareth, T.M.; Corrêa, J.A.F.; Pinto, A.C.S.M.; Palma, J.B.; Meca, G.; Bordin, K.; Luciano, F.B. Evaluation of gaseous allyl isothiocyanate against the growth of mycotoxigenic fungi and mycotoxin production in corn stored for 6 months. *J. Sci. Food Agric.* **2018**, *98*, 5235–5241. [CrossRef]
135. Nazareth, T.D.M.; Alonso-Garrido, M.; Stanciu, O.; Mañes, J.; Manyes, L.; Meca, G. Effect of allyl isothiocyanate on transcriptional profile, aflatoxin synthesis, and *Aspergillus flavus* growth. *Int. Food Res. J.* **2020**, *128*, 108786. [CrossRef]
136. Tracz, B.L.; Bordin, K.; Nazareth, T.D.M.; Costa, L.B.; Macedo, R.E.F.D.; Meca, G.; Luciano, F.B. Assessment of allyl isothiocyanate as a fumigant to avoid mycotoxin production during corn storage. *LWT* **2017**, *75*, 692–696. [CrossRef]
137. Azaiez, I.; Meca, G.; Manyes, L.; Luciano, F.B.; Fernández-Franzón, M. Study of the chemical reduction of the fumonisins toxicity using allyl, benzyl and phenyl isothiocyanate in model solution and in food products. *Toxicon* **2013**, *63*, 137–146. [CrossRef] [PubMed]
138. Azaiez, I.; Meca, G.; Manyes, L.; Fernández-Franzón, M. Antifungal activity of gaseous allyl, benzyl and phenyl isothiocyanate in vitro and their use for fumonisins reduction in bread. *Food Control.* **2013**, *32*, 428–434. [CrossRef]
139. Quiles, J.M.; Manyes, L.; Luciano, F.B.; Mañes, J.; Meca, G. Effect of the oriental and yellow mustard flours as natural preservative against aflatoxins B1, B2, G1 and G2 production in wheat tortillas. *Int. J. Food Sci. Technol.* **2015**, *52*, 8315–8321.
140. Lazzeri, L.; Curto, G.; Leoni, O.; Dallavalle, E. Effects of glucosinolates and their enzymatic hydrolysis products via myrosinase on the root-knot nematode *Meloidogyne incognita* (Kofoid et White) Chitw. *J. Agric. Food Chem.* **2004**, *52*, 6703–6707. [CrossRef]
141. Bhushan, S.; Gupta, S.; Sohal, S.K.; Arora, S. Assessment of insecticidal action of 3-Isothiocyanato-1-propene on the growth and development of *Spodoptera litura* (Fab.) (Lepidoptera: Noctuidae). *J. Entomol. Zool. Stud.* **2016**, *4*, 1068–1073.
142. Furlan, L.; Bonetto, C.; Finotto, A.; Lazzeri, L.; Malaguti, L.; Patalano, G.; Parker, W. The efficacy of biofumigant meals and plants to control wireworm populations. *Ind. Crops Prod.* **2010**, *31*, 245–254. [CrossRef]
143. Wu, H.; Zhang, G.A.; Zeng, S.; Lin, K.C. Extraction of allyl isothiocyanate from horseradish (*Armoracia rusticana*) and its fumigant insecticidal activity on four stored-product pests of paddy. *Pest. Manag. Sci.* **2009**, *65*, 1003–1008. [CrossRef]
144. Freitas, R.C.P.; Faroni, L.R.D.; Haddi, K.; Jumbo, V.L.O.; Oliveira, E.E. Allyl isothiocyanate actions on populations of *Sitophilus zeamais* resistant to phosphine: Toxicity, emergence inhibition and repellency. *J. Stored Prod. Res.* **2016**, *69*, 257–264. [CrossRef]
145. Messiha, N.K.; Ahmed, S.A.; El-Rokiek, K.G.; Dawood, M.G.; El-Masry, R.R. The physiological influence of allelochemicals in two *Brassicaceae* plant seeds on the growth and propagative capacity of *Cyprus rotundus* and *Zea mays* L. *World Appl. Sci. J.* **2013**, *26*, 1142–1149.
146. Smith, J.D.; Woldemariam, M.G.; Mescher, M.C.; Jander, G.; De Moraes, C.M. Glucosinolates from host plants influence growth of the parasitic plant *Cuscuta gronovii* and its susceptibility to aphid feeding. *Plant. Physiol.* **2016**, *172*, 181–197. [CrossRef]
147. Lazzeri, L.; Curto, G.; Dallavalle, E.; D'Avino, L.; Malaguti, L.; Santi, R.; Patalano, G. Nematicidal efficacy of biofumigation by defatted brassicaceae meal for control of Meloidogyne incognita (Kofoid et White) Chitw. on a full field zucchini crop. *Int. J. Agric. Sustain.* **2009**, *33*, 349–358.
148. Ntalli, N.; Caboni, P. A review of isothiocyanates biofumigation activity on plant parasitic nematodes. *Phytochem. Rev.* **2017**, *16*, 827–834. [CrossRef]
149. Curto, G.; Dallavalle, E.; Matteo, R.; Lazzeri, L. Biofumigant effect of new defatted seed meals against the southern root-knot nematode, Meloidogyne incognita. *Ann. Appl. Biol.* **2016**, *169*, 17–26. [CrossRef]
150. Meyer, S.L.F.; Zasada, I.A.; Orisajo, S.B.; Morra, M.J. Mustard seed meal mixtures: Management of Meloidogyne incognita on pepper and potential phytotoxicity. *J. Nematol.* **2011**, *43*, 7–15.
151. Laquale, S.; Candido, V.; Avato, P.; Argentieri, M.P.; D'addabbo, T. Essential oils as soil biofumigants for the control of the root-knot nematode Meloidogyne incognita on tomato. *Ann. Appl. Biol.* **2015**, *167*, 217–224. [CrossRef]
152. Ozone Secretariat. 2020. Available online: https://ozone.unep.org/sites/default/files/Handbooks/MP-Handbook-2020-English.pdf (accessed on 27 November 2020).
153. Lazzeri, L.; Baruzzi, G.; Malaguti, L.; Antoniacci, L. Replacing methyl bromide in annual strawberry production with glucosinolate-containing green manure crops. *Pest. Manag. Sci.* **2003**, *59*, 983–990. [CrossRef] [PubMed]

154. Curto, G.; Dallavalle, E.; Lazzeri, L. Life cycle duration of *Meloidogyne incognita* and host status of Brassicaceae and Capparaceae selected for glucosinate content. *Nematology* **2005**, *7*, 203–212. [CrossRef]
155. Lazzeri, L.; Leoni, O.; Manici, L.M.; Palmieri, S.; Patalano, G. Use of Seed Flour as Soil Pesticide. Patent US 7749549, 6 July 2010.
156. Lazzeri, L.; D'Avino, L.; Ugolini, L.; De Nicola, G.R.; Cinti, S.; Malaguti, L.; Bagatta, M.; Patalano, G.; Leoni, O.; Lazzeri, L. Bio-based products from Brassica carinata A. Braun oils and defatted meals by a second generation biorefinery approach. In Proceedings of the 19th European Biomass Conference, Berlin, Germany, 6–10 June 2011; pp. 1080–1092.
157. Ntalli, N.; Adamski, Z.; Doula, M.; Monokrousos, N. Nematicidal amendments and soil remediation. *Plants* **2020**, *9*, 429. [CrossRef]
158. Paudel, B.R.; Carpenter-Boggs, L.; Higgins, S. Influence of brassicaceous soil amendments on potentially beneficial and pathogenic soil microorganisms and seedling growth in Douglas-fir nurseries. *Appl. Soil Ecol.* **2016**, *105*, 91–100. [CrossRef]
159. Lazzeri, L.; Malaguti, L.; Cinti, S.; Ugolini, L.; De Nicola, G.R.; Bagatta, M.; Casadei, N.; D'avino, L.; Matteo, R.; Patalano, G. The *Brassicaceae* biofumigation system for plant cultivation and defence. An Italian twenty-year experience of study and application. *Acta Hortic.* **2013**, *1005*, 375–382.
160. Argento, S.; Melilli, M.G.; Branca, F. Enhancing Greenhouse Tomato-Crop Productivity by Using *Brassica macrocarpa* Guss. Leaves for controlling root-knot nematodes. *Agronomy* **2019**, *9*, 820. [CrossRef]
161. Ngala, B.M.; Haydock, P.P.; Woods, S.; Back, M.A. Biofumigation with *Brassica juncea*, *Raphanus sativus* and *Eruca sativa* for the management of field populations of the potato cyst nematode *Globodera pallida*. *Pest. Manag. Sci.* **2015**, *71*, 759–769. [CrossRef] [PubMed]
162. Mocali, S.; Landi, S.; Curto, G.; Dallavalle, E.; Infantino, A.; Colzi, C.; D'Errico, G.; Roversi, P.F.; D'Avino, L.; Lazzeri, L. Resilience of soil microbial and nematode communities after biofumigant treatment with defatted seed meals. *Ind. Crops Prod.* **2015**, *75*, 79–90.
163. Rudolph, R.E.; Sams, C.; Steiner, R.; Thomas, S.H.; Walker, S.; Uchanski, M.E. Biofumigation Performance of Four *Brassica* Crops in a Green Chile Pepper (*Capsicum annuum*) Rotation System in Southern New Mexico. *HortScience* **2015**, *50*, 247–253. [CrossRef]
164. Argento, S.; Raccuia, S.A.; Melilli, M.G.; Toscano, V.; Branca, F. Brassicas and their glucosinolate content for the biological control of root-knot nematodes in protected cultivation. *Acta Hortic.* **2013**, *1005*, 539–544.
165. Vandicke, J.; De Visschere, K.; Deconinck, S.; Leenknecht, D.; Vermeir, P.; Audenaert, K.; Haesaert, G. Uncovering the biofumigant capacity of allyl isothiocyanate from several *Brassicaceae* crops against *Fusarium* pathogens in maize. *J. Sci. Food Agric.* **2020**, *100*, 5476–5486. [CrossRef]
166. Piccinini, E.; Ferrari, V.; Campanelli, G.; Fusari, F.; Righetti, L.; Pagnotta, E.; Lazzeri, L. Effect of two liquid formulations based on *Brassica carinata* co-products in containing powdery mildew on melon. *Ind. Crops Prod.* **2015**, *75*, 48–53. [CrossRef]
167. Handiseni, M.; Jo, Y.; Lee, K.; Zhou, X. Screening brassicaceous plants as biofumigants for management of Rhizoctonia solani AG1-IA. *Plant. Dis.* **2016**, *100*, 758–763. [CrossRef] [PubMed]
168. Koron, D.; Sonjak, S.; Regvar, M. Effects of non-chemical soil fumigant treatments on root colonisation with arbuscular mycorrhizal fungi and strawberry fruit production. *Crop. Prot.* **2014**, *55*, 35–41. [CrossRef]
169. Neubauer, C.; Heitmann, B.; Müller, C. Biofumigation potential of Brassicaceae cultivars to *Verticillium dahliae*. *Eur. J. Plant. Pathol.* **2004**, *140*, 341–352.
170. Ugolini, L.; Martini, C.; Lazzeri, L.; D'Avino, L.; Mari, M. Control of postharvest grey mould (*Botrytis cinerea* Per.: Fr.) on strawberries by glucosinolate-derived allyl-isothiocyanate treatments. *Postharvest Biol. Technol.* **2014**, *90*, 34–39. [CrossRef]
171. Ríos, P.; González, M.; Obregón, S.; Carbonero, M.; Leal, J.; Fernández, P.; De Haro, A.; Sánchez, M. Brassica-based seedmeal biofumigation to control *Phytophthora cinnamomi* in the Spanish "dehesa" oak trees. *Phytopathol. Mediterr.* **2017**, *56*, 392–399.
172. Serrano- Pérez, P.; Palo, C.; Rodríguez -Molina, M.D.C. Efficacy of *Brassica carinata* pellets to inhibit mycelial growth and chlamydospores germination of *Phytophthora nicotianae* at different temperature regimes. *Sci. Hortic.* **2017**, *216*, 126–133. [CrossRef]
173. Ríos, P.; Obregón, S.; De Haro, A.; Fernández -Rebollo, P.; Serrano, M.; Sánchez, M. Effect of *Brassica* Biofumigant Amendments on Different Stages of the Life Cycle of *Phytophthora cinnamomi*. *J. Phytopathol. (1986)* **2016**, *164*, 582–594. [CrossRef]
174. Morales-Rodríguez, C.; Vettraino, A.M.; Vannini, A. Efficacy of biofumigation with *Brassica carinata* commercial pellets (BioFence) to control vegetative and reproductive structures of *Phytophthora cinnamomi*. *Plant. Dis.* **2016**, *100*, 324–330. [CrossRef] [PubMed]
175. Sukovata, L.; Jaworski, T.; Kolk, A. Efficacy of *Brassica juncea* granulated seed meal against *Melolontha* grubs. *Ind. Crops Prod.* **2015**, *70*, 260–265. [CrossRef]
176. Piccinini, E.; Ferrari, V.; Campanelli, G.; Fusari, F.; Righetti, L.; Matteo, R.; Lazzeri, L. Effect of two bio-based liquid formulations from *Brassica carinata* in containing red spider mite (*Tetranychus urticae*) on eggplant. *Ind. Crops Prod.* **2015**, *75*, 36–41. [CrossRef]
177. Zuluaga, D.L.; Van Ommen Kloeke, A.E.E.; Verkerk, R.; Röling, W.F.M.; Ellers, J.; Roelofs, D.; Aarts, M.G.M. Biofumigation using a wild *Brassica oleracea* accession with high glucosinolate content affects beneficial soil invertebrates. *Plant. Soil* **2015**, *394*, 155–163. [CrossRef]
178. Kirkegaard, J.A.; Matthiessen, J. Developing and refining the biofumigation concept. *Agroindustria* **2004**, *3*, 233–239.
179. Lu, P.; Gilardi, G.; Gullino, M.L.; Garibaldi, A. Biofumigation with Brassica plants and its effect on the inoculum potential of *Fusarium* yellows of *Brassica* crops. *Eur. J. Plant. Pathol.* **2010**, *126*, 387–402.

180. Regulation (EC) No 1907/2006 of the European Parliament and of the Council of 18 December 2006 concerning the Registration, Evaluation, Authorisation and Restriction of Chemicals (REACH). Available online: https://eur-lex.europa.eu/legal-content/en/ALL/?uri=CELEX%3A32006R1907 (accessed on 9 April 2021).
181. Matteo, R.; Back, M.A.; Reade, J.P.H.; Ugolini, L.; Pagnotta, E.; Lazzeri, L. Effectiveness of defatted seed meals from Brassicaceae with or without crude glycerin against black grass (*Alopecurus myosuroides* Huds.). *Ind. Crops Prod.* **2018**, *111*, 506–512. [CrossRef]
182. Brown, P.D.; Morra, M.J. Control of soil-borne plant pests using glucosinolate-containing plants. *Adv. Agron.* **1997**, *61*, 167–231.
183. Schnug, E.; Haneklaus, S. Glucosinolates—the agricultural story. In: Kopriva, S., ed. Glucosinolates. *Adv. Bot. Res.* **2016**, *80*, 281–302.
184. Gupta, S.K. Brassicas. In *Breeding Oilseed Crops for Sustainable Production*, 1st ed; Elsevier: Amsterdam, Netherlands, 2016; pp. 33–53.
185. Faulkner, K.; Mithen, R.; Williamson, G. Selective increase of the potential anticarcinogen 4-methylsulphinylbutyl glucosinolate in broccoli. *Carcinogenesis* **1998**, *19*, 605–609.
186. Mithen, R.; Faulkner, K.; Magrath, R.; Rose, P.; Williamson, G.; Marquez, J. Development of isothiocyanate enriched broccoli, and its enhanced ability to induce phase 2 detoxification enzymes in mammalian cells. *Theor. Appl. Genet.* **2003**, *106*, 727–734. [CrossRef] [PubMed]
187. Li, G.; Riaz, A.; Goyal, S.; Abel, S.; Quiros, C.F. Inheritance of Three Major Genes Involved in the Synthesis of Aliphatic Glucosinolates in *Brassica oleracea*. *J. Am. Soc. Hort. Sci.* **2001**, *126*, 427–431. [CrossRef]
188. Liu, Z.; Hirani, A.H.; McVetty, P.B.E.; Daayf, F.; Quiros, C.F.; Li, G. Reducing progoitrin and enriching glucoraphanin in *Brassica napus* seeds through silencing of the GSL-ALK gene family. *Plant. Mol. Biol.* **2012**, *79*, 179–189. [CrossRef]
189. Petersen, A.; Wang, C.; Crocoll, C.; Halkier, B.A. Biotechnological approaches in glucosinolate production. *J. Integr. Plant. Biol.* **2018**, *60*, 1231–1248. [CrossRef]
190. Brader, G.; Mikkelsen, M.D.; Halkier, B.A.; Tapio Palva, E. Altering glucosinolate profiles modulates disease resistance in plants. *Plant. J.* **2006**, *46*, 758–767. [CrossRef]
191. Wentzell, A.M.; Rowe, H.C.; Hansen, B.G.; Ticconi, C.; Halkier, B.A.; Kliebenstein, D.J. Linking metabolic QTL with network and cis-eQTLs controlling biosynthetic pathways. *PLoS Genet.* **2007**, *3*, e162. [CrossRef]
192. Neal, C.S.; Fredericks, D.P.; Griffiths, C.A.; Neale, A.D. The characterisation of AOP2: A gene associated with the biosynthesis of aliphatic alkenyl glucosinolates in *Arabidopsis thaliana*. *BMC Plant. Biol.* **2010**, *10*, 170. [CrossRef]
193. Bell, L.; Chadwick, M.; Puranik, M.; Tudor, R.; Methven, L.; Kennedy, S.; Wagstaff, C. The *Eruca sativa* Genome and Transcriptome: A Targeted Analysis of Sulfur Metabolism and Glucosinolate Biosynthesis Pre and Postharvest. *Front. Plant. Sci.* **2020**, *11*, 525102.
194. Falk, K.L.; Tokuhisa, J.G.; Gershenzon, J. The effect of sulfur nutrition on plant glucosinolate content: Physiology and molecular mechanisms. *Plant. Biol. (Stuttg)* **2007**, *9*, 573–581. [CrossRef] [PubMed]
195. Mithen, R.; Ho, E. Isothiocyanates for Human Health. *Mol. Nutr Food Res.* **2018**, *62*, e1870079. [CrossRef] [PubMed]
196. Traka, M.H.; Saha, S.; Huseby, S.; Kopriva, S.; Walley, P.G.; Barker, G.C.; Moore, J.; Mero, G.; van den Bosch, F.; Constant, H.; et al. Genetic regulation of glucoraphanin accumulation in Benefortè broccoli. *New Phytol.* **2013**, *198*, 1085–1095.
197. Ruffoni, B.; Pistelli, L.; Bertoli, A.; Pistelli, L. Plant cell cultures: Bioreactors for industrial production. *Adv. Exp. Med. Biol.* **2010**, *698*, 203–218.
198. Kastell, A.; Zrenner, R.; Schreiner, M.; Kroh, L.; Ulrichs, C.; Smetanska, I.; Mewis, I. Metabolic engineering of aliphatic glucosinolates in hairy root cultures of *Arabidopsis thaliana*. *Plant. Mol. Biol. Rep.* **2015**, *33*, 598–608. [CrossRef]
199. Wielanek, M.; Urbanek, H. Enhanced glucotropaeolin production in hairy root cultures of *Tropaeolum majus* L. by combining elicitation and precursor feeding. *Plant. Cell Tiss. Organ. Cult.* **2006**, *86*, 177–186. [CrossRef]
200. Hirani, A.H.; Li, G.; Zelmer, C.D.; McVetty, P.B.E.; Asif, M.; Goyal, A. Molecular genetics of glucosinolate biosynthesis in Brassicas: Genetic manipulation and application aspects. In *Crop Plant*; Goyal, A., Ed.; IntechOpen, Inc.: Palm Gardens, AB, Canada, 2012.
201. Neequaye, M.; Stavnstrup, S.; Lawrenson, T.; Hundleby, P.; Troncoso-Rey, P.; Saha, S.; Harwood, W.; Traka, H.M.; Mithen, R.; Østergaard, L. CRISPR-Cas9-mediated editing of myb28 impairs glucoraphanin accumulation of *Brassica oleracea* in the field. *bioRxiv Preprint* **2020**. [CrossRef]
202. Gao, M.; Li, G.; Yang, B.; Qiu, D.; Farnham, M.; Quiros, C. High-density *Brassica oleracea* linkage map: Identification of useful new linkages. *Theor. Appl. Genet.* **2007**, *115*, 277–287.
203. Issa, R.A. Identification of glucosinolate profile in *Brassica oleracea* for quantitative trait locus mapping. Ph.D. Thesis, University of Warwick, Coventry, UK, 2010.
204. Hirani, A.H. QTL mapping, gene identification and genetic manipulation of glucosinolates in *Brassica rapa* L. Ph.D. Thesis, University of Manitoba, Winnipeg, MB, Canada, 2011.
205. Lou, P.; Zhao, J.; He, H.; Hanhart, C.; Del Carpio, D.P.; Verkerk, R.; Custers, J.; Koornneef, M.; Bonnema, G. Quantitative trait loci for glucosinolate accumulation in *Brassica rapa* leaves. *New Phytol.* **2008**, *179*, 1017–1032.
206. Ramchiary, N.; Padmaja, K.L.; Sharma, S.; Gupta, V.; Sodhi, Y.S.; Mukhopadhyay, A.; Arumugam, N.; Pental, D.; Pradhan, A.K. Mapping of yield influencing QTL in *Brassica juncea*: Implications for breeding of a major oilseed crop of dryland areas. *Theor. Appl. Genet.* **2007**, *115*, 807–817. [CrossRef]
207. Hirani, A.H.; Geng, J.; Zhang, J.; Zelmer, C.D.; McVetty, P.B.E.; Daayf, F.; Li, G. Quantitative Trait Loci Mapping and Candidate Gene Identification for Seed Glucosinolates in *Brassica rapa* L. *Crop. Sci.* **2016**, *56*, 942–956. [CrossRef]

208. Zhang, J.; Wang, H.; Liu, Z.; Liang, J.; Wu, J.; Cheng, F.; Mei, S.; Wang, X. A naturally occurring variation in the BrMAM-3 gene is associated with aliphatic glucosinolate accumulation in *Brassica rapa* leaves. *Hortic. Res.* **2018**, *5*, 69. [CrossRef]
209. Quijada, P.A.; Udall, J.A.; Lambert, B.; Osborn, T.C. Quantitative trait analysis of seed yield and other complex traits in hybrid spring rapeseed (*Brassica napus* L.): Identification of genomic regions from winter germplasm. *Theor. Appl. Genet.* **2006**, *113*, 549–561. [CrossRef] [PubMed]
210. Hasan, M.; Friedt, W.; Pons-Kuhnemann, J.; Freitag, N.M.; Link, K.; Snowdon, R.J. Association of gene-linked SSR markers to seed glucosinolate content in oilseed rape (*Brassica napus* ssp. *napus*). *Theor. Appl. Genet.* **2008**, *116*, 1035–1049. [CrossRef] [PubMed]
211. Lionneton, E.; Aubert, G.; Ochatt, S.; Merah, O. Genetic analysis of agronomic and quality traits in mustard (*Brassica juncea*). *Theor. Appl. Genet* **2004**, *109*, 792–799. [CrossRef] [PubMed]
212. Ripley, V.L.; Roslinsky, V. Identification of an ISSR marker for 2-propenyl glucosinolate content in *Brassica juncea* L. and conversion to a SCAR marker. *Mol. Breed.* **2005**, *16*, 57–66. [CrossRef]
213. Howell, P.M.; Sharpe, A.G.; Lydiate, D.J. Homoeologous loci control the accumulation of seed glucosinolates in oilseed rape (*Brassica napus*). *Genome* **2003**, *46*, 454–460. [CrossRef]
214. Mahmood, T.; Ekuere, U.; Yeh, F.; Good, A.G.; Stringam, G.R. Molecular mapping of seed aliphatic glucosinolates in *Brassica juncea*. *Genome* **2003**, *46*, 753–760. [CrossRef]
215. Zou, Z.; Ishida, M.; Li, F.; Kakizaki, T.; Suzuki, S.; Kitashiba, H.; Nishio, T. QTL analysis using SNP markers developed by next-generation sequencing for identification of candidate genes controlling 4-methylthio-3-butenyl glucosinolate contents in roots of radish, *Raphanus sativus* L. *PLoS ONE* **2013**, *8*, e53541.
216. Javidfar, F.; Cheng, B. Construction of a genetic linkage map and QTL analysis of erucic acid content and glucosinolate components in yellow mustard (*Sinapis alba* L.). *BMC Plant. Biol.* **2013**, *13*, 142.
217. Cardellina, J.H. Biologically natural products in the search for new agrochemicals. In *Biologically active natural products: Potential use in agriculture*; Gulter, H.G., Ed.; American Chemical Society: Washington, WA, USA, 1988; pp. 305–311.
218. Gulter, H.G. Natural products and their potential in agriculture: A personal overview. In *Biologically active natural products: Potential use in agriculture*; Gulter, H.G., Ed.; American Chemical Society: Washington, WA, USA, 1998; pp. 1–2.
219. Emosairue, S.O.; Ukeh, D.A. Field trial of neem products for the control of okra flea beetles (*Podagrica* spp) in South Eastern Nigeria. *Afr. J. Plant. Prot.* **1996**, *6*, 27–33.
220. Tewari, S.N.; Nayak, M. Activity of four-plant leaf extracts against three fungal pathogens of rice. *Trop. Agric. (Trinidad)* **1991**, *68*, 373–375.
221. Al-Abed, A.S.; Quasem, J.R.; Abu-Blan, H.A. Antifungal effects of some common wild plant species on certain plant pathogenic fungi. *Dirasat (Pure Appl. Sci.)* **1993**, *20*, 149–158.
222. Amadioha, A.C. Control of powdery mildew of pepper (*Capsicum annum L*) by leaf extracts of papaya (*Asimina triloba*). *J. Herbs Spices Med. Plants* **1998**, *6*, 41–47. [CrossRef]
223. Amadioha, A.C. Controlling rice blast in vitro and in vivo with extracts of *Azadirachta indica*. *Crop. Prot.* **2000**, *19*, 287–290. [CrossRef]
224. Amadioha, A.C.; Obi, V.I. Fungitoxic activity of extracts from *Azadirachta indica* and *Xylopia aethiopica* on *Colletotrichum lindemuthianum* in Cowpea. *J. Herbs Spices Med. Plants* **1998**, *6*, 33–40. [CrossRef]
225. Mason, J.R.; Mathew, D.N. Evaluation of neem as a bird repellent chemical. *Int. J. Pest Manag* **1996**, *42*, 47–49. [CrossRef]
226. Metcalf, R.L.; Metclaf, R.A.; Rhodes, A.N. Cucurbitacins as Kairomones for diabroticide beetles. *Proc. Natl. Acad. Sci. USA* **1980**, *77*, 3769–3772. [CrossRef] [PubMed]
227. Neilson, J.K.; Larsen, L.M.; Sorenson, H.J. Cucurbitacins E and I in *Iberis amara*, feeding inhibitors for *Phyllotreta nemorum*. *Phytochemistry* **1977**, *16*, 1519–1522.
228. Singh, D.C. Scope of medicinal and aromatic plants in pest management. In Proceedings of the International Symposium, Allelopathy in sustainable Agriculture, Forestry and Environment, New Delhi, India, 6–8 September 1994; p. 68.
229. Fandohan, P.; Gbenou, J.D.; Gnonlonfin, B.; Hell, K.; Marasas, F.O.; Wingfield, M.G. Effect of essential oils on the growth of Fusarium verticillioides and fumonisin contamination in corn. *J. Agric. Food Chem.* **2004**, *52*, 6824–6829. [CrossRef]
230. Osbourn, A.E. Antimicrobial phytoprotectants and fungal pathogens: A commentary. *Fungal Genet. Biol.* **1999**, *26*, 163–168. [CrossRef] [PubMed]
231. Krishnamurthy, K.; Balconi, C.; Sherwood, J.E.; Giroux, M. Increased tolerance to fungal diseases of rice plants transformed with puroindoline genes. *Mol. Plant. Microbe Interact.* **2001**, *14*, 1255–1260. [CrossRef]
232. Ferreira, R.B.; Monteiro, S.; Freitas, R.; Santos, C.N.; Chen, Z.; Batista, L.M.; Duarte, J.; Borges, A.; Teixeira, A.R. The role of plant defence proteins in fungal pathogenesis. *Mol. Plant. Pathol.* **2007**, *8*, 677–700. [CrossRef]
233. Balconi, C.; Lanzanova, C.; Motto, M. Ribosome-inactivating proteins in cereals. In *Toxic Plant Proteins*; Martin, R.J., Ed.; Hartley, Springer series Plant Cell Monographs 18; Springer: Verlag: Berlin/Heidelberg, Germany, 2010; pp. 149–166.
234. Punja, Z. Genetic engineering of plants to enhance resistance to fungal pathogens—A review of progress and future prospects. *Can. J. Plant. Pathol.* **2001**, *23*, 216–235. [CrossRef]
235. Tarchevsky, I.A. Pathogen-Induced Plant Proteins (Review). *Appl. Biochem. Microbiol.* **2001**, *37*, 441–455. [CrossRef]
236. Keen, N.T. Plant disease resistance: Progress in basic understanding and practical application. *Adv. Bot Res.* **1999**, *30*, 291–328.

237. Balconi, C.; Stevanato, P.; Motto, M.; Biancardi, E. *Breeding for biotic stress resistance/tolerance in plants In Crop. Production for Agricultural Improvement*; Ashraf, M., Ozturk, M., Ahmad, M.S.A., Aksoy, A., Eds.; Springer Verlag: Berlin/Heidelberg, Germany, 2012; pp. 57–114.
238. Mesterházy, A.; Oláh, J.; Popp, J. Losses in the Grain Supply Chain: Causes and Solutions. *Sustainability* **2020**, *12*, 2342. [CrossRef]
239. Munkvold, G.P. Cultural and genetic approaches to managing mycotoxins in maize. *Annu. Rev. Phytopathol.* **2003**, *41*, 99–116. [CrossRef]
240. Balconi, C.; Berardo, N.; Locatelli, S.; Lanzanova, C.; Torri, A.; Redaelli, R. Evaluation of ear rot (*Fusarium verticillioides*) resistance and fumonisin accumulation in Italian maize inbred lines. *Phytopathol. Mediterr.* **2014**, *53*, 14–26.
241. Balconi, C.; Motto, M.; Mazzinelli, G.; Berardo, N. Ear secondary traits related to aflatoxin accumulation in commercial maize hybrids under artificial field inoculation. *World Mycotoxin J.* **2010**, *3*, 239–250. [CrossRef]
242. Castegnaro, M.; McGregor, D. Carcinogenic risk assessment of mycotoxins. *Rev. Med. Vet.* **1998**, *149*, 671–678.
243. CAST. *Mycotoxins: Risks in Plant, Animal, and Human Systems*; Task Force Report; Council for Agricultural Science and Technology: Ames, IA, USA, 2003; Volume 139.
244. Lanzanova, C.; Torri, A.; Motto, M.; Balconi, C. Characterization and interaction between b-32 maize ribosome-inactivating protein and fungal pathogens development in vivo and in vitro. *Maydica* **2011**, *56*, 83–93.
245. Balazs, E.; Schepers, J.S. The mycotoxin threat to world safety. *Int. J. Food Microbiol.* **2007**, *119*, 1–2. [CrossRef] [PubMed]
246. Berardo, N.; Lanzanova, C.; Locatelli, S.; Laganà, P.; Verderio, A.; Motto, M. Levels of total fumonisins in maize samples from Italy during 2006–2008. *Food Addit. Contam. Part. B* **2011**, *4*, 116–124. [CrossRef] [PubMed]
247. European Commission Regulation (EC) No 1881/2006 of 19 December 2006 Setting Maximum Levels for Certain Contaminants in Foodstuffs. Available online: https://eur-lex.europa.eu/legal-content/EN/ALL/?uri=CELEX%3A32006R1881 (accessed on 9 April 2021).
248. European Commission Regulation (EC) No 1126/2007 of 28 September 2007 Amending Regulation (EC) No 1881/2006 Setting Maximum Levels for Certain Contaminants in Foodstuffs as Regards Fusarium Toxins in Maize and Maize Products. Available online: https://eur-lex.europa.eu/legal-content/EN/ALL/?uri=CELEX%3A32007R1126 (accessed on 9 April 2021).
249. Hartings, H.; Lanzanova, C.; Lazzaroni, N.; Balconi, C. How maize trackles *Diabrotica v.* virgifera attack. *Maydica* **2016**, *61*, 1–8.
250. Herrera-Foessel, S.A.; Singh, R.P.; Huerta-Espino, J.; Crossa, J.; Yuen, J.; Djurle, A. Effect of leaf rust on grain yield and yield traits of durum wheats with race-specific and slow-rusting resistance to leaf rust. *Plant. Dis.* **2006**, *90*, 1065–1072. [CrossRef]
251. Savary, S.; Willocquet, L.; Pethybridge, S.; Esker, P.; McRoberts, N.; Nelson, A. The global burden of pathogens and pests on major food crops. *Nat. Ecol. Evol.* **2019**, *3*, 430–439. [CrossRef]

molecules

MDPI

Article

Phytochemical Investigation and Biological Activities of *Lantana rhodesiensis*

Fatimata Nea [1,2,*], Michel Boni Bitchi [1], Manon Genva [2], Allison Ledoux [3], Alembert Tiabou Tchinda [4], Christian Damblon [5], Michel Frederich [3], Zanahi Félix Tonzibo [1] and Marie-Laure Fauconnier [2]

1 Laboratory of Constitution and Reaction of Matter, UFR-SSMT, University Félix Houphouët-Boigny, 01 BP 582 Abidjan 01, Ivory Coast; bbonimichel@yahoo.fr (M.B.B.); tonzibz@yahoo.fr (Z.F.T.)
2 Laboratory of Chemistry of Natural Molecules, Gembloux Agro-Bio Tech, University of Liège, Passage des Déportés 2, 5030 Gembloux, Belgium; m.genva@uliege.be (M.G.); marie-laure.fauconnier@uliege.be (M.-L.F.)
3 Laboratory of Pharmacognosy, Center for Interdisciplinary Research on Medicines (CIRM), University of Liège, Avenue Hippocrate 15, 4000 Liège, Belgium; allison.ledoux@uliege.be (A.L.); m.frederich@uliege.be (M.F.)
4 Laboratory of Phytochemistry, Centre for Research on Medicinal Plants and Traditional Medicine, Institute of Medical Research and Medicinal Plants Studies, P.O. Box 13033 Yaoundé, Cameroon; talembert@gmail.com
5 MolSys Research Unit, Faculty of Sciences, University of Liège, 4000 Liège, Belgium; c.damblon@uliege.be
* Correspondence: Fatimata.Nea@uliege.be; Tel.: +32-(0)4-6521-2551

check for updates

Citation: Nea, F.; Bitchi, M.B.; Genva, M.; Ledoux, A.; Tchinda, A.T.; Damblon, C.; Frederich, M.; Tonzibo, Z.F.; Fauconnier, M.-L. Phytochemical Investigation and Biological Activities of *Lantana rhodesiensis*. *Molecules* **2021**, *26*, 846. https://doi.org/10.3390/molecules26040846

Academic Editors: Simona Fabroni, Krystian Marszałek and Aldo Todaro

Received: 8 January 2021
Accepted: 3 February 2021
Published: 5 February 2021

Abstract: *Lantana rhodesiensis* Moldenke is a plant widely used to treat diseases, such as rheumatism, diabetes, and malaria in traditional medicine. To better understand the traditional uses of this plant, a phytochemical study was undertaken, revealing a higher proportion of polyphenols, including flavonoids in *L. rhodesiensis* leaf extract and moderate proportion in stem and root extracts. The antioxidant activity of the extracts was also determined using three different assays: the radical 2,2-diphenyl-1-picrylhydrazyl (DPPH) scavenging activity, the FRAP method (Ferric-reducing antioxidant power) and the β-carotene bleaching test. The anti-malarial activity of each extract was also evaluated using asexual erythrocyte stages of *Plasmodium falciparum*, chloroquine-sensitive strain 3D7. The results showed that the leaf extract exhibited higher antioxidant and anti-malarial activities in comparison with the stem and root extracts, probably due to the presence of higher quantities of polyphenols including flavonoids in the leaves. A positive linear correlation was established between the phenolic compound content (total polyphenols including flavonoids and tannins; and total flavonoids) and the antioxidant activity of all extracts. Furthermore, four flavones were isolated from leaf dichloromethane and ethyl acetate fractions: a new flavone named rhodescine (5,6,3′,5′-tetrahydroxy-7,4′-dimethoxyflavone) (**1**), 5-hydroxy-6,7,3′,4′,5′-pentamethoxyflavone (**2**), 5-hydroxy-6,7,3′,4′-tetramethoxyflavone (**3**), and 5,6,3′-trihydroxy-7,4′-dimethoxyflavone (**4**). Their structures were elucidated by ^{1}H, ^{13}CNMR, COSY, HSQC, HMBC, and MS-EI spectral methods. Aside from compound **2**, all other molecules were described for the first time in this plant species.

Keywords: *Lantana rhodesiensis*; polyphenol content; flavonoid content; antioxidant activity; anti-malarial activity; flavones

1. Introduction

Lantana rhodesiensis (L. rhodesiensis) is an aromatic plant used in traditional medicine to treat many diseases, such as rheumatism, diabetes mellitus [1], malaria [2], cancer [3], congestive heart failure, and cardiac arrhythmia [4,5]. It is a woody herb or small shrub less than 2 m high, often with several stems, and without thorns, native to subtropical and tropical regions. *L. rhodesiensis* can be found in many African countries, such as Tanzania, Kenya, Rwanda, Ethiopia, Malawi, Cameroon, Sudan, Burkina Faso, and Côte d'Ivoire [1,6].

Several studies have already tried to correlate the traditional uses of this plant with its biological activities and chemical composition. As an example, aqueous extracts of *L. rhodesiensis* were screened for their hypoglycemic activities in alloxan-induced diabetic rats, with results confirming the antidiabetic activity of *L. rhodesiensis* when therapeutic doses were administered intra-peritoneally and orally [1]. In order to justify its traditional use for the treatment of cancer, the antiproliferative activity of *L. rhodesiensis* was evaluated. The results showed that *L. rhodesiensis* is not genotoxic and that this plant induces a strong antiproliferative effect against cancer cells in vitro. The high antioxidant activity of *L. rhodesiensis* methanol extracts [3] and decoctions [7] was also highlighted using DPPH method. In those studies, the methanol extracts contained high quantities of tannins and flavonoids and the decoctions were characterized by high total phenolic contents with low flavonoid quantities. *L. rhodesiensis* also showed significant repellency against *Anopheles gambiae sensu lato* Giles, the main vector of malaria in Africa [8]. Leaf essential oils from *L. rhodesiensis* have been extensively studied and shown to possess robust anti-inflammatory and antioxidant activity [9], which originates from their high content of phenolic compounds [10]. Studies have also shown that *L. rhodesiensis* contains triterpenes, steroids, phenols, alkaloids, polyphenols including flavonoids, and tannins [1–4,11]. Two polymethoxyflavones, 5,6,7,3′,4′,5′-hexamethoxyflavone and its analogue 5-hydroxy-6,7,3′,4′,5′-pentamethoxyflavone, were isolated from the whole plant of *L. rhodesiensis* [12].

The main purpose of the present research was to correlate the traditional medicine uses of *L. rhodesiensis* for treating rheumatism and malaria with the phytochemical composition of *L. rhodesiensis* extracts obtained from each plant organ and with their antioxidant and anti-malarial activities. The different plant organs were considered separately in order to determine the most active part of the plant. Finally, four major flavonoids were isolated from *L. rhodesiensis* leaves and their structures were determined, as well as their antioxidant activities.

2. Results and Discussion

2.1. Phytochemical Screening

2.1.1. Determination of Phytochemical Classes

The results of the qualitative phytochemical study of *L. rhodesiensis* organs (Table 1) showed that the leaf, stem, and root extracts of *L. rhodesiensis* contained polyphenols including flavonoids and tannins. Terpenes, sterols, saponins, and alkaloids were also detected in all the organs, while leuco-anthocyanins and anthocyanins were too low to be detectable. The results also highlighted considerable differences in the phytochemical classes found in the different plant organs, as the assays indicated higher levels of flavonoids and polyphenols in leaves than in stems and roots. Moreover, the results also indicated higher proportions of saponins in roots than in leaves and stems. The phytochemical classes detected in the leaf extract are in agreement with those already described in an aqueous leaf extract [1]. Moreover, the realized assays indicated a higher proportion of tannins in the methanolic extracts from the aerial parts (stems and leaves) in comparison with sterols/triterpenes, flavonoids, and saponins [3]. This is the first systematic phytochemical screening of *L. rhodesiensis* stems and roots.

Table 1. Phytochemical screening of *L. rhodesiensis* organs.

Phytochemical Classes		Test Performed	Leaves	Stems	Roots
Polyphenols		Iron chloride 2%	+++	++	+
Flavonoids		Cyanidin	+++	+	+
Terpenes/sterols		Lieberman and Bürchard	++	+	++
Tannins	catechin	Stiasny	++	+	+
	gallic	Stiasny	++	++	+
Saponins		Foam formation	+	+	++
Alkaloids		Dragendorff	+	+	+
Leuco-anthocyanins		Cyanidin	-	-	-
Anthocyanins		Cyanidin	-	-	-

Note: (-): not detectable, (+): low amounts, (++): high amounts, and (+++): very high amounts.

2.1.2. Polyphenolic Compound Quantification

Quercetin (coefficient of determination (R^2) = 0.9996) and gallic acid (R^2 = 0.9975) calibration curves were performed in order to determine the phenolic compound concentrations in the extracts. Total polyphenol contents (Table 2) ranged from 153.37 ± 0.61 to 273.27 ± 0.48 mg gallic acid equivalents (GAE)/g extract; the highest content was obtained with the leaf extract. For the total flavonoid assay, the contents ranged from 34.87 ± 0.34 to 110.54 ± 0.46 mg quercetin equivalents (QE)/g extract. Similarly, the leaf extract had the highest content, showing that *L. rhodesiensis* leaves are richer in polyphenols including flavonoids and tannins than the stems and roots. The lowest contents of total polyphenols and total flavonoids were observed in the root extract.

Table 2. Polyphenolic compound assay results. GAE: gallic acid equivalents, QE: quercetin equivalents (mean ± standard deviation of three independent tests).

	Polyphenolic Compound Contents	
	Total Polyphenols (mg GAE/g Extract)	**Total Flavonoids (mg QE/g Extract)**
Leaves	273.27 ± 0.48	110.54 ± 0.46
Stems	206.06 ± 0.87	52.95 ± 0.64
Roots	153.37 ± 0.61	34.87 ± 0.34

The determination of phenolic compounds in an aqueous extract of *L. rhodesiensis* leaves had been performed previously. The results obtained in that study showed that the amount of phenols (685.25 ± 30.77 mg GAE/g) was higher than that of tannins (323.61 ± 61.54 mg GAE/g) and flavonoids (187.33 ± 54.97 mg GAE/g) [1]. In addition, another study showed the total phenol (210.55 ± 7.5 mg GAE/g) and flavonoid (50.09 ± 1.9 mg QE/g) composition of the methanolic extract of leafy stems of *L. rhodesiensis* [11]. These results cannot be directly compared to those of the present study as the extracts, organs used, and the methods applied for the different tests are not the same. However, taking into account data from the literature, it can be said that the aqueous extract of *L. rhodesiensis* leaves is richer in phenolic compounds than the hydro-methanolic extract. On the other hand, the amount of total phenolic compounds in the methanolic extract of leaves and stems is lower than that of the leaf extract in our study.

A study using the same method to determine the total polyphenol content was carried out on leaf methanolic extracts of different *Lantana camara* varieties. Although it is not the same species, the results of two varieties (225.15 ± 12.52 and 232.99 ± 15.97 mg GAE/g extract) were found to be similar to those of the present study [13], highlighting the considerable interest in leaves from plants of the genus *Lantana* when searching for a source of polyphenolic compounds.

The protective effect of polyphenols has been attributed to their antioxidant properties, which can prevent molecular oxidative damage and cellular disorders leading to various pathologies such as cancer, Alzheimer's disease, type 2 diabetes, and cardiovascular and neurodegenerative diseases. Polyphenols are also capable of reducing other risk factors for cardiovascular disease involved in metabolic syndrome (hyperglycemia, high lipid levels, insulin resistance, abdominal obesity, and high blood pressure) [14].

2.2. Antioxidant Activity

The antioxidant activity of the leaf, stem, and root of *L. rhodesiensis* extracts was evaluated using three different methods on methanolic extracts at different concentrations (200–1000 µg/mL). Ascorbic acid was used as a standard and its activity was evaluated under the same conditions as the extracts.

The results of the DPPH radical scavenging test (Figure 1) show that the root extract had the lowest antioxidant activity (50% inhibition concentration, IC_{50} value: 561.36 ± 3.93 µg/mL), with higher antioxidant properties in the leaf (449.53 ± 0.56 µg/mL) and stem (512.81 ± 1.41 µg/mL) extracts. The ascorbic acid standard had an IC_{50} value of 122.09 ± 0.56 µg/mL.

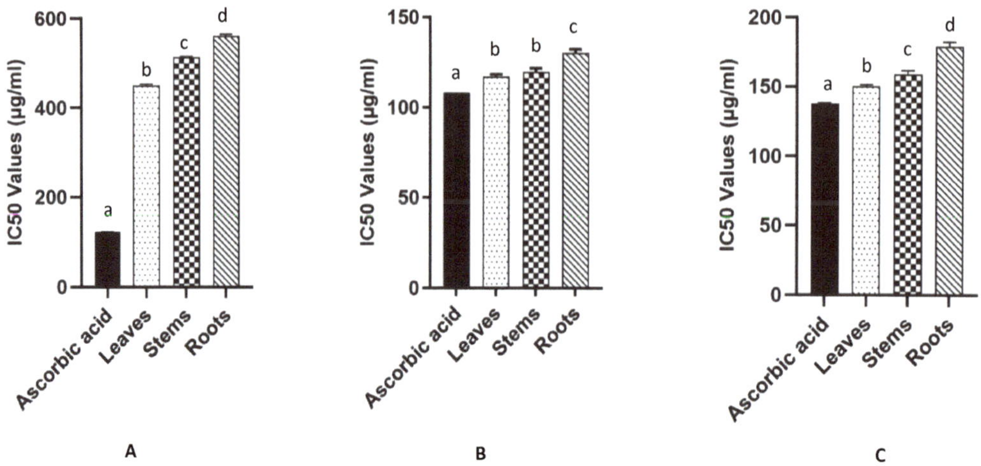

Figure 1. Antioxidant activity values of the methanolic extracts of *L. rhodesiensis* organs and ascorbic acid, (**A**). DPPH method, (**B**). FRAP (Ferric-reducing antioxidant power) method and (**C**). β-carotene method. Each result is the average of three values ($n = 3$). Histograms that do not share any letters are significantly different (p-value < 0.05).

The ability of phenolic compounds to reduce Fe^{3+} ions to Fe^{2+} was measured using the FRAP (Ferric-reducing antioxidant power) method. The results show that the leaf and stem extracts had low IC_{50} values (117.08 ± 1.1 µg/mL and 119.57 ± 2.17 µg/mL, respectively), similar to that of the ascorbic acid standard (108.01 ± 0.01 µg/mL), which confirms the ability of the extracts to reduce Fe^{3+} ions, to a greater extent than the root extract (130.04 ± 2.19 µg/mL). The same trend was highlighted with the β-carotene, test as the leaf and stem extracts had low IC_{50} values (150.18 ± 1.21 µg/mL and 158.91 ± 2.65 µg/mL, respectively), similar to that of ascorbic acid (IC_{50} = 137.55 ± 0.75 µg/mL), while the root extract IC_{50} was higher (178.92 ± 3.56 µg/mL).

Results obtained here are supported by a precedent study, showing that *L. rhodesiensis* methanolic extract dissolved in DMSO displayed a strong DPPH antioxidant activity, IC_{50} value of 5.96 ± 0.40 mg/mL [3]. The IC_{50} values obtained in the present study are different to that previous assay, but the methods used in those two studies were sensibly different (different plant parts, extracts preparation methods, dilution solvent, concentrations, volumes used, incubation time, etc.). In addition, a study has shown that using the same DPPH method, EtOH extract from the leaves of *Lantana montevidensis* showed lower antioxidant activity (IC_{50} = 290.5 ± 1.97 µg/mL) than aqueous extract (IC_{50} = 108.2 ± 3.4 µg/mL) [15]. Methanol extracts of leaves and flowers from *Lantana camara* were also already tested for their antioxidant potential, both extracts exhibiting high antioxidant and free radical scavenging activities with relatively stronger antioxidant activity in the case of whole flower extracts [16].

These three antioxidant tests show that the leaf and stem extracts of *L. rhodesiensis* have robust antioxidant activity, with interesting perspectives for their potential valorization as pharmaceuticals.

The therapeutic effects of medicinal plants are generally attributed to their phytochemicals. Specifically, many studies have correlated the antioxidant activity of plant extracts with the presence of phenolic compounds [17–20], as they are one of the main groups of molecules that act as primary antioxidants or free radical terminators [21]. The antioxidant potential of phenols is conferred by their hydroxyl (OH^-) group [18], which is directly linked to an aromatic hydrocarbon ring. This allows them to easily donate electrons to free radicals, and thus regulate their threat to living cells [22]. Generally, antioxidants (vitamins C, E, carotenoids, polyphenols) are important for good bone health. They neutralize reactive particles called free radicals that are associated with all inflammatory and painful phenomena [23]. The results obtained here show the high antioxidant properties of *L. rhodesiensis* extracts. More specifically, important quantities of phenols were highlighted in the leaf extract, showing higher antioxidant properties compared to the stem and root extracts. The differences between the values are in order with their phenolic content.

The results of the DPPH test showed a considerable difference between the antioxidant activity (IC_{50}) of the standard and each different organ extract studied, greater than the results of the other tests (FRAP and bleaching of β-carotene), where minor differences were highlighted. This may have been influenced by the method or test used for the evaluation of antioxidant activity, because each test has its specificities. The DPPH method is based on the measurement of antioxidant scavenging capacity towards the stable radical 2,2-diphenyl-1-picrylhydrazyl (DPPH). This method offers an easy and quick way to evaluate the anti-radical activities of antioxidants, since the radical compound is stable and does not have to be generated as in other radical scavenging tests [24]. The FRAP method is based on the ability of an antioxidant to transfer an electron to reduce any compound, including metals, carbonyl groups and radicals [25]. As for the β-carotene bleaching method, it is based on the ability of an antioxidant to neutralize free radicals generated by linoleic acid and to prevent the oxidation of β-carotene [26]. Indeed, phenolic compounds exert their antioxidant activity by several mechanisms, including the donation of hydrogen atoms to free radicals, or the trapping of other reactive species such as OH^-, NO^{2-}, N_2O_3, ONOOH, and HOCl. Some phenolic compounds, mainly di- and polyphenols, can react with O^{2-} or bind to transition metal ions (especially iron and copper). This often results in weakly active forms to promote free radical reactions [27,28].

Phenols play important roles in plants, such as protection against herbivores and insect pathogens. They are involved in cementing the material linking phenolic polymers to cell wall polysaccharides [29]. In addition, they play a role in the regulation of cell growth and division [13,30]. Flavonoids are the most common and most important group of naturally occurring phenolic compounds, probably because of their wide range of functions. Flavonoids generally act through a scanning or chelation process. Flavonoids act as antioxidants by breaking radical chains in more stable products in the membranes of liver microsomes. They also play an important role in instinctive protection against oxidative stress [21,31–33]. In the present study, a positive linear correlation was established between the content of phenolic compounds (total polyphenols including flavonoids and tannins; and total flavonoids) and the antioxidant activity of all extracts (Figure 2). The Pearson's correlation coefficient (r) and the coefficient of determination (R^2) were higher ($r = 0.9978$, $R^2 = 0.9956$) between the total polyphenolic content and DPPH activity than those of the total polyphenolic content and bleaching activity of β-carotene ($r = 0.9688$, $R^2 = 0.9386$), followed by the total polyphenolic content and FRAP activity ($r = 0.9230$, $R^2 = 0.852$). The correlation between the total flavonoid content and antioxidant capacity (DPPH test) was even higher ($r = 0.9879$, $R^2 = 0.9759$). A moderate correlation ($r = 0.8902$, $R^2 = 0.7924$) was observed for the total flavonoid content and bleaching activity of β-carotene. For FRAP activity, the correlation was lower ($r = 0.8153$, $R^2 = 0.6648$).

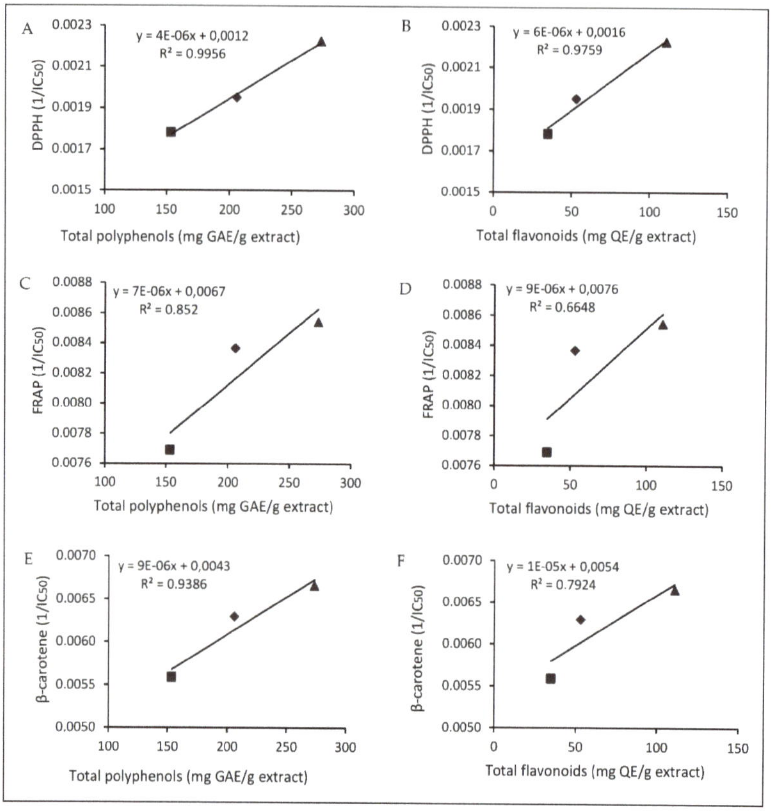

Figure 2. (**A**). Correlation between the total polyphenolic content and DPPH activity; (**B**). correlation between the total flavonoid content and DPPH activity; (**C**). correlation between the total polyphenolic content and FRAP activity; (**D**). correlation between the total flavonoid content and FRAP activity; (**E**). correlation between the total polyphenolic content and bleaching activity of β-carotene; (**F**). correlation between the total flavonoid content and bleaching activity of β-carotene; r: correlation coefficient and R^2: determination coefficient; roots (■), stems (◆), leaves (▲).

The correlation between total polyphenols and antioxidant activity was the strongest, indicating that a high phenolic content correlates with higher antioxidant activity. Phenolic compounds are produced differently depending on the plant species [34]. In addition, environmental factors, such as the drying technique, storage conditions, and the plant organ used as the source and the moisture content are parameters that could influence the phytochemical content of a plant [35,36]. Furthermore, the extraction process appears to affect the total phenolic content and antioxidant activity of the plants [3,37–41].

A positive linear correlation was also established between the three different methods used to evaluate antioxidant activity in this study. The Pearson correlation coefficient (r) for the DPPH and β-carotene assays (0.9558) was higher than that of the DPPH and FRAP assays (0.9140). However, the correlation between the FRAP method and the bleaching of β-carotene had a coefficient of 0.9929. These results indicate that the antioxidant activity values tested by the three different methods are highly correlated. Those results were expected as several others studies on plant extracts have confirmed the relationship between antioxidant activities and polyphenolic compounds [42–44].

2.3. Anti-Malarial Activity

The hydro-methanolic extracts from the different *L. rhodesiensis* organs were tested on a chloroquino-sensitive strain (3D7) of *Plasmodium falciparum* in order to evaluate their in vitro anti-malarial activity. Artemisinin was used as a positive control. The concentration that inhibited 50% of the strain (IC_{50}) was determined using sigmoidal curves for each extract (Table 3). The hydro-methanolic leaf extract was found to be active against *Plasmodium falciparum* strain 3D7, while the stem and root extracts were inactive. These results highlight, for the first time, the possible value of *L. rhodesiensis* leaves in traditional medicine for the treatment of malaria. *L. camara* leaves, a plant of the same genus, has been shown to have an IC_{50} value similar to that found in this study [45,46], highlighting the interest of this plant genus for the treatment of malaria and encouraging further studies.

Table 3. Results of the anti-malarial activity of the different extracts obtained by non-sequential extraction (50% inhibition concentration, IC_{50}).

Extract ($MeOH/H_2O$)	3D7, IC_{50} (μg/mL)
Leaves	12.5 ± 2.5
Stems	>100
Roots	>100
Artemisinin	0.004 ± 0.001

Some studies argue that major phytochemical groups such as flavonoids, tannins, saponins, coumarins, alkaloids, triterpenes, sesquiterpenes and steroids [47–49] may be responsible for the anti-malarial activity observed in some plants. As an example, *L. camara* aqueous and ethanolic leaf extracts have shown antimalarial activity close to that of the standard drug chloroquine. In that study, alkaloids, cardiac glycosides, saponins, carbohydrates, flavonoids, steroids, tannins, and terpenoids were present in the different extracts [50]. These phytochemical groups with anti-malarial potential are present in the leaves of *L. rhodesiensis*, as previously described.

2.4. Determination of the HPLC-PDA (Photodiode Array Detector) Polyphenol Profile Leaf, Stem, and Root Extracts

The *L. rhodesiensis* leaf, stem, and root extracts were analyzed by HPLC-PDA. The major phenolic compounds were identified in each extract (Figure 3) based on their retention index and their PDA spectrum with comparison to a library. Results showed that all extracts were characterized by high quantities of isomers of acteoside, a phenolic molecule well known for its wide range of biological properties including anti-inflammatory, antioxidant and hepatoprotective activities [51–55].

2.5. Structural Elucidation

Compound **1** was obtained as yellow needles. The protonated mass, measured by LC/MS in positive mode electrospray ionization, was 346.9 $[M + H]^+$, corresponding to the formula $C_{17}H_{14}O_8$. The ^{1}H- and ^{13}C-NMR data for compound **1** were quite similar to those of compound **4**. In the ^{1}H NMR spectrum of compound **1** (Table 1), a signal at δ_H 7.12 (2H, s) was attributed to two protons (H-2′, H-6′) of the B ring indicating oxygenation at C-3′, C-4′ and C-5′. Two singlets at δ_H 6.81 (1H) and at δ_H 6.59, were assigned to the H-8 and H-3 protons, respectively. These data as well as the intense signals at d 3.99 and 3.95 (both 3H, s), relative to two OCH_3 groups, suggested presence of a tetrahydroxyflavone with two additional methoxyl group substitutions [56,57]. The ^{13}CNMR spectrum of compound **1** (Table 1) shows values between 130–155 suggesting an oxygenated A-ring. After careful analysis of 2D NMR, the hydrogen group at C-5′ in 3 was replaced by a hydroxy group in 1. So, in the HMBC spectrum, cross-peaks disclosing the bonding site of each methoxyl were observed: δ_H 3.99 correlated with δ_C 154.4 (C-7), and δ_H 3.95 correlated with δ_C 148.5 (C-4′). Correlations were also observed between H-3/C-1′, C-2, C-4 and C-10, H-8/C-6, C-7, C-9 and C-10, H-2′ and H-6′/ C-2, C-1′, C-2′, C-4′, C-5′, C-6′. Consequently, the structure of compound **1** was determined to be the new 5,6,3′,5′-tétrahydroxy-7,4′-dimethoxyflavone, named rhodescine (Figure 1).

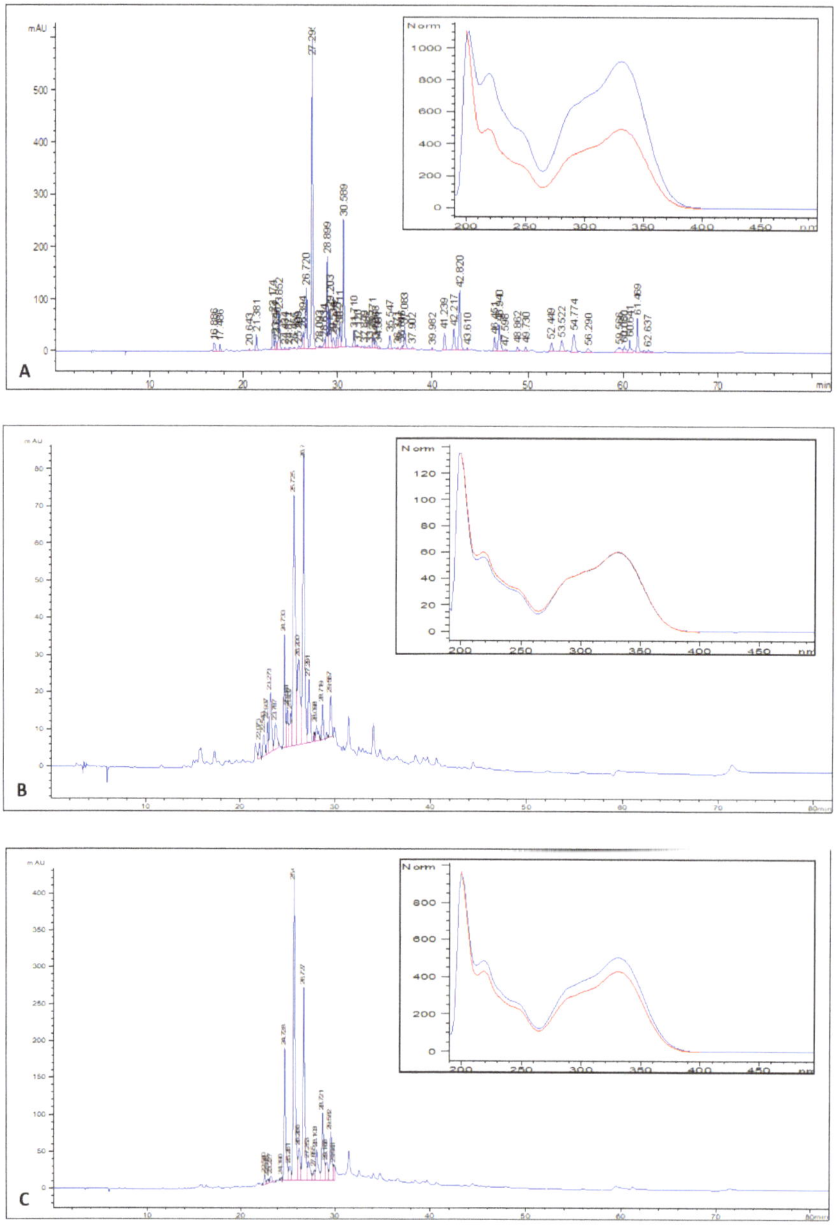

Figure 3. HPLC-PDA (photodiode array detector) chromatograms of the chemical profiles of polyphenols occurring in the studied extracts of leaves (**A**), stems (**B**), and roots (**C**). For each extract, the insert presents the PDA spectrum of the major peak and the PDA spectrum of the acteoside reference (in red).

Compound **2**, a white amorphous powder, possessed a molecular formula of $C_{20}H_{20}O_8$ based on the protonated ion peak at m/z 388.9 by LC/MS, indicating seven degrees of unsaturation. The ^{1}H-NMR (proton nuclear magnetic resonance) (Table 4) displayed resonances for three singlets at δ_H 7.12, 6.64, and 6.61 ppm, suggesting aromatic ring hydrogens, and five singlets between δ_H 3.8 and 4.1 ppm, integrating for the three protons characteristic of a methoxy group. The ^{13}C-NMR (carbon-13 nuclear magnetic resonance, J-modulated) data exhibited in total 20 carbon resonances attributed to one ketone carbonyl carbon (δ_C 182.8), 11 quaternary carbons (singlet, δ_C 165–106), four carbons (doublets, δ_C 105–90) suggesting C-H bonds and five carbons (singlet, δ_C 56–62) characteristic of carbons linked to a methoxy group. Thus, the structure of compound **1** was established as 5-hydroxy-6,7,3′,4′,5′-pentamethoxyflavone (Figure 4). This result was compared to [58].

Table 4. ^{1}H and ^{13}C-NMR spectroscopic data for compounds **1–4** (500 MHz, MeOD).

	1		**2**		**3**		**4**	
	δ_H m (*J* in Hz)	**δ_C**	**δ_H m (*J* in Hz)**	**δ_C**	**δ_H m (*J* in Hz)**	**δ_C**	**δ_H m (*J* in Hz)**	**δ_C**
2		165.1		164.1		165.3		165.0
3	6.59, s	102.5	6.64, s	105.3	6.76, s	104.1	6.67, s	102.4
4		182.8		182.8		182.6		182.9
5		150.9		153.4		151.8		150.6
6		130.0		132.7		131.7		130.0
7		154.4		159.1		159.4		154.4
8	6.81, s	90.4	6.61, s	91.0	6.89, s	92.7	6.85, s	90.6
9		150.6		152.7		154.9		150.6
10		105.2		106.2		105.4		105.2
1′		121.4		126.5		122.0		122.4
2′	7.12, s	101.7	7.12, s	103.9	7.14, d (8.5)	111.1	7.52, d (2.13)	109.2
3′		138.5		153.7		152.5		150.7
4′		148.5		141.5		151.0		148.1
5′		145.7		153.7	7.55, d (2.2)	108.8	6.94, d (8.3)	115.4
6′	7.12, s	107.3	7.12, s	103.9	7.68, dd (8.5–2.2)	119.9	7.54, dd (8.3–2.13)	120.4
6-OCH$_3$			4.00, s	61.1	3.85, s	61.1		
7-OCH$_3$	3.99, s	55.5	3.93, s	56,5	4.00, s	57.0	4.02, s	55.6
3′-OCH$_3$			3.98, s	56,5	3.93, s	56.8		
4′-OCH$_3$	3.95, s	55.5	3.94, s	60.9	3.96, s	56.6	3.99, s	55.3
5′-OCH$_3$			3.98	56,5				

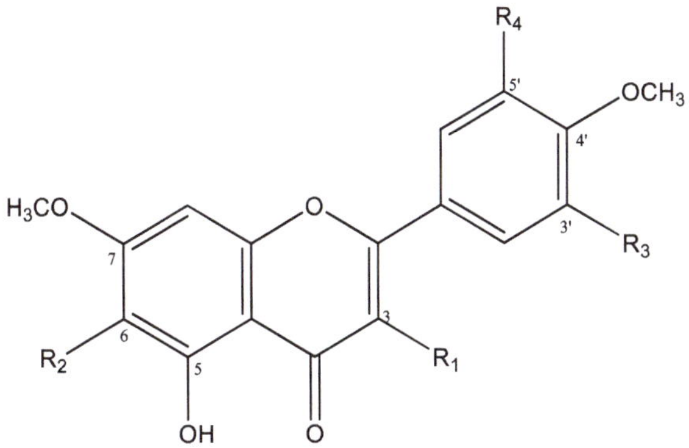

Compounds	R$_1$	R$_2$	R$_3$	R$_4$
1	H	OH	OH	OH
2	H	OCH$_3$	OCH$_3$	OCH$_3$
3	H	OCH$_3$	OCH$_3$	H
4	H	OH	OH	H

Figure 4. Molecular structure of purified flavones.

Compound **3** was found as colorless needles. Its molecular formula was $C_{19}H_{18}O_7$ according to the protonated ion peak at CNMRm/z 358.1 $[M + H]^+$. The ^{1}H-NMR spectrum exhibited signals for five aromatic ring hydrogens (δ_H 6.76 (1H, s), 6.89 (1H, s), 7.14 (1H, d, J = 8.5), 7.55 (1H, d, J = 2.2) and 7.68 (1H, dd, J = 8.5; 2.1)) and four methoxy groups [δ_H 3.85 (3H, s), 4.00 (3H, s), 3.93 (3H, s) and 3.96 (3H, s)] (Table 4). The ^{13}C-NMR spectrum of compound **2** exhibited 19 carbon resonances (Table 4). The ^{13}C-NMR data for compound **3** were quite similar to those of compound **2**. However, on the ^{13}C-NMR spectrum of compound **3**, there were four signals characteristic of the carbons of the methoxy group. Compound **3** was identified as 5-hydroxy-6,7,3′,4′-tétraméthoxyflavone (Figure 4). The data were compared to [59].

Compound **4** was obtained as yellow needles. The high-resolution mass spectrum of compound **3** in positive mode electrospray ionization generated a protonated ion peak at CNMRm/z 331.0862. This is compatible with the elemental composition $C_{17}H_{14}O_7$. The ^{1}H-NMR spectrum showed signals for five aromatic rings (δ_H 6.67 (1H, s), 6.85 (1H, s), 6.94 (1H, d, J = 8.3), 7.54 (1H, dd, J = 8.3–2.13), and 7.52 (1H, d, J = 2.13)) and two methoxy groups (δ_H 4.00 (3H, s) and 3.97 (3H, s)) (Table 4). The ^{13}C-NMR spectrum of compound **3** exhibited in total 17 carbon resonances (Table 4) attributed to one ketone carbonyl carbon (δ_C 182.9), 10 quaternary carbons (δ_C 167–104), five carbons (δ_C 122–92) suggesting C-H bonds and two carbons of methoxy groups (δ_C 58.8 and 57.5). Compound **4** was characterized as 5,6,3′-trihydroxy-7,4′-dimethoxyflavone. The results were compared to [57].

All data of compounds **2**, **3**, and **4** were in good agreement with the respective literature data.

Compound **2** (5-hydroxy-6,7,3′,4′,5′-pentamethoxyflavone) has been previously reported in extracts obtained from the whole *L. rhodesiensis* plant. This molecule shows interesting anti-proliferative and pro-apoptotic properties [12]. Compounds **3** and **4** were already reported in various plant organs from other genera [57,59–65], but were observed here for the first time in *L. rhodesiensis* (Figure 4).

In order to explain if the high antioxidant properties of *L. rhodesiensis* leaf extracts originates from the presence of those molecules in high proportions, the antioxidant activities of purified compounds **1**, **2**, **3** and **4** were evaluated in the present study. The results showed that at a concentration of 1 mg/mL, compound **1** (97.92 ± 0.20%) inhibited DPPH better than compounds **2** (0.57 ± 0.04%), **3** (1.98 ± 0.64%) and **4** (61.77 ± 3.53%). The inhibition by compound **1** was similar to that of the standard drug used (ascorbic acid; %I − 98.50 ± 0.56).

3. Materials and Methods

3.1. Plant Materials

Leaves, stems, and roots of *Lantana rhodesiensis* (*L. rhodesiensis*) were collected from the north of Côte d'Ivoire at Kapélé (9°25′60″ N, 5°42′0″ W). Sample collection occurred in the morning, from 9:00 to 12:00 a.m. The plant material was identified by Professor Ake Assi and a voucher specimen (N° UCJ017435) was deposited at the Centre National de Floristique (CNF, Abidjan, Côte d'Ivoire). Each plant organ was dried during one week at room temperature (25°C) and was subsequently ground into a fine powder using mechanic ball mill, type BB-27 (E2ME). The final particle size was from a few tens to a few hundred micrometers and the moisture content was 8.5 ± 0.18% for leaves; 7.52 ± 0.64% for stems; and 6.01 ± 0.09% for roots. To determine the moisture content, the sample (organ powder) was weighed to the nearest 10 mg and dried in a drying oven at 70 °C. After 48 h, the moisture content was determined as followed:

$$\%\ Moisture = \frac{m1 - m2}{m1} \times 100$$

where $m1$ is mass of the organ powder before drying and $m2$ is mass of the organ powder after drying (n = 3).

3.2. Reagent and Solvents

All reagents and solvents were either HPLC or analytical grade. Moreover, 2,2-diphenyl-1-picrylhydrazyl (DPPH), gallic acid (98%), quercetin (98%), artemisinin (98%), Folin–Ciocalteu reagent, ascorbic acid, β-carotene, and linoleic acid were purchased from Sigma-Aldrich (St. Louis, MO, USA). Anhydrous sodium sulfate, potassium ferricyanide, trichloroacetic acid (TCA), ferric chloride, sodium chloride, potassium chloride, disodium hydrogen phosphate, potassium dihydrogen phosphate, and hydrochloric acid were bought from VWR Chemicals (Radnor, PA, USA). Hexane, dichloromethane ($CHCl_3$), ethyl acetate (EtOAc) and methanol (technical and HPLC) were purchased from VWR International (Fontenay-sous-Bois, Val-de-Marne, France).

3.3. Determination of Phytochemical Classes

The different groups of compounds (sterols/terpenes, polyphenols, flavonoids, tannins, alkaloids, saponins, leuco-anthocyanins, and anthocyanins) present in *L. rhodesiensis* leaf, stem, and root powders or extracts were identified using the methods described by Bekro et al., Bidie et al., and Nineza Claire and Nkengurutse Jacques [61–63].

To highlight sterols and terpenes, the reagent of Liebermann was used. Five mL of each organ extracts were evaporated on a water bath (100 °C). The residue was dissolved in 1 mL of acetic anhydride and 0.5 mL of concentrated sulfuric acid was added. The appearance of a purple and violet ring at the interphase, turning blue and then green, indicated a positive reaction. The positive standard used is the cholesterol.

In order to highlight polyphenols, the reaction with ferric chloride ($FeCl_3$) was used. To 2 mL of each extract, a drop of 2% ferric chloride alcoholic solution was added. In the presence of polyphenol derivatives, ferric chloride causes the appearance of a dark blue-blackish or green coloration. The control is carried out with the alcoholic solution of gallic acid.

To highlight flavonoids, the "cyanidin" reaction was used. Two mL of each extract were evaporated and the residue was taken up in 5 mL of hydrochloric alcohol diluted twice. Then, three magnesium shavings were added and a pinkish-orange or purplish coloration was observed. By adding three drops of isoamyl alcohol, the coloration was intensified. This confirmed the presence of flavonoids. An alcoholic solution of quercetin was used as a control.

The leuco-anthocyanins were characterized by performing the same reaction as for the identification of flavonoids without the addition of magnesium shavings by heating for 15 min in a water bath. The appearance of a cherry-red or purplish coloration indicates the presence of leuco-anthocyanins.

To characterize anthocyanins, 5 mL of sulfuric acid and then 5 mL of ammonium hydroxide are added to 5 mL of the extracts. If the coloration is accentuated by acidification and then changes to purplish blue in basic medium, the presence of anthocyanins can be concluded.

Catechic tannins were identified by Stiasny reagent (formol 30%, concentrated HCl: 1/0.5). Five mL of each extract was evaporated. After 15 mL of Stiasny reagent were added to the residue. The mixture was then kept in a water bath at 80 °C for 30 min. The observation of a large flaky precipitate characterized the catechin tannins. The obtained solution was filtered and the collected filtrate was saturated with sodium acetate. The addition of three drops of 2% $FeCl_3$ caused the appearance of an intense blue-black coloration, indicating the presence of gallic tannins. An alcoholic solution of gallic acid was used as a control.

In order to highlight alkaloids, the Dragendorff (iodobismuthate) reagent was used. Six mL of each extract were evaporated to dryness. The residue was taken up again in 6 mL of alcohol at 60 °C. The addition of two drops of the Dragendorff reagent to the alcohol solution caused a precipitate or an orange coloration and indicated a positive reaction.

To highlight the saponins we used the method of foam appearance by agitation. A height of persistent foam, higher than 1 cm indicates the presence of saponosides.

3.4. Extract Preparation for the Determination of Total Phenolic and Flavonoid Contents and Tests for Biological Activity

A methanol:water (50:50, v/v) extract was obtained by stirring 100 g of each sample with 1.5 L of the solvent mixture at 25 °C and 150 rpm for 48 h. The extract was then filtered twice through cotton and once through WATTMAN 3 mm filter paper. The solvent was then evaporated at 40°C using a rotary evaporator and the residue was subsequently lyophilized. The obtained powder was used to carry out the biological tests. The extracts were prepared on the basis of the method described by MacDonald et al. [18].

3.5. Determination of the Total Phenolic Content

The total phenolic content of the extracts was evaluated using the Folin-Ciocalteu method according to the Shahidi and Naczk procedure [66] with MacDonald et al. modifications [18]. Briefly, a gallic acid calibration curve was established (0, 50, 100, 150, 200, 250 mg/L) in methanol:water (50:50, v/v). Leaf, stem and root extracts were prepared in methanol:water (50:50, v/v) at a concentration of 3 mg/mL. Then, 0.5 mL of each sample or phenolic standard was mixed with 2.5 mL of Folin–Ciocalteu reagent (diluted 1:10 with distilled water) and 2 mL of aqueous sodium carbonate solution (1 M). The tubes were allowed to stand for 15 min at room temperature before the absorbance of the mixture was measured at 765 nm using a spectrophotometer. All determinations were performed in triplicate.

The total phenolic content was calculated as gallic acid equivalents (GAE) by the following:

$$T = C \times \frac{V}{M}$$

where T is the total phenolic content in mg/g of the extracts as GAE, C is the concentration of gallic acid established from the calibration curve in mg/mL, V is the volume of the extract solution in mL and M is the weight of the extract in g.

3.6. Determination of the Total Flavonoid Content

The total flavonoid content of the extract was determined as previously described [31,42,67]. Different concentrations (0.01–0.2 mg/mL) of quercetin, the standard molecule, were prepared in methanol. Organ extracts were also diluted in methanol (3 mg/mL). Then, 0.5 mL of methanolic samples and standards was added to 0.5 mL of aluminum chloride 10% (w/v). The same volume of sodium acetate (1 M) was added to the solution, which was then brought up to 3500 µL with distilled water. After incubation at room temperature for 30 min, the absorbance was measured at 415 nm. All determinations were carried out in triplicate. The total flavonoid content (TFC) is presented as mg of quercetin equivalents (QE) per gram of the extract.

The total flavonoid content was calculated as quercetin equivalents (QE) by the following:

$$T = C \times \frac{V}{M}$$

where T is the total flavonoid content in mg/g of the extracts as QE, C is the concentration of quercetin established from the calibration curve in mg/mL, V is the volume of the extract solution in mL and M is the weight of the extract in g.

3.7. Antioxidant Activity

3.7.1. DPPH Radical Scavenging Assay

The antioxidant activities were measured in terms of hydrogen-donating or radical-scavenging ability, using the stable radical 2,2-diphenyl-1-picrylhydrazyl (DPPH) as a reagent [67]. To do so, various concentrations (200, 400, 600, 800, 1000 µg/mL) of each *L. rhodesiensis* organ extract and ascorbic acid were prepared in methanol. Then, 50 µL of each sample concentration was added to 2 mL of a 0.004% (w/v) DPPH methanolic solution. After 30 min of incubation at room temperature in the dark, absorbance was measured at 517 nm using an Ultrospec UV-visible dual beam spectrophotometer (GE Healthcare,

Cambridge, UK). A blank sample containing the same amount of methanol and DPPH solution was used as the negative control. All determinations were performed in triplicate.

The inhibition percentage (%I) of the DPPH radical by the samples was calculated according to the formula [68]:

$$\%I = \frac{Ab - Aa}{Ab} \times 100$$

where *Ab* is the absorbance of the blank sample and *Aa* is the absorbance of the test sample.

The inhibition percentage was plotted versus the sample concentration to obtain the IC_{50} index.

3.7.2. Reducing Power

The reducing power of the extracts and a standard (ascorbic acid) was determined by mixing 1 mL of the extract or standard at different concentrations in methanol (200 to 1000 μg/mL) with 1 mL of phosphate buffer (0.2 M, pH 6.6) and 1 mL of potassium ferricyanide [$K_3Fe(CN)_6$] solution (1%, *w*/*v*). The mixture was incubated at 50 °C for 20 min. After incubation, 1 mL of trichloroacetic acid (TCA) (10% *v*/*v*) was added to the solution to stop the reaction. This solution was then centrifuged at 3000 g for 10 min at room temperature. The supernatant was recovered and mixed with distilled water (1.0 mL) and 0.1% $FeCl_3$ (150 μL). Then, the absorbance was measured at 700 nm. Higher absorbance of the reaction mixture (according to the blank) indicates greater reducing power. This determination was made according to a published protocol [69] with some modifications. The inhibition percentage (%I) was calculated according to the formula:

$$\%I = \frac{Ab - Aa}{Ab} \times 100$$

where *Ab* is the absorbance of the blank sample and *Aa* is the absorbance of the test sample.

The inhibition percentage was plotted versus the sample concentration to obtain the IC_{50} index.

3.7.3. β-Carotene Blanching Test

The β-carotene/linoleic acid test evaluates the inhibitory effect of a compound or a mixture on β-carotene oxidation in the presence of molecular oxygen (O_2) and gives an estimation of the antioxidant potential of the sample.

As previously described [70], a mixture of β-carotene and linoleic acid was prepared by adding together 0.5 mg of β-carotene, 25 μL of linoleic acid and 200 mg of Tween-40 in 1 mL of chloroform. The chloroform was then completely evaporated under vacuum and 100 mL of oxygenated water was subsequently added to the residue and mixed to form a clear yellowish emulsion. Then, 350 μL of various sample concentrations (200, 400, 600, 800, 1000 μg/mL) in methanol (extracts and ascorbic acid) was added to 2.5 mL of the above emulsion and mixed. The test tubes were incubated in a water bath at 50 °C for 2 h together with a negative control (blank) containing pure methanol instead of sample. The absorbance values were measured at 470 nm.

The antioxidant activity (percentage inhibition, % I) of the samples was calculated as follows:

$$\%I = \frac{A\ (\beta - carotene\ after\ 2\ h\ assay)}{A\ (initial\ \beta - carotene)} \times 100$$

where *A (β-carotene after* 2 *h assay)* is the absorbance value of β-carotene remaining in the samples, after the 2 h assay whereas *A (initial β-carotene)* is the absorbance value of β-carotene in the freshly prepared standard solution. The activity was calculated as 50% inhibition concentration (IC_{50}). All experiments were repeated three times on independent samples and the data are expressed as mean ± standard deviation (SD).

3.8. Anti-Malarial Activity

The anti-malarial activity was determined as previously described [71]. The asexual erythrocyte stages of *P. falciparum*, chloroquine-sensitive strain 3D7 were maintained in continuous in vitro culture, according the procedure of Trager and Jensen. The host erythrocytes were A+ human red blood cells obtained from a patient from Schiphol in the Netherlands (BEI Reagent Search) [72]. Crude extract solutions were prepared in DMSO (Sigma-Aldrich, Darmstadt, Germany, D-4540) at 10 mg/mL (or 1 mg extract diluted in 100 μL DMSO). The extract solutions were diluted 10 times in ready-to-use culture medium to give a 1 mg/mL solution. In a 96-well plate, each test sample was applied in a series of eight two-fold dilutions and tested in triplicate. Parasitemia was 2% and hematocrit was 1%, as described by Murebwayire et al. [73]. Infected red blood cells were used as a positive growth control and unparasitized red blood cells were used as a negative (blank) control. Artemisinin 98% (Sigma-Aldrich, Machelen, Belgium) at an initial concentration of 100 ng/mL was used as a positive control in all experiments. The plate was incubated for 48 h at 37 °C in a hermetically sealed culture dish impregnated with a GENbox microaer gasbag (bioMerieux, 96125) to generate a microaerobic medium. It was then kept at –20°C for 24 h after the 48-h incubation and thawed at 37 °C for 45 min. Then, 20 μL of each homogenized well was transferred to a new 96-well plate and 100 μL of a solution consisting of 1 mL Triton X-100 (Sigma, X100), 10 mg saponin (Merck, A18820), 1 g lithium L-lactate (Sigma, L2250), and 200 mg APAD (Sigma, A5251)/100 mL TRIS pH 8 buffer (Sigma, Darmstadt, Germany, T6664) was added. The new 96-well plate was incubated for 15 min at 37 °C, then 20 μL of a solution mixture prepared from 1 mL of a NTB solution (nitro-blue tetrazolium chloride; Sigma, Darmstadt, Germany, N6639) (2 mg/mL) in distilled water and 1 mL of a PES solution (phenazine ethosulfate; Sigma, P4544) (0.1 mg/mL) in TRIS pH 8 buffer were added protected from light and incubated for 30 to 45 min at 37 °C. Parasite growth was estimated by the determination of lactate dehydrogenase (LDH) activity, using the colorimetric method described in 1993 by Makler et al. [74,75]. Absorbance was measured with a spectrophotometer (Stat Fax 2100, Fisher, Illkirch, France) at 630 nm. The intensity of coloration is proportional to the amount of enzyme present in the reaction medium and, thus, to the amount of parasites. The IC_{50} values were calculated from the graphs.

3.9. Statistical Analysis of Biological Data

Data are expressed as means ± S.D; for the statistical analysis, ANOVA followed by Tukey's test was used GraphPad Prism. Statistically significant differences were considered for *p*-values < 0.05. Pearson's method was used to determine correlations.

3.10. Extraction, Isolation, and Characterization of Compounds

To better understand the origin of the reported biological activities of the leaf hydromethanolic extract of *L. rhodesiensis*, the four flavones present in highest concentration in the methanolic extract were isolated and characterized. To do so, 1.0 kg of dry *L. rhodesiensis* leaves was ground and mixed with 15 L MeOH/H_2O (50/50, *v*/*v*). The extract solution was concentrated under vacuum to give a dark-brown residue (122 g). Eighty-five grams of the residue was suspended in MeOH/H_2O (850 mL). This solution was successively partitioned with solvents of increasing polarity such as hexane, dichloromethane ($CHCl_3$) and ethyl acetate (EtOAc) (1.0 L: 2 × 500 mL each). The EtOAc fraction (22.7 g) was purified over a silica gel CC with stepwise $CHCl_3$-EtOAc solvents (30:0 to 70:100) and stepwise EtOAc-MeOH solvents (95:0 to 5:100) to obtain eight subfractions (FAE.1-FAE.8), after combining the eluates on the basis of TLC (Thin-layer chromatography) analysis. Subfraction FAE.2 (2.5 g) was separated using preparative HPLC with an ACN-H_2O + 0.1% HFO (20–40% ACN) solvent system for 25 min to afford compound **1** (4.8 mg). In addition, the $CHCl_3$ fraction (10.7 g) was purified on a silica gel chromatographic column (CC) with a gradient of $CHCl_3$-EtOAc solvents (70:0 to 30:100) and stepwise EtOAc-MeOH solvents (90:0 to 10:100) to obtain 24 eluates. Deposits were obtained from subfractions FD1p, FD2p,

FD3p, FD5p, FD6p, FD7p, FD12p, and FD13p. Other subfractions (FDM.1-FDM.10) were obtained after eluates were combined according to the TLC analysis. The subfractions FD1p, FD2p and FD3p were separated using preparative HPLC with an ACN-H_2O + 0.1% H_3PO_4 (20–100% ACN) solvent system for 25 min to afford compound **2** (16.7 mg) and compound **3** (7.0 mg). Subfraction FDM.5 was also subjected to preparative HPLC with an ACN-H_2O + 0.1% H_3PO_4 (35–45% ACN) solvent system to yield compound **4** (13.0 mg)

The structures of the compounds were established by spectral analysis, mainly HR ESI-MS, Q-TOF, ^{1}H, ^{13}C and 2D-NMR (COSY, HSQC, and HMBC), as well as by comparing their spectroscopic data with those reported in the literature.

Compound **1** (yellow needles):

^{1}H-NMR (MeOH-d4, 500 MHz) δ: 7.12 (2H, s, H-2′, H-6′), 6.81(1H, s, H-8), 6.59 (1H, s, H-3), 3.99, 3.95 (3H each, both s, OMe-7, OMe-4′). ^{13}C-NMR (MeOH-d4, 126 MHz) δ: 182.8 (C-4), 165.1 (C-2), 154.4 (C-7), 150.6 (C-9), 150.9 (C-5), 148.5 (C-4′), 145.7 (C-5′), 138.5 (C-3′), 102.5 (C-3), 130.0 (C-6), 121.4 (C-1′), 107.3 (C-6′), 105.2 (C-10), 101.7 (C-2′), 90.4 (C-8), 55.5, 55.5 (OMe-7, OMe-4′). HR-ESI-MS m/z 346.9 ((M + H)$^+$, 100%).

Compound **2** (white, amorphous powder):

^{1}H-NMR (MeOH-d4, 500 MHz) δ: 7.12 (2H, s, H-2′, H-6′), 6.64*, 6.61* (1H, each, s, H-3, H-8), 4.00, 3.98, 3.94, 3.93 (3H, 6H, 3H, 3H, s, OMe). ^{13}C-NMR (MeOH-d4, 126 MHz) δ: 180.8 (C-4), 164.17 (C-2), 159.09 (C-7), 153.69 (C-3′, C-5′), 153.48 (C-5), 152.73 (C-9), 141.53 (C-4′), 132.76 (C-6), 126.58 (C-1′), 106.19 (C-10), 105.38 (C-3), 103.96 (C-2′, C-6′), 91.00 (C-8), 61.15 (OMe-6), 60.98 (OMe-4′), 56,50 (OMe-3′, OMe-5′, OMe-7); HR-ESI-MS m/z: = 388.9 (M + H)$^+$. An asterisk (*) means that the values may be interchanged.

Compound **3** (colorless needles):

^{1}H-NMR (MeOH-d4, 500 MHz, d, ppm, J/Hz): 3.96 (3H, s, OMe-4′), 3.93 (3H, s, OMe-3′), 3.85 (3H, s, OMe-6), 4.00 (3H, s, OMe-7), 6.76 (1H, s, H-3), 6.89 (1H, s, H-8), 7.55 (1H, d, J = 2.2, H-5′), 7.14 (1H, d, J = 8.5, H-2′), 7.68 (1H, dd, J = 8.5; 2.1, H-6′). ^{13}C-NMR (MeOH-d4, 126 MHz, d) 56.6 (OMe-4′), 56.8 (OMe-3′), 57.0 (OMe-7), 61.1 (OMe-6), 92.4 (C-8), 104.7 (C-3), 106.8 (C-10), 110.6 (C-5′), 112.8 (C-2′), 124.8 (C-1′), 121.7 (C-6′), 133.8 (C-6), 151.0 (C-4′), 154.1 (C-3′), 151.8 (C-5), 154.9 (C-9), 161.0 (C-7), 166.0 (C-2), 184.2 (C-4). Mass spectrum Q-TOF, $C_{19}H_{18}O_7 m/z$ 358.1 (M + H)$^+$, Wiley library score 97.65%.

Compound **4** (yellow needles):

^{1}H-NMR (MeOH-d4, 500 MHz) δ: 7.52 (1H, d, H-2′), 7.54 (1H, dd, H-6′), 6.94 (1H, d, H-5′), 6.85 (1H, d, H-8), 6.67 (1H, d, H-3), 4.00, 3.97 (3H, s, OMe-7, OMe-4′). ^{13}C-NMR (MeOH-d4, 126 MHz) δ: 182.9 (C-4), 165.0 (C-2), 154.4 (C7), 150.7 (C-3′), 150.6 (C-5, C-9), 148.1 (C-4′), 130.0 (C-6), 122.4 (C-l′), 120.4 (C-6′), 115.4 (C-5′), 109.2 (C-2′), 105.2 (C-10), 102.4 (C-3), 90.6 (C-8), 55.6 (OMe-7), 55.3 (OMe-4′). HR-ESI-MS m/z 331.0862 ((M + H)$^+$, 100%), Q-TOF $C_{17}H_{14}O_7$.

The antioxidant potential of all purified compounds was determined based on the DPPH method.

3.11. General Procedure for the Determination of Compounds

^{1}H and ^{13}CNMR (proton and carbon nuclear magnetic resonance) spectra were recorded in MeOH-d4 on a Bruker NEO 500 MHz spectrometer equipped with a cryoprobe. 2D experiments were performed using standard Bruker microprograms. The heteronuclear multiple bond correlation (HMBC) spectrum analysis allowed us to confirm the position of carbonyl, methoxy and C-H correlations. The heteronuclear single quantum coherence (HSQC) spectrum analysis allowed us to identify C-H correlations. LC/MS was carried out on a Thermo Scientific LTQ orbitrap XL mass spectrometer with an ESI source in positive mode with an RP select B LiChrospher 60 (250 mm × 4.6 mm, 5 μm) column. A chromatography Interchim puriFlash 4250 equipped with a Büchi fraction collector C-660 unit was used to accomplish the preparative isolation. An Agilent Technologies G1311B 1260 quant pump apparatus equipped with a PDA detector and a C18 column (Agilent, Santa Clara, CA, USA, Eclipse XDB-C18; 3.5 μm; 4.6 × 150 mm) were employed for analytical HPLC. HPLC-PDA determination of the chemical profiles of polyphenols in the studied extracts of leaves, stems, and roots were performed using a gradient of methanol and 0.05% trifluoroacetic acid (1 mL/min). Analytical TLC was used during the extraction and purification procedures in order to confirm the presence of polyphenol and flavonoid molecules in the different fractions. TLC was performed on pre-coated Silica gel 60 F254 (Merck, Hohenbrunn, Germany) plates. After development with EtOAc/formic acid/HOAc/H_2O (100:11:11:26), the dried plates were sprayed with NP-PEG [natural product reagent (1% diphenylboryloxyethylamine in MeOH) and polyethylene glycol 4000 (5% polyethylene glycol 4000 in EtOH)]. The plates were dried again and examined under ultraviolet light (366 nm).

4. Conclusions

In the present study, different *L. rhodesiensis* organs were submitted to hydro-methanolic extraction and the antioxidant and anti-malarial activities of those extracts were evaluated. The present study showed variable results depending on the plant organ. Leaf and stem extracts showed an interesting phenolic compound content correlated with robust antioxidant and anti-malarial activities, while the root extract displayed lower activities. As antioxidant molecules are able to neutralize reactive particles called free radicals that are associated with inflammatory and painful phenomena, the antioxidant activities of *L. rhodesiensis* extracts support the claim regarding the traditional use of this plant for the treatment of various affections, such as rheumatism.

Hence, *L. rhodesiensis* is a potential source for isolating new exogenous antioxidant and anti-malarial molecules. Moreover, this is the first report on the in vitro anti-malarial activity of *L. rhodesiensis*. Four compounds were isolated from the *L. rhodesiensis* leaf extract. Compounds **1**, **3**, and **4** were reported for the first time in this plant. Compound **1**, which displayed the highest number of free hydroxyl groups on the benzene rings among all the purified molecules, had a high antioxidant potential, whereas compound 4 displayed an average potential. This study reported one new flavone isolated from the leaves of *L. rhodesiensis* (compound **1**). Further anti-malarial tests supported by bioassay-guided isolation of the active compounds in the leaf extract are suggested. Moreover, as the biological activities were highlighted here using in vitro assays, it is necessary to confirm them in vivo. In the next part of our study, we plan to evaluate the antioxidant and antimalarial activity of the isolated flavonoid compounds.

L. rhodesiensis is a plant that is widely present in tropical and sub-tropical regions. However, it should be noted that if the local population uses it extensively for its biological properties, it would be important to cultivate it in order to avoid its loss. Moreover, it would be interesting to study the variability in extract compositions and in the biological activities of plants grown in different locations and seasons, as it is known that the culture conditions can widely impact the production of secondary metabolites by plants.

Author Contributions: Conceptualization, F.N., Z.F.T. and M.-L.F.; data curation, F.N., M.B.B., M.G., A.L., A.T.T., C.D. and M.F.; formal analysis, F.N.; funding acquisition, Z.F.T. and M.-L.F.; investigation, F.N., M.B.B. and A.L.; methodology, F.N.; project administration, Z.F.T. and M.-L.F.; resources, F.N. and M.-L.F.; software, F.N., M.B.B., M.G. and A.T.T.; supervision, M.F., Z.F.T. and M.-L.F.; validation, C.D., Z.F.T. and M.-L.F.; visualization, F.N., M.B.B., M.G. and A.T.T.; writing—original draft, F.N., M.B.B. and M.G.; writing—review and editing, M.G., C.D., M.F., Z.F.T. and M.-L.F. All authors have read and agreed to the published version of the manuscript.

Funding: This research was funded by the Education, Audiovisual, and Culture Executive Agency (EACEA), through EOHUB project 600873EPP-1-2018-1ES-EPPKA2-KA.

Data Availability Statement: Data is contained within the article.

Acknowledgments: The authors wish to thank the members of the Laboratory of Constitution and Reaction of Matter of the University Felix Houphouet Boigny, Abidjan, Cocody (Côte d'Ivoire), and all of the laboratory staff at the Chemistry of Natural Molecules, University of Liege, Gembloux Agro-Bio Tech (Belgium) for their availability and scientific contribution, particularly Thomas Bertrand,

Danny Trisman, and Franck Michels. This research was supported by the "Fondation Universitaire de Belgique" (Belgian University Foundation).

Conflicts of Interest: The authors declare no conflict of interest.

Sample Availability: Samples of the compounds are available from the authors in very small amount.

References

1. Ngugi, P.M.; Kimuni, N.; Ngeranwa, N.; Orinda, O.; Njagi, M.; Maina, D.; Agyirifo, S.; Gathumbi, K.; King'e, W.; Eliud, E.N. Antidiabetic and Safety of Lantana rhodesiensis in Alloxan Induced Diabetic Rats. *J. Dev. Drugs* **2015**, *4*, 1–10.
2. Fratkin, E. Traditional medicine and concepts of healing among Samburu pastoralists of Kenya. *J. Ethnobiol.* **1996**, *16*, 63–97.
3. Sawadogo, W.R.; Maciuk, A.; Banzouzi, J.T.; Champy, P.; Figadere, B.; Guissou, I.P.; Nacoulma, O.G. Mutagenic effect, antioxidant and anticancer activities of six medicinal plants from Burkina Faso. *Nat. Prod. Res.* **2012**, *26*, 575–579. [CrossRef]
4. Jean, B.M.; Nâg-Tiero, M.R.; Martin, K.; Germaine, N.O.; Norma, A.-A. A review on Lantana rhodesiensis Moldenke: Traditional uses, phytochemical constituents and pharmacological activities. *Int. J. Phytomed.* **2017**, *9*, 1–9. [CrossRef]
5. Edeoga, H.O.; Okwu, D.E.; Mbaebie, B.O. Phytochemical constituents of some Nigerian medicinal plants. *Afr. J. Biotechnol.* **2005**, *4*, 685–688. [CrossRef]
6. Ruffo, C.K.; Birnie, A.; Tengnäs, B. *Edible Wild Plants of Tanzania*; Regional Land Management Unit/Sida: Nairobi, Kenya, 2002; ISBN 0-947643-51-6.
7. Jean, B.M.; Alfonso, R.-M.; Hervé, C.T.; Véronique, M.; Germaine, N.O. Phytochemistry investigation and antioxidant activity of four edible Verbenaceae of Burkina Faso. *Int. J. Phytomed.* **2017**, *9*, 655. [CrossRef]
8. Seyoum, A.; Killeen, G.F.; Kabiru, E.W.; Knols, B.G.J.; Hassanali, A. Field efficacy of thermally expelled or live potted repellent plants against African malaria vectors in western Kenya. *Trop. Med. Int. Health* **2003**, *8*, 1005–1011. [CrossRef]
9. Nea, F.; Amenan, E.; Leon, E.; Kenne, T.; Genva, M.; Saive, M.; Felix, Z.; Fauconnier, M. A new chemotype of Lantana rhodesiensis Moldenke essential oil from Côte d'Ivoire: Chemical composition and biological activities. *Ind. Crops Prod.* **2019**, *141*, 111766. [CrossRef]
10. Zheng, W.; Wang, S.Y. Antioxidant activity and phenolic compounds in selected herbs. *J. Agric. Food Chem.* **2001**, *49*, 5165–5170. [CrossRef]
11. Bangou, M.J.; Kiendrebeogo, M.; Compaoré, M.; Coulibaly, A.Y.; Meda, N.-T.R.; Abarca, N.A.; Zeba, B.; Millogo-Rasolodimby, J.; Nacoulma, O.G. Enzyme Inhibition Effect and Polyphenolic Content of Medicinal Plant Extracts from Burkina Faso. *J. Biol. Sci.* **2011**, 31–38. [CrossRef]
12. Sawadogo, W.R.; Cerella, C.; Al-mourabit, A.; Moriou, C.; Teiten, M.; Guissou, I.P.; Dicato, M.; Diederich, M. Cytotoxic, Antiproliferative and Pro-Apoptotic Effects of 5-Hydroxyl-6,7,3′,4′,5′-Pentamethoxyflavone Isolated from Lantana ukambensis. *Nutrients* **2015**, *7*, 10388–10397. [CrossRef]
13. Kumar, S.; Sandhir, R.; Ojha, S. Evaluation of antioxidant activity and total phenol in different varieties of Lantana camara leaves. *BMC Res. Notes* **2014**, *7*, 1–9. [CrossRef]
14. Amiot, M.J.; Riollet, C.; Landrier, J.F. Polyphénols et syndrome métabolique. *Med. Mal. Metab.* **2009**, *3*, 476–482. [CrossRef]
15. Barros, L.M.; Duarte, A.E.; Waczuk, E.P.; Roversi, K.; Da Cunha, F.A.B.; Rolon, M.; Coronel, C.; Gomez, M.C.V.; De Menezes, I.R.A.; Da Costa, J.G.M.; et al. Safety assessment and antioxidant activity of lantana montevidensis leaves: Contribution to its phytochemical and pharmacological activity. *EXCLI J.* **2017**, *16*, 566–582.
16. Mansoori, A.; Singh, N.; Dubey, S.K.; Thakur, T.K.; Alkan, N.; Das, S.N.; Kumar, A. Phytochemical Characterization and Assessment of Crude Extracts From *Lantana camara* L. for Antioxidant and Antimicrobial Activity. *Front. Agron.* **2020**, *2*, 1–14. [CrossRef]
17. Roby, M.H.H.; Sarhan, M.A.; Selim, K.A.-H.; Khalel, K.I. Evaluation of antioxidant activity, total phenols and phenolic compounds in thyme (*Thymus vulgaris* L.), sage (*Salvia officinalis* L.), and marjoram (*Origanum majorana* L.) extracts. *Ind. Crops Prod.* **2013**, *43*, 827–831. [CrossRef]
18. McDonald, S.; Prenzler, P.D.; Antolovich, M.; Robards, K. Phenolic content and antioxidant activity of olive extracts. *Food Chem.* **2001**, *73*, 13–84. [CrossRef]
19. Skotti, E.; Anastasaki, E.; Kanellou, G.; Polissiou, M.; Tarantilis, P.A. Total phenolic content, antioxidant activity and toxicity of aqueous extracts from selected Greek medicinal and aromatic plants. *Ind. Crops. Prod.* **2014**, *53*, 46–54. [CrossRef]
20. Velioglu, Y.S.; Mazza, G.; Gao, L.; Oomah, B.D. Antioxidant Activity and Total Phenolics in Selected Fruits, Vegetables, and Grain Products. *J. Agric. Food Chem.* **1998**, *46*, 4113–4117. [CrossRef]
21. Oriakhi, K.; Oikeh, E.I.; Ezeugwu, N.; Anoliefo, O.; Aguebor, O. Comparative Antioxidant Activities of Extracts of Vernonia amygdalina and Ocimum gratissimum Leaves. *J. Agric. Sci.* **2014**, *6*, 13–20. [CrossRef]
22. Uyoh, E.A.; Chukwurah, P.N.; David, I.A.; Bassey, A.C. Evaluation of Antioxidant Capacity of Two Ocimum Species Consumed Locally as Spices in Nigeria as a Justification for Increased Domestication. *Am. J. Plant Sci.* **2013**, *04*, 222–230. [CrossRef]
23. Arulselvan, P.; Fard, M.T.; Tan, W.S.; Gothai, S.; Fakurazi, S.; Norhaizan, M.E.; Kumar, S.S. Role of Antioxidants and Natural Products in Inflammation. *Oxid. Med. Cell. Longev.* **2016**, *2016*, 5276130. [CrossRef]
24. Brand-Williams, W.; Cuvelier, M.E.; Berset, C. Use of a Free Radical Method to Evaluate Antioxidant Activity. *Leb. Wiss. Technol.* **1995**, *28*, 25–30. [CrossRef]

25. Miguel, M.G. Antioxidant activity of medicinal and aromatic plants. A review. *Flavour Fragr. J.* **2010**, *25*, 291–312. [CrossRef]
26. Kulisic, T.; Radonic, A.; Katalinic, V.; Milos, M. Use of different methods for testing antioxidative activity of oregano essential oil. *Food Chem.* **2004**, *85*, 633–640. [CrossRef]
27. Taubert, D.; Breitenbach, T.; Lazar, A.; Censarek, P.; Harlfinger, S.; Berkels, R.; Klaus, W.; Roesen, R. Reaction rate constants of superoxide scavenging by plant antioxidants. *Free Radic. Biol. Med.* **2003**, *35*, 1599–1607. [CrossRef]
28. Zin, Z.M.; Abdul Hamid, A.; Osman, A.; Saari, N. Antioxidative activities of chromatographic fractions obtained from root, fruit and leaf of Mengkudu (*Morinda citrifolia* L.). *Food Chem.* **2006**, *94*, 169–178. [CrossRef]
29. Wallace, G.; Fry, S.C. Phenolic components of the primary cell wall. *Inter. Rev. Cytol.* **1994**, *151*, 229–267.
30. Binns, A.N.; Chen, R.H.; Wood, H.N.; Lynn, D.G. Cell division promoting activity of naturally occurring dehydrodiconiferyl glucosides: Do cell wall components control cell division? *Proc. Natl. Acad. Sci. USA* **1987**, *84*, 980–984. [CrossRef]
31. Martial-didier, A.K.; Hubert, K.K.; Eugène, K.; Parfait, J.; Kablan, T. Phytochemical Properties and Proximate Composition of Papaya (*Carica papaya* L. var solo 8) Peels. *Turk. J. Agric. Food Sci. Technol.* **2017**, *5*, 676–680. [CrossRef]
32. Arbaayah, H.; Umi Kalsom, Y. Antioxidant properties in the oyster mushrooms (*Pleurotus* spp.) and split gill mushroom (*Schizophyllum commune*) ethanolic extracts. *Mycosphere* **2013**, *4*, 661–673. [CrossRef]
33. Gil, M.I.; Tomás-Barberán, F.A.; Hess-Pierce, B.; Kader, A.A. Antioxidant capacities, phenolic compounds, carotenoids, and vitamin C contents of nectarine, peach, and plum cultivars from California. *J. Agric. Food Chem.* **2002**, *50*, 4976–4982. [CrossRef] [PubMed]
34. Tavassoli, S.; Djomeh, Z.E. Total phenols, antioxidant potential and antimicrobial activity of methanol extract of rosemary (*Rosmarinus officinalis* L.). *Glob. Vet.* **2011**, *7*, 337–341.
35. Bergonzi, M.C.; Bilia, A.R.; Gallori, S.; Guerrini, D.; Vincieri, F.F. Variability in the content of the constituents of Hypericum perforatum L. and some commercial extracts. *Drug Dev. Ind. Pharm.* **2001**, *27*, 491–497. [CrossRef]
36. Wang, S.Y.; Zheng, W. Effect of plant growth temperature on antioxidant capacity in strawberry. *J. Agric. Food Chem.* **2001**, *49*, 4977–4982. [CrossRef] [PubMed]
37. Casassa, I.A.F.; Sari, I.A.S.; Avagnina, E.S.; Díaz, T.G.M.; Jofré, L.Q.V.; Fanzone, L.B.M.; Catania, I.A.C. Influence de deux techniques de maceration sur la composition phenolique et aromatique et les caractéristiques orgnoleptiques de vins cv. merlot. *Rev. Internet Vitic. Oenol.* **2007**, *25*, 1–12.
38. Yang, D.J.; Hwang, L.S.; Lin, J.T. Effects of different steeping methods and storage on caffeine, catechins and gallic acid in bag tea infusions. *J. Chromatogr. A* **2007**, *1156*, 312–320. [CrossRef]
39. Teixeira, B.; Marques, A.; Ramos, C.; Serrano, C.; Matos, O.; Neng, N.R.; Nogueira, J.M.F.; Saraiva, J.A.; Nunes, M.L. Chemical composition and bioactivity of different oregano (*Origanum vulgare*) extracts and essential oil. *J. Sci. Food Agric.* **2013**, *93*, 2707–2714. [CrossRef]
40. Venditti, E.; Bacchetti, T.; Tiano, L.; Carloni, P.; Greci, L.; Damiani, E. Hot vs. cold water steeping of different teas: Do they affect antioxidant activity? *Food Chem.* **2010**, *119*, 1597–1604. [CrossRef]
41. Bouyahya, A.; Abrini, J.; Bakri, Y.; Dakka, N. Screening phytochimique et évaluation de l'activité antioxydante et antibactérienne des extraits d'Origanum compactum. *Phytotherapie* **2017**, *15*, 379–383. [CrossRef]
42. Meda, A.; Lamien, C.E.; Marco, R.; Millogo, J.; Nacoulma, O.G. Determination of the total phenolic, flavonoid and proline contents in Burkina Fasan Honey, as well as their radical scavenging activity. *Food Chem.* **2005**, *91*, 571–577. [CrossRef]
43. Gursoy, N.; Sarikurkcu, C.; Cengiz, M.; Solak, M.H. Antioxidant activities, metal contents, total phenolics and flavonoids of seven Morchella species. *Food Chem. Toxicol.* **2009**, *47*, 2381–2388. [CrossRef]
44. Compaoré, M.; Lamien-Meda, A.; Mogoşan, C.; Lamien, C.E.; Kiendrebeogo, M.; Voştinaru, O.; Vlase, L.; Ionescu, C.; Nacoulma, O.G. Antioxidant, diuretic activities and polyphenol content of *Stereospermum kunthianum* Cham. (Bignoniaceae). *Nat. Prod. Res.* **2011**, *25*, 1777–1788. [CrossRef]
45. Jonville, M. Etude de la Composition Chimique et des Potentialités Antipaludiques de Plantes Utilisées en Médecine Traditionnelle au Cambodge et dans L'archipel des Mascareignes. Ph.D. Thesis, Université de Liège, Liège, Belgium, 2011.
46. Clarkson, C.; Maharaj, V.J.; Crouch, N.R.; Grace, O.M.; Pillay, P.; Matsabisa, M.G.; Bhagwandin, N.; Smith, P.J.; Folb, P.I. In vitro antiplasmodial activity of medicinal plants native to or naturalised in South Africa. *J. Ethnopharmacol.* **2004**, *92*, 177–191. [CrossRef]
47. Bashige-Chiribagula, V.; Bakari-Amuri, S.; Mbuyi-Kalonji, S.; Kahumba-Byanga, J.; Duez, P.; Lumbu-Simbi, J.B. Étude ethnobotanique, phytochimique et évaluation de l'activité antiplasmodiale de 13 plantes réputées antipaludéennes dans la commune du Kenya (Lubumbashi, RDC). *Phytotherapie* **2017**, 1–10. [CrossRef]
48. Esseh, K.; Afanyibo, Y.-G.; Ahama-Esseh, K.Y.S.; Idoh, K.; Koudouvo, K.; Agbonon, A.; Gbeassor, M. Screening Phytochimique, Étude Toxicologique, Évaluation des Activités Antiplasmodiale et Antiradicalaire de la Tige Feuillée de Senna occidentalis Linn (Fabaceae). *Eur. Sci. J. ESJ* **2019**, *15*, 411–433.
49. Nassirou, R.; Ibrahim, M.; Ilagouma, A.; Mahamadou, A.; Mamoudou, M.; Abdoulaye, A.; Oukem-Boyer, O.; Ikhiri, K. Évaluation in vitro de l'activité antiplasmodiale d'extraits de plantes issues de la pharmacopée traditionnelle du Niger. *J. Appl. Biosci.* **2015**, *89*, 8291–8300. [CrossRef]
50. Baba, G.; Adewumi, A.A.J.; Aina, V.O. Phytochemical Characterization and in-vivo Anti- Malaria Activity of Lantana camara Leaf Extract. *Br. J. Pharmacol. Toxicol.* **2011**, *2*, 277–282.
51. Chiou, W.F.; Lin, L.C.; Chen, C.F. Acteoside protects endothelial cells against free radical-induced oxidative stress. *J. Pharm. Pharmacol.* **2004**, *56*, 743–748. [CrossRef] [PubMed]

52. Chen, C.; Lin, Y.; Chien, M.; Hou, W.; Hu, M. Antioxidant and antihypertensive activities of acteoside and its analogs. *Biochemistry* **2012**, *53*, 421–429.
53. Lee, H.D.; Kim, J.H.; Pang, Q.Q.; Jung, P.M.; Cho, E.J.; Lee, S. Antioxidant activity and acteoside analysis of abeliophyllum distichum. *Antioxidants* **2020**, *9*, 1–10. [CrossRef] [PubMed]
54. Jang, T.W.; Choi, J.S.; Park, J.H. Protective and inhibitory effects of acteoside from Abeliophyllum distichum Nakai against oxidative DNA damage. *Mol. Med. Rep.* **2020**, *22*, 2076–2084. [CrossRef] [PubMed]
55. Lee, K.J.; Woo, E.R.; Choi, C.Y.; Shin, D.W.; Lee, D.G.; You, H.J.; Jeong, H.G. Protective Effect of Acteoside on Carbon Tetrachloride-Induced Hepatotoxicity. *Life Sci.* **2004**, *74*, 1051–1064. [CrossRef]
56. Nair, A.G.R.; Kotiyal, J.P.; Bhardwaj, D.K. Myricetin 7,4′-dimethyl ether and its 3-galactoside from Rhus lancea. *Phytochemistry* **1983**, *22*, 318–319. [CrossRef]
57. Nagao, T.; Abe, F.; Kinjo, J.; Okabe, H. Antiproliferative constituents in plants 10. Flavones from the leaves of Lantana montevidensis Briq. and consideration of structure-activity relationship. *Biol. Pharm. Bull.* **2002**, *25*, 875–879. [CrossRef] [PubMed]
58. Rwangabo, P.C.; Claeys, M.; Pieters, L.; Corthout, J.; Vanden Berghe, D.A.; Vlietinck, A.J. Umuhengerin, a new antimicrobially active flavonoid from lantana trifolia. *J. Nat. Prod.* **1988**, *51*, 966–968. [CrossRef]
59. Akkal, S.; Benayache, F.; Bentamene, A.; Medjroubi, K.; Seguin, E.; Tillequin, F. Flavonoid aglycones from Centaurea napifolia. *Chem. Nat. Compd.* **2003**, *39*, 165–166. [CrossRef]
60. Servi, H.; Goren, N. Chemistry of endemic Tanacetum mucroniferum Hub.-Mor. & Grierson extracts and three new sesquiterpene lactones. *Turk. J. Chem.* **2019**, *43*, 352–358.
61. Abu-Niaaj, L.; Katampe, I. Isolation and characterization of flavones from Artemisia monosperma. *Pharmacogn. J.* **2018**, *10*, 1018–1023. [CrossRef]
62. Itoh, T.; Ohguchi, K.; Iinuma, M.; Nozawa, Y.; Akao, Y. Inhibitory effects of polymethoxy flavones isolated from Citrus reticulate on degranulation in rat basophilic leukemia RBL-2H3: Enhanced inhibition by their combination. *Bioorg. Med. Chem.* **2008**, *16*, 7592–7598. [CrossRef]
63. Yang, L.N.; Xing, J.G.; He, C.H.; Wu, T. The phenolic compounds from *Dracocephalum moldavica* L. *Biochem. Syst. Ecol.* **2014**, *54*, 19–22. [CrossRef]
64. Belkacem, S.; Belbache, H.; Boubekri, C.; Mosset, P.; Rached-Mosbah, O.; Marchioni, E.; Benayache, S.; Benayache, F. Chemical Constituents from Centaurea parviflora Desf. *Res. J. Pharm. Biol. Chem. Sci.* **2014**, *5*, 1275–1279.
65. Polatoğlu, K.; Karakoç, Ö.C.; Demirci, F.; Gökçe, A.; Gören, N. Chemistry and biological activities of Tanacetum chiliophyllum var. oligocephalum extracts. *J. AOAC Int.* **2013**, *96*, 1222–1227. [CrossRef]
66. Shahidi, F.; Naczk, M. Food Phenolics: Sources, Chemistry, Effects and Applications. *Food Chem.* **1995**, *57*, 481–482.
67. Barkat Malika. Imène Laib Composition chimique et activité antioxydante de l'huile essentielle des fleurs sèches de *Lavandula officinalis*. *Rev. Génie Ind.* **2011**, *6*, 46–54.
68. Rebey, I.B.; Wannes, W.A.; Kaab, S.B.; Bourgou, S.; Tounsi, M.S.; Ksouri, R.; Fauconnier, M.L. Bioactive compounds and antioxidant activity of *Pimpinella anisum* L. accessions at different ripening stages. *Sci. Hortic.* **2019**, *246*, 453–461. [CrossRef]
69. Bettaieb Rebey, I.; Bourgou, S.; Saidani Tounsi, M.; Fauconnier, M.L.; Ksouri, R. Phytochemical composition and antioxidant activity of Lavandula dentate extracts Etude de la composition chimique et de l'activité antioxydante des différents extraits de la Lavande dentée (*Lavandula dentata*). *J. New Sci. Agric. Biotechnol.* **2017**, *39*, 2096–2105.
70. Lamia, S.A.; Moussa, B.; Marie-laure, F.; Georges, L. Chemical Composition and Antioxidant Activity of Thymus fontanesii Essential Oil from Algeria. *Nat. Prod. J.* **2018**, *8*, 1–7.
71. Ledoux, A.; St-gelais, A.; Cieckiewicz, E.; Jansen, O.; Illien, B.; Di Giovanni, N.; Marvilliers, A.; Hoareau, F. Antimalarial Activities of Alkyl Cyclohexenone Derivatives Isolated from the Leaves of Poupartia borbonica. *J. Nat. Prod.* **2017**, *80*, 1750–1757. [CrossRef] [PubMed]
72. Delemarre, J.M.; Van Der Kaay, H.J. Plasmodium falciparum malaria contracted in the natural way in the Netherlands. *Ned. Tijdschr. Geneeskd.* **1979**, *123*, 1981–1982.
73. Murebwayire, S.; Frédérich, M.; Hannaert, V.; Jonville, M.C.; Duez, P. Antiplasmodial and antitrypanosomal activity of Triclisia sacleuxii (Pierre) Diels. *Phytomedicine* **2008**, *15*, 728–733. [CrossRef] [PubMed]
74. Makler, M.T.; Ries, J.M.; Williams, J.A.; Bancroft, J.E.; Piper, R.C.; Gibbins, B.L.; Hinrichs, D.J. Parasite lactate dehydrogenase as an assay for Plasmodium falciparum drug sensitivity. *Am. J. Trop. Med. Hyg.* **1993**, *48*, 739–741. [CrossRef] [PubMed]
75. Makler, M.T.; Hinrichs, D.J. Measurement of the lactate dehydrogenase activity of Plasmodium falciparum as an assessment of parasitemia. *Am. J. Trop. Med. Hyg.* **1993**, *48*, 205–210. [CrossRef] [PubMed]

molecules

Article

In Vivo and In Vitro Assays Evaluating the Biological Activity of Taurine, Glucose and Energetic Beverages

Marcos Mateo-Fernández [1,*], Fernando Valenzuela-Gómez [1,*], Rafael Font [2], Mercedes Del Río-Celestino [2], Tania Merinas-Amo [1] and Ángeles Alonso-Moraga [1]

1 Department of Genetics, University of Córdoba, 14071 Córdoba, Spain; tania.meram@gmail.com (T.M.-A.); ge1almoa@uco.es (Á.A.-M.)
2 Agri-Food Laboratory, Avda. Menéndez Pidal, s/n, 14080 Córdoba, Spain; rafaelm.font@juntadeandalucia.es (R.F.); mercedes.rio.celestino@juntadeandalucia.es (M.D.R.-C.)
* Correspondence: mmateo@ceslopedevega.com (M.M.-F.); fernando.valenzuela@unican.es (F.V.-G.)

Abstract: Taurine is one of the main ingredients used in energy drinks which are highly consumed in adolescents for their sugary taste and stimulating effect. With energy drinks becoming a worldwide phenomenon, the biological effects of these beverages must be evaluated in order to fully comprehend the potential impact of these products on the health due to the fact nutrition is closely related to science since the population consumes food to prevent certain diseases. Therefore, the aim of this study was to evaluate the biological effects of taurine, glucose, classic Red Bull® and sugar-free Red Bull® in order to check the food safety and the nutraceutical potential of these compounds, characterising different endpoints: (i) Toxicology, antitoxicology, genotoxicology and life expectancy assays were performed in the *Drosophila melanogaster* model organism; (ii) The in vitro chemopreventive activity of testing compounds was determined by assessing their cytotoxicity, the proapoptotic DNA-damage capability to induce internucleosomal fragmentation, the strand breaks activity and the modulator role on the methylation status of genomic repetitive sequences of HL-60 promyelocytic cells. Whereas none tested compounds showed toxic or genotoxic effect, all tested compounds exerted antitoxic and antigenotoxic activity in *Drosophila*. Glucose, classic Red Bull® and sugar-free Red Bull® were cytotoxic in HL-60 cell line. Classic Red Bull® induced DNA internucleosomal fragmentation although none of them exhibited DNA damage on human leukaemia cells. In conclusion, the tested compounds are safe on *Drosophila melanogaster* and classic Red Bull® could overall possess nutraceutical potential in the in vivo and in vitro model used in this study. Besides, taurine could holistically be one of the bioactive compounds responsible for the biological activity of classic Red Bull®.

Keywords: classic Red Bull®; taurine; glucose; *Drosophila melanogaster*; HL-60 cell line

check for updates

Citation: Mateo-Fernández, M.; Valenzuela-Gómez, F.; Font, R.; Del Río-Celestino, M.; Merinas-Amo, T.; Alonso-Moraga, Á. In Vivo and In Vitro Assays Evaluating the Biological Activity of Taurine, Glucose and Energetic Beverages. *Molecules* **2021**, *26*, 2198. https://doi.org/10.3390/molecules26082198

Academic Editors: Simona Fabroni, Krystian Marszałek and Aldo Todaro

Received: 9 March 2021
Accepted: 9 April 2021
Published: 11 April 2021

1. Introduction

Energy drinks are caffeinated soft drinks which adolescents consume for the taste and stimulating effect. Energy drink intake is increasing all over the world as well as the scientific studies related to their effects [1,2]. Manufacturers of these beverages encourage their consumption with statements claiming a diversity of benefits [3], stating that these products are appropriate for consumers and free of danger. However, among scientists the food safety of these beverages is controversial [4]. With energy drinks becoming a worldwide phenomenon, the biological effects of these beverages must be evaluated in order to fully comprehend the potential impact of these products on health.

Taurine (TAU) is one of the main ingredients used in energy drinks such as Classic Red Bull® (CRB) and Sugar Free Red Bull® (SFRB). TAU is a natural amino acid that can be synthesised from methionine and cysteine mainly within the liver, kidney, astrocytes or testis [5]. However, dietary taurine supplementation may be required to achieve optimal taurine status in newborn infants [6] playing an important role in human during development, bile acid conjugation, osmoregulation, detoxification of xenobiotics, cell membrane stabilization, modulation of cellular calcium flux, modulation of neuronal excitability, antioxidant and anti-inflammatory properties among others [7]. Therefore, TAU is an essential amino-acid end-product. TAU has been extensively studied because of the variable evidence for the beneficial effects of this amino acid food supplementation [8].

Glucose (GLU) is also added in energy drinks such as CRB [9] and high content of GLU has been demonstrated to produce an oxidative stress in different trials and it is also related to several diseases [10–12]. Therefore, it is needed to evaluate the glucose role in the biological activity of energy drinks, comparing CRB to SFRB.

Nutrition is closely related to science since population intakes food to prevent certain diseases. A nutraceutical is any substance that is a food or a part of a food and provides medical or health benefits, including the prevention and treatment of disease. A nutraceutical substance should not be toxic and it should be able to prevent toxicity, avoid genetic oxidative damage, modulate the epigenome marks and induce cell death in a programmed manner in tumour cells, and maintain the epigenome marks in normal cells [13].

Toxicity, antitoxicity, genotoxicity, antigenotoxicity and life and healthspan in vivo assays were carried out in order to characterise the safety and nutraceutical potential of CRB, SFRB, TAU and GLU, establishing their effects on larvae viability, somatic mutations and how they affect life expectancy in *Drosophila melanogaster* model, respectively. *Drosophila* model organism has been used as a reliable model to evaluate the toxicity, genotoxicity and other degenerative process [14,15] and the Somatic Mutation And Recombination Test (SMART), based on the genetic alterations produced in the cells of imaginal discs of the larvae, has been shown to be able to detect genotoxic activity of various compounds with different chemical structures and complex mixtures [16,17].

Regarding in vitro assays, the human leukaemia cell line HL-60 is widely investigated as a model for inducible cell differentiation. This phenomenon might affect the cell ability to proliferate, and thus their immortality, with the appearance of apoptosis [18]. Compounds capable of inducing differentiation and apoptosis in tumour cell lines are candidates to act as chemopreventive agents against cancer [19,20]. Apoptosis is featured by the degradation of genomic DNA into internucleosomal fragments [21]. In addition, apoptosis may be assessed in single cell gel electrophoresis (SCGE) since this assay is capable of measuring DNA breaks and genotoxicity in single cells [22]. Chemopreventive therapies are also related to DNA methylation status in transposable sequences [23] and it is known that environmental exposures to nutritional and chemical factors could alter the epigenome pattern [24].

The aim of this study was to evaluate the biological effects of CRB, SFRB, TAU and GLU in order to check the food safety and the nutraceutical potential of these compounds, characterising different endpoints: (i) Assays of toxicology, antitoxicology, genotoxicology and life expectancy were performed in individuals chronically treated with the tested compounds in the *Drosophila melanogaster* model organism; (ii) The in vitro chemopreventive activity of testing compounds was determined by assessing their cytotoxicity, the proapoptotic DNA-damage capability to induce internucleosomal fragmentation, the strand breaks activity and the modulator role on the methylation status of genomic repetitive sequences of HL-60 promyelocytic cells.

2. Results

2.1. Toxicity/Antitoxicity

Toxicity and antitoxicity results are shown in Table 1. TAU exerted toxic effects in *Drosophila* larvae at all assayed concentrations, except for the lowest one and GLU provided this toxic effect when larvae were treated with 87.5 and 350 mM GLU. CRB showed a similar pattern to GLU being toxic the two-highest concentrations. Contrarily, SFRB were toxic at all assayed concentrations, except for the two-highest concentrations. Although all tested compounds were toxic in some extent, the lethal dose 50 (LD_{50}) standardized parameter was not reached at any of the significant toxic concentration.

Table 1. Toxicity and antitoxicity levels of CRB, SFRB, TAU and GLU in *D. melanogaster*.

CRB (mg/mL)	Survival (%)		SFRB (mg/mL)	Survival (%)		TAU (mM)	Survival (%)		GLU (mM)	Survival (%)	
	Simple Treatment (1)	Combined Treatment (2)		Simple Treatment	Combined Treatment		Simple Treatment	Combined Treatment		Simple Treatment	Combined Treatment
0	100	100	0	100	100	0	100	100	0	100	100
H_2O_2	-	38.46	H_2O_2	-	58	H_2O_2	-	38.46	H_2O_2	-	60
1.02	97	49.23 * (4)	0.16	77.65 *	65	0.25	94	40.76	2.7	98	60
4.06	93	47.69 *	0.625	80.3 *	87 *	1	78 *	57.69 *	11	92.65	84.35 *
8.13	88	51.53 *	1.25	74 *	106.68 *	2	77 *	56.9 *	21.86	90.32	88.65 *
32.5	75 * (3)	40	5	93	69 *	8	76 *	40	87.5	70.3 *	79 *
130	59 *	39	20	95.32	53.65	32	78 *	42.3	350	65 *	72 *

(1) Data are expressed as percentage of survival adults with respect to 300 untreated 72-h-old larvae from three independent experiments. (2) Combined treatments using standard medium and 0.15 M hydrogen peroxide. (3) Asterisks (*) indicate significant differences (one tail) with respect to the untreated control group and (4) the hydrogen peroxide control group: * Chi-square value higher than 5.02 [17].

Regarding antitoxicity assay, all substances showed some degree of capability to protect larvae population against the oxidative stress of genotoxicant hydrogen peroxide. SFRB, TAU and GLU provided antioxidant effects in a Gaussian-like manner, being antitoxic all the tested GLU concentrations, except for the lowest one. As regards CRB, this beverage exerted antitoxic activities at the three-lowest concentrations in a dose negative response manner.

2.2. Genotoxicity/Antigenotoxicity

Genotoxicity and antigenotoxicity results performing the SMART assay are shown in Table 2 and representative photograph of different types of hairs are depicted in Figure 1. The positive result obtained in the positive control after applying the binomial Kastenbaum-Bowman test, validated the accuracy of the SMART.

Table 2. Genotoxicity and antigenotoxicity of CRB, SFRB, TAU and GLU in the *Drosophila* wing spot test.

Compound	Clones Per Wings (Number of Spots) (1)					Mann-Whitney Test (2)	IP (%) (3)
	Wings	Small Single Spots	Large Simple Spots	Twin Spots	Total Spots		
	Number	(1–2 Cells)	(>2 Cells)	m = 5	m = 2		
		m = 2	m = 5				
H_2O	41	0.147 (6)	0.048 (2)	0	0.195 (8)		
H_2O_2 (0.15 M)	40	0.375 (15)	0.05 (2)	0	0.425 (17)+		
			SIMPLE TREATMENT				
CRB (mg/mL)							
[4.06]	40	0.2 (8)	0	0	0.2 (8)i	λ	
[130]	40	0.3 (12)	0.025 (1)	0	0.325 (13)i	λ	
SFRB (mg/mL)							
[0.625]	36	0.055 (2)	0.027 (1)	0	0.083 (3)-		
[20]	40	0.125 (5)	0	0	0.125 (5)-		
TAU (mM)							
[1]	37	0.027 (1)	0.08 (3)	0	0.108 (4)-		
[32]	40	0.15 (6)	0.05 (2)	0	0.2 (8)i	λ	
GLU (mM)							
[11]	40	0.125 (5)	0.05 (2)	0	0.175 (7)i	λ	
[350]	40	0.1 (4)	0.025 (1)	0	0.125 (5)-		

Table 2. *Cont.*

Compound	Clones Per Wings (Number of Spots) [1]					Mann-Whitney Test [2]	IP (%) [3]
	Wings	Small Single Spots	Large Simple Spots	Twin Spots	Total Spots		
	Number	(1–2 Cells) m = 2	(>2 Cells) m = 5	m = 5	m = 2		
		COMBINED TREATMENT (*mwh/flr*^{3})					
CRB (mg/mL)							
[4.06]	40	0.275 (11)	0.1 (4)	0	0.375 (15) β	λ	
[130]	44	0.11 (5)	0.023 (1)	0	0.13 (6) *		69.4
SFRB (mg/mL)							
[0.625]	38	0.079 (3)	0.053 (2)	0.026 (1)	0.158 (6) *		62.8
[20]	40	0.2 (8)	0.025 (1)	0.075 (3)	0.3 (12) *		29.4
TAU (mM)							
[1]	40	0.125 (5)	0	0	0.125 (5) *		70.6
[32]	38	0.236 (9)	0.026 (1)	0	0.263 (10) *		38.1
GLU (mM)							
[11]	40	0.075 (3)	0.025 (1)	0	0.1 (4) *		76.5
[350]	40	0.175 (7)	0.025 (1)	0	0.2 (8) *		52.9

Statistical diagnosis according to Frei and Wurgler [25]: + (positive), − (negative) and i (inconclusive) vs. negative control; * (positive), Δ (negative) and β (inconclusive) vs. respective positive control; m: multiplication factor. Kastenbaum-Bowman Test without Bonferroni correction, probability levels: α = β = 0.05. [1] No. of spots in parentheses. [2] Mann-Whitney test was used when appropriate to resolve inconclusive results. Lambda symbol (λ) means that there are not significant differences with respect to the negative control. [3] Inhibition percentage values were included when appropriate.

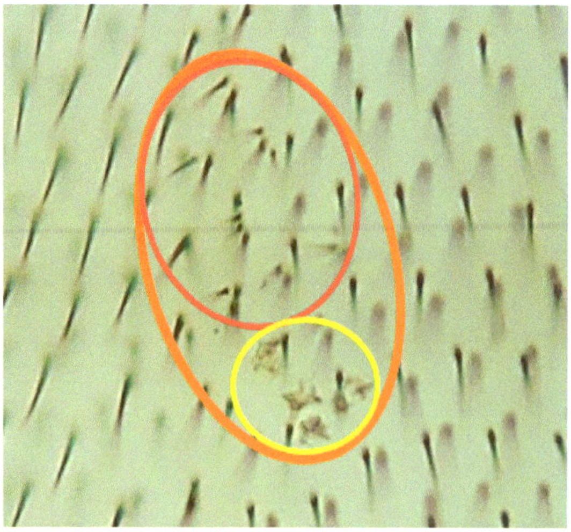

Figure 1. Representative homemade photograph was taken showing the different kind of hairs scored in the tested compounds: twin spot (**orange circle**); *mwh* mutation (**red circle**); *flr3* mutation (**yellow circle**); left being wild type single trichome hairs.

A Kastenbaum-Bowman test revealed negative results for both tested concentrations of SFRB, 1 mM TAU and 350 mM GLU. This test provided inconclusive results in both concentrations of CRB, 32 mM TAU and 11 mM GLU. The inconclusive results were resolved applying the Mann-Whitney test obtaining negative statistically values. Therefore, none of the tested compounds were genotoxic at any concentrations assayed. On the other hand, the IP values were calculated to determine the antigenotoxic potential of each compound when the clones per wings of the total spots from antigenotoxicity assay were different from the positive control (0.425) after applying the Kastenbaum-Bowman test (positive results). All tested compounds resulted in protecting *Drosophila* larvae against hydrogen peroxide in a negative dose-dependent manner, except for 4 mg/mL CRB which provided 0.375 clones per wing being statistically similar to the positive control. The IP values obtained were as follows: 69.4% for 130 mg/mL CRB; 62.8% and 29.4% for 0.625 and 20 mg/mL SFRB, respectively; 70.6% and 38.1% for 1 and 32 mM TAU, respectively and 76.5% and 52.9% for 11 and 350 mM GLU, respectively.

2.3. Lifespan

Figure 2 and Table 3 show the lifespan results obtained in *Drosophila melanogaster* experimental model treated with CRB, SFRB, TAU and GLU, resulting in an significant decreasing (roughly 18%) of the life expectancy in flies treated with 20 mg/mL of SFRB. However, CRB, GLU and TAU did not significantly differ from their concurrent control.

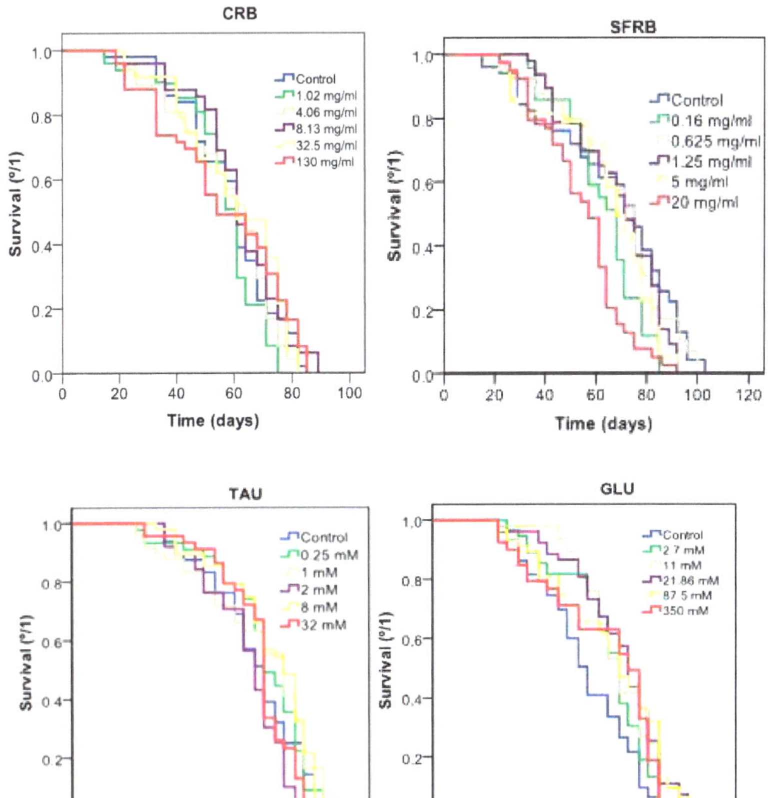

Figure 2. Effect of CRB, SFRB, TAU and GLU supplementation on the lifespan of *Drosophila melanogaster*.

Table 3. Effects of CRB, SFRB, TAU and GLU treatments on the *Drosophila melanogaster* mean lifespan and healthspan.

	Mean Lifespan	Mean Lifespan Difference	Healthspan (80th Percentile)	Healthspan Difference
	(days)	(%) [a]	(days)	(%) [a]
CRB (mg/mL)				
Control	58.797 ± 2.27	0	37.553 ± 1.96	0
1.02	56.191 ± 2.22	−4.4	34 ± 3.327	−9.3
4.06	57.671 ± 2.548	−2	26.625 ± 2.23 *	−29
8.13	61.59 ± 2.233	4.76	35.4 ± 3.637	−5.6
32.5	59.742 ± 2.637	1.6	32.4 ± 3.04	−13.6
130	56.425 ± 3.06	−4	27.5 ± 2.1 **	−26.6
SFRB (mg/mL)				
Control	67.08 ± 3.6	0	32.4 ± 1.78	0
0.16	62.75 ± 2.5	−6.4	33.75 ± 0.975	4.16
0.625	68.46 ± 3.4	2.2	36.7 ± 3.3	13.27
1.27	68.22 ± 2.7	1.8	36.8 ± 2.21	13.58
5	64.37 ± 3.24	−4	25.75 ± 0.94 **	−20.5
20	54.7 ± 2.7 ***	−18.35	30.25 ± 1.5	−6.6
TAU (mM)				
Control	64.97 ± 2.426	0	37.78 ± 1.94	0
0.25	67.94 ± 2.603	4.57	34.98 ± 3.954	−7.3
1	66.81 ± 3.215	2.81	30.22 ± 1.878	−20
2	61.83 ± 2.25	−4.8	35.357 ± 0.9	−6.4
8	70.142 ± 2.385	7.95	36.8 ± 2.832	−2.6
32	66.04 ± 2.055	1.6	36.57 ± 3.55	−3.2
GLU (mM)				
Control	59.67 ± 2.92	0	32.63 ± 1.49	0
2.7	61.25 ± 2.12	2.65	31.14 ± 1.53	−4.5
11	64.9 ± 2.05	8.76	41 ± 3.4	25.65
21.86	63.21 ± 2.08	5.9	30.66 ± 3.62	−5.8
87.5	60.5 ± 2.5	1.39	29.9 ± 1.96	−8.4
350	59 ± 3.1	−1.2	21.5 ± 0.76 ***	−34

[a] The difference was calculated by comparing treated flies with the concurrent water control. Positive numbers indicate lifespan increase and negative numbers indicate lifespan decrease. Data are expressed as mean value ±SE. * $p \leq 0.05$, ** $p \leq 0.01$, *** $p \leq 0.001$ significances obtained with the log-rank (Mantel-Cox) test.

Regarding healthspan showed in Table 3, 4.06 and 130 mg/mL CRB were able to decrease the quality of life around 27–29% with respect to the control. In addition, 5 mg/mL SFRB also reduced 20% the healthspan in *Drosophila*. By contrast, 350 mM GLU was able to increase the quality of life in 34%.

2.4. Cytotoxicity

Our cytotoxicity assays reported that all tested substances were chemopreventive compounds reaching IC_{50} at lowest tested concentrations, except for the TAU where IC_{50} was not found as it is depicted in Figure 3.

2.5. Internucleosomal DNA Fragmentation

According to internucleosomal DNA fragmentation assay, 32.5 mg/mL CRB was able to induce the typical ladder pattern showed in apoptotic cells. Nevertheless, SFRB, TAU and GLU did not induce DNA fragmentation in HL-60 cell line (Figure 4).

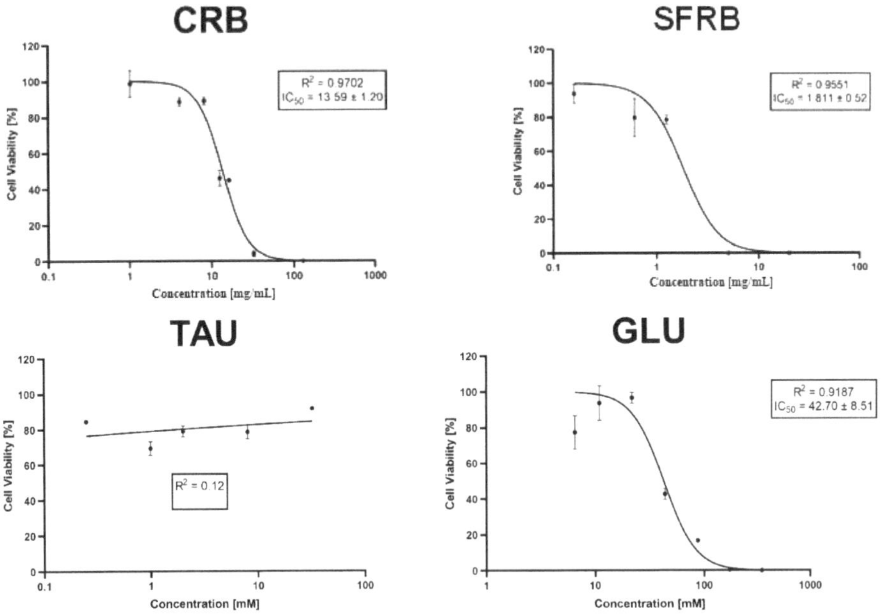

Figure 3. Cytotoxic effects of CRB, SFRB, TAU and GLU. Viability curves at 72 h of treatment. Data are expressed as mean ± SD, including IC_{50} value.

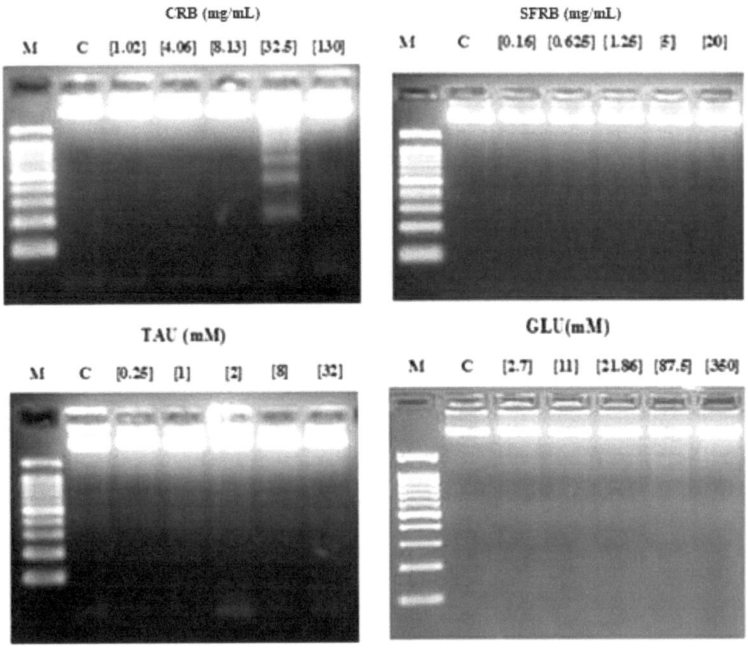

Figure 4. Internucleosomal DNA fragmentation after 5 h of treatment with CRB, SFRB, TAU and GLU. Letters M and C mean weight size marker and negative control respectively.

2.6. Comet Assay

Figure 5 shows representative photographs of Tail Moment (TM) obtained in (A) negative control (TM = 0), (B) treated HL-60 cells with GLU (TM = 5) and (C) positive control (with a hedgehog pattern). Our SCGE assay revealed that CRB and GLU significantly induce TM values higher than the concurrent control producing DNA damage in HL-60 cell line as well as 0.25 and 2 mM TAU. However, SFRB was not able to induce DNA strand breaks as it is shown in Figure 6. The concentrations used in this SCGE assay were determined according to the results obtained in the previous cytotoxicity assay.

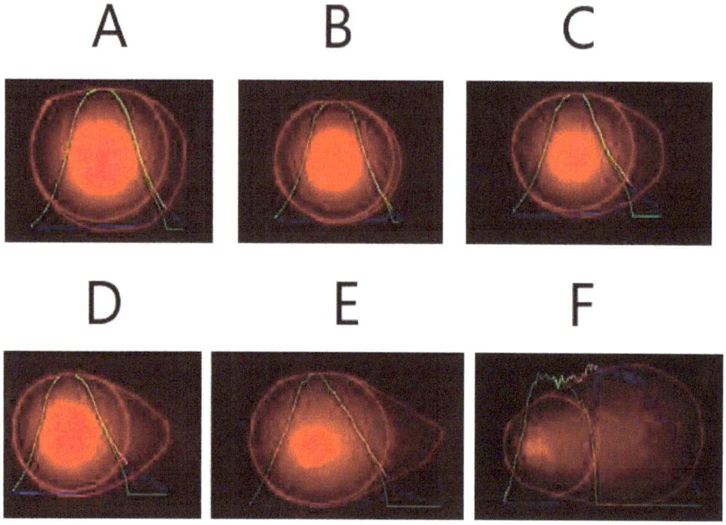

Figure 5. Representative photographs obtained in our SCGE assay for each treatment once Tail Moment (TM) parameter was analysed using OpenComet plugin for ImageJ Software. (**A**) negative control (TM = 0.81), (**B**) treated HL-60 cells with SFRB (TM = 0.125), (**C**) treated HL-60 cells with TAU (TM = 2.49), (**D**) treated HL-60 cells with RB (TM = 3.57), (**E**) treated HL-60 cells with GLU (TM = 5.51) and (**F**) positive control showing the hedgehog pattern (TM = 82.74).

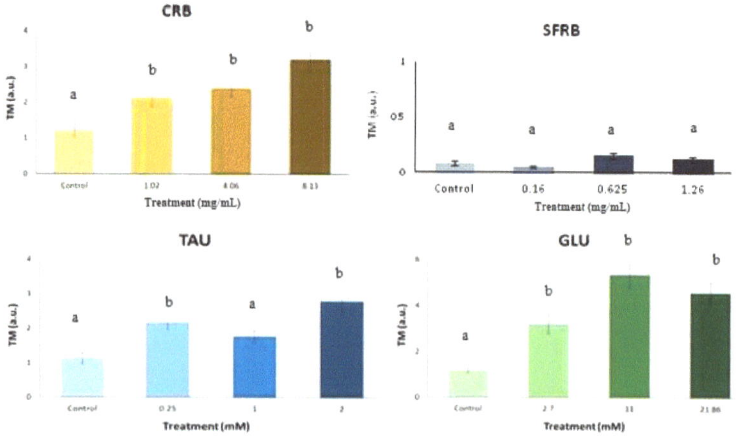

Figure 6. Alkaline comet assay (pH < 13) of HL-60 cells after 5 h-treatment with different concentrations of CRB, SFRB, TAU and GLU. DNA migration is reported as mean TM. The plot shows mean TM values and standard errors. Different letters mean different values after one-way ANOVA and post hoc Tukey's test.

2.7. Methylation Status

Figure 7 shows the relative normalised methylation status (RMS) of the three repetitive sequences (LINE-1, Alu M1, and Sat-α) in HL-60 cell line treated with the tested compounds. CRB and TAU hypomethylated LINE and Sat-α repetitive elements showing a similar pattern in satellite sequences. However, while CRB was able to reduce the methylation status of LINE sequences at the lowest concentration assayed, TAU reduced this methylation status at the highest one. Both tested concentrations of TAU significantly hypermethylated satellite alpha sequences. The highest assayed concentration of SFRB hypermethylated Alu and LINE repetitive elements, whereas the lowest one reduced the methylation status of Sat-α sequences. Finally, 11 mM GLU hypermethylated Alu sequences and hypomethylated LINE elements and whereas 350 mM GLU reduced the methylation status of LINE repetitive element in HL-60 cell line, this concentration hypermethylated satellite alpha sequences.

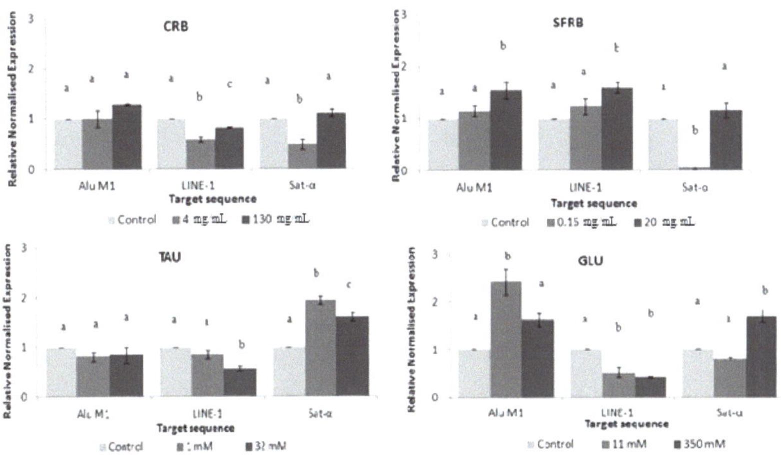

Figure 7. Relative normalised expression data of each repetitive element. Different letters are associated with different means applying One-Way ANOVA test and post hoc Tuckey's test.

3. Discussion

The arrival of the "genomic era" has raised the concept of the nutrigenomic studies, which aim is to expose the relationship between nutrition and genome to supply a scientific basis for improved public health through dietary habits [26]. aking into account that TAU is one of the main bioactive additives of CRB and SFRB, its evaluation is necessary to understand the physiological effects of both energy drinks.

We performed in vivo assays ascertaining on toxicity/antitoxicity and genotoxicity/antigenotoxicity in the SMART model as well as on lifespan and healthspan using the same *Drosophila* genetic background. We also carried out in vitro assays assessing on the cytotoxicity, internucleosomal DNA fragmentation, DNA damage and methylation status using HL-60 human leukaemia cell line. None information has been found in scientific databases about CRB and SFRB related with our assays. Contrarily, there are some data available about TAU and GLU. Therefore, it is the first time that CRB and SFRB are studied with this scope so our results will provide some important data about the biological effects of these worldwide consumed compounds.

The effect of energy beverages, glucose and taurine on *D. melanogaster* in vivo model *Drosophila* is considered an accurate in vivo model to study human diseases and further substantial contributions in this sense are expected [27]. To our knowledge, this is the first attempt to characterise the genotoxic effect of CRB and SFRB using the *Drosophila* in vivo model and HL-60 in vitro model. Experimental doses of TAU mimicking the concentration used in CRB were tested as well as GLU in this study.

The toxicity results obtained suggest that CRB, TAU and GLU are significantly toxic at most assayed concentrations, showing the beverage and GLU a positive dose-dependent manner. Contrarily, SFRB significantly resulted in being toxic at the three lowest concentrations in a negative dose-dependent manner. CRB (10 mg/kg/day, during 4 weeks) was demonstrated to induce distorted architecture of pancreatic sections of rats and the level of glucose in blood was increased [28] increase the oxidative stress altering the superoxide dismutase in liver, kidney, testes and brain but had not effect in heart resulting toxic the consumption of CRB in rats [29]. Moreover, the toxicity of CRB was reported in rabbit treated during 21 days at doses lower than the manufacturer's recommended equivalent for human intake [30]. Our CRB results are consistent with the toxicity mentioned above up to a point since de classical toxicity parameter LD_{50} has been not reached at any tested concentration. Bigard [31] demonstrated that although TAU was toxic in encephalopathy diseases, its toxicity was low besides, energy drinks were related to adverse effects due to their components. Besides, a great deal of incidents related to energy drinks toxicity is underestimated by the National Poison Data System [32]. Our results agree with those obtained since TAU could provide the toxicity of CRB as well as GLU which was demonstrated to be toxic in several in vivo trials triggering oxidative stress [33] being in agreement with our results.

Nevertheless, the standardized lethal dose 50 (LD_{50}) was not reached by any concentration and substance assayed in this study thus they are in agreement with TAU security demonstrated by Shao and Hathcock [34] both in human and animals trials using lots of concentrations. Moreover, we used a wide range of concentrations taking into account the *Drosophila* consumption and weight in order to be able to extrapolate the results to humans. If we consider a person whose daily intake was about 2 L of CRB, the concurrent concentration would be 4 mg/mL for RB and 1 mM TAU in *Drosophila* organism model [35]. If these concentrations are fixed, CRB would not be toxic at all. Regarding TAU, the toxicity could be due to the organism model used since Massie et al. [36] demonstrated that TAU could be act as a larvicide in a mechanism which only would occur in insect, although LD_{50} was not found as it was mentioned before. Therefore, the toxicity observed in our study could be explained in this sense. We hypothesise that the lack of toxicity of 4 mg/mL CRB could be due to others additive such as GLU or caffeine [37], among others.

Ceriello et al. [10] revealed that GLU produces an oxidative stress which is in agreement with the certain toxicity found in our GLU results. In addition, in this study was demonstrated that this oxidative stress in human endothelial cells triggers the induction of antioxidant enzymes which could be the responsible for the antioxidant activity of GLU found in *Drosophila* at all concentrations except for the lowest one. Our results provided evidences of the antioxidant activity of TAU at lowest concentrations. This result is in agreement with the hypothesis that TAU is able to block the ROS generation related to the toxicity produced by oxidative stress [38]. TAU was able to reduce the stress oxidative and apoptosis led by temperature in pufferfish, enhancing the activity of antioxidant enzymes [39]. Similar results to TAU antioxidant and antiapoptotic effects were found by Abdel-Daim et al. [40] in rats. However, TAU was demonstrated to be a low antioxidant compound in in vivo (rats) assays some years ago [41]. Therefore, the antioxidant effects of CRB in *Drosophila* may be due to the TAU and GLU biological synergic activity.

Research using *Drosophila* has provided seminal insights into gene function which are relevant to human health [42]. The genomic stability (lack of genotoxicity) observed in *Drosophila* with all the compounds assayed confirmed their safety. The inconclusive results obtained in some tested concentrations mean that the null hypothesis is accepted thus it is assumed that there is no difference in the mutation frequency between control and treated series. However, the alternative hypothesis is also accepted what means that the mutation frequency is not significantly 2-fold lower than the postulated increased frequency. No significant differences were found between the tested concentrations and the negative control when inconclusive results were resolved applying the Mann-Whitney U-test [43].

Direct studies on genotoxicity of TAU, CRB and SFRB are barely available because the epidemiological data always shows protective anticancer health properties such as antioxidant or antigenotoxic activity. These results agree with the previously reported by Cozzi et al. [44] who determined that 1 mM TAU was not able to induce neither chromosome aberrations nor sister chromatid exchanges in CHO cells. Lymphocytes cells treated with 20–50 mM TAU showed the similar number of sister chromatid exchange values than the concurrent control [45]. TAU and/or caffeine could account for the absence of genotoxicity of CRB [37]. GLU is involved in DNA damage processes lead by glycated products provoking base modification, strand breaks and apurinic/apyrimidinic sites in DNA [46]. Sucrose and GLU were determined as genotoxic compounds in the colon epithelium of rats at higher concentration (30% GLU) than the human intake using the mutation frequencies of the *E. coli* lambda and the level of bulky DNA adducts tests [47]. On basis to literature, the lack of genotoxicity found in our GLU results may be due to the fact that the mutagenicity showed by GLU is related to Maillard reaction products when the sugar is heated in presence of other compound.

It was demonstrated the antigenotoxic effect of TAU on aluminum sulphate-induced DNA damage in human peripheral lymphocytes protecting the DNA against the stress oxidative induced by this coagulant [48]. In addition, 5–20 mM TAU was reported as DNA-protected concentrations against oxidative damage in calf thymus DNA [49] and reduced genotoxicity activity of some drugs (benznidazole, cyclophosphamide and metronidazole, among others) in mice [50]. Our glucose results suggested that this sugar is able to protect the DNA of *Drosophila* against oxidative damage caused by hydrogen peroxide in large extent. All information available in scientific database on glucose DNA protection is related to cause this damage instead of protecting it and the information found about CRB and SFRB is scarce in connection with our assays.

D. melanogaster is an excellent model for the study of aging because adults show many similarities with the cellular senescence observed in mammals [51]. This is the reason why this particular model is frequently used to understand the relationship between nutrient metabolism and aging mechanisms [52]. As far as we know, the anti-ageing and anti-degenerative effects of CRB were assayed for the first time using *D. melanogaster* in our study. We demonstrated that none of the tested substances was able to increase the life expectancy of fruit flies. However, CRB decreased the quality of life of this model organism. TAU had not any influence on quality of life of *Drosophila* although an increase of hemoglobin levels was associated with high amount of TAU in human patients thus the quality of life could be improved [53]. Furthermore, it is used as antioxidant in order to improve the quality of life reducing the adverse effects of some disease such as diabetes mellitus [54]. The absence of lifespan improvement was reported long time ago as TAU (0.05 to 0.20 M) had no influence on adult lifespan of *Drosophila* [36]. In fact, 1.6% TAU was demonstrated to decrease the lifespan of *Drosophila*. Conversely, subsequent reports on *Drosophila* lifespan demonstrated that TAU increased the estimated mean values of survival at 8–24 mM [53]. Furthermore, Smith et al. [55] reported that TAU could also increase the lifespan of *Drosophila* [56]. TAU reduced the negative effect of tunicamycin on *C. elegans* lifespan restoring its normal values [57]. A multi-country epidemiological meta-analysis revealed that dietary TAU intake is negatively correlated with mortality from ischemic heart disease [57,58]. Nevertheless, Ito et al. [59] demonstrated that tissue TAU depletion shortens lifespan concomitant with acceleration in tissue aging. Despite a lifespan extension was not induced by GLU in our assays, it is well-known that GLU restriction can extend the lifespan of normal cells [52] including *Drosophila* [60] and the information found in scientific databases is restricted in this sense. Although CRB did not increase the lifespan, its safety has been demonstrated again. Conversely, SFRB decreased the lifespan of flies at the highest tested concentration in spite of the caloric restriction which flies were undergone to.

The in vitro evaluation of the anti-cancer properties of nutraceutical compounds or foods is the first step of a large pathway to obtain suitable conclusions to be extrapolated to human [37]. We determined the potential chemopreventive and genotoxic effect of CRB, SFRB, TAU and GLU on a human cancer cell model (HL-60 cell line).

CRB, SFRB and GLU showed cytotoxic effects following a positive dose-dependent response, being the inhibitory concentration 50 (IC_{50}) 13.59 mg/mL, 1.81 mg/mL and 42.70 mM, respectively. On the other hand, TAU did not show cytotoxic activity in HL-60 cells at any assayed concentration, and only the 2 mM concentration induced a decrease of cell proliferation. Jeon et al. [61] reported that TAU increased cell proliferation although they were not cancerous and Heidari et al. [62] demonstrated the protective effects of TAU against isoniazid and its intermediary metabolite hydrazine cytotoxicity in rat hepatocytes. These reports support our findings, suggesting that TAU is not cytotoxic since it may even protect HL-60 cells against toxic damage. Among the other tested compounds, they showed a similar pattern inhibiting around 100% of the cell growth at the second-highest concentration, which does not occur with none of the tested concentrations of TAU. In addition, our result fit in with those obtained by Chen et al. [63] who demonstrated that TAU up-regulate the taurine-upregulated gene 1 (TUG1) which serves an oncogenic role in the development of several tumors. Therefore, TAU would not be the responsible for the cytotoxic activity showed by CRB, being GLU one of the bioactive compounds involved in that effect. It has been previously described that GLU deprivation induces cell death in human breast carcinoma cells [64] and a high-rate glycolysis is related to promote cancer cell survival [65]. Therefore, our GLU result is not in agreement with these assumptions being a noteworthy finding. Nevertheless, 30 mM GLU cytotoxicity was demonstrated in human umbilical endothelial cells inducing stress oxidative by the generation of free radicals [66] being in concordance with the idea based on the hyperglycemia cytotoxicity in a positive dose-dependent response [67] in normal cells. On the other hand, fructose has been considered as a chemopreventive additive [68] while there is controversial information according to caffeine cytotoxic properties. We hypothesise that CRB sugar content could provide the chemopreventive potential of this energy drink and the artificial sugar sweetened could act similarly in SFRB.

The degradation of genomic DNA into internucleosomal fragments was proposed as a major mechanism affecting cancer cell apoptosis. Figure 4 indicates that DNA fragmentation was observed in the sec highest concentration of CRB (32.5 mg/mL) whereas none of the tested concentrations of SFRB, TAU and GLU showed this ladder pattern. Thus, these three compounds are not able to induce apoptosis in HL-60 cell line and these results fit in with the studies of Chen et al. [69] and Takatani et al. [70] who proved that TAU has no effect on induction of cell apoptosis. Chang et al. [71] demonstrated that DNA protecting compounds were able to prevent apoptosis which is in agreement with our results. ROS are essential mediators of apoptosis which eliminates cancerous and other life-threatening cells. Excessive DNA protecting substances could interfere with this mechanism [72]. Our findings suggest that the cytotoxicity observed in CRB may be conducted by proapoptotic mechanisms in some extent. This type of death cell is not induced in HL-60 cell line treated with TAU because of the lack of cytotoxicity showed by this additive. According to GLU, it has been demonstrated to induce DNA fragmentation in normal cells such as proximal tubular cells treated with 10% GLU under hyperglycemia conditions [73] and in cultured human umbilical vein endothelial cells fed with 30 mM GLU [74]. However, our results fit in with those obtained by Cao et al. [65] and Vaughn and Deshmukh [75] who reported that hyperglycemia environment stabilised tumour cells and block apoptosis mechanisms. We suggest the fructose as one of the main responsible for this proapoptotic mechanism induced by CRB since fructose was able to induce DNA fragmentation [76] once any tested concentrations of caffeine did not reach IC_{50} in a similar assay [37].

We performed alkaline SCGE in order to detect DNA damage [77], which are widely used to determine whether cells are triggering apoptotic and/or necrotic pathways [78]. It is assumed that apoptosis occurs when treatments induces a TM > 30 (hedgehog pattern) whereas control cells remain lower than 2 (no tails). On the contrary, necrosis shows a short comet-tail pattern since the majority of the damaged DNA remains in the comet head [79]. Our results showed that the damage induced by CRB and TAU in HL-60 cells was characterised by necrosis (short tails, TM < 3.2, Figure 6). These results agree with our cytotoxicity and DNA fragmentation assays, demonstrating that CRB, TAU and GLU induced cell death in HL-60, probably mediated by a necrotic pathway, when appropriate. The apoptotic way observed in HL-60 cell line treated with CRB also are in agreement with these comet results since the concentration which induced apoptosis was not assayed on comet assay due to the fact that it was highly cytotoxic and none cell was found. Our results fit in with Mochizuki et al. [80] who reported that tumour necrosis factor alpha (TNF-α) increases the amount of the TAU transporter and its affinity, resulting in an increase of the intracellular TAU level. These tested compounds exhibited the same DNA damage pattern: class 1; TM between 1 and 5 according to Fabiani et al. [81]. The lack of in vitro genotoxicity induced by TAU was also observed previously by Ahmad et al. [82] where the tail length of cell treated with TAU was similar to the concurrent control in rats. In normal cells such as human endothelial cells, 30 mM GLU was demonstrated to induce single strand breaks [83] and necrosis mechanisms in peripheral blood monocytes [84].

The genome instability is triggering by a globally hypermethylated status and the repetitive elements are highly methylated in somatic normal cells. However, cancer cells are generally hypomethylated, especially transposable elements, contributing to genome instability [85,86]. LINE-1(Long interspersed nuclear elements), Alu M4 and Sat-α repetitive elements are studied in the present work thus 32% of the genome has been analysed searching for methylation levels on transposable elements [87]. LINE-1 is non-random distributed by accumulating in G-positive bands (AT-rich regions) whilst Alu elements are included in non-coding GC-rich regions. On the other hand, Sat-α repetitive elements are AT-rich regions present in centromere [88,89].

Our results of DNA methylation status showed a different response with respect to the repetitive sequences screened. The methylation status of Alu sequences was significantly increased at 20 mg/mL SFRB and 11 mM GLU. CRB and GLU generally hypomethylated LINE-1 as well as 4 mg/mL of CRB and 0.16 mg/mL of SFRB hypomethylated Sat-α repetitive element, whereas SFRB hypermethylated LINE sequences at the highest concentration. In addition, TAU induced hypomethylation in LINE-1 at 32 mM. Contrarily, both tested concentrations of TAU hypermethylated Sat-α in a significant negative dose-dependent manner and this results fit in with Lleu and Huxtable [90] who proved that TAU increases the degree of phospholipid methylation in vivo in cerebral cortical synaptosomes of developing rats. This hypermethylation could be considered as a benefit since Sat-α represents the main DNA component of every human centromere [88,91]. Although the information related to the modification of the methylation status of repetitive elements induced by GLU is rather scarce on scientific database, its ability to modify the methylation pattern has been proved since hyperglycemia is associated with epigenetic changes in the promoter of the nuclear factor kB subunit p65 in aortic endothelial cells in mice and induces proinflamatory cytokines modulating the histones [92,93]. Glucose restriction is also related to the induction of DNA methylation changes in normal and immortalised cells acting as a therapy in cancer cells [94].

It has been demonstrated that the expression of satellite sequences is associated with a hypomethylation triggering cancer cells. Therefore, methylation process in satellite sequences is a potential mechanism for silencing its satellite expression in transformed cells [95]. On the other hand, human therapies against cancer are based on hypomethylation agents since this therapy is highly related to gene silencing thus this fact could activate tumour suppressor genes and be a positive highlight [96]. These are the reasons why CRB, TAU and GLU could be proposed as hypomethylated agents since they induce a global hypomethylation in some extent and hypermethylated satellite sequences. Moreover, TAU and GLU could be suggested as the responsible compound for the CRB properties. The hypermethylation showed by SFRB could be considered as a benefit since LINE-1 has been related to C-met oncogene what would be silenced [97].

4. Materials and methods

4.1. Samples

The energy drinks selected for this study, CRB and SFRB, was bought in a local market, lyophilised (SCAI, University of Córdoba, Córdoba, Spain) and stored at room temperature in a dark and dry atmosphere until being used. Furthermore, two single bioactive compound of these beverages were analysed, TAU (2-aminoethanesulfonic acid, Cat. No. 107-35-7, Sigma-Aldrich, St. Louis, MO, USA) and GLU (Cat. No. 50-99-7, Applichem Panreac, Barcelona, Spain). In order to make comparable with a 330 mL daily intake of energy drinks consumption in human (70 kg human body weight), the concentration ranges of compounds used in the different assays were calculated regarding the average daily food intake of *D. melanogaster* (1 mg/day) and the average body weight of *D. melanogaster* individuals (1 mg) [98].

4.2. Fly Stocks

Two *Drosophila melanogaster* strains with genetic markers that affect the wing-hair phenotype were used: (i) *mwh/mwh*, carrying the recessive mutation *mwh* (multiple wing hairs) [99], and (ii) flr^3*/In (3LR) TM3*, $rip^p se^p bx^{34e} e^s Bd^S$, where the flr^3 (flare) [100] marker is a homozygous recessive lethal mutation which is viable in homozygous somatic cells once larvae start developing and producing deformed trichomas.

Two strains of *D. melanogaster*, *mwh* (multiple wing hairs) that and flr^3 (flare) have been used. Both strains carry hair markers genes on the third chromosomes, being different in the shape and the number of hairs per cell as follows: (i) mwh/mwh, carrying the recessive mutation mwh that produces multiple trichomas per cell instead of one per cell in homozygosis [99]; (ii) *flr3/In (3LR) TM3*, $rip^p sepbx^{34e} e^s Bd^S$ is a lethal recessive marker in homozygosis that produces deformed trichomas but is viable in homozygous somatic cells once larvae start the development [100].

4.3. Cell Culture Conditions

Promyelocytic human leukaemia (HL-60) immortal cells originated from a female suffering myeloid leukaemia [101] were grown as suspension cultures in RPMI-1640 medium (R5886, Sigma-Aldrich, St. Louis, MO, USA) supplemented with 10% heat-inactivated foetal bovine serum (S01805, Linus, Madrid, Spain), L-glutamine 2 mM (G7513, Sigma-Aldrich, St. Louis, MO, USA) and 1× antibiotic-antimycotic solution from a 100× stock solution containing 10,000 units of penicillin, 10 mg of streptomycin and 25 µg of amphotericin B per mL (A5955, Sigma-Aldrich, St. Louis, MO, USA). Cells were incubated at 37 °C in a 80% humidified atmosphere 5% CO_2. Cultures were plated at 2.5×10^4 cells/mL density in 10 mL culture bottles and passed every 2 days.

4.4. In Vivo Assays

4.4.1. Toxicity and Antitoxicity Assays

Toxicity was assayed according to our standard protocols. CRB and SFRB were tested at five concentrations: 1, 4, 8.12, 32.5 and 130 mg/mL and 0.16, 0.625, 1.25, 5 and 20 mg/mL, respectively. TAU and GLU were tested at 0.25, 1, 2, 8 and 32 mM and 2.7, 11, 21.86, 87.5 and 350 mM, respectively. Negative (H_2O) and positive (0.15 M H_2O_2) toxicant concurrent controls were also assayed. Tested groups consisted of larvae fed with *Drosophila* Instant Medium (Formula 4–24, Carolina Biological Supply, Burlington, NC, USA) supplemented with the assessed compounds concentrations. Emerging adults of all groups were counted and toxicity was determined as the percentage of hatched individuals in each treatment compared with the negative control. Antitoxicity was assessed using the same procedure and experimental concentrations as in toxicity assays, but in combined treatments with

0.15 M H_2O_2 and comparing the percentage of emerging adults with the positive toxicant control [102].

4.4.2. Genotoxicity and Antigenotoxicity Assays

The wing spot test was performed to evaluate genotoxicity assays following the standard procedure [14]. Briefly, trans-heterozygous larvae for *mwh* and *flr*^{3} genes were obtained by crossing four day-old virgin *flr*^{3} females with *mwh* males in a 2:1 ratio. Four days after fertilization, females were allowed to lay eggs in fresh yeast medium (25 g yeast and 4 mL sterile distilled water) for 8 h in order to obtain synchronized larvae. After 72 h, larvae were collected, washed with distilled water, and clustered in groups of 100 individuals. Each group was fed with a mixture containing 0.85 g *Drosophila* Instant Medium (Formula 4–24, Carolina Biological Supply) and 4 mL water supplemented with the different compounds and concentrations assayed and negative (H_2O) and positive (0.15 M H_2O_2) controls until pupae hatching (10–12 days). Adult flies were collected and stored in 70% ethanol until the wings were removed and mounted on slides using Faure's solution. Mutant spots were assessed in both dorsal and ventral surfaces of the wings in a bright light microscope at 400× magnification. The frequencies of each type of mutant clone per wing (single, large or twin spot) were compared with the concurrent negative control [43]. All inconclusive and positive results ($p > 0.05$) were analysed with the nonparametric U-test of Mann, Whitney and Wilcoxon ($\alpha = \beta = 0.05$).

Anter et al. [13] described the antigenotoxicity method conducted in *D. melanogaster*. The same compounds and concentrations were assayed in combined treatment with hydrogen peroxide (0.15 M) acting as concurrent genotoxicant. Single and twin spots per wing were also recorded and compared with the concurrent negative control as described before. Finally, the inhibition percentages (IP) for the combined treatments were calculated as described by Abraham [103]: IP = [(genotoxin alone—combined treatment)/genotoxin alone] × 100.

4.4.3. Chronic Treatments: Lifespan and Healthspan Assays

Flies used in the lifespan assays show the same genetic background those used in the genotoxicity assays in order to compare both result. The treated adults consisted of the F1 progeny from *mwh* and *flr*^{3} parental strains produced by a 24 h eggs-lying in yeast medium. The same compounds and concentrations as in the toxicity/genotoxicity experiments were assayed. Lifespan assays were carried out at 25 °C according to the procedure described by Fernandez-Bedmar et al. [16]. Briefly, synchronized 72 ± 12-h-old trans-heterozygous larvae were washed in distilled water, collected and transferred in groups of 100 individuals into test vials containing 0.85 g *Drosophila* Instant Medium and 4 mL of the different concentrations of the compounds to be assayed. Emerged adults from pupae were collected under CO_2 anaesthesia and placed in groups of 25 individuals of the same sex into sterile vials containing 0.21 g *Drosophila* Instant Medium and 1 mL of different concentrations of the compounds to be tested. Flies were chronically treated during all their life. The number of survivors was determined twice a week in three different replicates.

4.5. In Vitro Assays

4.5.1. Cytotoxicity Assay

Trypan blue exclusion test was used to determine the cell viability exerted by the assayed compounds, according to our standard procedures [13]. HL-60 cells were placed in 96 well plates (2×10^4 cells/mL) and cultured for 72 h supplemented with 5 different concentrations of CRB, SFRB, TAU and GLU which were selected to assess the cytotoxic doses ranging the inhibitory concentration 50 (IC_{50}). After culture, cells were stained with a 1:1 volume ratio of Trypan blue dye (T8154, Sigma-Aldrich, St. Louis, MO, USA) and counted in a Neubauer chamber at 100× magnification. The survival percentage of each treatment compared with the control was recorded in three independent replicates. Data are expressed as mean ± standard deviation of the mean (SEM). When appropriate, the

IC_{50} values were analysed using simple linear or nonlinear regression fitting curve with the normalised response using GraphPad Software Prism 9 (San Diego, CA, USA).

4.5.2. DNA Fragmentation Status

DNA fragmentation induction was determined as described by Anter et al. (2014) [104]. Briefly, HL-60 cells (1×10^6/mL) were co-cultured with 5 different concentrations of CRB, SFRB, TAU and GLU similar to those used in cytotoxicity assays, during 5 h. After treatment, genomic DNA was extracted using a commercial kit (Blood Genomic DNA Extraction Mini Spin Kit, Canvax Biotech, Córdoba, Spain). Then, DNA was incubated overnight with RNase at 37 °C and quantified in a spectrophotometer (Nanodrop® ND-1000, NanoDrop Technologies, Inc., Wilmington, DE, USA). Finally, 1200 ng DNA were electrophoresed in a 2% agarose gel for 120 min at 50 V, stained with ethidium bromide and visualized under UV light. The apoptosis process is recognised by the appearance of internucleosomal DNA fragments that are multiple of 200 base pairs.

4.5.3. Clastogenicity: SCGE (Comet Assay)

DNA integrity was assayed by SCGE as described by Olive and Banáth [78] with minor modifications. HL-60 cells (5×10^5) in exponential growing phase were incubated in 1.5 mL of culture medium supplemented with CRB, SFRB, TAU and GLU (the three lowest concentrations selected in the cytotoxicity assays) for 5 h in P-12 plates. After treatment, cells were washed twice and adjusted to 6.25×10^5 cells/mL in PBS. Electrophoresis gels were prepared pouring a 1:4 dilution (cells in liquid low-melting-point agarose at 40 °C, A4018, Sigma) into slides. Gels were covered with a coverslip and allowed to solidify at RT for 30 min. Once the slides solidified, the coverslips were carefully removed and slides were bathed in freshly prepared lysing solution (2.5 M NaCl, 100 mM Na-EDTA, 10 mM Tris, 250 mM NaOH, 10% DMSO and 1% Triton X-100; pH = 13) for 1 h at 4 °C. Thereafter, slides were equilibrated in alkaline electrophoresis buffer (300 mM NaOH and 1 mM Na-EDTA, pH = 13) for 20–30 min at 4 °C. Once equilibrated, the slides were underwent electrophoresis (12 V, 400 mA for 8 min) in the dark and were immediately neutralized in cold neutral solution (0.4 M Tris-HCl buffer, pH 7.5) for 10 min. Finally, slides were dried overnight at RT in the dark. Gels were stained with 7 µL propidium iodide and photographed in a DM2500 microscope (Leica Microsystems GmBH, Wetzlar, Germany) at 400× magnification. At least 50 single cells from each treatment were analysed using the Open Comet™ software (OpenComet, GNU General Public License version 3.0, GPLv3) [105]. A one-way ANOVA and post-hoc Tukey's test with SPSS Statistics for Windows, Version 19.0 (2010, IBM Corporation, Armonk, NY, USA) was applied to determine the effect of the tested compounds on HL-60 cell DNA integrity from the analyses of Tail Moment (TM) data.

4.5.4. Methylation Status of HL-60 Cells

HL-60 cells were treated with different concentrations of CRB, SFRB, TAU and GLU (as selected in SMART assay) for 5 h. Then, DNA was extracted similarly to previously described DNA fragmentation assay. After that, the DNA was converted with bisulphite (EZ DNA Methylation-Gold™ Kit, Zymo Research, Irvine, CA, USA). Bisulphite-modified DNA was used for fluorescence-based real-time quantitative Methylation-Specific PCR (qMSP) using 5 µM of each forward and reverse primer (Isogen Life Science BV, Utrecht, The Netherlands), 2 µL of iTaq™ Universal SYBR® Green Supermix (Bio-Rad Laboratories, Inc., Hercules, CA, USA); it contains antibody-mediated hot-start iTaq DNA polymerase, dNTPs, MgCl2, SYBR® Green I dye, enhancers, stabilizers and a blend of passive reference dyes including ROX and fluorescein) and 25 ng of bisulphite converted genomic DNA.

PCR conditions included initial denaturalisation at 95 °C for 3 min and amplification which consisted of 45 cycles at 95 °C for 10 s, 60 °C for 15 s and 72 °C for 15 s, taking picture at the end of each elongation cycle. After that, melting curve was determined increasing 0.5 °C each 0.05 s from 60 °C to 95 °C and taking pictures.

QMSP was carried out in 48 well plates in MiniOpticon Real-Time PCR System (MJ Mini Personal Thermal Cycler, Bio-Rad) and were analysed by Bio-Rad CFX Manager 3.1 Software. The housekeeping Alu-C4 was used as a reference to correct for total DNA input. Alu-C4 and the target repetitive elements Alu-M1, LINE-1 and Sat-α were obtained from Isogen Life Science (Utrecht, The Netherlands) and their sequences are shown in Table 4. Each sample was analysed in triplicate [106].

Table 4. Primers information [85].

Primer	Forward Primer Sequence 5′ to 3′ (N)	Reverse Primer Sequence 5′ to 3′ (N)
ALU-C4	GGTTAGGTATAGTGGTTTATATTTGTAATTTTAGTA (−36)	ATTAACTAAACTAATCTTAAACTCCTAACCTCA (−33)
ALU-M1	ATTATGTTAGTTAGGATGGTTTCGATTTT (−29)	CAATCGACCGAACGCGA (−17)
LINE-1-M1	GGACGTATTTGGAAAATCGGG (−21)	AATCTCGCGATACGCCGTT (−19)
SAT-α-M1	TGATGGAGTATTTTTAAAATATACGTTTTGTAGT (−34)	AATTCTAAAAATATTCCTCTTCAATTACGTAAA (−33)

The results of each CT were obtained from each qMSP. Data were normalised with the housekeeping Alu C4 using the Nikolaidis et al. [107] and Liloglou et al. [108] comparative CT method (ΔΔCT). One-way ANOVA and post-hoc Tukey's test are used to evaluate the differences between the tested compounds, repetitive elements and concentrations.

5. Conclusions

In summary, this study provides a new corpus data from nutraceutical potential and food safety assays evaluating the toxi/antitoxicity, geno/antigenotoxicity and lifespan in *Drosophila* and the cytotoxicity and clastogenicity, as well as the methylation status of HL-60 cell line. The results shown that CRB, SFRB, TAU and GLU could be considered safe on *Drosophila* taking the LD_{50} parameter into account as well as genotoxicity was not found. On the one hand, according to nutraceutical potential assays related to the protective activity in in vivo assays, both energy drinks, TAU and GLU behaved as antioxidant in some extent against hydrogen peroxide in *Drosophila*. However, promising results were not found in the lifespan assay on *D. melanogaster* even SFRB significantly reduced de life expectancy. On the other hand, chemopreventive potential was shown by all tested compounds, except for TAU in vitro in the HL-60 cell line. CRB induced DNA fragmentation in the second-highest tested concentration although none of the tested compounds displayed significant apoptotic TM values in SCGE test. Therefore, necrotic pathway is the main mechanism from death cell occurs when HL-60 cell line is treated with CRB, GLU and TAU. Finally, TAU and GLU increased the methylation status of satellite sequences in HL-60 cell line providing nutraceutical benefits. On the other hand, CRB globally decreased the methylation pattern which fit in with the fact that some cancer therapies are based on globally hypomethylation. In conclusion, the tested compounds are safe on *Drosophila melanogaster* and CRB could overall possess nutraceutical potential in the in vivo and in vitro model used in this study. Besides, TAU could holistically be one of the bioactive compounds responsible for the biological activity of CRB.

Author Contributions: M.M.-F. and F.V.-G.: These authors contributed equally to this work sharing the authorship as first authors; Methodology, T.M.-A.; investigation, R.F. and M.D.R.-C.; Conceptualization, Á.A.-M. All authors have read and agreed to the published version of the manuscript.

Funding: No external funding was received.

Institutional Review Board Statement: Not applicable.

Informed Consent Statement: Not applicable.

Data Availability Statement: The data presented in this study are available on request from the corresponding author.

Conflicts of Interest: The authors declare no conflict of interest.

Sample Availability: Samples of the compounds are not available from the authors.

References

1. Ehlers, A.; Marakis, G.; Lampen, A.; Hirsch-Ernst, I. Risk assessment of energy drinks with focus on cardiovascular parameters and energy drink consumption in Europe. *Food Chem. Toxicol.* **2019**, *130*, 109–121. [CrossRef]
2. Vercammen, K.A.; Koma, J.W.; Bleich, S.N. Trends in energy drink consumption among U.S. adolescents and adults, 2003–2016. *Am. J. Prevent. Med.* **2019**, *56*, 827–833. [CrossRef]
3. Kammerer, M.; Jaramillo, J.A.; García, A.; Calderín, J.C.; Valbuena, L.H. Effects of energy drink major bioactive compounds on the performance of young adults in fitness and cognitive tests: A randomized controlled trial. *J. Int. Soc. Sports Nutr.* **2014**, *11*, 1–7. [CrossRef] [PubMed]
4. Alsunni, A.A. Energy drink consumption: Beneficial and adverse health effects. *Int. J. Health Sci.* **2015**, *9*, 468. [CrossRef]
5. Bouckenooghe, T.; Remacle, C.; Reusens, B. Is taurine a functional nutrient? *Curr. Opin. Clin. Nutr. Metab. Care* **2006**, *9*, 728–733. [CrossRef] [PubMed]
6. Verner, A.; Craig, S.; Mc Guire, W. Effect of taurine supplementation on growth and development in preterm or low birth weight infants. *Cochrane Database Syst. Rev.* **2007**, *4*, 1–16. [CrossRef] [PubMed]
7. Higgins, J.P.; Tuttle, T.D.; Higgins, C.L. Energy beverages: Content and safety. In *Mayo Clinic Proceedings*; Elsevier: Amsterdam, The Netherlands, 2010; pp. 1033–1041.
8. Lee, S.E.; Lee, S.Y.; Jo, E.J.; Kim, M.Y.; Yang, M.S.; Chang, Y.S.; Kim, S.H. A case of taurine-containing drink induced anaphylaxis. *As. Pac. Allergy* **2013**, *3*, 70–73. [CrossRef]
9. Cavka, A.; Stupin, M.; Panduric, A.; Plazibat, A.; Cosic, A.; Rasic, L.; Debeljak, Z.; Martinovic, G.; Drenjancevic, I. Adrenergic system activation mediates changes in cardiovascular and psychomotoric reactions in young individuals after red bull © energy drink consumption. *Int. J. Endocrinol.* **2015**, *10*, 751530. [CrossRef]
10. Ceriello, A.; dello Russo, P.; Amstad, P.; Cerutti, P. High glucose induces antioxidant enzymes in human endothelial cells in culture: Evidence linking hyperglycemia and oxidative stress. *Diabetes* **1996**, *45*, 471–477. [CrossRef]
11. Piwkowska, A.; Rogacka, D.; Audzeyenka, I.; Jankowski, M.; Angielski, S. High glucose concentration affects the oxidant-antioxidant balance in cultured mouse podocytes. *J. Cell. Biochem.* **2011**, *112*, 1661–1672. [CrossRef]
12. Karunakaran, U.; Park, K.G. A systematic review of oxidative stress and safety of antioxidants in diabetes: Focus on islets and their defense. *Diabet. Metab. J.* **2013**, *37*, 106–112. [CrossRef]
13. Anter, J.; Fernández-Bedmar, Z.; Villatoro-Pulido, M.; Demyda-Peyras, S.; Moreno-Millán, M.; Alonso-Moraga, Á.; Muñoz-Serrano, A.; de Castro, M.D. A pilot study on the DNA-protective, cytotoxic, and apoptosis-inducing properties of olive-leaf extracts. *Mutat. Res.* **2011**, *723*, 165–170. [CrossRef] [PubMed]
14. Graf, U.; Würgler, F.E.; Katz, A.J.; Frei, H.; Juon, H.; Hall, C.B.; Kale, P.G. Somatic mutation and recombination test in *Drosoplila melanogaster*. *Environ. Mutagen.* **1984**, *6*, 153–188. [CrossRef]
15. Siddique, Y.H.; Anjum, S.; Jyoti, S. Evaluation of the toxic potential of aspartame in third instar larvae of transgenic *Drosoplila melanogaster* (hsp70-lacZ) Bg9. *All Res. J. Biol.* **2017**, *8*, 16–24.
16. Fernández-Bedmar, Z.; Anter, J.; de La Cruz-Ares, S.; Muñoz-Serrano, A.; Alonso-Moraga, Á.; Pérez-Guisado, J. Role of citrus juices and distinctive components in the modulation of degenerative processes: Genotoxicity, antigenotoxicity, cytotoxicity, and longevity in Drosoplila. *J. Toxicol. Environ. Health A* **2011**, *74*, 1052–1066. [CrossRef] [PubMed]
17. Mateo-Fernández, M.; Alves-Martínez, P.; Río-Celestino, D.; Font, R.; Merinas-Amo, T.; Alonso-Moraga, Á. Safety and nutraceutical potential of caramel colour class iv using *In vivo* and *In vitro* assays. *Foods* **2019**, *8*, 392. [CrossRef] [PubMed]
18. Anazetti, M.C.; Melo, P.S.; Duran, N.; Haun, M. Comparative cytotoxicity of dimethylamide-crotonin in the promyelocytic leukemia cell line (HL60) and human peripheral blood mononuclear cells. *Toxicology* **2003**, *188*, 261–274. [CrossRef]
19. Fesus, L.; Szondy, Z.; Uray, I. Probing the molecular program of apoptosis by cancer chemopreventive agents. *J. Cell. Biochem. Suppl.* **1995**, *22*, 151–161. [CrossRef]
20. Hong, W.K.; Sporn, M.B. Recent advances in chemoprevention of cancer. *Science* **1997**, *278*, 1073–1077. [CrossRef] [PubMed]
21. Gaido, M.L.; Cidlowski, J.A. Identification, purification, and characterization of a calcium-dependent endonuclease (NUC18) from apoptotic rat thymocytes. NUC18 is not histone H2B. *J. Biol. Chem.* **1991**, *266*, 18580–18585. [CrossRef]
22. Collins, A.R. The comet assay for DNA damage and repair: Principles, applications, and limitations. *Mol. Biotechnol.* **2004**, *26*, 249–261. [CrossRef]
23. Roman-Gomez, J.; Jimenez-Velasco, A.; Agirre, X.; Castillejo, J.A.; Navarro, G.; San Jose-Eneriz, E.; Garate, L.; Cordeu, L.; Cervantes, F.; Prosper, F.; et al. Repetitive DNA hypomethylation in the advanced phase of chronic myeloid leukemia. *Leuk. Res.* **2008**, *32*, 487–490. [CrossRef]
24. Jirtle, R.L.; Skinner, M.K. Environmental epigenomics and disease susceptibility. *Nat. Rev. Genet.* **2007**, *8*, 253–262. [CrossRef]
25. Frei, H.; Wurgler, F.E. Statistical methods to decide whether mutagenicity test data from *Drosoplila* assays indicate a positive, negative, or inconclusive result. *Mutat. Res.* **1988**, *203*, 297–308. [CrossRef]

26. Dang, T.S.; Walker, M.; Ford, D.; Valentine, R.A. Nutrigenomics: The role of nutrients in gene expression. *Periodontology* **2014**, *64*, 154–160. [CrossRef] [PubMed]
27. Beckingham, K.M.; Armstrong, J.D.; Texada, M.J.; Munjaal, R.; Baker, D.A. *Drosoplila melanogaster*-the model organism of choice for the complex biology of multi-cellular organisms. *Gravit. Space Res.* **2007**, *18*, 17–30.
28. El Desouky, A.A.; Abo Zaid, A.; El Saify, G.H.; Noya, D.A. Ameliorative effect of omega-3 on energy drinks-induced pancreatic toxicity in adult male albino rats. *Egypt. J. Histol.* **2019**, *42*, 324–334. [CrossRef]
29. Okoro, E.; Asagba, S.; Okoro, I. Effects of energy drink and alcohol on some clinical and oxidative stress parameters in rats. *Biokemistri* **2020**, *31*, 2.
30. Taiwo, O.I.; Adesokan, A.A. Toxicodynamic effects of "Red Bull" energy drink in a randomised controlled study on local strains of adult rabbits. *J. Biol. Life Sci* **2018**, *9*, 46–64.
31. Bigard, A. Risks of energy drinks in youths. *Arch. Pediatr. Org. Off. Soc. Franc. Pediatr.* **2010**, *17*, 1625–1631. [CrossRef]
32. Seifert, S.M.; Schaechter, J.L.; Hershorin, E.R.; Lipshultz, S.E. Health effects of energy drinks on children, adolescents, and young adults. *Pediatrics* **2011**, *127*, 511–528. [CrossRef]
33. Robertson, R.P.; Harmon, J.S. Diabetes, glucose toxicity, and oxidative stress: A case of double jeopardy for the pancreatic islet β cell. *Free Rad. Biol. Med.* **2006**, *41*, 177–184. [CrossRef]
34. Shao, A.; Hathcock, J.N. Risk assessment for the amino acids taurine, l-glutamine and l-arginine. *Reg. Toxicol. Pharmacol.* **2008**, *50*, 376–399. [CrossRef]
35. Deshpande, S.A.; Carvalho, G.B.; Amador, A.; Phillips, A.M.; Hoxha, S.; Lizotte, K.J.; Ja, W.W. Quantifying Drosoplila food intake: Comparative analysis of current methodology. *Nat. Method* **2014**, *11*, 535–540. [CrossRef]
36. Massie, H.R.; Williams, T.R.; DeWolfe, L.K. Changes in taurine in aging fruit flies and mice. *Exp. Gerontol.* **1989**, *24*, 57–65. [CrossRef]
37. Mateo-Fernández, M.; Merinas-Amo, T.; Moreno-Millán, M.; Alonso-Moraga, Á.; Demyda-Peyrás, S. *In vivo* and *in vitro* genotoxic and epigenetic effects of two types of cola beverages and caffeine: A multiassay approach. *BioMed Res. Int.* **2016**, *2016*, 1–15. [CrossRef] [PubMed]
38. Schaffer, S.W.; Azuma, J.; Mozaffari, M. Role of antioxidant activity of taurine in diabetes this article is one of a selection of papers from the NATO advanced research workshop on translational knowledge for heart health. *Can. J. Physiol. Pharmacol.* **2009**, *87*, 91–99. [CrossRef] [PubMed]
39. Cheng, C.H.; Guo, Z.X.; Wang, A.L. The protective effects of taurine on oxidative stress, cytoplasmic free-Ca^{2+} and apoptosis of pufferfish (*Takifugu obscurus*) under low temperature stress. *Fish Shellfish Immunol.* **2018**, *77*, 457–464. [CrossRef]
40. Abdel-Daim, M.M.; Dessouki, A.A.; Abdel-Rahman, H.G.; Eltaysh, R.; Alkahtani, S. Hepatorenal protective effects of taurine and N-acetylcysteine against fipronil-induced injuries: The antioxidant status and apoptotic markers expression in rats. *Sci. Total Environ.* **2019**, *650*, 2063–2073. [CrossRef] [PubMed]
41. Aruoma, O.; Halliwell, B.; Hoey, B.M.; Butler, J. The antioxidant action of taurine, hypotaurine and their metabolic precursors. *Biochem. J.* **1988**, *256*, 251–255. [CrossRef] [PubMed]
42. Fortini, M.E.; Skupski, M.P.; Boguski, M.S.; Hariharan, I.K. A survey of human disease gene counterparts in the *Drosoplila* genome. *J. Cell Biol.* **2000**, *150*, F23–F30. [CrossRef]
43. Frei, H.; Wurgler, F.E. Optimal experimental design and sample size for the statistical evaluation of data from somatic mutation and recombination tests (SMART) in *Drosoplila*. *Mutat. Res.* **1995**, *334*, 247–258. [CrossRef]
44. Cozzi, R.; Ricordy, R.; Bartolini, F.; Ramadori, L.; Perticone, P.; De Salvia, R. Taurine and ellagic acid: Two differently-acting natural antioxidants. *Environ. Mol. Mut.* **1995**, *26*, 248–254. [CrossRef]
45. Ergun, M.A.; Soysal, Y.; Kismet, E.; Akay, C.; Dundaroz, R.; Ilhan, M.N.; Imirzalioglu, N. Investigating the *in vitro* effect of taurine on the infant lymphocytes by sister chromatid exchange. *Pediatr. Int.* **2006**, *48*, 284–286. [CrossRef] [PubMed]
46. Mullokandov, E.; Franklin, W.; Brownlee, M. DNA damage by the glycation products of glyceraldehyde 3-phosphate and lysine. *Diabetologia* **1994**, *37*, 145–149. [CrossRef] [PubMed]
47. Hansen, M.; Baunsgaard, D.; Autrup, H.; Vogel, U.B.; Møller, P.; Lindecrona, R.; Wallin, H.; Poulsen, H.E.; Loft, S.; Dragsted, L.O. Sucrose, glucose and fructose have similar genotoxicity in the rat colon and affect the metabolome. *Food Chem. Toxicol.* **2008**, *46*, 752–760. [CrossRef] [PubMed]
48. Türkez, H.; Geyikoğlu, F. The anti-genotoxic effect of taurine on aluminum sulphate-induced DNA damage in human peripheral lymphocytes. *IUFS J. Biol.* **2010**, *69*, 25–32.
49. Messina, S.A.; Dawson, R., Jr. Attenuation of oxidative damage to DNA by tairome and taurine analogs. In *Taurine 4*; Springer: Boston, MA, USA, 2002; Volume 483, pp. 355–367.
50. Bosquesi, P.L.; Scarim, C.B.; de Oliveira, J.R.S.; de Oliveira Vizioli, E.; dos Santos, J.L.; Chin, C.M. Protective effect of taurine in the induction of genotoxicity by mutagenic drugs. *J. Pharm. Pharmacol.* **2018**, *6*, 1–9.
51. Fleming, J.E.; Reveillaud, I.; Niedzwiecki, A. Role of oxidative stress in *Drosoplila* aging. *Mutat. Res.* **1992**, *275*, 267–279. [CrossRef]
52. Li, Y.; Liu, L.; Tollefsbol, T.O. Glucose restriction can extend normal cell lifespan and impair precancerous cell growth through epigenetic control of hTERT and p16 expression. *FASEB J.* **2010**, *24*, 1442–1453. [CrossRef]
53. Yang, X.; Zhang, Z.; Feng, Y.; Ren, H.; Liu, F.; Zu, T. Effect of taurine on lifespan and antioxidant in *Drosophila*. In Proceedings of the Biomedical Engineering and Biotechnology (iCBEB), International Conference, Macau, Macao, 28–30 May 2012; pp. 206–209.

54. Franconi, F.; Loizzo, A.; Ghirlanda, G.; Seghieri, G. Taurine supplementation and diabetes mellitus. *Curr. Opin. Clin. Nutr. Metab. Care* **2006**, *9*, 32–36. [CrossRef]
55. Smith, L.; Habib, I.; Shirkey, S.; Talon, B.; Milne, A.; Nadolski, J. Sexual dimorphism in the effect of a taurine supplemented diet on life span in adult *Drosoplila melanogaster*. *Int. J. Zool. Res.* **2011**, *7*, 34. [CrossRef]
56. Suh, H.J.; Shin, B.; Han, S.H.; Woo, M.J.; Hong, K.B. Behavioral changes and survival in melanogaster: Effects of Ascorbic acid, taurine, and caffeine. *Biol. Pharm. Bull.* **2017**, *40*, 1873–1882. [CrossRef] [PubMed]
57. Kim, H.M.; Do, C.H.; Lee, D.H. Taurine reduces ER stress in *C. Elegans*. *J. Biomed. Sci.* **2010**, *17*, S26. [CrossRef] [PubMed]
58. Yamori, Y.; Liu, l.; Ikeda, K.; Miura, A.; Mizushima, S.; Miki, T.; Nara, Y. Distribution of twenty-four hour urinary taurine excretion and association with ischemic heart disease mortality in 24 populations of 16 countries: Results from the WHO-CARDIAC study. *Hypertens. Res.* **2001**, *24*, 453–457. [CrossRef]
59. Ito, T.; Yoshikawa, N.; Inui, T.; Miyazaki, N.; Schaffer, S.W.; Azuma, J. Tissue depletion of taurine accelerates skeletal muscle senescence and leads to early death in mice. *PLoS ONE* **2014**, *9*, e107409. [CrossRef]
60. Troen, A.M.; French, E.E.; Roberts, J.F.; Selhub, J.; Ordovas, J.M.; Parnell, L.D.; Lai, C.-Q. Lifespan modification by glucose and methionine in *Drosoplila melanogaster* fed a chemically defined diet. *Age* **2007**, *29*, 29–39. [CrossRef]
61. Jeon, S.H.; Lee, M.Y.; Kim, S.J.; Joe, S.G.; Kim, G.B.; Kim, I.S.; Kim, N.S.; Hong, C.U.; Kim, S.Z.; Kim, J.S.; et al. Taurine increases cell proliferation and generates an increase in [Mg^{2+}] i accompanied by ERK 1/2 activation in human osteoblast cells. *FEBS Lett.* **2007**, *581*, 5929–5934. [CrossRef]
62. Heidari, R.; Babaei, H.; Eghbal, M.A. Cytoprotective effects of taurine against toxicity induced by isoniazid and hydrazine in isolated rat hepatocytes. *Arch. Ind. Hygiene Toxicol.* **2013**, *64*, 201–210. [CrossRef]
63. Chen, L.; Zhao, H.; Wang, C.; Hu, N. TUG1 knockdown enhances adriamycin cytotoxicity by inhibiting glycolysis in adriamycin-resistant acute myeloid leukemia HL60/ADR cells. *RSC Adv.* **2019**, *9*, 10897–10904. [CrossRef]
64. Lee, Y.J.; Galoforo, S.S.; Berns, C.M.; Chen, J.C.; Davis, B.H.; Sim, J.E.; Corry, P.M.; Spitz, D.R. Glucose deprivation-induced cytotoxicity and alterations in mitogen-activated protein kinase activation are mediated by oxidative stress in multidrug-resistant human breast carcinoma cells. *J. Biol. Chem.* **1998**, *273*, 5294–5299. [CrossRef] [PubMed]
65. Cao, X.; Fang, L.; Gibbs, S.; Huang, Y.; Dai, Z.; Wen, P.; Zheng, X.; Sadee, W.; Sun, D. Glucose uptake inhibitor sensitizes cancer cells to daunorubicin and overcomes drug resistance in hypoxia. *Cancer Chemother. Pharmacol.* **2007**, *59*, 495–505. [CrossRef] [PubMed]
66. Ciszewicz, M.; PoLubinska, A.; Antoniewicz, A.; Suminska-Jasinska, K.; BrĘborowicz, A. Sulodexide suppresses inflammation in human endothelial cells and prevents glucose cytotoxicity. *Trans. Res.* **2009**, *153*, 118–123. [CrossRef] [PubMed]
67. Xian-pei, H.; Ke-ji, C.; Zheng-feng, H.; Wei-dong, H.; Ke-dan, C.; Wen-lie, C.; Xu-zheng, C.; Hai-xia, Z.; Ling, C.; Liu-qing, Y.; et al. Glucose endothelial cytotoxicity and protection by Dan Gua-Fang, a Chinese herb prescription in huVEC in hyperglycemia medium. *J. Diab. Compl.* **2009**, *23*, 297–303. [CrossRef] [PubMed]
68. Jing, Q.; Xin, S.M.; Cheng, Z.J.; Zhang, W.B.; Zhang, R.; Qin, Y.W.; Pei, G. Activation of p38 mitogen-activated protein kinase by oxidized LDL in vascular smooth muscle cells mediation via pertussis toxin–sensitive g proteins and association with oxidized LDL-Induced cytotoxicity. *Circ. Res.* **1999**, *84*, 831–839. [CrossRef] [PubMed]
69. Chen, Y.X.; Zhang, X.R.; Xie, W.F.; Li, S. Effects of taurine on proliferation and apoptosis of hepatic stellate cells *in vitro*. *Hepatobil. Pancreat. Dis. Int.* **2004**, *3*, 106–109.
70. Takatani, T.; Takahashi, K.; Uozumi, Y.; Shikata, E.; Yamamoto, Y.; Ito, T.; Matsuda, T.; Schaffer, S.; Fujio, Y.; Azuma, J. Taurine inhibits apoptosis by preventing formation of the Apaf-1/caspase-9 apoptosome. *Am. J. Physiol. Cell Physiol.* **2004**, *287*, C949–C953. [CrossRef] [PubMed]
71. Chang, D.J.; Ringold, G.M.; Heller, R.A. 1992 Cell killing and induction of manganous superoxide dismutase by tumor necrosis factor-(α is mediated by lipoxygenase metabolites of arachidonic acid. *Biochem. Biophys. Res. Commun.* **1995**, *188*, 538–546. [CrossRef]
72. Salganik, R.I. The benefits and hazards of antioxidants: Controlling apoptosis and other protective mechanisms in cancer patients and the human population. *J. Am. College Nutr.* **2001**, *20*, 464S–472S. [CrossRef]
73. Ishii, N.; Ogawa, Z.; Suzuki, K.; Numakami, K.; Saruta, T.; Itoh, H. Glucose loading induces DNA fragmentation in rat proximal tubular cells. *Metabolism* **1996**, *45*, 1348–1353. [CrossRef]
74. Baumgartner-Parzer, S.M.; Wagner, L.; Pettermann, M.; Grillari, J.; Gessl, A.; Waldhäusl, W. High-glucose–triggered apoptosis in cultured endothelial cells. *Diabetes* **1995**, *44*, 1323–1327. [CrossRef]
75. Vaughn, A.E.; Deshmukh, M. Glucose metabolism inhibits apoptosis in neurons and cancer cells by redox inactivation of cytochrome c. *Nat. Cell Biol.* **2008**, *10*, 1477. [CrossRef]
76. Mateo-Fernández, M.; Moreno-Ariza, V.; Balongo-Escobar, M.; Luque-Bravo, L.; Fernández-Bedmar, Z.N.; Martínez-Jurado, M.; Demyda-Peyrás, S.; Merinas-Amo, T. Biological effects of classic and diet soda drinks assessed in *in vivo* and *in vitro* models. *Toxicol. Lett.* **2015**, *238*, S65. [CrossRef]
77. Forchhammer, L.; Ersson, C.; Loft, S.; Möller, L.; Godschalk, R.W.L.; Van Schooten, F.J.; Jones, G.D.D.; Higgins, J.A.; Coke, M.; Mistry, V.; et al. Inter-laboratory variation in DNA damage using a standard comet assay protocol. *Mutagenesis* **2012**, *27*, 665–672. [CrossRef]
78. Olive, P.L.; Banáth, J.P. The comet assay: A method to measure DNA damage in individual cells. *Nat. Protoc.* **2006**, *1*, 23–29. [CrossRef]

79. Fairbairn, D.W.; O'Neill, K.L. Necrotic DNA degradation mimics apoptotic nucleosomal fragmentation comet tail length. *In Vitro Cell. Dev. Biol. Anim.* **1995**, *31*, 171–173. [CrossRef]
80. Mochizuki, T.; Satsu, H.; Shimizu, M. Tumor necrosis factor α stimulates taurine uptake and transporter gene expression in human intestinal Caco-2 cells. *FEBS Lett.* **2002**, *517*, 92–96. [CrossRef]
81. Fabiani, R.; Rosignoli, P.; De Bartolomeo, A.; Fuccelli, R.; Morozzi, G. Genotoxicity of alkene epoxides in human peripheral blood mononuclear cells and HL60 leukaemia cells evaluated with the comet assay. *Mutat. Res.* **2012**, *747*, 1–6. [CrossRef] [PubMed]
82. Ahmad, M.K.; Khan, A.A.; Ali, S.N.; Mahmood, R. Chemoprotective effect of taurine on potassium bromate-induced DNA damage, DNA-protein cross-linking and oxidative stress in rat intestine. *PLoS ONE* **2015**, *10*, e0119137. [CrossRef] [PubMed]
83. Lorenzi, M.; Montisano, D.F.; Toledo, S.; Barrieux, A. High glucose induces DNA damage in cultured human endothelial cells. *J. Clin. Investig.* **1986**, *77*, 322. [CrossRef] [PubMed]
84. Morohoshi, M.; Fujisawa, K.; Uchimuraa, I.; Numano, F. Glucose-dependent interleukin 6 and tumor necrosis factor production by human peripheral blood monocytes *in vitro*. *Diabetes* **1996**, *45*, 954–959. [CrossRef]
85. Weisenberger, D.J.; Campan, M.; Long, T.I.; Kim, M.; Woods, C.; Fiala, E.; Ehrlich, M.; Laird, P.W. Analysis of repetitive element DNA methylation by MethyLight. *Nucl. Acids Res.* **2005**, *33*, 6823–6836. [CrossRef] [PubMed]
86. López-Serra, L.; Esteller, M. Proteins that bind methylated DNA and human cancer: Reading the wrong words. *Br. J. Cancer* **2008**, *98*, 1881–1885. [CrossRef] [PubMed]
87. Martínez, J.G.; Pérez-Escuredo, J.; Castro-Santos, P.; Marcos, C.Á.; Pendás, J.L.L.; Fraga, M.F.; Hermsen, M.A. Hypomethylation of LINE-1, and not centromeric SAT-α, is associated with centromeric instability in head and neck squamous cell carcinoma. *Cell. Oncol.* **2012**, *35*, 259–267. [CrossRef] [PubMed]
88. Waye, J.; Willard, H. Structure, organization, and sequence of alpha satellite DNA from human chromosome 17: Evidence for evolution by unequal crossing-over and an ancestral pentamer repeat shared with the human X chromosome. *Mol. Cell. Biol.* **1986**, *6*, 3156–3165. [CrossRef]
89. Grover, D.; Majumder, P.P.; Rao, C.B.; Brahmachari, S.K.; Mukerji, M. Nonrandom distribution of alu elements in genes of various functional categories: Insight from analysis of human chromosomes 21 and 22. *Mol. Biol. Evol.* **2003**, *20*, 1420–1424. [CrossRef]
90. Lleu, P.L.; Huxtable, R. Phospholipid methylation and taurine content of synaptosomes from cerebral cortex of developing rat. *Neurochem. Int.* **1992**, *21*, 109–118. [CrossRef]
91. Consortium, I.H.G.S. Initial sequencing and analysis of the human genome. *Nature* **2001**, *409*, 860.
92. El-Osta, A.; Brasacchio, D.; Yao, D.; Pocai, A.; Jones, P.L.; Roeder, R.G.; Cooper, M.E.; Brownlee, M. Transient high glucose causes persistent epigenetic changes and altered gene expression during subsequent normoglycemia. *J. Exp. Med.* **2008**, *205*, 2409–2417. [CrossRef]
93. Yun, J.M.; Jialal, I.; Devaraj, S. Epigenetic regulation of high glucose-induced proinflammatory cytokine production in monocytes by curcumin. *J. Nutr. Biochem.* **2011**, *22*, 450–458. [CrossRef] [PubMed]
94. Li, S.; Chen, K.; Li, X.; Zhang, X.; Liu, S.V. A new cultivation system for studying chemical effects on the lifespan of the fruit fly. *Exp. Gerontol.* **2010**, *45*, 158–162. [CrossRef] [PubMed]
95. Ting, D.T.; Lipson, D.; Paul, S.; Brannigan, B.W.; Akhavanfard, S.; Coffman, E.J.; Contino, G.; Deshpande, V.; Lafrate, A.J.; Letovsky, S.; et al. Aberrant overexpression of satellite repeats in pancreatic and other epithelial cancers. *Science* **2011**, *331*, 593–596. [CrossRef] [PubMed]
96. Wild, L.; Flanagan, J.M. Genome-wide hypomethylation in cancer may be a passive consequence of transformation. *Biochim. Biophys. Acta Rev. Cancer* **2010**, *1806*, 50–57. [CrossRef]
97. Wilson, A.S.; Power, B.E.; Molloy, P.L. DNA hypomethylation and human diseases. *Biochim. Biophys. Acta Rev. Cancer* **2007**, *1775*, 138–162. [CrossRef]
98. Ja, W.W.; Carvalho, G.B.; Mak, E.M.; De La Rosa, N.N.; Fang, A.Y.; Liong, J.C.; Brummel, T.; Benzer, S. Prandiology of Drosophila and the CAFE assay. *Proc. Natl. Acad. Sci. USA* **2007**, *104*, 8253–8256. [CrossRef]
99. Yan, J.; Huen, D.; Morely, T.; Johnson, G.; Gubb, D.; Roote, J.; Adler, P.N. The multiple-wing-hairs gene encodes a novel GBD-FH3 domain-containing protein that functions both prior to and after wing hair initiation. *Genetics* **2008**, *180*, 219–228. [CrossRef]
100. Ren, N.; Charlton, J.; Adler, P.N. The flare gene, which encodes the AIP1 protein of *Drosoplila*, functions to regulate F-actin disassembly in pupal epidermal cells. *Genetics* **2007**, *176*, 2223–2234. [CrossRef]
101. Gallagher, R.; Collins, S.; Trujillo, J.; McCredie, K.; Ahearn, M.; Tsai, S.; Metzgar, R.; Aulakh, G.; Ting, R.; Ruscetti, F. Characterization of the continuous, differentiating myeloid cell line (HL-60) from a patient with acute promyelocytic leukemia. *Blood* **1979**, *54*, 713–733. [CrossRef]
102. Tasset-Cuevas, I.; Fernandez-Bedmar, Z.; Lozano-Baena, M.D.; Campos-Sánchez, J.; de Haro-Bailón, A.; Muñoz-Serrano, A.; Alonso-Moraga, A. Protective effect of borage seed oil and gamma linolenic acid on DNA: *In vivo* and *in vitro* studies. *PLoS ONE* **2013**, *8*, e56986. [CrossRef]
103. Abraham, S.K. Antigenotoxicity of coffee in the Drosophila assay for somatic mutation and recombination. *Mutagenesis* **1994**, *9*, 383–386. [CrossRef]
104. Anter, J.; Tasset, I.; Demyda-Peyras, S.; Ranchal, I.; Moreno-Millan, M.; Romero-Jimenez, M.; Muntané, J.; Luque de Castro, M.D.; Serran-Muñoz, A.; Alonso-Moraga, A. Evaluation of potential antigenotoxic, cytotoxic and proapoptotic effects of the olive oil by-product alperujo, hydroxytyrosol, tyrosol and verbascoside. *Mutat. Res. Genet. Toxicol. Environ. Mutagen.* **2014**, *772*, 25–33. [CrossRef] [PubMed]

105. Gyori, B.M.; Venkatachalam, G.; Thiagarajan, P.S.; Hsu, D.; Clement, M.V. OpenComet: An automated tool for comet assay image analysis. *Redox Biol.* **2014**, *2*, 457–465. [CrossRef] [PubMed]
106. Merinas-Amo, T.; Tasset-Cuevas, I.; Díaz-Carretero, A.M.; Alonso-Moraga, A.; Calahorro, F. Role of choline in the modulation of degenerative processes: *In vivo* and *In vitro* studies. *J. Med. Food* **2017**, *20*, 223–234. [CrossRef] [PubMed]
107. Nikolaidis, G.; Raji, O.Y.; Markopoulou, S.; Gosney, J.R.; Bryan, J.; Warburton, C.; Walshaw, M.; Sheard, J.; Field, J.K.; Liloglou, T. DNA methylation biomarkers offer improved diagnostic efficiency in lung cancer. *Cancer Res.* **2012**, *72*, 5692–5701. [CrossRef]
108. Liloglou, T.; Bediaga, N.G.; Brown, B.R.; Field, J.K.; Davies, M.P. Epigenetic biomarkers in lung cancer. *Cancer Lett.* **2014**, *342*, 200–212. [CrossRef]

molecules

MDPI

Article

Screening of Wood/Forest and Vine By-Products as Sources of New Drugs for Sustainable Strategies to Control *Fusarium graminearum* and the Production of Mycotoxins

Mathilde Montibus [1], Xavier Vitrac [2], Véronique Coma [3], Anne Loron [3], Laetitia Pinson-Gadais [4], Nathalie Ferrer [4], Marie-Noëlle Verdal-Bonnin [4], Julien Gabaston [5], Pierre Waffo-Téguo [5], Florence Richard-Forget [4] and Vessela Atanasova [4,*]

1 Institut Technologique FCBA, Allée de Boutaut, BP 227, F-33028 Bordeaux, France; mathilde.montibus@fcba.fr
2 Laboratoire Phenobio SAS, ZA Les Pins Verts, 22 Allée de Migelane, F-33650 Saucats, France; xavier.vitrac@phenobio.fr
3 Laboratoire de Chimie des Polymères Organiques, Université de Bordeaux, CNRS, Bordeaux INP, UMR 5629, 16 Avenue Pey-Berland, F-33600 Pessac, France; veronique.coma@u-bordeaux.fr (V.C.); anne.loron@enscbp.fr (A.L.)
4 INRAE, UR1264 Mycology and Food Safety (MycSA), F-33882 Villenave d'Ornon, France; laetitia.pinson-gadais@inrae.fr (L.P.-G.); nathalie.ferrer@inrae.fr (N.F.); marie-noelle.verdal-bonnin@inrae.fr (M.-N.V.-B.); florence.forget@inrae.fr (F.R.-F.)
5 Faculté des Sciences Pharmaceutiques, Unité de Recherche Œnologie, EA 4577, USC 1366 INRAE, Equipe Molécules d'Intérêt Biologique-ISVV, Université de Bordeaux, F-33882 Villenave d'Ornon, France; julien.gabaston@gmail.com (J.G.); pierre.waffo-teguo@u-bordeaux.fr (P.W.-T.)
* Correspondence: vessela.atanasova@inrae.fr; Tel.: +33-557-122-497

Citation: Montibus, M.; Vitrac, X.; Coma, V.; Loron, A.; Pinson-Gadais, L.; Ferrer, N.; Verdal-Bonnin, M.-N.; Gabaston, J.; Waffo-Téguo, P.; Richard-Forget, F.; et al. Screening of Wood/Forest and Vine By-Products as Sources of New Drugs for Sustainable Strategies to Control *Fusarium graminearum* and the Production of Mycotoxins. *Molecules* **2021**, *26*, 405. https://doi.org/10.3390/molecules26020405

Received: 23 December 2020
Accepted: 12 January 2021
Published: 14 January 2021

Publisher's Note: MDPI stays neutral with regard to jurisdictional claims in published maps and institutional affiliations.

Abstract: *Fusarium graminearum* is a fungal pathogen that can colonize small-grain cereals and maize and secrete type B trichothecene (TCTB) mycotoxins. The development of environmental-friendly strategies guaranteeing the safety of food and feed is a key challenge facing agriculture today. One of these strategies lies on the promising capacity of products issued from natural sources to counteract crop pests. In this work, the in vitro efficiency of sixteen extracts obtained from eight natural sources using subcritical water extraction at two temperatures was assessed against fungal growth and TCTB production by *F. graminearum*. Maritime pine sawdust extract was shown to be extremely efficient, leading to a significant inhibition of up to 89% of the fungal growth and up to 65% reduction of the mycotoxin production by *F. graminearum*. Liquid chromatography/mass spectrometry analysis of this active extract revealed the presence of three families of phenolics with a predominance of methylated compounds and suggested that the abundance of methylated structures, and therefore of hydrophobic compounds, could be a primary factor underpinning the activity of the maritime pine sawdust extract. Altogether, our data support that wood/forest by-products could be promising sources of bioactive compounds for controlling *F. graminearum* and its production of mycotoxins.

Keywords: *Fusarium graminearum*; type B trichothecenes; natural extracts; ecological strategies; biofungicides

1. Introduction

Fusarium graminearum was recently ranked as the fourth most important fungal plant pathogen regarding scientific and economic criteria [1]. This fungus infests cereal crops and can produce type B trichothecenes (TCTB) mycotoxins that are of major concern due to their toxicity to animals and humans [2]. TCTB include deoxynivalenol (DON) and its acetylated forms 3-acetyl-4-deoxynivalenol and 15-acetyl-4-deoxynivalenol (3ADON and 15ADON), and nivalenol (NIV) and its acetylated form 4-acetylnivalenol or fusarenon X (FX). DON is the most commonly found TCTB on crops. DON is regulated in many countries worldwide, with a strict limitation set at 1750 $\mu g\ kg^{-1}$ for unprocessed maize, durum wheat and oats in the European Union (European Commission in the Commission Regulation No 1126/2007), at 2000 μg kg^{-1} for unprocessed wheat in Canada and at 1000 μg kg^{-1} for cereals in China, for example [3].

Combined with good agricultural practices, the use of fungicides is a key factor in the integrated management strategies aiming to tackle this economical and public health issue. Their repeated use over decades has, however, disrupted natural biological systems and resulted in the development of fungal resistance [4]. In addition, the European regulation REACH and the French Ecophyto plans have been established to decrease the use of synthetic fungicides over the next few years. There is therefore a major need for bio-based and non-toxic alternatives, efficient to prevent *Fusarium* growth and its production of mycotoxins.

In the context of biopesticide formulations to control phytopathogenic fungi such as *F. graminearum*, a wide range of plant extracts has showed a broad spectrum of activity. Several of these extracts were characterized by a richness in terpenes [5,6] and phenolic compounds, for which antifungal properties have been widely described, including their capacity to reduce the mycelial growth of *F. graminearum* and its mycotoxin production [7–9]. Caffeic acid [10] or ferulic acid [11] have been the most extensively studied for their ability to inhibit the yield of TCTB while tetrahydrocurcumin was shown to minimize the accumulation of fumonisins produced by *Fusarium proliferatum* [12]. Compared to pure molecules, natural crude or partially purified extracts from plants could be of interest due to the potential benefit resulting from synergic effects between different active substances [13]. Another advantage is that a mixture of active molecules with different fungal physiological targets can limit the development of resistance.

Forests and wood by-products are currently poorly valorized, although they could be promising sources of extracts characterized by a high abundance of phenolic compounds or terpenes. Thus, extracts of seven Amazonian woods including *Acacia mangium* (acacia), *Paraserianthes falcataria* (sengon) and *Swietenia mahagoni* (mahoni) barks have been shown to efficiently reduce wood rot diseases [14,15]. Similarly, the capacity of spruce bark extracts to inhibit the development of *Plasmopara viticola*, the causal agent of downy mildew on grapevine, has been reported [16]. Extracts from Norway spruce barks have been characterized for their potential to reduce the mycelial growth of different fungi as well as bacteria [17].

One of the guiding ideas of the present work was to exploit the richness in potential bioactive compounds of wood/forest and vine by-products and provide a use for this neglected biomass. Indeed, vineyard management often implies cane and vine-shoot pruning, leaf trimming and grub up vineyards and was reported to generate annually about 5 tons of solid waste per hectare [18]. In the same way, wood/forest wastes are a versatile raw material that can be used for various applications. According to a Parliament briefing, in 2012, 52.9 million tons of wood waste were produced in the European Union [19].

Driven by environmental issues, the objective of this work was to investigate the biological activity of natural extracts obtained from wood/forest and vine by-products using a subcritical water extraction procedure. After characterization of the total phenolic content and the antioxidant activity of each extract, their capacity to restrain the fungal growth of *F. graminearum* and to inhibit the production of TCTB were evaluated. Additionally, to better understand composition/bioactivity relationships, a deeper study of the phenolic composition of the most active extract was carried out using liquid chromatography/mass spectrometry.

2. Results

2.1. Characterization of Wood/Forest and Vine By-Product Extracts

2.1.1. Plant Extraction Yield

A subcritical water extraction method was used to produce sixteen extracts from eight wood/forest and vine by products. Subcritical water is defined as hot water at sufficient pressure to maintain the liquid state at a critical temperature between 100 °C (the boiling point of water) and 374 °C (the critical point of water) under the critical pressure (1–22.1 MPa). As the temperature increases, the dielectric constant of the water changes, so that the polar, medium polar, weakly polar, and nonpolar compounds can be extracted differentially. In this study, two temperatures, 125 °C and 175 °C, were used for the extraction, leading to extracts designed as extract 125 °C and extract 175 °C.

The obtained extraction yields associated with the different plant materials are presented in Table 1. Our results showed how the extraction yields varied with samples, from 1.1% for locust chips 125 °C, to 7.9% for maritime pine needles 175 °C. Extracts obtained from the same plant source but with different extraction conditions resulted in similar extraction yields with the exception of vine root extracts (extraction yield of vine roots 175 °C was 3.7 times higher than that of vine roots 125 °C). The highest extraction yields were observed for the two maritime pine needle extracts (7–8%), vine roots 175 °C (7%), the two vine cane extracts (5–6%), and the two maritime pine bark extracts (5%). Lower extraction yields, below 5%, were obtained for the two locust chip, chestnut chip, oak chip, maritime pine sawdust extracts and the vine root 125 °C extract.

Table 1. Origin, extraction yield, concentration, and total phenolic content of natural extracts.

Plant Species	Natural Extract	Extraction Yield (%) [1]	Concentration (g L^{-1}) [2]	Total Phenolic Content (mg g^{-1})
		Forest and Wood By-Products		
Pinus pinaster	Maritime pine barks 175 °C	5.0	10.0	334.8 ± 4.7
	Maritime pine barks 125 °C	4.7	9.5	346.6 ± 6.2
	Maritime pine sawdust 175 °C	2.3	4.6	121.3 ± 17.5
	Maritime pine sawdust 125 °C	1.8	3.7	65.8 ± 0.3
	Maritime pine needles 175 °C	7.9	15.9	127.8 ± 3.2
	Maritime pine needles 125 °C	7.3	14.6	96.7 ± 2.4
Quercus robur	Oak chips 175 °C	2.5	5.1	351.1 ± 11.1
	Oak chips 125 °C	2.5	5.1	277.9 ± 5.7
Castanea sativa	Chestnut chips 175 °C	2.6	5.2	533.1 ± 7.4
	Chestnut chips 125 °C	2.1	4.2	551.6 ± 27.4
Robinia pseudoacacia	Locust chips 175 °C	1.6	3.1	340.5 ± 4.2
	Locust chips 125 °C	1.1	2.1	320.6 ± 4.8
		Vine By-Products		
Vitis vinifera	Vine canes 175 °C	5.3	10.5	94.2 ± 1.7
	Vine canes 125 °C	5.7	11.3	91.7 ± 3.7
	Vine roots 175 °C	7.1	14.1	120.9 ± 2.7
	Vine roots 125 °C	1.9	3.8	65.6 ± 1.7

[1] Extraction yield % = weight of extracted plant residues/weight of plant raw sample × 100; [2] Concentration: g of dry matter in one-liter of natural extract.

2.1.2. Total Phenolic Content and Antioxidant Activity of Wood/Forest and Vine By-Product Extracts

The total phenolic content of the natural extracts considered in the present study are gathered in Table 1. The antioxidant activity values can be retrieved in Figure 1. According to their phenolic concentration, the extracts can be classified into three groups:

Low total phenolic concentration (<200 mg g^{-1}) for maritime pine sawdust, maritime pine needle, vine cane, and vine root extracts obtained at 125 and 175 °C.

Intermediate total phenolic concentration (ranging from 200 to 400 mg g^{-1}) for maritime pine bark, oak chip, and locust chip extracts obtained at 125 and 175 °C.

High total phenolic concentration (>500 mg g^{-1}) for chestnut chip extracts obtained at 125 and 175 °C.

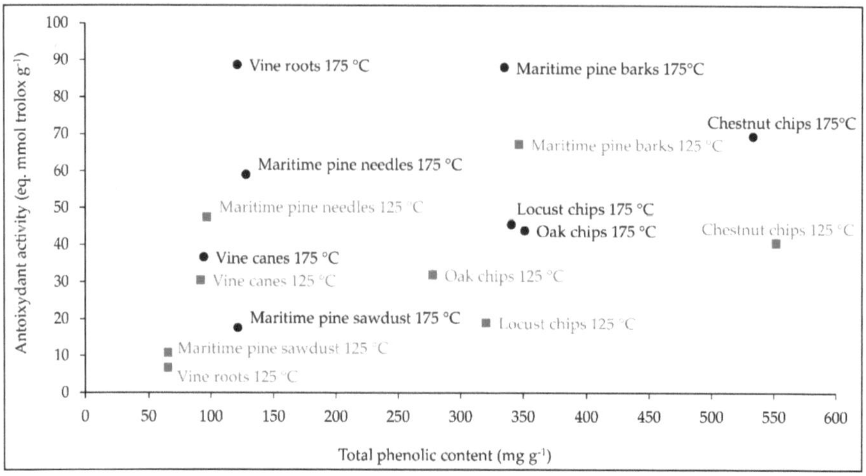

Figure 1. Correlation between total phenolic content and antioxidant activity of extracts obtained at 175 °C (black circles) and extracts obtained at 125 °C (gray squares). Data are means using three biological replicates.

Overall, total phenolic concentration and antioxidant activity of extracts obtained from the same plant source at the two temperatures were quite similar, with the exception of vine root and locust chip extracts where the antioxidant activity of the extracts 175 °C was 13 and 2 times higher than that in the extracts 125 °C, respectively (Figure 1).

As shown in Figure 1, a significant positive correlation was observed between antioxidant activity and total phenolic content ($r > 0.351$, $p < 0.015$). However, for some extracts, such as vine roots 175 °C and maritime pine needles 175 °C, the high antioxidant activity (88.64 and 59.05 eq. mmol trolox g^{-1}, respectively) was not associated with a high phenolic content.

2.2. Screening of the Wood/Forest and Vine By-Product Extracts for Their Antifungal Activity Against *F. graminearum* CBS 185.32

A first screening assay was implemented to assess whether the wood/forest and vine by-products considered in the present study could be promising sources of subcritical water extractable antifungal compounds. The effect of crude extracts obtained from the same initial quantity of raw materials on the radial growth of *F. graminearum* CBS 185.32 strain was investigated in Potato Dextrose Agar (PDA) medium. The extract activities were expressed as the percentage of growth inhibition/activation compared to the corresponding control (Figure 2). Our data indicated that the maritime pine sawdust 175 °C, maritime pine sawdust 125 °C, vine cane 125 °C, and maritime pine bark 125 °C extracts led to a significant inhibition of the fungal growth with inhibition percentages ranging between 35 and 89%. The highest inhibition rate was obtained for the maritime pine sawdust source and the 175 °C extraction procedure. None of the other extracts had any statistically significant effect on the fungal growth. Indeed, the inhibitions observed in PDA media with the maritime pine bark 125 °C, vines cane 125 °C, maritime pine needle 175 °C and vine root 175 °C extracts, ranging between 20–33%, were not statistically significant. In contrast, several extracts including oak chips 175 °C, locust chips 175 °C and maritime pine needles 175 °C were shown to increase non-significantly the fungal growth by 34, 29, and 27% compared to the control, respectively.

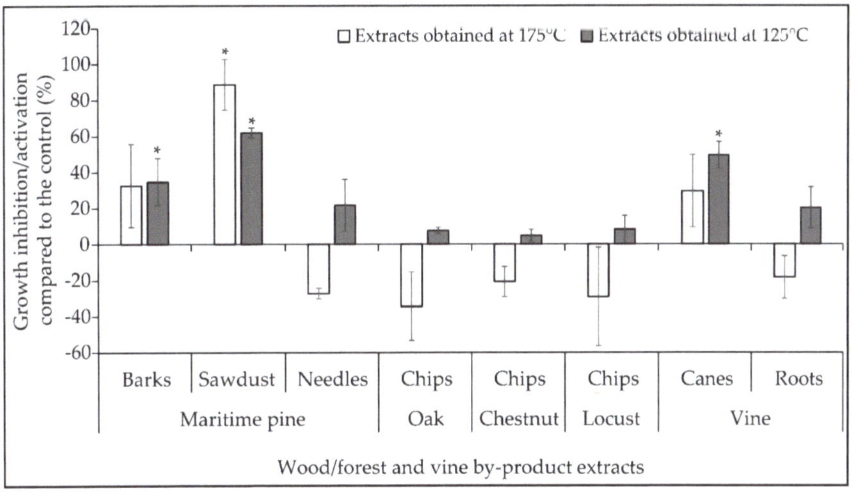

Figure 2. Percentage of inhibition/activation of the radial growth of *Fusarium graminearum* by the natural extracts obtained at 175 °C (white bars) and extracts obtained at 125 °C (gray bars). Data are means ± standard deviation using three biological replicates. Asterix (*) indicates significant differences when compared with control (Student's t test, control versus treated, $p < 0.05$). The inhibitory activity of extracts was calculated according to the following formula: percentage of mycelial growth inhibition % = [(area in control culture − area in extract-amended culture)/area in control culture] × 100.

No correlation was evidenced between the antifungal activity of the extracts and the extract yield, neither with the antioxidant activity nor the total phenolic content.

We further investigated the effect of the most promising extracts, i.e., maritime pine sawdust 175 °C, maritime pine sawdust 125 °C, vines cane 125 °C, and maritime pine bark 125 °C extracts, on the production of TCTB by *F. graminearum* CBS 185.32.

2.3. Impact of Selected Wood/Forest and Vine By-Product Extracts on the Mycelial Biomass of *F. graminearum* and the Production of TCTB in Liquid Cultures

2.3.1. Comparative Efficiencies of Maritime Pine Sawdust 175 °C, Maritime Pine Sawdust 125 °C, Vine Cane 125 °C and Maritime Pine Bark 125 °C Extracts to Affect the Fungal Growth and the Production of TCTB by F. graminearum CBS 185.32

The effect of the maritime pine sawdust 175 °C, maritime pine sawdust 125 °C, vine cane 125 °C and maritime pine bark 125 °C extracts on TCTB production by *F. graminearum* CBS 185.32 was studied in the appropriate toxin-inducing liquid Minimal Synthetic medium (MS) [8,10]. Each natural extract was tested at five concentrations: 100, 200, 300, 500, and 1000 mg L^{-1}; results are gathered in Figure 3a,b.

With the exception of maritime pine bark 125 °C extract at 100 and 500 mg L^{-1} and vine cane 125 °C extract at 100 mg L^{-1}, the different extracts induced a significant increase in the biomass ranging between 9 and 69% (Figure 3a). The highest activation rate (69%) was observed for the maritime pine bark 125 °C extract at 1000 mg L^{-1}.

DON and 15ADON were the major TCTB produced by the *F. graminearum* CBS185.32 strain in MS medium. Because the DON/15ADON ratio was not significantly affected by the extracts, the sum of the two mycotoxins was used to express the amount of TCTB produced.

The maritime pine bark 125 °C and maritime pine sawdust 125 °C extracts at all the tested concentrations did not affect the TCTB production, with the exception of maritime pine barks 125 °C at 500 mg L^{-1} that induced a significant inhibition of 36%. The supplementation of liquid cultures with vine cane 125 °C extract at 100, 300, and 500 mg L^{-1} led to a significant reduction close to 46% in TCTB accumulation. The maritime pine sawdust 175 °C extract at all the tested concentrations was characterized by the highest potential to reduce the TCTB production; inhibitions of 53, 65, and 61% were observed with 200, 500, and 1000 mg L^{-1}, respectively.

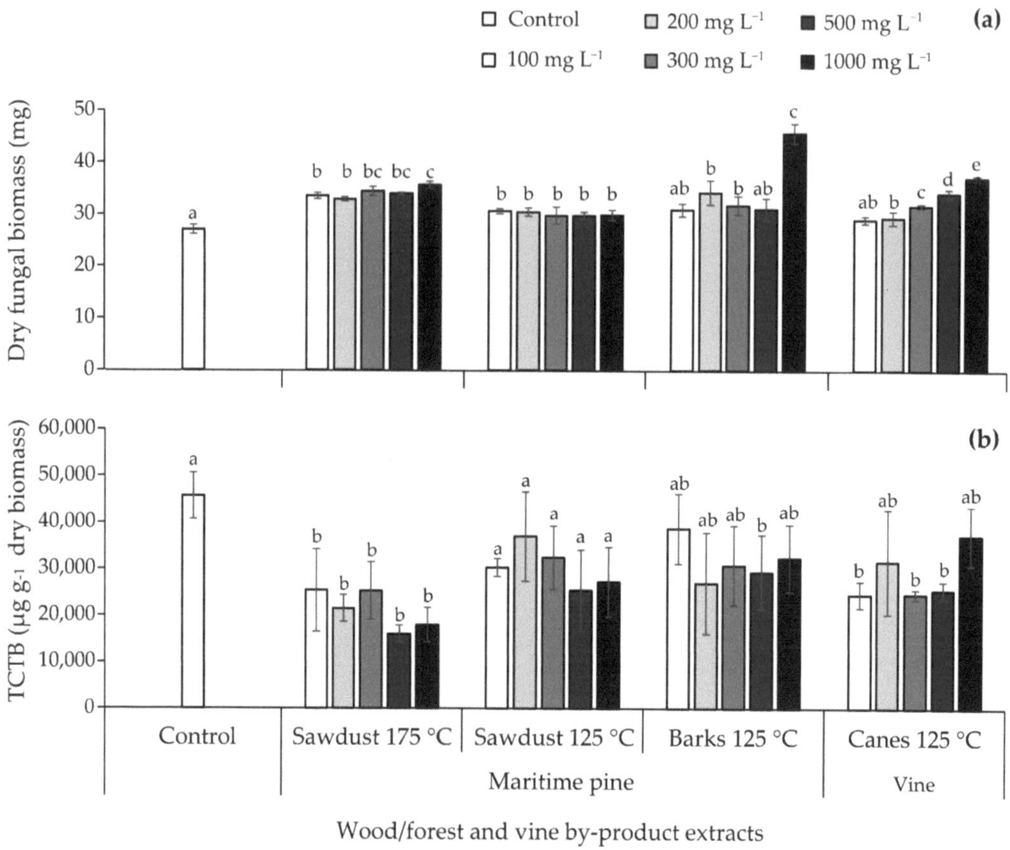

Figure 3. Effect of natural extracts obtained at 125 °C or 175 °C at 100, 200, 300, 500, and 1000 mg L^{-1} on (**a**) fungal biomass and (**b**) type B trichothecenes (TCTB) production by *Fusarium graminearum* CBS 185.32. Values are expressed as means ± standard deviation using four biological replicates. Differences between control and tested extract concentrations in terms of fungal biomass and TCTB were determined separately for each extract with multiple comparisons using the Dunn–Sidak method. For each extract, different letters indicate significant differences ($p < 0.05$).

Overall, our data indicated that among the four extracts selected for their capacity to restrain the radial fungal growth in PDA medium, the maritime pine sawdust 175 °C extract was the most effective to reduce the TCTB yield with an inhibition reaching more than 50%. The 200 and 500 mg L^{-1} concentrations were selected for the next step of our work.

2.3.2. Effect of the Maritime Pine Sawdust 175 °C Extract on the Fungal Biomass and TCTB Yield by a Panel of *F. graminearum* Strains

To confirm the promising potential of the maritime pine sawdust 175 °C extract, its effect on the TCTB production was investigated in liquid MS medium using six additional strains of *F. graminearum*. Results on the fungal biomass and the TCTB accumulation are reported in Figure 4a,b, respectively.

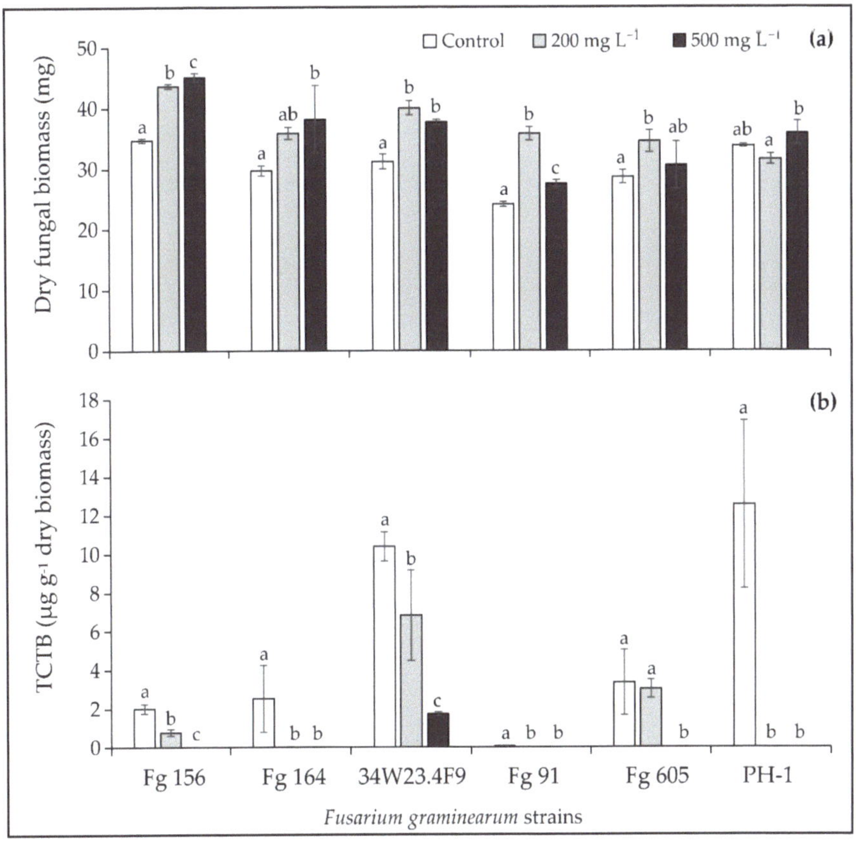

Figure 4. Effect of the maritime pine sawdust extract obtained at 175 °C at 200 and 500 mg L^{-1} on (**a**) fungal biomass and (**b**) type B trichothecenes (TCTB) production by six *Fusarium graminearum* (Fg) strains. Values are expressed as means ± standard deviation using four biological replicates. Differences between control and treated conditions in terms of fungal biomass and TCTB were determined separately for each fungal strain with multiple comparisons and the Dunn–Sidak method. For each fungal strain, different letters indicate significant differences between tested conditions ($p < 0.05$).

Whatever the *F. graminearum* strain, the maritime pine sawdust 175 °C extract at the two tested concentrations had no effect or significantly activated the fungal biomass, which corroborates the results obtained with the *F. graminearum* CBS 185.32 strain (Figure 4a).

As it clearly appears in Figure 4b, the TCTB amounts yielded by the six studied strains were significantly reduced by the treatments with the maritime pine sawdust 175 °C extract. At 500 mg L^{-1}, the extract led to a complete inhibition of the mycotoxin accumulation by Fg 156, Fg 164, Fg 91, Fg 605, and PH-1, and induced an inhibition of 83% for the 34W23.4F9 strain. Whereas the TCTB yield was abrogated with 200 mg L^{-1} for the PH-1 strain, the decrease in TCTB accumulation was less drastic for the other strains (99, 61, 56 and 34% for Fg 164, Fg 156, Fg 91 and 34W23.4F9, respectively). Besides, no significant inhibition of TCTB production was observed for the Fg 605 strain in the broths supplemented with 200 mg L^{-1} of maritime pine sawdust 175 °C extract.

In conclusion, the maritime pine sawdust 175 °C extract can substantially inhibit the TCTB biosynthesis by *F. graminearum*, with an inhibition efficiency that varies according to the considered strain.

2.4. Characterization of the Phenolic Composition of the Maritime Pine Sawdust 175 °C Extract

The soluble phenolic composition of the maritime pine sawdust 175 °C extract was characterized using liquid chromatography/diode-array detector coupled with mass spectrometry (LC-DAD/MS). Eleven peaks characterized by a high intensity were unequivocally identified by comparison with retention times, mass and UV-visible spectra of reference standards and quantified using their respective external calibration curves (Table 2, Figure 5). According to identification data, phenolic compounds were shown to belong to three main groups: phenolic acids/aldehydes/alcohols, lignans, and flavonoids.

Table 2. Phenolic compounds identified in the active maritime pine sawdust extract obtained at 175 °C.

Compounds	Retention Time (min)	λmax (nm)	$(M-H)^-$	Concentration ($mg\ g^{-1}$ of Extract)
		Phenolic Acids/Aldehydes/Alcohols		
Protocatechuic acid	1.8	260	153	0.25 (± 0.01)
Vanillic acid	3.7	262	167	0.53 (± 0.06)
Caffeic acid	3.8	325	179	0.15 (± 0.01)
Coniferyl alcohol	4.6	264	179	4.89 (± 0.22)
Vanillin	4.7	280	151	0.64 (± 0.01)
Ferulic acid	5.1	325	193	0.35 (± 0.01)
Coniferyl aldehyde	5.9	340	177	0.83 (± 0.01)
Unknown [1]	10.6	253	329	1.90 (± 0.06)
Lignans				
Nortrachelogenin	6.9	281	373	7.30 (± 0.25)
Pinoresinol [2]	7.7	281	357	2.84 (± 0.29)
Flavonoids				
Pinobanksin [3]	9.2	289	271	1.01 (± 0.01)
Pinocembrin	11.8	289	255	0.41 (± 0.02)
Total				21.09 (± 0.40)

[1] Quantified using coniferyl alcohol as standard. [2] Quantified using nortrachelogenin as standard. [3] Quantified using pinocembrin as standard. Standard deviation using three replicates in brackets.

The most representative group was the phenolic acids/aldehydes/alcohols with seven identified molecules (Table 2). Regarding the concentration values, coniferyl alcohol was the predominant compound with a concentration reaching 4.89 $mg\ g^{-1}$, followed by coniferyl aldehyde, vanillin, vanillic acid, ferulic acid, protocatechuic acid, and caffeic acid. One additional peak eluted at the end of the chromatogram (10.5 min) was characterized by a UV-visible spectrum with λmax at 253 and 300 nm, i.e., with a slight hypsochromic effect in comparison with coniferyl alcohol (λmax at 264 and 300 nm), suggesting that this unidentified compound could belong to the same family. The compound eluting in this peak presented a deprotonated molecular ion at m/z 329. As we could not identify this compound, its concentration was estimated to be the equivalent of coniferyl alcohol (1.90 $mg\ g^{-1}$).

The highest concentrations of phenolic compounds were measured in the group of lignans. Two lignans were identified and quantified in the active fraction: nortrachelogenin and pinoresinol at 7.30 $mg\ g^{-1}$ and 2.84 $mg\ g^{-1}$, respectively. The nortrachelogenin was the most abundant identified compound in the active extract.

Among the flavonoids, two compounds, the pinobanksin and its precursor (the pinocembrin), were identified and quantified at 1.01 and 0.41 $mg\ g^{-1}$, respectively.

Regarding the chemical structure of identified molecules, it is interesting to note that seven (coniferyl alcohol, coniferyl aldehyde, vanillin, vanillic acid, ferulic acid, nortrachelogenin, and pinoresinol) of the eleven identified compounds were mono- or di-methylated. The sum of the concentrations of methylated compounds was 17.38 $mg\ g^{-1}$, which represents 82% of the quantified compounds.

To summarize, 11 compounds belonging to three different groups of phenolics were unequivocally identified in the active maritime pine sawdust 175 °C extract. The group of phenolic acids contained the largest number of molecules and was estimated to represent 0.95% ($w\ w^{-1}$) of the dried biomass of the extract. The main compound in this group was coniferyl alcohol representing 0.49% ($w\ w^{-1}$) of the dried biomass. The group of lignans was represented by two compounds: nortrachelogenin and pinoresinol, contributed to 1.01% ($w\ w^{-1}$) of the dried biomass. The lowest phenolic concentrations were associated with flavonoids (about 0.14% ($w\ w^{-1}$)). The mono- and di-methylated phenolic molecules were shown to predominate qualitatively and quantitatively in the active extract.

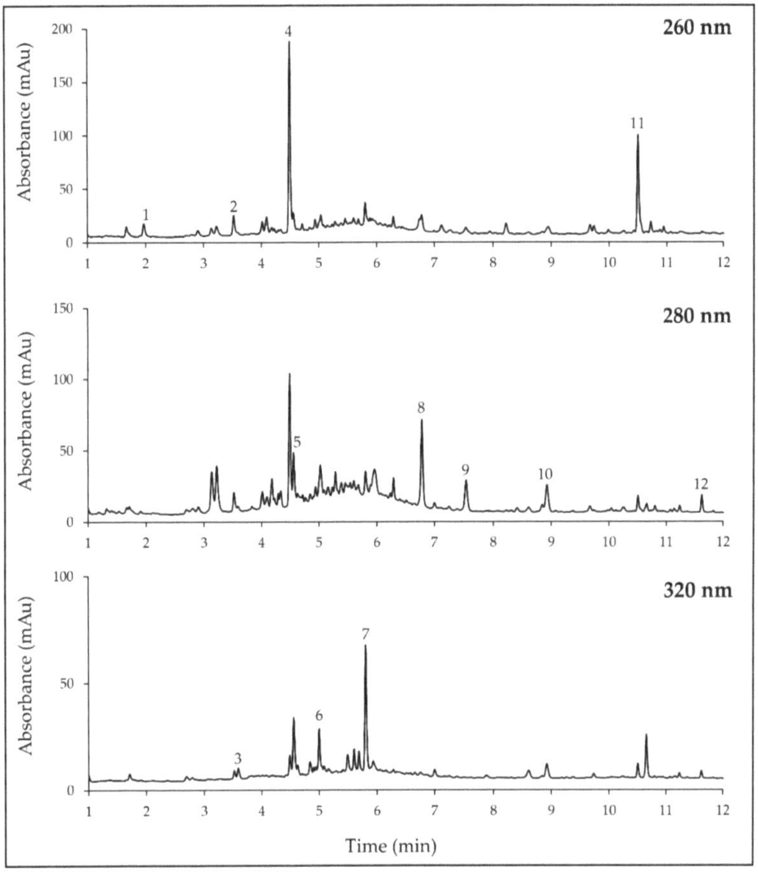

Figure 5. Chromatograms at 260, 280, and 320 nm of the active maritime pine sawdust extract obtained at 175 °C. 1, protocatechuic acid; 2, vanillic acid; 3, caffeic acid; 4, coniferyl alcohol; 5, vanillin; 6, ferulic acid; 7, coniferyl aldehyde; 8, nortrachelogenin; 9, pinoresinol; 10, pinobanksin; 11, unknown; 12, pinocembrin.

3. Discussion

The aim of this study was to characterize natural extracts obtained from wood/forest and vine by-products using an environment-friendly extraction procedure for their capacity to reduce *F. graminearum* growth and TCTB accumulation, and to provide first insights regarding the compounds involved in their bioactivity.

3.1. Variations in Antioxidant Activity and Total Phenolic Composition of Wood/Forest and Vine By-Product Extracts

In the first part of this work, sixteen extracts were obtained from eight natural sources using a dynamic subcritical water extraction method and two experimental conditions, 125 and 175 °C. Characterization of the total phenolic composition and antioxidant activity of these extracts led to values ranging from 65.6 to 551.6 mg g^{-1} and from 6.7 to 88 eq. mmol trolox g^{-1}, respectively. Concerning total phenolic composition, values associated with maritime pine barks (more than 300 mg g^{-1}) were higher than the results previously reported for barks, between 22 and 62 mg g^{-1} depending on the extraction method [20]. Regarding the chestnut source, our results were fully consistent with previous data indicating a total phenolic content of approximatively 500 mg g^{-1} [21,22]. Concerning the antioxidant activity, the highest values (ranging from 66 to 88 eq. mmol Trolox g^{-1}) were obtained for chestnut chip 175 °C, maritime pine bark 125 °C, maritime pine bark 175 °C and vine root 175 °C extracts. For pine bark extracts, antioxidant potential was higher than that obtained by Legault at al. [23] for different bark extracts using the same oxygen radical absorbance capacity method (ORAC). However, the comparison of our data regarding total phenolic composition and the antioxidant activity with previously published results have to be considered with caution as these values are dependent on method, condition, and origin of the raw material.

In the present study, the antioxidant activity of the extracts was positively correlated with total phenolic contents and seemed to be influenced by phenolic composition. The same trend has frequently been reported in previous studies focusing on natural extracts [24,25]. The high levels of antioxidant activity of chestnut chip 175 °C, maritime pine bark 125 °C, and maritime pine bark 175 °C extracts could be related to their phenolic composition. Maritime pine bark extracts are acknowledged to contain high levels of procyanidins (condensed tannins) [20], whereas chestnut chip extracts have been reported to contain high amounts of both procyanidins and ellagitannins (hydrolysable tannins) [26]. All these extract constituents are polymerized structures containing many hydroxyl functional groups which are supposed to contribute to the antioxidant activity of the extracts. Indeed, Yokozawa et al. [27] postulated that an increase in molecular weight, number of hydroxyl and galloyl groups, and ortho-hydroxyl structure improved the antioxidant activity of molecules, being important features for the scavenging of free radicals. On the contrary, when the free hydroxyl group was methoxylated or glycosylated, the antioxidant activity was decreased or even abolished [27]. Accordingly, the lower antioxidant activity of the active maritime pine sawdust 175 °C extract could be explained, on the one hand, by the presence of a multitude of methylated compounds at high amounts, and on the other hand, by its relatively low concentrations in total phenolics (121.3 mg g^{-1}).

However, in the present study, several extracts including maritime pine needle 175 °C and vine root 175 °C extracts were characterized by a high antioxidant activity but a low total phenolic content. The latter finding could be explained by the fact that the presence of other non-phenolic compounds could contribute to the overall antioxidant potential. Pine extracts could contain some lipophilic metabolites such as terpenoids and resin acids that certainly contribute to the antioxidant activity [28,29].

3.2. The Maritime Pine Sawdust 175 °C Extract Is a Strong Inhibitor of Fungal Growth and TCTB Production by *F. graminearum* CBS 185.32

Among the sixteen natural extracts studied in the present work, the two maritime pine sawdust, the vine cane 125 °C and the maritime pine bark 125 °C extracts were shown as the most promising for their capacity to restrain the *F. graminearum* CBS 185.32 radial growth. Several studies have previously highlighted the antifungal or antibacterial activity of extracts obtained from various parts of the pine tree [16,30–32]. However, to our knowledge, the present work is the first report demonstrating the antifungal activity of maritime pine sawdust extract against *F. graminearum*. For example, extracts from Palestian Aleppo pine seeds, barks, and cones were reported to reduce by 80–95% the growth of *Shigella*, *Escherichia coli* and *Staphylococcus aureus* [32]. In the study of Jung et al. [31], the effect of pine needle extracts was shown against several fungi and furfural was demonstrated as the active molecule responsible for the growth inhibition of *Alternia mali*. Spruce bark extract, which is considered as a wood industry by-product, was also identified as a potential inhibitor of downy mildew (*Plasmopara viticola*) [16]. While the antimicrobial activity of pine extracts has been the subject of several studies, to our knowledge, their effect on mycotoxin production by different mycotoxigenic fungi has never been reported previously. For the first time, the present study evidenced the capacity of pine sawdust extract 175 °C to efficiently limit the production of TCTB by a panel of *F. graminearum* strains.

Several hypotheses can be raised to explain the activities of the pine sawdust 175 °C extract. Numerous studies have shown positive relations between antifungal efficiency of natural extracts, contents in total phenolics and antioxidant potential, i.e., free radical scavenging capacity [33,34]. In addition, several authors have suggested that the efficiency of phenolic acids in reducing TCTB production could be linked to their antioxidant properties [7,35,36]. However, in the present study, no correlation was found between the antifungal effect of plant extracts, their total phenolic contents and antioxidant activities. Such an outcome was previously observed for 22 extracts from African tropical wood species [37]. The previous results suggest that in addition to the free radical scavenging activity, other mechanisms could explain the bioactivity of the extract. In accordance with the literature data [38], the capacity of natural compounds to disrupt the membrane integrity, perturb the ionic homeostasis and/or to directly interact with key fungal enzymes are additional properties that could explain their effect. Besides, lipophilic properties of natural extracts/compounds have been shown as primary factors sustaining the perturbations they can induce to fungal membranes.

We have demonstrated that the active pine sawdust 175 °C extract was characterized by a qualitative and quantitative predominance of mono- and di-methylated phenolic compounds. Previous studies indicated that the main classes of natural antimicrobial phenolic compounds were methylated phenolics [39,40]. For example, 5,6,7,8-tetramethoxyflavone and 4′-hydroxy-5,6,7,8-tetramethoxyflavone characterized by four methoxyl groups were active against *Candida albicans* and *Staphylococcus aureus* [40]. In the study of Fitzgerald et al. [39], the structure-antifungal activity analysis of vanillin and some structural analogues has demonstrated that 4-anisaldehyde (one methoxyl group) showed lower Minimal Inhibitory Concentration (MIC) values than 4-hydroxybenzaldehyde (one hydroxyl group). Similarly, Koh et al. [41] have reported that pterostilbene (two methoxyl and one hydroxyl groups), but not resveratrol (three hydroxyl groups), detained potent fungicidal and sporicidal activities against *Leptosphaeria maculans* and could protect canola seedlings from blackleg infection. Recent studies, dealing with the efficiency of hydroxycinnamic acids to inhibit fungal growth and toxin biosynthesis by *Fusarium avenaceum, Aspergillus westerdijkiae* and *Penicillium verrucosum*, supported the key importance of methoxyl groups [42,43]. Since it is widely acknowledged that substitution with methoxyl groups contributes to the lipophilicity of a compound, we can suppose that the predominance of mono- and di-methylated phenolics in the maritime pine 175 °C extract partially explains its high efficiency.

The analysis of the phenolic composition of the maritime pine sawdust 175 °C extract has evidenced the occurrence of some phenolic compounds in which antifungal efficiency has been previously illustrated. According to Peng et al. [44], the pinocembrin flavonoid is efficient to control the growth of *Penicillium italicum* by interfering with energy homeostasis and inducing cell membrane damage in the pathogen. Nortrachelogenin has also been described as an efficient antifungal agent against *Candida albicans* [45]. According to the former study, high concentrations of nortrachelogenin were shown to induce membrane disruption, whereas low concentrations were responsible for a mitochondria and metacaspase-dependent apoptotic mechanism. In addition, by studying structure-activity relationships of phenolic acids, Shalaby et al. [46] postulated that the highest toxicity of ferulic acid (one methoxyl group) could be explained by its limited metabolization by *Cochliobolus heterostrophus*. The previous authors indicated that the lowest toxicity of coumaric and caffeic acids (one and two hydroxyl groups, respectively) could be related to their capacity to induce the expression of fungal genes encoding phenolic acid degrading enzymes. Besides, the afore published studies suggested that the inhibition of the trichothecene and enniatin mycotoxin production by ferulic acid could be explained by the reduction of the expression of some key biosynthetic genes suggesting the occurrence of a transcriptional control exerted by phenolic acid [9,11,42].

Finally, the antifungal and anti-mycotoxin activities of maritime pine sawdust 175 °C extract can certainly not be ascribed to the action of a single active molecule, but to the result of synergistic interactions between several compounds. Besides, other molecules, not identified in this study, in synergy or not with phenolics, certainly contribute to the bioactivity of the extract.

4. Materials and Methods

4.1. Chemicals and Standards

For LC/MS analysis, LC-MS-grade acetonitrile from VWR (Fontenay-Sous-Bois, France) and formic acid from Fisher Scientific (Loughborough, U.K.) were purchased while water was purified by an Elga water purification system (High Wycombe, U.K.). Reagents were purchased from Scharlau (Barcelona, Spain), Sigma Aldrich (St Louis, MO, USA), and Fischer Scientific (Loughborough, U.K.).

Vanillic acid, protocatechuic acid, caffeic acid, ferulic acid, gallic acid, vanillin, coniferyl aldehyde, and pinocembrin were purchased from Sigma-Aldrich (St Louis, MO, USA) and coniferyl alcohol was obtained from Extrasynthese (Genay, France). Nortrachelogenin was isolated and purified from pine extract in the Molécules d'Intérêt Biologique laboratory (ISVV, Villenave d'Ornon, France) [47].

Standard solution of TCTB were purchased from Romer Labs (Tulln, Austria).

4.2. Natural Sources and Plant Material Preparation Prior to Extraction

The eight natural sources used in this study were collected in the Nouvelle Aquitaine Région, France in 2017. All sources came from agricultural by-products (vine canes and vine roots), and wood/forest by-products (maritime pine needles, barks and sawdust; oak chips; chestnut chips; and locust chips). The ethnobotanical data of used plant species are summarized in Table 1. Wood/forest by-products were supplied by the LESBATS society and vine by-products were collected by the Laboratory PHENOBIO. Upon receipt, the natural sources were dried at 40 °C during 7–10 days, ground into fine powder to pass through a 0.5 mm sieve, and stored at −20 °C.

4.3. Preparation of Natural Extracts and Determination of Extraction Yield and Extract Concentration

Dynamic subcritical water device used in this study included a High Performance Liquid Chromatography (HPLC) pump, a 300 mL extraction vessel, a heating device, a pressure controller, and a collector. For extraction of each raw material, 50 g was loaded into the high-pressure vessel. The vessel was placed in an oven set at 120 °C, and water was heated with a heating jacket at the chosen temperature (125 °C or 175 °C). The outlet valve of the extraction vessel was then closed and the system was pressurized to the desired pressure of 15×10^5 Pa at a constant flow rate of 15 mL min^{-1}. The liquid-to-solid ratio was maintained at the value of 5, to obtain 250 mL of extract. After the extraction vessel, the extract passed through a cooling system, was collected in a sampling vessel and then stored at −20 °C for further analysis.

The concentration of natural extracts was determined by lyophilizing an aliquote of 50 mL of each extract and weighing a dry sample. Extract concentration was expressed as a gram of dry matter per liter of extract. Extraction yield was calculated using the following formula: extraction yield % = (weight of extracted plant/weight of plant raw sample) × 100.

4.4. *Fusarium* Strains, Media, and Culture Conditions

Seven *F. graminearum* strains were used in this study; their origin and characteristics are summarized in Table 3.

When inoculum was required, fungal strains were grown on PDA at 25 °C for one week in the dark. Spore suspensions were prepared in carboxymethyl cellulose medium (CMC) (15 g L^{-1} carboxymethyl cellulose, 1 g L^{-1} yeast extract, 0.5 g L^{-1} $MgSO_4{\cdot}7H_2O$, 1 g L^{-1} NH_4NO_3, 1 g L^{-1} KH_2PO_4). Four agar blocks (5 mm in diameter) carrying mycelia were introduced into 20 mL of CMC medium and incubated in darkness at 25 °C and at 180 rpm in a Multitron incubator shaker (INFORS AG, Bottmingen, Switzerland). After 2–3 days, culture broths were filtered in order to remove the mycelia. Spores were collected after centrifugation (2300× *g* for 10 min) and washed two times with sterile distilled water. Spores were suspended in 2% tween (for experimentations in PDA medium) or in MS medium (for experimentation in liquid MS medium), and concentration was determined using a spores counting method.

Table 3. Origin and characteristics of *Fusarium graminearum* strains used in this study.

Strain	Source	Host	Country	Chemotype
Fg 605	INRAE MycSA collection, France [1]	Unknown	Unknown	DON/15ADON
Fg 156	INRAE MycSA collection, France [1]	Wheat	France	DON/15ADON
Fg 164	INRAE MycSA collection, France [1]	Wheat	France	DON/15ADON
Fg 91	INRAE MycSA collection, France [1]	Corn	France	NIV/FX
34W23.4F9	AgResearch, Hamilton, New Zealand [48]	Unknown	New Zealand	DON/15ADON
PH-1	Fungal Genetic Stock Center, USA	Corn	USA	DON/15ADON
CBS 185.32	Westerdijk Fungal Biodiversity Institute, The Netherlands [2]	Corn	Unknown	DON/15ADON

[1] INRAE MycSA strains are stored in the International Center for Microbial Resources-Filamentous Fungi, Marseille, France (https://www6.inrae.fr/cirm_eng/Filamentous-Fungi/Strains-catalogue). [2] Westerdijk Fungal Biodiversity Institute, previously Centraalbureau voor Schimmelcultures (CBS). Fg, *Fusarium graminearum*; DON, deoxynivalenol; 15ADON, 15-acetyl-4-deoxynivalenol; NIV, nivalenol; FX, fusarenon X.

To screen the natural extracts at their initial concentration (Table 1) for their antifungal activity, a solid PDA medium and the CBS 185.32 strain were used. One milliliter of each natural extract was mixed with one milliliter of spore suspension at 10^4 spores mL^{-1}. Three times 20 μL of each mixture were equidistantly placed in a Petri dich (90 mm diameter). Three replications were performed for each individual treatment. Appropriate positive (Amphotericin B at 250 mg L^{-1}) and negative controls were done. Cultures were incubated in the dark for 4 days at 25 °C. The extent of *F. graminearum* growth on each sample was photographed at day 2 and 3, and the fungal surface area was determined using the ImageJ software package developed by the National Institutes of Health (Bethesda, Maryland, USA). The inhibitory activity of extracts was calculated according to the following formula: percentage of mycelial growth inhibition % = [(area in control culture − area in extract-amended culture)/area in control culture] × 100.

Liquid-culture tests were done in MS medium as described by Gauthier at al. [8] Medium was supplemented or not with natural extracts at 100, 200, 300, 500, and 1000 mg L^{-1} depending of experimentation. Petri plates containing 8 mL of MS medium were inoculated with 2 × 10^4 spores mL^{-1} and incubated in darkness at 25 °C for 10 days. Cultures were stopped by centrifugation (3000× *g* for 10 min) and 4 mL of each supernatant were stored at −20 °C before TCTB analysis. Fungal biomass production was measured by weighing the mycelia after 48 h of freeze-drying. For each trial, it was ensured that supplementation with extracts did not modify the pH values of the treated batches compared to controls. Four replications were done for each condition.

4.5. Extraction and TCTB Analysis

TCTB were extracted as described by Gauthier et al. [8] Quantification of mycotoxins was performed on a 1100 Series HPLC chain, equipped with an autosampler, an Agilent photodiode array detector and the ChemStation software (Agilent, Les Ulis, France). Separation was achieved on a Kinetex XB-C18 − 100 Å column (150 × 4.6 mm; 2.6 µm) (Phenomenex, Le Pecq, France) maintained at 45 °C. Mobile phase consisted of water acidified with *ortho*-phosphoric acid at pH 2.6 (solvent A) and acetonitrile (solvent B). TCTB were separated using a gradient elution as follows: 7 to 30% B in 10 min, 30–90% B in 5 min, 90% B for 5 min, 90 to 7% B for 2 min, 7% B for 5 min. The flow was kept at 1 mL min^{-1}. The injection volume was 5 µL. The UV-visible spectra were recorded from 190 to 400 nm and peak areas were measured at 230 nm. Quantification was performed by external calibration. Toxin yields were expressed in µg g^{-1} of dry biomass.

4.6. Determination of Total Phenolic Content

The total phenolic content was spectrophotometrically measured according to 96-well microplates modified with the Folin–Ciocalteu method as previously described [49]. A microplate spectrophotometer (BMG Labtech, Champigny sur Marne, France) was used for measurement. Each well was filled with 20 µL of distilled water or natural extracts, followed by 100 µL of the Folin–Ciocalteu commercial solution diluted to 1/10 in water (Sigma-Aldrich, St Quentin Fallavier, France) and 80 µL of 7.5% (*w*/*v*) Na_2CO_3 solution. Prior to measuring the absorbance at 760 nm, the mixture was incubated for 20 min under dark conditions at room temperature. Gallic acid (0.03–0.5 mg L^{-1}) was used as standard for calibration. The results were given as the mean of three determinations.

4.7. Determination of Free Radical Scavenging Potential by the Oxygen Radical Absorbance Capacity (ORAC)

The oxygen radical absorbance capacity analysis was carried out using 96-well fluorescence microplates and a FLUOstar Optima microplate reader (BMG LabTech, Champigny sur Marne, France). The reaction was carried out in a phosphate buffer (75 mM, pH 7.4); 30 µL of the extract solution, 180 µL of fluorescein (117 nM final concentration), and 90 µL of dihydrochloride (AAPH) at 40 mM were successively added to each well. The mixture was shaken and left to stand for 1.5 h at 37 °C. Fluorescence was recorded every minute during this period at excitation and emission wavelengths of 485 and 530 nm respectively. A blank sample (phosphate buffer replaced the sample) and Trolox® calibration solutions (1–40 µM) were also performed simultaneously on the same microplate. The area under the curve (AUC) was calculated for each extract sample by integrating its relative fluorescence curve. By subtracting the AUC of the blank, the net AUC of the extracts was calculated and correlated with Trolox® concentrations.

4.8. LC/MS Analysis

Analysis were performed using a 1290 Series UHPLC from Agilent Technologies (Santa Clara, CA, USA) equipped with an auto sampler module, a binary pump with degasser, a column heater/selector and an UV-visible diode-array detector. The UHPLC apparatus was coupled to an Esquire 6000 ion trap mass spectrometer using an electrospray ionization source (Bruker Daltonics, Billerica, MA, USA).

Chromatographic elution was carried out on a 100 mm × 2.1 mm i.d., 1.8 µm, Zorbax SB-C18 column, with a 2.1 mm × 5 mm i.d. guard column (Agilent Technologies, Santa Clara, CA, USA). A binary solvent system composed of water acidified with 0.1% formic acid (solvent A) and acetonitrile acidified with 0.1% formic acid (solvent B) was used at a flow rate of 0.4 mL min^{-1} with the following gradient: 0–1.7 min, 10% B; 1.7–5.1 min, 10–30% B; 5.1–6.8 min, 30% B; 6.8–8.5 min, 30–35% B; 8.5–11.9 min, 35–60% B; 11.9–15.3 min, 60–100% B; 15.3–17 min, 100% B; 17–17.3 min, 100–10% B. Parameters of mass spectrometry were set in negative mode with a range of m/z 100–1200 with nitrogen as the drying gas at 10 L min^{-1}, nebulizer pressure of 40 psi at 365 °C, capillary voltage at 3100 V, capillary exit voltage at −118.3 V, skimmer voltage at −40 V, and the trap drive at 58.1.

Maritime pine sawdust 175 °C extract was dissolved in methanol/water (3:7, v/v) at 10 mg mL^{-1}, filtered on 0.45 μm and injected at 1 μL. The phenolic compounds were characterized by using commercial standards and the polyphenol library developed in Molécules d'Intérêt Biologique laboratory (ISVV, Villenave d'Ornon, France) [47,50]. Identification was performed by comparing retention time, UV data (λmax) as well as mass spectrometry data with pseudomolecular ion and product ions. Quantification was carried out for each compound with its own calibration curve at the maximum wavelength. Pure standards were injected at several concentrations (0, 1, 5, 20, 50, 100 μg mL^{-1}) in independent triplicate to obtain calibration and equation curves. All data were processed with Bruker Data Analysis 3.2 software (Bruker Daltonics, Billerica, MA, USA).

4.9. Expression of Results and Statistical Analyses

Results of fungal growth and TCTB production were reported as mean values ± standard deviation of three or four biological replications depending of the experimentation.

The significance of the effect of the extracts on radial fungal growth in PDA medium was tested with Student's t test (control versus treated).

For experimentations in liquid MS medium, one-way ANOVAs for TCTB and fungal biomass were carried out. Differences between control and tested extract concentrations in terms of fungal biomass and TCTB were determined separately for each extract or each strain with multiple comparison tests using the Dunn–Sidak method.

The Pearson correlation coefficient (r) was used to study the relationship between the antioxidant activity and the total phenolic content and between the antifungal efficiency of extracts and the total phenolic content, the antioxidant activity and the plant extraction yields.

The level of significance was set at $p = 0.05$.

Statistical analyses were performed using XLSTAT 2019 software (Addinsoft, Rennes, France).

5. Conclusions

In the present study, sixteen plant extracts were produced from eight wood/forest and vine by-products using a subcritical water extraction method, i.e., without the use of toxic solvents. On the one hand, the obtained findings showed that this green technology is appropriate for valorizing forest and agricultural by-products by producing extracts possessing various antioxidant properties and different levels of total phenolic content. Such extracts could be promising sources of bioactive molecules for different fields such as pharmaceutical, cosmetic, or food industries. On the other hand, our findings showed that several wood/forest and vine by-product extracts could be promising sources of subcritical water extractable inhibitors of *F. graminearum* radial growth. In addition, one extract from maritime pine sawdust was demonstrated to possess a high anti-mycotoxin capacity. This extract shows a great potential to be used for the development of biofungicide to combat *F. graminearum* and its TCTB production on agricultural commodities.

The phenolic characterization of the maritime pine sawdust 175 °C extract allowed us to make first hypotheses regarding its composition in active molecules and suggested that methylated compounds could be essential for its antifungal and anti-mycotoxin properties. However, it is most likely that biological activities are the result of a mixture of compounds rather than individual molecules, and that additive, synergistic, and/or antagonist effects between compounds of the same or different groups may be existing.

Further studies are necessary to determine the chemical identity of the active molecules responsible for the observed antifungal and anti-mycotoxin effectiveness. This essential step will allow providing the required basis to potentiate the properties of the maritime pine sawdust extract.

Author Contributions: Conceptualization, F.R.-F. and V.A.; Data curation, L.P.-G., F.R.-F. and V.A.; Formal analysis, V.A.; Funding acquisition, M.M. and V.A.; Investigation, X.V., A.L., N.F., M.-N.V.-B., J.G. and P.W.-T.; Methodology, V.A.; Project administration, F.R.-F. and V.A.; Resources, M.M., X.V., P.W.-T., F.R.-F. and V.A.; Supervision, M.M., L.P.-G. and V.A.; Validation, X.V., J.G., F.R.-F. and V.A.; Visualization, L.P.-G. and V.A.; Writing—original draft, M.M., X.V., V.C., A.L., L.P.-G., J.G., F.R.-F. and V.A.; Writing—review and editing, M.M., V.C., A.L., L.P.-G., F.R.-F. and V.A. All authors have read and agreed to the published version of the manuscript.

Funding: This research was funded by INRAE and the Nouvelle Aquitaine Region, France (BioAXoM project, N°2017-1R20203, 2007-2021). LC/MS experiments were performed at MetaboHUB-Bordeaux facility and supported by MetaboHUB (ANR-11-INBS-0010).

Data Availability Statement: The data presented in this study are available on request from the corresponding author.

Acknowledgments: Authors gratefully acknowledge the LESBATS society for supplying wood/forest by-products. Authors are thankful to Chloé Agro and Jérémy Esteves for technical support.

Conflicts of Interest: The authors declare no conflict of interest.

Sample Availability: Not available.

References

1. Dean, R.; Van Kan, J.A.L.; Pretorius, Z.A.; Hammond-Kosack, K.E.; Di Pietro, A.; Spanu, P.D.; Rudd, J.J.; Dickman, M.; Kahmann, R.; Ellis, J.; et al. The Top 10 fungal pathogens in molecular plant pathology. *Mol. Plant Pathol.* **2012**, *13*, 414–430. [CrossRef] [PubMed]
2. Pestka, J.J. Deoxynivalenol: Mechanisms of action, human exposure, and toxicological relevance. *Arch. Toxicol.* **2010**, *84*, 663–679. [CrossRef] [PubMed]
3. Mahdjoubi, C.K.; Arroyo-Manzanares, N.; Hamini-Kadar, N.; García-Campaña, A.M.; Mebrouk, K.; Gámiz-Gracia, L. Multi-mycotoxin occurrence and exposure assessment approach in foodstuffs from Algeria. *Toxins* **2020**, *12*, 194. [CrossRef] [PubMed]
4. Chen, Y.; Zhou, M.-G. Characterization of *Fusarium graminearum* isolates resistant to both carbendazim and a new fungicide. *Phytopathology* **2009**, *99*, 441–446. [CrossRef]
5. Cowan, M.M. Plant products as antimicrobial agents. *Clin. Microbiol. Rev.* **1999**, *12*, 19. [CrossRef]
6. Nazzaro, F.; Fratianni, F.; Coppola, R.; Feo, V.D. Essential oils and antifungal activity. *Pharmaceuticals* **2017**, *10*, 86. [CrossRef]
7. Ponts, N.; Pinson-Gadais, L.; Boutigny, A.L.; Barreau, C.; Richard-Forget, F. Cinnamic-derived acids significantly affect *Fusarium graminearum* growth and in vitro synthesis of type B trichothecenes. *Phytopathology* **2011**, *101*, 929–934. [CrossRef]
8. Gauthier, L.; Bonnin-Verdal, M.-N.; Marchegay, G.; Pinson-Gadais, L.; Ducos, C.; Richard-Forget, F.; Atanasova-Penichon, V. Fungal biotransformation of chlorogenic and caffeic acids by *Fusarium graminearum*: New insights in the contribution of phenolic acids to resistance to deoxynivalenol accumulation in cereals. *Int. J. Food Microbiol.* **2016**, *221*, 61–68. [CrossRef]
9. Ferruz, E.; Atanasova-Pénichon, V.; Bonnin-Verdal, M.-N.; Marchegay, G.; Pinson-Gadais, L.; Ducos, C.; Lorán, S.; Ariño, A.; Barreau, C.; Richard-Forget, F. Effects of phenolic acids on the growth and production of T-2 and HT-2 Toxins by *Fusarium langsethiae* and *F. sporotrichioides*. *Molecules* **2016**, *21*, 449. [CrossRef]
10. Atanasova-Penichon, V.; Legoahec, L.; Bernillon, S.; Deborde, C.; Maucourt, M.; Verdal-Bonnin, M.-N.; Pinson-Gadais, L.; Ponts, N.; Moing, A.; Richard-Forget, F. Mycotoxin biosynthesis and central metabolism are two interlinked pathways in *Fusarium graminearum*, as demonstrated by the extensive metabolic changes induced by caffeic acid exposure. *Appl. Environ. Microbiol.* **2018**, *84*, e01705-17. [CrossRef]
11. Boutigny, A.-L.; Barreau, C.; Atanasova-Penichon, V.; Verdal-Bonnin, M.-N.; Pinson-Gadais, L.; Richard-Forget, F. Ferulic acid, an efficient inhibitor of type B trichothecene biosynthesis and *Tri* gene expression in *Fusarium* liquid cultures. *Mycol. Res.* **2009**, *113*, 746–753. [CrossRef]
12. Coma, V.; Portes, E.; Gardrat, C.; Richard-Forget, F.; Castellan, A. In vitro inhibitory effect of tetrahydrocurcuminoids on *Fusarium proliferatum* growth and fumonisin B-1 biosynthesis. *Food Addit. Contam. A* **2011**, *28*, 218–225. [CrossRef]
13. Valette, N.; Perrot, T.; Sormani, R.; Gelhaye, E.; Morel-Rouhier, M. Antifungal activities of wood extractives. *Fungal Biol. Rev.* **2017**, *31*, 113–123. [CrossRef]
14. Rodrigues, A.M.S.; Stien, D.; Eparvier, V.; Espindola, L.S.; Beauchêne, J.; Amusant, N.; Leménager, N.; Baudassé, C.; Raguin, L. The wood preservative potential of long-lasting Amazonian wood extracts. *Int. Biodeter. Biodegr.* **2012**, *75*, 146–149. [CrossRef]
15. Rosdiana, N.A.; Dumarçay, S.; Gérardin, C.; Chapuis, H.; Santiago-Medina, F.J.; Sari, R.K.; Syafii, W.; Gelhaye, E.; Raharivelomanana, P.; Mohammed, R.; et al. Characterization of bark extractives of different industrial Indonesian wood species for potential valorization. *Ind. Crops Prod.* **2017**, *108*, 121–127. [CrossRef]
16. Gabaston, J.; Richard, T.; Biais, B.; Waffo-Teguo, P.; Pedrot, E.; Jourdes, M.; Corio-Costet, M.-F.; Mérillon, J.-M. Stilbenes from common spruce (Picea abies) bark as natural antifungal agent against downy mildew (*Plasmopara viticola*). *Ind. Crops Prod.* **2017**, *103*, 267–273. [CrossRef]

17. Burčová, Z.; Kreps, F.; Greifová, M.; Jablonský, M.; Ház, A.; Schmidt, Š.; Šurina, I. Antibacterial and antifungal activity of phytosterols and methyl dehydroabietate of Norway spruce bark extracts. *J. Biotechnol.* **2018**, *282*, 18–24. [CrossRef] [PubMed]
18. Zacharof, M.-P. Grape winery waste as feedstock for bioconversions: Applying the biorefinery concept. *Waste Biomass Valor.* **2017**, *8*, 1011–1025. [CrossRef]
19. Bourgignon, D.; European Parliament. European Parliamentary Research Service. Understanding Waste Streams: Treatment of Specific Waste. Briefing. 2015. Available online: https://www.europarl.europa.eu/thinktank/en/document.html?reference=EPRS_BRI(2015)564398 (accessed on 12 January 2021).
20. Chupin, L.; Motillon, C.; Charrier-El Bouhtoury, F.; Pizzi, A.; Charrier, B. Characterisation of maritime pine (*Pinus pinaster*) bark tannins extracted under different conditions by spectroscopic methods, FTIR and HPLC. *Ind. Crops Prod.* **2013**, *49*, 897–903. [CrossRef]
21. Vázquez, G.; Fontenla, E.; Santos, J.; Freire, M.S.; González-Álvarez, J.; Antorrena, G. Antioxidant activity and phenolic content of chestnut (*Castanea sativa*) shell and eucalyptus (*Eucalyptus globulus*) bark extracts. *I Ind. Crops Prod.* **2008**, *28*, 279–285. [CrossRef]
22. Aires, A.; Carvalho, R.; Saavedra, M.J. Valorization of solid wastes from chestnut industry processing: Extraction and optimization of polyphenols, tannins and ellagitannins and its potential for adhesives, cosmetic and pharmaceutical industry. *J. Waste Manag.* **2016**, *48*, 457–464. [CrossRef] [PubMed]
23. Legault, J.; Girard-Lalancette, K.; Dufour, D.; Pichette, A. Antioxidant potential of bark extracts from boreal forest conifers. *Antioxidants* **2013**, *2*, 77–89. [CrossRef] [PubMed]
24. Aspé, E.; Fernández, K. Comparison of phenolic extracts obtained of *Pinus radiata* bark from pulp and paper industry and sawmill industry. *Maderas Cienc. Technol.* **2011**, *13*, 243–252. [CrossRef]
25. Jerez, M.; Selga, A.; Sineiro, J.; Torres, J.L.; Núñez, M.J. A comparison between bark extracts from *Pinus pinaster* and *Pinus radiata*: Antioxidant activity and procyanidin composition. *Food Chem.* **2007**, *100*, 439–444. [CrossRef]
26. De Vasconcelos, M.C.; Bennett, R.N.; Rosa, E.A.; Ferreira-Cardoso, J.V. Composition of European chestnut (*Castanea sativa Mill.*) and association with health effects: Fresh and processed products. *J. Sci. Food Agric.* **2010**, *90*, 1578–1589. [CrossRef]
27. Yokozawa, T.; Chen, C.P.; Dong, E.; Tanaka, T.; Nonaka, G.-I.; Nishioka, I. Study on the inhibitory effect of tannins and flavonoids against the 1,1-diphenyl-2-picrylhydrazyl radical. *Biochem. Pharmacol.* **1998**, *56*, 213–222. [CrossRef]
28. Conde, E.; Díaz-Reinoso, B.; Moure, A.; Hemming, J.; Willför, S.M.; Domínguez, H.; Parajó, J.C. Extraction of phenolic and lipophilic compounds from pinus pinaster knots and stemwood by supercritical CO_2. In Proceedings of the III Iberoamerican Conference on Supercritical Fluids—PROSCIBA, Cartagena de Indias, Colombia, 1–5 April 2013.
29. Willför, S.; Hemming, J.; Reunanen, M.; Holmbom, B. Phenolic and lipophilic extractives in scots pine knots and stemwood. *Holzforschung* **2003**, *57*, 359–372. [CrossRef]
30. Maimoona, A.; Naeem, I.; Saddiqe, Z.; Jameel, K. A review on biological, nutraceutical and clinical aspects of French maritime pine bark extract. *J. Ethnopharmacol.* **2011**, *133*, 261–277. [CrossRef]
31. Jung, K.-H.; Yoo, S.-K.; Moon, S.-K.; Lee, U.-S. Furfural from pine needle extract inhibits the growth of a plant pathogenic fungus, *Alternaria mali*. *Mycobiology* **2007**, *35*, 39–43. [CrossRef]
32. Salim, H.; Rimawi, W.H.; Shaheen, S.; Mjahed, A. Phytochemical analysis and antibacterial activity of extracts from Palestinian Aleppo pine seeds, bark and cones. *Asian. J. Chem.* **2019**, *31*, 143–147. [CrossRef]
33. Ferreira-Santos, P.; Genisheva, Z.; Botelho, C.; Santos, J.; Ramos, C.; Teixeira, J.A.; Rocha, C.M.R. Unravelling the biological potential of *Pinus pinaster* bark extracts. *Antioxidants* **2020**, *9*, 334. [CrossRef] [PubMed]
34. Kim, H.; Lee, B.; Yun, K.W. Evaluation of antimicrobial activity and total phenolic content of three *Pinus* species. *J. Ecol. Environ.* **2013**, *36*, 57–63. [CrossRef]
35. Montibus, M.; Pinson-Gadais, L.; Richard-Forget, F.; Barreau, C.; Ponts, N. Coupling of transcriptional response to oxidative stress and secondary metabolism regulation in filamentous fungi. *Crit. Rev. Microbiol.* **2015**, *41*, 295–308. [CrossRef] [PubMed]
36. Reverberi, M.; Ricelli, A.; Zjalic, S.; Fabbri, A.A.; Fanelli, C. Natural functions of mycotoxins and control of their biosynthesis in fungi. *Appl. Microbiol. Biotechnol.* **2010**, *87*, 899–911. [CrossRef]
37. Huang, Z.; Hashida, K.; Makino, R.; Kawamura, F.; Shimizu, K.; Kondo, R.; Ohara, S. Evaluation of biological activities of extracts from 22 African tropical wood species. *J. Wood Sci.* **2009**, *55*, 225–229. [CrossRef]
38. Atanasova-Penichon, V.; Barreau, C.; Richard-Forget, F. Antioxidant secondary metabolites in cereals: Potential involvement in resistance to *Fusarium* and mycotoxin accumulation. *Front. Microbiol.* **2016**, *7*. [CrossRef]
39. Fitzgerald, D.J.; Stratford, M.; Gasson, M.J.; Narbad, A. Structure−function analysis of the vanillin molecule and its antifungal properties. *J. Agric. Food Chem.* **2005**, *53*, 1769–1775. [CrossRef]
40. Bastos, M.; Lima, M.; Conserva, L.M.; Andrade, V.S.; Rocha, E.M.; Lemos, R.P. Studies on the antimicrobial activity and brine shrimp toxicity of *Zeyheria tuberculosa* (*Vell.*) *Bur.* (*Bignoniaceae*) extracts and their main constituents. *Ann. Clin. Microbiol. Antimicrob.* **2009**, *8*, 16. [CrossRef]
41. Koh, J.C.O.; Barbulescu, D.M.; Salisbury, P.A.; Slater, A.T. Pterostilbene is a potential candidate for control of blackleg in *Canola*. *PLoS ONE* **2016**, *11*, e0156186. [CrossRef]
42. Gautier, C.; Pinson-Gadais, L.; Verdal-Bonnin, M.-N.; Ducos, C.; Tremblay, J.; Chéreau, S.; Atanasova, V.; Richard-Forget, F. Investigating the efficiency of hydroxycinnamic acids to inhibit the production of enniatins by *Fusarium avenaceum* and modulate the expression of enniatins biosynthetic genes. *Toxins* **2020**, *12*, 735. [CrossRef]

43. Boonmee, S.; Atanasova, V.; Chéreau, S.; Marchegay, G.; Hyde, K.D.; Richard-Forget, F. Efficiency of hydroxycinnamic phenolic acids to inhibit the production of ochratoxin A by *Aspergillus westerdijkiae* and *Penicillium verrucosum*. *Int. J. Mol. Sci.* **2020**, *21*, 8548. [CrossRef] [PubMed]
44. Peng, L.; Yang, S.; Cheng, Y.J.; Chen, F.; Pan, S.; Fan, G. Antifungal activity and action mode of pinocembrin from propolis against *Penicillium italicum*. *Food Sci. Biotechnol.* **2012**, *21*, 1533–1539. [CrossRef]
45. Lee, H.; Woo, E.-R.; Lee, D.G. (-)-Nortrachelogenin from *Partrinia scabiosaefolia* elicits apoptotic response in *Candida albicans*. *FEMS Yeast Res.* **2016**, fow013. [CrossRef] [PubMed]
46. Shalaby, S.; Larkov, O.; Lamdan, N.-L.; Goldshmidt-Tran, O.; Horwitz, B.A. Plant phenolic acids induce programmed cell death of a fungal pathogen: MAPK signaling and survival of *Cochliobolus heterostrophus*. *Environ. Microbiol.* **2016**, *18*, 4188–4199. [CrossRef] [PubMed]
47. Gabaston, J.; Richard, T.; Cluzet, S.; Palos Pinto, A.; Dufour, M.-C.; Corio-Costet, M.-F.; Mérillon, J.-M. *Pinus pinaster* knot: A source of polyphenols against *Plasmopara viticola*. *J. Agric. Food Chem.* **2017**, *65*, 8884–8891. [CrossRef]
48. Cooney, J.M.; Lauren, D.R.; di Menna, M.E. Impact of competitive fungi on trichothecene production by *Fusarium graminearum*. *J. Agric. Food Chem.* **2001**, *49*, 522–526. [CrossRef]
49. Yammine, S.; Delsart, C.; Ghidossi, R.; Vitrac, X.; Mietton Peuchot, M.; Ghidossi, R. Characterisation of polyphenols and antioxidant potential of red and white pomace by-product extracts using subcritical water extraction. *OENO One* **2020**, *54*. [CrossRef]
50. Gabaston, J.; Leborgne, C.; Waffo-Téguo, P.; Pedrot, E.; Richard, T.; Mérillon, J.-M.; Valls Fonayet, J. Separation and isolation of major polyphenols from maritime pine (*Pinus pinaster*) knots by two-step centrifugal partition chromatography monitored by LC-MS and NMR spectroscopy. *J. Sep. Sci.* **2020**, *43*, 1080–1088. [CrossRef]

molecules

MDPI

Article

Cheminformatics-Based Identification of Potential Novel Anti-SARS-CoV-2 Natural Compounds of African Origin

Samuel K. Kwofie [1,2,3], Emmanuel Broni [1], Seth O. Asiedu [4], Gabriel B. Kwarko [2], Bismark Dankwa [4], Kweku S. Enninful [4], Elvis K. Tiburu [1,2] and Michael D. Wilson [3,4,*]

[1] Department of Biomedical Engineering, School of Engineering Sciences, College of Basic and Applied Sciences, University of Ghana, Legon P.O. Box LG 54, Accra, Ghana; skkwofie@ug.edu.gh (S.K.K.); ebroni002@st.ug.edu.gh (E.B.); etiburu@ug.edu.gh (E.K.T.)

[2] West African Centre for Cell Biology of Infectious Pathogens, Department of Biochemistry, Cell and Molecular Biology, University of Ghana, Legon P.O. Box LG 54, Accra, Ghana; gabrielbrako@ymail.com

[3] Department of Medicine, Loyola University Medical Center, Maywood, IL 60153, USA

[4] Department of Parasitology, Noguchi Memorial Institute for Medical Research, University of Ghana, Legon P.O. Box LG 581, Accra, Ghana; sethasieduosei@gmail.com (S.O.A.); bdankwa@noguchi.ug.edu.gh (B.D.); kenninful@noguchi.ug.edu.gh (K.S.E.)

* Correspondence: MWilson@noguchi.ug.edu.gh; Tel.: +233-276-018892

Abstract: The coronavirus disease 2019 (COVID-19) pandemic caused by the severe acute respiratory syndrome virus 2 (SARS-CoV-2) has impacted negatively on public health and socioeconomic status, globally. Although, there are currently no specific drugs approved, several existing drugs are being repurposed, but their successful outcomes are not guaranteed. Therefore, the search for novel therapeutics remains a priority. We screened for inhibitors of the SARS-CoV-2 main protease and the receptor-binding domain of the spike protein from an integrated library of African natural products, compounds generated from machine learning studies and antiviral drugs using AutoDock Vina. The binding mechanisms between the compounds and the proteins were characterized using LigPlot+ and molecular dynamics simulations techniques. The biological activities of the hit compounds were also predicted using a Bayesian-based approach. Six potential bioactive molecules NANPDB2245, NANPDB2403, fusidic acid, ZINC000095486008, ZINC0000556656943 and ZINC001645993538 were identified, all of which had plausible binding mechanisms with both viral receptors. Molecular dynamics simulations, including molecular mechanics Poisson-Boltzmann surface area (MM/PBSA) computations revealed stable protein-ligand complexes with all the compounds having acceptable free binding energies <−15 kJ/mol with each receptor. NANPDB2245, NANPDB2403 and ZINC000095486008 were predicted as antivirals; ZINC000095486008 as a membrane permeability inhibitor; NANPDB2403 as a cell adhesion inhibitor and RNA-directed RNA polymerase inhibitor; and NANPDB2245 as a membrane integrity antagonist. Therefore, they have the potential to inhibit viral entry and replication. These drug-like molecules were predicted to possess attractive pharmacological profiles with negligible toxicity. Novel critical residues identified for both targets could aid in a better understanding of the binding mechanisms and design of fragment-based *de novo* inhibitors. The compounds are proposed as worthy of further in vitro assaying and as scaffolds for the development of novel SARS-CoV-2 therapeutic molecules.

Keywords: SARS-CoV-2; coronavirus; African natural products; molecular docking; virtual screening; molecular dynamics; SARS-CoV-2 inhibitors

check for updates

Citation: Kwofie, S.K.; Broni, E.; Asiedu, S.O.; Kwarko, G.B.; Dankwa, B.; Enninful, K.S.; Tiburu, E.K.; Wilson, M.D. Cheminformatics-Based Identification of Potential Novel Anti-SARS-CoV-2 Natural Compounds of African Origin. *Molecules* **2021**, *26*, 406. https://doi.org/10.3390/molecules26020406

Academic Editors: Simona Fabroni, Krystian Marszałek, Aldo Todaro and Theodore Tselios

Received: 20 October 2020
Accepted: 8 January 2021
Published: 14 January 2021

1. Introduction

Coronavirus disease 2019 (COVID-19) is caused by severe acute respiratory syndrome coronavirus 2 (SARS-CoV-2) [1,2]. As of 8 December 2020, the novel coronavirus disease 2019 (COVID-19) has spread globally, with 66,729,375 confirmed cases, including 1,535,982 deaths [3]. The disease outbreak, declared a pandemic, has led to nearly three billion people in about 82 countries under partial or full lockdowns due to the infections [4]. The mild symptoms include fever, dry cough, runny nose, sore throat, and difficulty in breathing [5]. Chemosensory dysfunctions such as ageusia (loss of taste) and anosmia (loss of smell) have also been reported [6–9]. In more severe cases, the symptoms include severe muscle pain, cardiovascular shock, arrhythmia, acute respiratory distress syndrome (ARDS), and hyper inflammation [5,10]. The infection is spread mainly from one person to others via droplets produced from the respiratory systems of infected people, often during coughing or sneezing [5].

SARS-CoV-2, a positive-sense single-stranded RNA virus, is a member of the coronaviridae family, which are enveloped nonsegmented viruses with large surface spike proteins [11]. Most of the RNA encodes for RNA synthesis material, viral polymerase (RdRp), and nonstructural polyproteins [12,13], while the remainder encodes four structural proteins comprising spike (S), envelope (E), membrane (M), nucleocapsid (N), and other helper proteins [12].

There is evidence that the SARS-CoV-2 infects the host's cells that co-express both angiotensin-converting enzyme 2 (ACE2) and transmembrane protease serine 2 (TMPRSS2). The ACE2 binds to the spike protein, facilitating the viral cell entry, and TMPRSS2 proteolytically cleaves it, resulting in fragments that activate cell-cell and virus-cell fusion and interferes with antibody-mediated neutralization [14]. This suggests that TMPRSS2 might impact the spread of the virus via two independent mechanisms [14]. The S1 subunit of the spike protein binds to the ACE2 receptor, and the S2 subunit is responsible for the fusion between the virion and host receptors [15,16]. Upon binding to ACE2, the virion releases its RNA into the cell and takes over the host cell machinery to produce copies of itself, which are then shed via exocytosis to infect fresh cells [12,17–19].

The viral main protease (M^{pro}) and the papain-like proteases are critical in the translation of viral RNA to polyproteins. The viral spike protein also possesses a variable receptor-binding domain (RBD), which, if inhibited, should prevent viral attachment, fusion and entry, and thus make it also another attractive target [15].

Although, there are no FDA-approved drugs for the treatment and prevention of COVID-19, several antiviral compounds including remdesivir, favipiravir and darunavir; anti-HCV; nucleotide inhibitors sofosbuvir, IDX-184 and ribavirin; and kinase inhibitor imatinib are being repurposed [20–24]. Some of these repurposed drugs are currently undergoing various phases of clinical trials. Others such as dexamethasone and remdesivir have been given provisional approval for COVID-19 treatment [25,26]. The outcomes of these efforts are not assured, therefore emphasizing the urgent need to identify novel molecules that are therapeutically efficacious and maximally safe to human hosts.

Considering the urgency that the current situation of the pandemic demands, the conventional approach to drug discovery is time-consuming, but computational methods offer cost-effective and faster alternatives of discovering novel compounds. A network-based approach was used to predict repurposed drugs or combination therapies for SARS-CoV-2 by quantifying the interplay between drugs and human coronaviruses (HCoVs)-host interactome [27]. Additionally, a total of 1000 compounds were prioritized for downstream analysis as potential inhibitors via screening using the deep docking of 1.3 billion compounds present in the ZINC15 library against SARS-CoV-2 M^{pro} [28]. Camostat mesylate, which is an inhibitor of serine protease TMPRSS2, was shown to block cell entry of SARS-CoV-2 [17]. Other efforts geared towards the unraveling of potential antivirals targeting SARS-CoV-2 have been reported [24,29–32]. Nevertheless, issues of therapeutic potency persist, since potential promising molecules produced disappointing results during trials [33]. It is exigent to identify novel molecules that can disrupt critical mechanisms and biomolecular pathways involving SARS-CoV-2. These molecules must be therapeutically efficacious and safe to human hosts with negligible side effects.

Pharmacoinformatics methods have been applied in the development of antivirals such as Boceprevir, Saquinavir and Rupintrivir [34]. The viral M^{pro}, alongside the papain-like proteases, is critical in the translation of viral RNA to polyproteins. The M^{pro} has been suggested as a drug target, and its inhibition can obstruct viral replication [35,36]. Furthermore, the cleavage specificity of the main protease is absent in human proteases, suggesting potential inhibitors are unlikely to be toxic to the human host [36]. The enzyme spike protein is also another attractive drug target [1], which plays an essential role in facilitating viral entry into target cells [17]. The spike proteins contain a variable receptor-binding domain (RBD) located in the S1 subunit, which binds to the ACE2 receptor found most abundantly in the lungs and organs, including the heart, kidneys, and gastrointestinal tract. The spike protein S2 subunit is responsible for the fusion between viral and host receptors [15,16]. Inhibiting the viral spike protein will occlude viral attachment, fusion and entry [15].

The structural and chemical diversity of natural product-derived compounds serve as rich sources of scaffolds for the discovery of novel drug leads [37,38]. Natural products have traditionally played a prominent role in treatment [39]. Currently, natural products and their derivatives represent over one-third of all new molecular entities (NMEs) approved by the US FDA [40]. Cheminformatics databases, including the Traditional Chinese Database [41], African Natural Product Database (AfroDB) [42], and North African Natural Product Database [43], collectively contain over 30,000 freely accessible unique natural product compounds. Natural products including saikosaponins and raoulic acid have been reported to exhibit antiviral properties against coronaviruses [44].

The recent strides in understanding the genome of SARS-CoV-2 and the available plethora of diverse structural genomic data combined with advanced high-performance biocomputing are key in expediting the identification of COVID-19 drugs. Therefore, this study sought to identify promising novel molecules with the potential of disrupting the critical mechanisms involving the SARS-CoV-2, including viral attachment, entry and replication processes using cheminformatics. The study seeks to virtually screen an integrated library made up of diverse African natural compounds [45] and recently prioritized hits from a deep docking study of 1.3 billion compounds [28] together with drugs currently undergoing clinical trials against SARS-CoV-2 M^{pro} and RBD of the spike protein. This is to identify potential polypharmacological antiviral compounds targeting both proteins [46,47] with novel scaffolds to augment the design of next-generation inhibitors against the cell entry and replication of SARS-CoV-2. Moreover, we sought to gain novel insights into the mechanisms of binding between the receptors and compounds using molecular dynamics (MD) simulations, including molecular mechanics Poisson-Boltzmann surface area (MM/PBSA) methods, by evaluating their binding free energies [48]. A Bayesian-based technique was also used to predict the antiviral activity of the compounds as a measure to characterize their anti-SARS-CoV-2 propensities.

2. Results and Discussion

2.1. Description of Binding Sites of M^{pro} and RBD Structures

The binding sites of the M^{pro} and RBD of the spike protein structures were characterized using the Computed Atlas of Surface Topography of proteins (CASTp) version 3.0 (available at http://sts.bioe.uic.edu/castp/calculation.html). CASTp utilizes theoretical and algorithmic results of computational geometry to analytically predict pockets and cavities while excluding shallow depressions from the calculations [49]. The predicted binding cavities were analyzed using Chimera version 1.12 and PyMOL. The volumes and areas of the sites were also determined using Chimera version 1.12 (Table 1). Predicted binding sites with very small volumes and areas such that no ligands could fit were not considered for downstream virtual screening. The surface representations of the two proteins with three putative binding pockets are shown (Figure 1).

2.1.1. Binding Site Analysis of M^{pro}

The three-dimensional structure of the M^{pro} SARS-CoV-2 solved using X-ray diffraction at a resolution of 1.31 Å is a homodimer with a sequence length of 306 along one chain and shares a 96% sequence identity to the SARS-CoV M^{pro} [36]. The plausible binding sites of the SARS-CoV-2 M^pro^ were predicted via CASTp (Table 1) using the M^pro^ structure with a Protein Data Bank (PDB) ID 5R82.

Table 1. Predicted binding sites in the viral main protease (M^pro^) and receptor-binding domain (RBD) via CASTp, including the dimensions of the volumes and areas.

Pocket	Pocket Area (Å^2)	Volume (Å^3)	Residues Lining Pockets
		M^pro^	
1	557.1	920.6	Thr24, Thr25, Thr26, Leu27, His41, Cys44, Thr45, Ser46, Met49, Leu50, Phe140, Leu141, Asn142, Gly143, Ser144, Cys145, His163, His164, Met165, Glu166, Leu167, Pro168, Asp187, Arg188, Gln189, Thr190, Gln192
2	380.0	465.1	Met6, Ala7, Phe8, Pro9, Gly11, Lys12, Val13, Gln127, Phe150, Ile152, Asp153, Tyr154, Val157, Phe291, Asp295, Arg298, Gln299, Val303, Thr304
3	107.5	151.3	Phe3, Arg4, Lys5, Trp207, Leu282, Ser284, Glu288, Phe291
4	193.5	225.2	Pro108, Gly109, Gln110, Pro132, Ile200, Thr201, Val202, Asn203, Glu240, His246, Ile249, Thr292, Pro293, Phe294
5	86.6	131.4	Glu14, Gly15, Met17, Val18, Trp31, Ala70, Gly71, Val73, Asn95, Lys97
		RBD	
A	148.3	169.2	Arg454, Phe456, Arg457, Lys458, Asp467, Ser469, Glu471, Ile472, Tyr473, Pro491
B	52.3	90.3	Phe342, Asn343, Leu368, Ser371, Ser373, Phe374
C	146.8	160.6	Glu340, Val341, Ala344, Arg346, Phe347, Ala348, Asn354, Arg355, Lys356, Ala397, Ser399, Val511

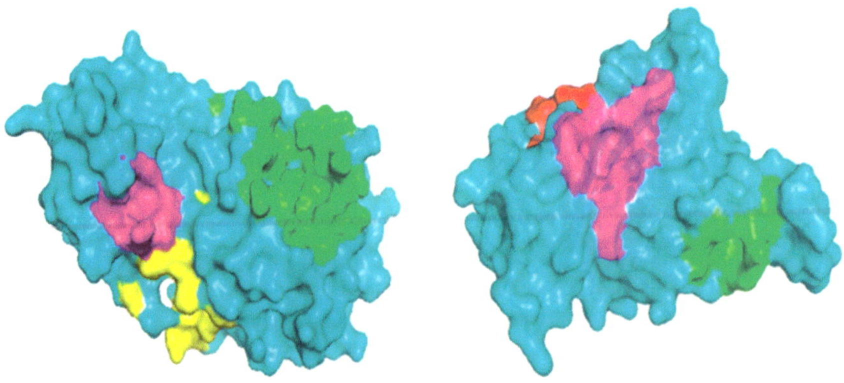

(**a**) (**b**)

Figure 1. Surface representations of the protein structures generated using PyMOL. (**a**) The structure of severe acute respiratory syndrome virus 2 main protease (SARS-CoV-2 M^pro^) is colored cyan, and the predicted binding pockets 1, 2, and 3 are colored green, yellow and magenta, respectively. (**b**) The receptor-binding domain (RBD) of the spike protein is colored cyan, and the predicted binding pockets A, B, and C are colored green, red and magenta, respectively.

Ser1, His41, Met49, Gly143, Phe140, Ser144, Cys145, His163, His164, Glu166, Pro168 and Gln189 have been determined in previous studies as lining the active sites of M^pro^ [36,50], consistent with the prediction by CASTp as Pocket 1 (Table 1). A computational study to identify potential SARS-CoV-2 M^pro^ inhibitors reported that the top compound docked into the M^pro^-binding cavity lined by residues His41, Met49, Tyr54, Phe140, Leu141, Asn142, Ser144, Cys145, His163, Met165, Glu166, Leu167, Pro168, Asp187, Arg188, Gln189 and Gln192 [28].

Other studies have also reported that residues of M^{pro} forming hydrogen bonds include Gly143, His163, His164, Glu166, Gln189 and Thr190, with Cys145 forming a covalent bond [29].

Additionally, the region around Glu288-Asp289-Glu290 has been reported to be involved in a likely enzyme dimerization in SARS-CoV [51]. Mutating Glu288, Asp289, Glu290, Arg298 and Gln299 led to a high decrease in enzymatic activities of the protease, and this region overlaps with pockets 2 and 3 (Table 1). Additionally, mutations in Asn214, Leu282 and Cys300 significantly decreased the activity of the protease. Furthermore, the replacement of Ser284, Thr285, Ile286 or Phe291 by Ala produced mutated proteases with higher enzymatic activities than the wild type [51]. Leu282, Ser284 and Phe291 were predicted as residues lining pocket 3 (Table 1).

A molecular docking study of compounds against M^{pro} identified possible binding sites of the protein. Remdesivir was reported to dock in the region lined by residues Gln107, Pro108, Pro132, Ile200, Glu240 and His246 [52]. This binding site was also identified by other computational studies [53,54]. Herein, these residues were predicted as lining pocket 4 (Table 1). Pockets 1, 2, 3, 4 and 5 were selected as the most plausible binding cavities for the main protease (Table 1) and were considered for the virtual screening process.

2.1.2. Binding Site Analysis of RBD

The 3D structure of the RBD of the spike protein was obtained from the PDB database with ID 6M0J [55]. The RBD was solved using X-ray diffraction at a resolution of 2.45 Å. The receptor-binding domain is located on chain E of the spike protein with a residue count of 229 [55]. A previous study identified a significant difference between the C-terminus residues of the RBDs of SARS-CoV and SARS-CoV-2, although this did not affect the capability to engage ACE2 [56]. The binding sites of both RBD proteins of SARS-CoV-2 and SARS-CoV are highly conserved [57].

Lys417, Gly446, Tyr449, Tyr453, Leu455, Phe456, Phe486, Asn487, Tyr489, Gln493, Gly496, Gln498, Thr500, Asn501, Gly502 and Tyr505 are contacting residues of the RBD at the SARS-CoV-2 RBD-ACE2 interface [55]. An in silico study to repurpose FDA-approved drugs as SARS-CoV-2 spike protein inhibitors via virtual screening revealed the binding of compounds occurred at the SARS-CoV-2 RBD-ACE2 interface [58].

In a recent computational study to identify anti-SARS-CoV-2 spike protein molecules, the compounds were reported to interact with at least one of the following residues: Leu335, Cys336, Phe337, Phe338, Gly339, Val341, Phe342, Asn343, Ala344, Thr345, Lys356, Asp364, Val367, Leu368, Tyr369, Asn370, Ser373, Phe374, Phe377, Lys378, Cys379, Ala397, Asn422, Gly431, Trp436, Leu441, Arg509, Val510, Val511, Phe515 and Asn643 [59]. Cys336 is a critical residue for the RBD, since it forms a disulfide bond with Cys361, which helps stabilize the β sheet structure [55]. A NAG glycan has also been reported to be linked to Asn343 of the RBD [55], which was identified as a critical residue in our study. These residues overlap with pockets B and C, as predicted (Table 1). For the RBD, pockets A, B and C (Table 1) and the RBD-ACE2 interface were considered as plausible binding sites.

2.2. Virtual Screening Studies

2.2.1. Molecular Docking Studies

Molecular docking is an essential technique in computer-aided drug discovery [60]. With the structure of the receptor known, compounds of interest are screened in silico to guide the selection of potential leads. AutoDock Vina used for docking employs an empirical and knowledge-based scoring function to predict the binding affinity of compounds [61]. The grid boxes for both receptors were set to cover all the binding sites (Table 1). Compounds with binding energies of −7.5 kcal/mol or less for both receptors were selected for the downstream analysis. This threshold was used, since −7.0 kcal/mol has been reported to significantly discriminate between putative specific and nonspecific protein-ligands for viruses [62]. A total of 1462 ligands from the African natural compounds (ANC) library were successfully screened against both the M^{pro} and RBD, whilst 940 compounds from the ML study were screened against both receptors, including 43 known antivirals.

Molecular Docking Studies of M^{pro}

A grid box with a dimension of 37.58 * 64.78 * 62.77 Å^3 and center 49.33, 49.36 and 49.56 Å was specified for M^{pro}. Out of 1462 successfully screened ANC compounds, 65 had a binding affinity ≤−7.5 kcal/mol. A total of 112 compounds from the ML library and 14 known drugs met the threshold. Ledipasvir had the highest binding affinity of −9.6 kcal/mol to the M^{pro} (Table S1). Velpatasvir and imatinib also had binding affinities of −8.9 and −8.5 kcal/mol, respectively (Table S1). Other studies have proposed ledipasvir and velpatasvir as potential anti-SARS-CoV-2 M^{pro} molecules [63,64]. After using AutoDock to screen FDA-approved drugs against the M^{pro}, velpatasvir was reported to possess a binding affinity of −9.1 kcal/mol [63]. Imatinib, currently in phase 3 clinical trials for adults hospitalized with SARS-CoV-2 (https://clinicaltrials.gov/ct2/show/record/NCT04394416), was also reported to demonstrate the highest binding affinity (−11.46 kcal/mol) to the M^{pro} in a recent study [65]. These results are consistent with the docking outcomes reported herein. The differences in binding affinity values could be due to the various software packages used and the different preparation protocols the protein and ligand structures were subjected to.

For the ANC library, ZINC000095486008, NANPDB2403 (retusolide B), NANPDB2245 (helioscopinolide B) and NANPDB2510 (jolkinolide E) demonstrated binding affinities of −8.2, −8.1, −8.0 and −7.9 kcal/mol, respectively, with the M^{pro} (Table 2). ZINC001657931232 had the highest binding affinity of −8.4 kcal/mol to the M^{pro} among the ML compounds. ZINC001181689720 and ZINC001460974086 also had binding affinities of −8.3 and −8.2 kcal/mol, respectively (Table S1). Remdesivir, hydroxychloroquine and chloroquine had low binding affinities of −6.8, −5.9 and −5.5 kcal/mol with the M^{pro}, respectively (Table 2). In the quest of finding potential SARS-CoV-2 inhibitors, antiviral and antimalarial drugs were virtually screened against the M^{pro} (PDB ID: 6LU7) and S-protein. Remdesivir, hydroxychloroquine, and chloroquine were reported to possess binding affinities of −6.5, −5.3 and −5.1 kcal/mol with the M^{pro}, respectively, consistent with the results obtained herein [66]. Although, the binding affinities are relatively low, it does not exclude them as anti-SARS-CoV-2 M^{pro} molecules.

Molecular Docking Studies of RBD

For the RBD, the grid box was set with dimensions of 42.54 * 42.73 * 42.44 Å^3 and centered at 41.19, 47.74, and 55.37 Å. A total of 26 compounds from the ANC library, 22 from the ML-based study and 9 known drugs demonstrated binding affinities ≤−7.5 kcal/mol, thereby complying with the threshold. Ledipasvir showed the highest binding affinity (−9.9 kcal/mol), followed by velpatasvir and imatinib, which also had binding affinities of −8.5 and −8.1 kcal/mol, respectively (Table S1). A recent study reported that ledipasvir and velpatasvir have high binding affinities of −8.4 and −7.9 kcal/mol with the spike glycoprotein, respectively [67].

From the ANC library, ZINC000095486008, NANPDB2403 (retusolide B), NANPDB2245 (helioscopinolide B) and NANPDB2510 (jolkinolide E) had binding affinities of −7.8, −7.8, −7.7 and −7.6 kcal/mol, respectively (Table 2). For the ML library, ZINC001657931232, ZINC001181689720 and ZINC001460974086 had the highest binding affinities to the M^{pro} with binding affinities of −7.8, −7.5 and −7.6 kcal/mol, respectively. Remdesivir, hydroxychloroquine and chloroquine also had low binding affinities of −6.3, −5.5 and −4.9 kcal/mol, respectively (Table 2).

Shortlisted Compounds for Downstream Analysis

Thirteen compounds from the ANC library, fourteen from the ML library and nine known antivirals met the threshold for both RBD and M^{pro} (Table 2). In total, 36 compounds, including experimental drugs had binding affinities of ≤−7.5 kcal/mol with both receptors (Table 2) and were shortlisted for further analysis.

Table 2. The binding energies and intermolecular interactions between selected compounds and M^{pro} as well as RBD. Compounds are from the African Natural Compounds (ANC) database and Machine Learning Study (ML). In addition, antivirals and experimental drugs are included.

Compound	Source	Binding Energy (kcal/mol)		Hydrogen Bonds [Bond Length (Å)]		Hydrophobic Bonds	
		M^{pro}	RBD	M^{pro}	RBD	M^{pro}	RBD
Selected Hits							
NANPDB2403	ANC	−8.1	−7.8	Leu287 (3.22)	-	Thr199, Tyr237, Tyr239, Leu271, Leu272, Leu286.	Leu335, Cys336, Phe338, Phe342, Asn343, Asp364, Val367, Leu368, Ser371,
NANPDB2245	ANC	−8.0	−7.7	Arg131 (2.92)	Asn343 (2.96)	Lys137, Thr199, Tyr237, Tyr239, Leu271, Leu272, Leu286, Leu287, Asp289	Leu335, Cyc336, Phe338, Gly339, Asp364, Val367, Leu368, Ser371, Phe374
ZINC000055656943	ML	−8.0	−8.0	Asp197 (2.80)	-	Arg131, Thr198, Thr199, Tyr237, Tyr239, Leu272, Leu287	Leu335, Cys336, Phe338, Phe342, Asp364, Val367, Leu368, Ser371, Phe374,
ZINC000095486008	ANC	−8.2	−7.8	Lys5 (3.1), Glu288 (3.02)	Cys336 (2.96), Phe338 (3.26), Gly339 (3.3)	Lys137, Asp197, Thr199, Tyr239, Leu272, Leu286, Leu287, Asp289, Glu290	Pro337, Phe342, Asn343, Val367, Leu368, Ser371, Phe374, Trp436
ZINC001645993538	ML	−7.7	−7.5	Thr199 (314)	-	Lys137, Asp197, Tyr239, Leu272, Leu286, Leu287, Glu288, Asp289	Cys336, Phe338, Asp364, Val367, Leu368, Ser371, Phe374
Known Antivirals and Experimental Drugs							
Oxymetholone		−7.8	−7.7	Thr25 (2.81), Glu166 (2.9, 3.00)	Cys336 (3.0), Asn343 (3.09)	His41, Ser46, Thr45, Asn142, Gly143, Cys145, His164, Met165	Leu335, Phe338, Gly339, Phe342, Asp364, Val367, Leu368, Ser371, Phe374
Dexamethasone		−7.6	−6.7	Asp197 (2.9, 3.24), Met276 (3.01), Leu287 (3.29, 3.32)	Arg355 (2.99, 3.01), Thr430 (3.18), Glu516 (2.76)	Lys137, Thr198, Thr199, Tyr239, Leu271, Gly275, Leu286, Leu287, Asp289	Pro426, Phe429, Pro463, Phe464, Phe515
Remdesivir		−6.8	−6.3	Lys137 (3.19), Thr199 (2.82, 3.05), Leu287 (3.09), Asp289 (2.84)	Gly496 (2.84, 2.96), Asn501 (2.9)	Arg131, Asp197, Thr198, Tyr237, Asn238, Tyr239, Leu271, Leu272, Asn274, Gly275, Met276, Leu286	Arg403, Tyr453, Leu455, Ser494, Tyr495, Phe497, Tyr505
Hydroxychloroquine		−5.9	−5.5	Asp197 (3.05, 3.22), Thr199 (3.26)	Thr345 (2.97), Asn354 (3.04), Ala397 (2.72), Ser399 (2.99),	Arg131, Thr198, Tyr237, Tyr239, Leu272, Met276, Ala285, Leu286, Leu287, Asp289	Glu340, Val341, Ala344, Arg346, Phe347, Ala348, Arg355, Lys356, Asp398
Chloroquine		−5.5	−4.9	Tyr239 (3.2)	-	Arg131, Asp197, Thr198, Thr199, Tyr237, Leu272, Leu286, Leu287, Asp289,	Arg403, Tyr449, Tyr453, Ser494, Tyr495, Gly496, Phe497, Asn501, Tyr505

2.2.2. Characterization of the Protein-Ligand Interactions

The nature of the active site and the functional groups on the ligands are critical for stabilization within the binding pocket of a receptor [68]. Studies into these interactions are key in determining whether a ligand is considered as a promising lead. The protein-ligand interactions of the 36 shortlisted compounds were studied using LigPlot+ [69] and PyMOL. After analyzing their binding interactions, five compounds comprising NANPDB2403, NANPDB2245, ZINC000055656943, ZINC000095486008 and ZINC001645993538 were selected as hits (Table 2). The interactions of the known antivirals and experimental drugs with the respective targets were compared. Characterizing the binding interactions enabled the identification of certain critical residues within the active pockets of the respective protein targets.

Characterization of the M^{pro}-Ligand Interactions

Considering the molecular interactions of the M^{pro}, all the 36 compounds that met the ≤ -7.5 kcal/mol threshold, except for ZINC000544552417 and ZINC000621286015, formed at least one hydrogen bonding with the M^{pro} residues. Oxymetholone docked into pocket 1 (Figure 2d, and Tables 1 and 2), the known active site of the M^{pro} [36,50]. It formed a hydrogen bond with Thr25 (bond length 2.81 Å); two with Glu166 (bond lengths 2.9 Å and 3.00 Å); and hydrophobic bonds with His41, Ser46, Thr45, Asn142, Gly143, Cys145, His164 and Met165. A recent study also revealed that tipranavir, which is a nonpeptidic protease inhibitor used in combination with ritonavir to treat HIV, interacted with Gln192 and Met165 (both formed hydrogen bonds) and Gln189, Asp187, Met49, Arg188, Ser46, Cys44, Thr25 and His41 in a different conformation from that of the α-ketoamide inhibitor [70]. It was further identified that raltegravir demonstrated a high binding affinity to the M^{pro} than the co-crystallized α-ketoamide, with interactions from His164, Arg188, Gln192, Glu166, Met49, Met165, Phe140, Pro168 and Leu167 [70]. Considering our results and references to existing literature, we suggest that Gly143, Cys145 and Glu166 are critical residues for binding.

Interestingly, ledipasvir, NANPDB2403, NANPDB2245 (Figure 3a), ZINC0095486008 (Figure 3b), and ZINC001645993538 (Figure 2c) were found to interact with the main protease in a different binding cavity with binding energies of −9.6, −8.1 and −8.0, respectively (Table 2). The residues lining this cavity include Arg131, Lys137, Thr199, Tyr237, Tyr239, Leu271, Leu272, Gly275, Leu286, Leu287, Glu288, Asp289 and Glu290. This binding pocket is located between pockets 2, 3 and 4. Other in silico studies also identified this binding cavity [71,72]. Ledipasvir formed hydrogen bonding with Met276 of a bond length 2.92 Å and hydrophobic contacts with Lys5, Gly124, Tyr126, Gln127, Lys137, Gly138, Ser139, Thr199, Tyr237, Tyr239, Leu272, Gly275, Asn277, Gly278, Leu286, Leu287 and Glu290. NANPDB2245 formed a hydrogen bond with Arg131 (bond length 2.92 Å) and nine hydrophobic contacts with M^{pro} residues comprising Lys137, Thr199, Tyr237, Tyr239, Leu271, Leu272, Leu286, Leu287 and Asp289 (Table 2 and Figure 2a). ZINC000095486008 formed two hydrogen bonds with Lys5 (bond length 3.1 Å) and Glu288 (bond length 3.02 Å), and interacted with Lys137, Asp197, Thr199, Tyr239, Leu272, Leu286, Leu287, Asp289, and Glu290 via hydrophobic contacts (Table 2 and Figure 2b).

Characterization of the RBD-Ligand Interactions

A total of 27 out of the 36 compounds formed hydrogen bonds of varying lengths with the RBD (Figure S1 and Table S1). Ledipasvir interacted via hydrogen bonding with Gly339 and formed hydrophobic contacts with Leu335, Cys336, Pro337, Phe338, Phe342, Asn343, Ala363, Asp364, Leu368, Ser371, Ala372, Ser373, Phe374, Ser375, Trp436, Asn437 and Tyr508 (Table S1).

Oxymetholone formed hydrogen bonding with Cys336 (bond length of 3.0 Å) and Asn343 (bond length of 3.09 Å); and hydrophobic contacts with Leu335, Phe338, Gly339, Phe342, Asp364, Val367, Leu368, Ser371 and Phe374 (Table 2). ZINC000095486008 also interacted with the RBD via three hydrogen bonds with Cys336, Phe338 and Gly339 of bond lengths 2.96, 3.26 and 3.3 Å, respectively (Table 2 and Figure S3B). NANPDB2245 interacted via nine hydrophobic contacts Leu335, Cys336, Phe338, Gly339, Asp364, Val367, Leu368, Ser371 and

Phe374, and formed one hydrogen bond with Asn343 (Figure S3A). Therefore, we suggest Cys336, Ser373 and Phe374 as potential critical residues.

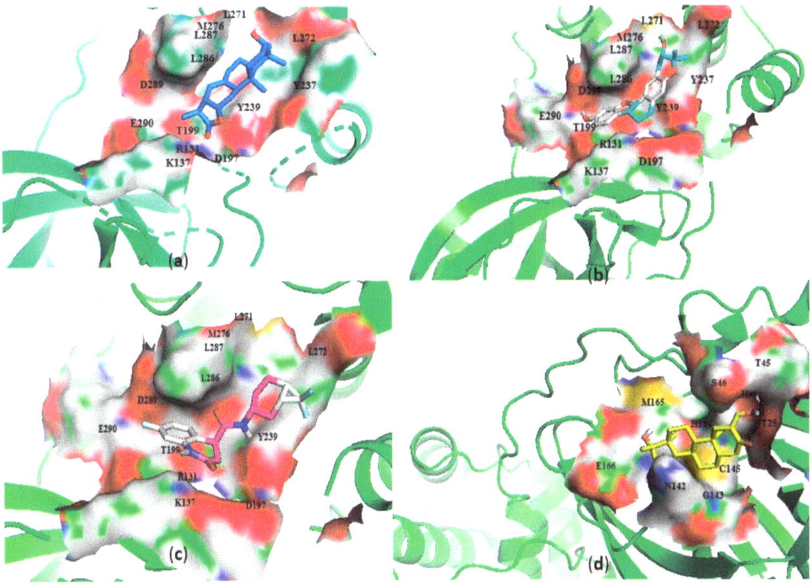

Figure 2. Cartoon representation of the main protease (M^{pro}) in complex with: (**a**) NANPDB2245 (helioscopinolide B), (**b**) ZINC000095486008, (**c**) ZINC001645993538, and (**d**) oxymetholone. The binding sites are shown as a surface representations, with the ligands shown as sticks.

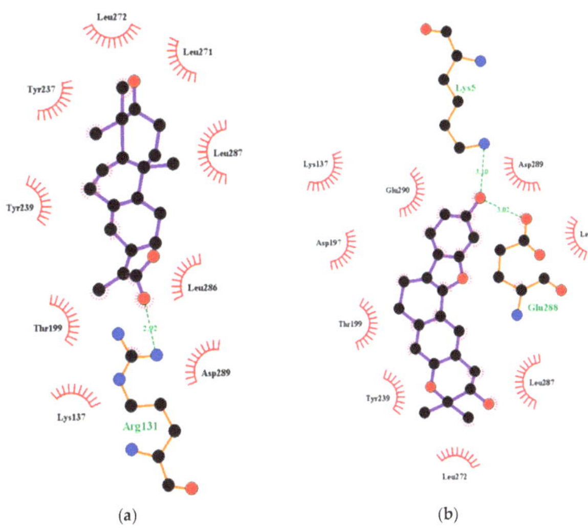

Figure 3. Two-dimensional diagrams of the M^{pro}–ligand interactions generated using LigPlot+. (**a**) Interaction profiles of the M^{pro}–NANPDB2245, and (**b**) M^{pro}–ZINC000095486008 complexes. Ligands are colored in purple, hydrogen bonds are represented as green dash lines, and hydrophobic contacts are represented as red spoke arcs.

The residues Leu455, Phe486 and Gln493 of RBD have been reported to interact with Lys31 (hotspot 31), whereas residues Asn487 and Ser494 of the RBD are described to interact with Lys353 (hotspot 353) of the human ACE2 [73,74]. Compounds KT185, KT203, GSK1838705A, BMS195614 and RS504393 were reported to bind to the RBD by interacting with Leu455, Phe486, Asn487, Gln493 and Ser494 [75]. These compounds were predicted to interact and block key RBD residues responsible for recognizing hotspot 31 and hotspot 353 of SARS-CoV-2 [75].

2.2.3. Predictions of Biological Activities

The biological activities for the 36 shortlisted compounds were elucidated. All the 13 ANC compounds shortlisted and two of the machine learning compounds were predicted to be antivirals, with a probable activity (Pa) > 0.3 and Pa > Pi (probable inactivity). Furthermore, the propensity of the compounds to be cell adhesion molecule inhibitors and membrane permeability inhibitors was considered in this study. The mechanisms of entry into the host cell are key in the survival of viruses, including coronaviruses [76]. More so, the first step in coronavirus entry is the adhesion to the host cell surface. The inhibition of SARS-CoV-2 spike proteins is a critical antiviral strategy, since it serves as the major barrier to block the infection [15,77]. A total of seven ANC and one machine learning compound were predicted to be cell adhesion inhibitors or membrane permeability inhibitors. The stimulation of human nuclear factor kappa B (NF-κB) transcription factors has been linked to the mediation of the induction of antiviral and inflammatory responses [78]. NANPDB2403 was predicted to be a cell adhesion inhibitor with a Pa of 0.760 and Pi of 0.005; anti-influenza activity with a Pa of 0.543 and Pi of 0.021; an anti-rhinovirus with a Pa of 0.499 and Pi of 0.02; and an anti-herpes activity with a Pa of 0.366 and Pi of 0.068. NANPDB2245 was predicted to possess anti-herpes activity with a Pa of 0.446 and Pi of 0.019. The ZINC000095486008 was also predicted to have anti-rhinovirus activities with a Pa of 0.511 and Pi of 0.021; an anti-influenza with a Pa of 0.431 and Pi of 0.037; and anti-herpes with a Pa of 0.431 and Pi of 0.037. ZINC000095486008 possesses potential anti-Ebola activity with a free binding energy of −114.650 kcal/mol against the VP24 protein target [45]. Since the Pa > Pi, these compounds can be considered as prospective antivirals, necessitating further experimental testing.

2.3. Existing Drugs Proposed as Potential Frontline Treatment Options

After molecular docking, nine known antivirals and experimental drugs comprising ledipasvir, velpatasvir, imatinib, dactinomycin, dolutegravir, bictegravir, oxymetholone, raltegravir and sirolimus were predicted to have high binding affinities (≤ -7.5 kcal/mol) to both M^{pro} and RBD of the spike glycoprotein (Table S1). A recent study also proposed the use of ledipasvir and velpatasvir for the treatment of SARS-CoV-2 [64]. Since all these drugs are FDA-approved, more attention must be focused on exploring their therapeutic potentials against SARS-CoV-2.

2.3.1. Similarity Search of Hits

A structural similarity search of the hits was conducted via DrugBank. The search revealed that fusidic acid is structurally similar to NANPDB2245 and NANPDB2403, with similarity scores of 0.729 and 0.717, respectively. The structural similarity search also revealed that betulinic acid is similar to oxymetholone, with a score of 0.712. Oxymetholone used to treat HIV/AIDS wasting syndrome had binding energies of −7.8 and −7.7 kcal/mol against M^{pro} and RBD, respectively (Table 2).

2.3.2. Fusidic Acid and Betulinic Acid as Potential Anti-SARS-CoV-2 Compounds

Fusidic acid and betulinic acid were virtually screened against both the M^{pro} and RBD. Fusidic acid had binding affinities of −6.9 and −7.2 kcal/mol against the M^{pro} and RBD, respectively. Betulinic acid also demonstrated good binding affinities of −7.7 and −7.4 kcal/mol with the M^{pro} and RBD, respectively. A recent in silico study proposed the use of betulinic acid for the treatment of SARS-CoV-2, since it had a good binding affinity to the main protease [79].

The interaction profiles between the receptors and these two compounds were also investigated. Fusidic acid formed hydrogen bonds with Lys137 (bond length of 2.8 Å), Leu271 (bond length of 2.94 Å) and Leu272 (bond length of 3.05 Å), and hydrophobic contacts with Arg131, Asp197, Thr199, Tyr239, Gly275, Met276, Leu286 and Asp289 (Table S1). For the RBD, fusidic acid formed hydrogen bonding with Ser371 (bond length of 2.7 Å) and Ser373 (bond length of 2.92 Å); and hydrophobic bonding with Leu335, Cys336, Gly339, Phe342, Asn343, Val367, Leu368, Phe374 and Trp436, which lined binding pocket B (Table 1 and Table S1). Fusidic acid is a natural product-derived bacteriostatic antibiotic classified under both approved and investigational drug categories in DrugBank. Fusidic acid suppresses bacterial growth and enhances the clearance of infections by the immune system by inhibiting translocation during the synthesis of protein. Fusidic acid was reported to inhibit the replication of feline infectious peritonitis virus (FIPV) in vitro by significantly reducing the viral titer [80]. Since fusidic acid is already an FDA-approved drug for humans, this offers the opportunity to further explore its therapeutic potential against SARS-CoV-2.

Additionally, betulinic acid formed a hydrogen bond with Asp289 of a bond length 2.79 Å and interacted with Arg131, Lys137, Asp197, Thr198, Thr199, Tyr237, Tyr239, Leu271, Leu272, Gly275, Met276, Leu286 and Leu287 via hydrophobic bonds. For the RBD, betulinic acid formed two hydrogen bonds with Phe515 (lengths 2.73 Å and 3.16 Å) and one with Thr430 (2.71 Å). Betulinic acid also interacted with the RBD via hydrophobic interactions with residues Pro426, Asp428, Phe429, Lys462, Pro463, Phe464 and Ser514. Betulinic acid was also reported to possess inhibitory activity against SARS-CoV M^{pro} with an IC_{50} value of 10 μM [81]. Based on modelization studies, the inhibition by betulinic acid was attributed to the multiple hydrogen bonds formed between betulinic acid and the M^{pro} [81]. Betulinic acid is an HIV-1 inhibitor with an EC_{50} value of 1.4 μM [82]. Other studies have also shown that derivatives of betulinic acid can interfere with HIV-1 virus entry in cells [83,84]. Since the S2 subunit of the spike protein of SARS-CoV and the glycoprotein 41 (Gp41) of HIV-1 are similar, blocking of the entry or fusion of the SARS-CoV viral particles to the human cell membrane was proposed as anti-SARS-CoV mechanisms [81].

2.4. Molecular Mechanics/Poisson-Boltzmann Surface Area (MM/PBSA) Calculations

Molecular dynamics (MD) simulations are performed after docking to assess the predicted binding modes of the top-ranking compounds as a filter in silico or to guide chemical synthesis for hit optimization [85]. At a quantitative level, simulation-based methods provide substantially more accurate estimates of ligand binding affinities (free energies) [86]. To understand the biophysical basis of recognition of inhibitors, MM/PBSA is employed for each system to identify stable MD trajectories, and the results are evaluated based on the total binding free energy of the ligand-receptor complex [48,87]. Binding free energy (ΔG_{bind}) is used to quantify the affinity of a ligand to its target and is the free energy difference between the ligand-bound state (complex) and the corresponding unbound states of proteins and ligands. Assessing the ΔG_{bind} of a series of ligands against a particular target can unravel those ligands with higher binding affinities with the target. Thus, the ΔG_{bind} calculations are important to gain in-depth knowledge about the binding modes of the hits in drug design [88]. The various contributing energy terms were computed in this study (Table 3). Binding free energies between the hits and the respective targets were calculated after a 10 ns production MD run of the respective complexes using GROMACS. The binding free energies between fusidic acid, oxymetholone and remdesivir with the respective targets were also calculated.

2.4.1. MM/PBSA-Binding Free Energy Computational Analysis of M^{pro}

Talampicillin was found to have a binding energy of −11.17 kcal/mol and MM/PBSA-binding free energy of −2.8 kcal/mol (−11.7152 kJ/mol) [89]. Talampicillin formed a network of bonds with residues His41, Met49, Gly143, Cys145, Met165, Glu166, Leu167, Pro168,

Gln189 and Gln192 [89]. ZINC000015988935 also had binding energy of −12.39 kcal/mol and free binding energy of −4.62 kcal/mol (−19.33008 kJ/mol). It also showed a p-sulfur interaction with Met49; seven hydrogen bonds with the residues Arg188, Asp187, Gln189, Ser144, Cys145 and Glu166; an alkyl interaction with Met165; and a p-alkyl interaction with Cys145 [89].

Table 3. Contributing energy terms of the Molecular Mechanics/Poisson-Boltzmann Surface Area (MM/PBSA) computations for receptor–ligand complexes. Values are shown as average ± standard deviations in kJ/mol. The terms consist of van der Waals, electrostatic, polar solvation, solvent-accessible surface area (SASA) and binding energies.

Compound	van der Waals Energy (kJ/mol)	Electrostatic Energy (kJ/mol)	Polar Solvation Energy (kJ/mol)	SASA Energy (kJ/mol)	Binding Energy (kJ/mol)
			M^{pro}		
NANPDB2245	−85.61 +/− 11.970	−6.274 +/− 7.537	46.495 +/− 10.814	−10.829 +/− 1.110	−56.223 +/− 11.988
NANPDB2403	−77.965 +/− 12.063	−6.624 +/− 7.992	36.397 +/− 13.775	−9.939 +/− 1.139	−58.132 +/− 13.000
ZINC000095486008	−98.620 +/− 15.067	−20.464 +/− 14.240	84.718 +/− 29.042	−12.692 +/− 1.538	−47.058 +/− 20.877
ZINC000055656943	−18.966 +/− 26.649	−2.907 +/− 8.983	7.468 +/− 57.684	−2.692 +/− 4.061	−17.097 +/− 45.262
ZINC001645993538	−84.952 +/− 12.296	−20.470 +/− 13.867	62.338 +/− 24.852	−10.702 +/− 1.140	−53.785 +/− 18.652
Oxymetholone	−60.820 +/− 13.039	−3.207 +/− 5.288	27.787 +/− 20.226	−8.485 +/− 1.967	−44.724 +/− 17.562
Remdesivir	−114.276 +/− 18.798	−19.410 +/− 12.604	89.056 +/− 41.414	−13.726 +/− 2.248	−58.356 +/− 31.051
			RBD		
NANPDB2245	−30.310 +/− 43.669	−2.337 +/− 4.496	14.435 +/− 40.707	−3.930 +/− 5.769	−22.142 +/− 39.775
NANPDB2403	−79.080 +/− 14.764	−2.714 +/− 7.624	39.552 +/− 18.265	−10.898 +/− 1.698	−53.140 +/− 20.905
ZINC000095486008	−119.217 +/− 10.410	−8.227 +/− 7.728	77.567 +/− 12.472	−15.298 +/− 1.031	−65.174 +/− 10.495
ZINC000055656943	−58.972 +/− 54.205	−11.991 +/− 12.656	34.870 +/− 57.870	−7.003 +/− 6.417	−43.096 +/− 39.685
ZINC001645993538	−109.967 +/− 10.090	−0.990 +/− 6.308	63.10 +/− 8.655	−13.921 +/− 0.760	−61.778 +/− 9.594
Oxymetholone	−109.874 +/− 9.028	−15.240 +/− 7.816	74.123 +/− 15.363	−13.752 +/− 0.876	−64.742 +/− 14.235
Remdesivir	−100.708 +/− 18.622	−11.616 +/− 11.476	80.060 +/− 24.762	−12.206 +/− 1.981	−44.471 +/− 19.222

Herein, the binding free energies of hits ranged from −17.097 to −61.090 kJ/mol (Table 3). Fusidic acid demonstrated the highest binding affinity with a binding free energy of −61.090 kJ/mol, while ZINC000055656943 had the highest binding free energy of −17.097 kJ/mol. NANPDB2245, NANPDB2403, ZINC000095486008, ZINC001645993538, remdesivir and oxymetholone had binding free energies of −56.223, −58.132, −47.058, −53.785, −58.356 and −44.724 kJ/mol, respectively, with M^{pro} (Table 3).

2.4.2. MM/PBSA-Binding Free Energy Computational Analysis of RBD

For the RBD, ZINC000095486008 demonstrated the highest affinity with a binding free energy of −65.174 kJ/mol. NANPDB2245 demonstrated the highest binding free energy (−22.142 kJ/mol) with the RBD. NANPDB2403, fusidic acid, ZINC000055656943, ZINC0016 45993538, remdesivir and oxymetholone also had binding free energies of −53.140, −55.858, −43.096, −61.778, −44.471 and −64.742 kJ/mol, respectively (Table 3). The low binding free energies exhibited by these compounds make them promising anti-SARS-CoV-2 molecules worthy of in vitro studies.

2.5. Other Contributing Energy Terms

The nonpolar component of the solvation free energy is estimated by the molecular solvent-accessible surface area (SASA) [90]. The SASA analysis measures the interaction between complexes and solvents. The M^{pro}-ligand complexes had SASA energies ranging from −2.692 to −13.726 kJ/mol. M^{pro}-Remdesivir demonstrated the lowest SASA energy of −13.726 kJ/mol, while M^{pro}-ZINC000055656943 had a SASA energy of −2.692 kJ/mol. M^{pro}-ZINC000095486008, M^{pro}-fusidic acid, M^{pro}-NANPDB2245, M^{pro}-ZINC001645993538, M^{pro}-NANPDB2403 and M^{pro}-oxymetholone also had −12.692, −12.623, −10.829, −10.702, −9.939, and −8.485 kJ/mol, respectively. For the RBD-ligand complexes, the SASA energy values observed were between −3.930 and −15.298 kJ/mol (Table 3). ZINC000095486008 demonstrated the lowest SASA energy value of −15.298 kJ/mol.

The energy terms van der Waals, electrostatic and polar solvation energies are useful for analyzing free binding energies. Studies have shown that electrostatic and van der Waals forces contribute predominantly and continuously to the binding energy, along with simulations that favor the binding of complexes [91,92]. All the M^{pro}-ligand complexes demonstrated very low van der Waals energies ranging from −18.966 to −114.276 kJ/mol (Table 3). M^{pro}-Remdesivir demonstrated the lowest van der Waals energy of −114.276 kJ/mol, while ZINC000055656943 showed the highest van der Waals energy of −18.966 kJ/mol (Table 3). Fusidic acid, ZINC000095486008, NANPDB2245, ZINC001645993538, NANPDB2403 and oxymetholone demonstrated van der Waals energies of −99.476, −98.620, −85.615, −84.952, −77.965 and −60.820 kJ/mol, respectively (Table 3). For the RBD-ligand complexes, the van der Waals energies ranged between −30.310 and −119.217 kJ/mol. RBD-ZINC000095486008 demonstrated the lowest van der Waals energy of −119.217 kJ/mol, while RBD-NANPDB2245 had −30.310 kJ/mol (Table 3).

Compounds with low electrostatic energies and high polar energies have been reported to be active against receptors [93]. The electrostatic energies ranged between −2.907 to −20.464 kJ/mol for the M^{pro}-ligand complexes and −2.337 to −16.338 kJ/mol for the RBD-ligand complexes. (Table 3). High polar solvation energies were also observed for all protein-ligand complexes. For the M^{pro}-ligand complexes, the polar solvation energies ranged between 7.468 and 89.056 kJ/mol (Table 3). M^{pro}-Remdesivir demonstrated the highest polar energy of 89.056 kJ/mol, while M^{pro}-ZINC000055656943 had the lowest (7.468 kJ/mol). For the RBD-ligand complexes, polar solvation energies ranging from 14.435 to 80.060 kJ/mol were observed (Table 3). RBD-Remdesivir demonstrated the highest polar solvation energy of 80.060 kJ/mol.

2.5.1. Energy Decomposition per Residue

MM/PBSA computations are used to decompose calculated free energies either by per-residue or pair-wise decompositions [90,94]. The per-residue decomposition involves the decomposition of each residue by including the interactions in which one residue atom is involved. Alternatively, pair-wise decomposition interactions can be decomposed by specific residue pairs by including only those interactions in which one atom from each of the analyzed residues is participating [90,94]. These techniques provide useful insight into important interactions of key residues in free energy contribution. Residues contributing binding free energy greater than 5 kJ/mol or less than −5 kJ/mol are worthy of consideration as key residues for the binding of a ligand to a protein [95].

Per-Residue Energy Decomposition of M^{pro}-Ligand Complexes

Generally, amino acid residues within the range of 230 to 290 were observed to contribute energies beyond the ±5 kJ/mol threshold. For the M^{pro}-ligand complexes, Tyr237, Tyr239 and Leu272 were common residues that contributed energies greater than 5 kJ/mol or lesser than 5 kJ/mol (Figure 4 and Figure S5A–G). For the M^{pro}-NANPDB2245 complex, Tyr237, Tyr239, Leu271, Leu272 and Ala285 were involved in the protein-ligand interaction with individual residue energies of 9.4613, 7.4232, 5.8197, 8.5205, and 5.4298 kJ/mol, respectively (Figure S5B). Lys236, Tyr237, Asn238, Tyr239 and Leu272 contributed energies of 9.4957, 10.5023, 5.0114, 9.6197 and 10.2874 kJ/mol, respectively, in the M^{pro}-fusidic acid complex (Figure S5D). For the M^{pro}-Remdesivir, only Tyr237 was observed to contribute energy beyond the ±5 kJ/mol threshold (−5.5271 kJ/mol) (Figure S5A).

The M^{pro}-ZINC000095486008 complex had the greatest number of residues contributing energies to the interaction (Figure 4). Lys137 contributed the highest energy (33.1602 kJ/mol), followed by Leu287 (16.6352 kJ/mol). Thr199, Tyr239, Met276, Leu286, Glu288 and Asp289 also contributed individual energies of 5.9548, 10.5721, 5.6664, 5.7757, 7.8710 and 9.4673 kJ/mol, respectively (Figure 4). In a previous study, ZINC000095486008 in a complex with Ebola virus viral protein 24 (EBOV VP24) also had high interaction energies per residue [45]. Only Lys137 contributed beyond the threshold (10.7331 kJ/mol) in the M^{pro}-ZINC000055656943 complex (Figure S5E). The interaction between ZINC001645993538 and M^{pro} caused residues Tyr237, Tyr239, Leu271, Leu272, Gln273 and Gly275 to contribute energies of 9.7242, 7.5793, 5.6107, 15.8221, 8.4805 and 5.0989 kJ/mol, respectively (Figure S5F). The interaction with oxymetholone also involved Ser46 and Gln189 with energy contributions 8.2579 and 12.2857 kJ/mol, respectively (Figure S5G).

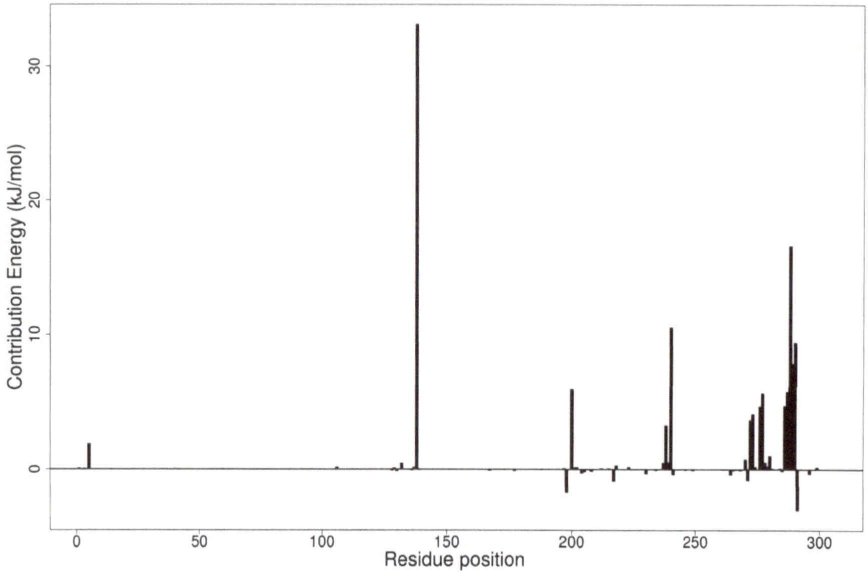

Figure 4. Molecular mechanics/Poisson-Boltzmann surface area (MM/PBSA) plot of binding free energy contribution per residue of the M^{pro}-ZINC000095486008 complex.

Per-Residue Energy Decomposition of the RBD-Ligand Complexes

For the RBD-ligand complexes, residues within the range 340–375 were observed to be involved in interactions contributing energies beyond ±5 kJ/mol (Figure S5H–O). Only Tyr505 showed energy beyond ±5 kJ/mol for the RBD-Remdesivir complex (−5.9154 kJ/mol) (Figure S5H). Residues from RBD complexes of NANPDB2245, ZINC000095486008 and ZINC001645993538 were observed to contribute individual energies beyond the ±5 kJ/mol thresholds (Figure S5I,L,N). For the RBD-NANPDB2245 complex, Asn343 and Ser373 were observed to contribute 5.3285 and 5.286 kJ/mol, respectively (Figure S5I). Leu368 and Phe374 contributed −5.0598 and −5.3661 kJ/mol, respectively, in the RBD-ZINC0000 95486008 complex (Figure S5L). Additionally, Leu335, Asp364 and Val367 contributed energies of −6.3651, 5.5776 and 7.2196 kJ/mol in the RBD-ZINC001645993538 interactions, respectively, (Figure S5N). These residues need to be investigated to ascertain their critical roles in RBD-ligand binding.

2.6. Molecular Dynamics

To further understand the dynamic behavior of the hits within the active sites of the protein structures, 100 ns MD simulations were performed for the unbound protein structures and two selected protein-ligand complexes (NANPDB2403 and ZINC95486008) in duplicates. The root mean square deviation (RMSD), the root mean square fluctuations (RMSF), and the radius of gyration (Rg) were assessed. In addition, graphs for the 10 ns MD simulations were provided in Figure S4.

2.6.1. Root Mean Square Deviation of the Complexes for 100 ns MD Simulations

The RMSD is a plausible measure of protein stability. For the unbound protein, its RMSD sharply rose to 0.2 nm and plateaued around this figure during the simulation (Figure 5a). A similar occurrence was observed for the duplication run. Both runs for the Mpro-NANPDB2403 complex showed similar RMSD, with little deviations (Figure 5a). The duplicate run of the M^{pro}-ZINC000095486008 complex was observed to have a stable RMSD with an average value of 0.2 nm until about 30 ns, where a rise was observed until the end of the 100 ns period (Figure 5a). In all the structures, the RMSDs were observed to fluctuate within the range of 0.13–0.32 nm (except the duplicate run of M^{pro}-

ZINC000095486008) (Figure 5a). A recent study virtually screened FDA-approved antiviral drugs against the M^{pro} and performed 100 ns molecular dynamics simulations of protein-ligand complexes [96]. The study reported an RMSD range of 1.5 Å (0.15 nm) to 3 Å (0.3 nm), with an average RMSD of 2.25 Å (0.225 nm) for all complexes [96], consistent with the RMSD range reported (Figure 6a) and other studies [97,98].

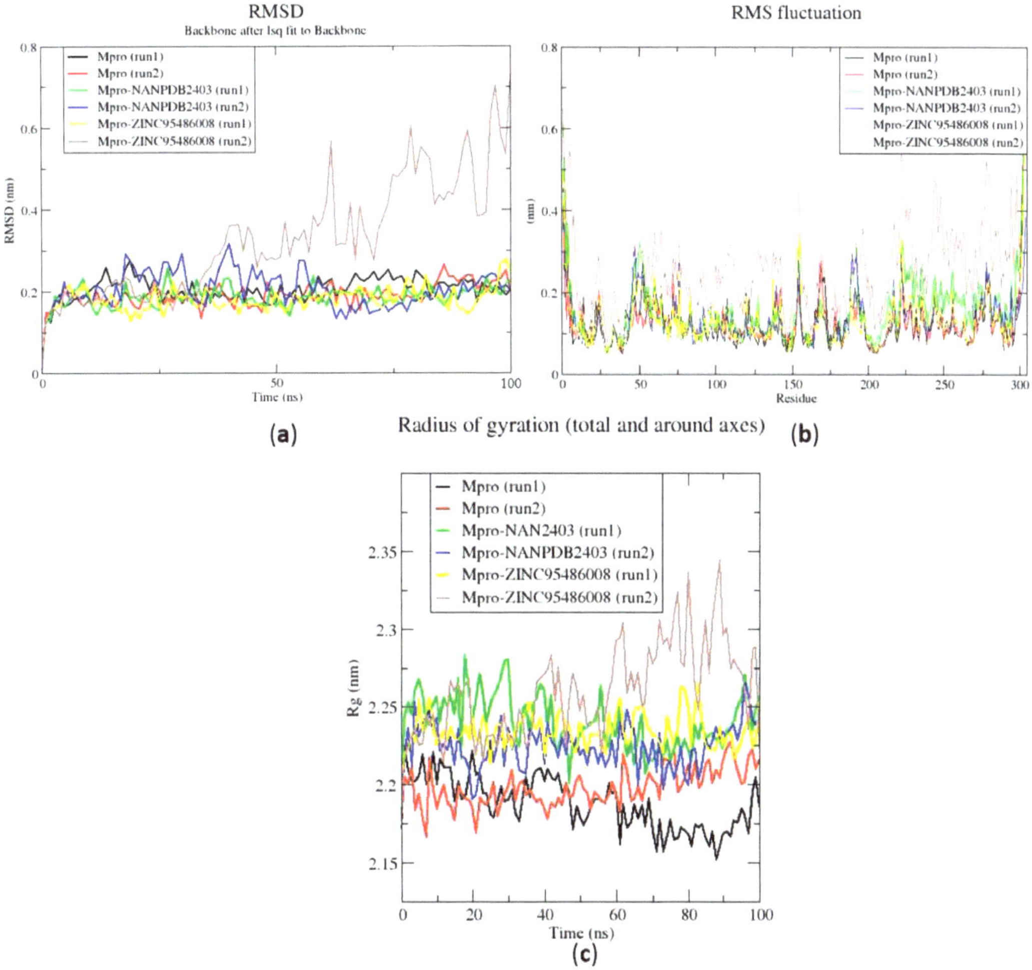

Figure 5. Root mean square deviation (RMSD), root mean square fluctuations (RMSF), and the radius of gyration (Rg) plots of the extended molecular dynamics (MD) simulations of the M^{pro}-ligand complexes generated over 100 ns using GROMACS. (**a**) RMSD versus a time graph of the M^{pro}-ligand, (**b**) analysis of RMSF trajectories of residues of the M^{pro}-ligand, and (**c**) the Rg versus a time graph of the M^{pro}-ligand complexes. In all the three graphs, the unbound M^{pro}, unbound M^{pro} duplicate run, Mpro-NANPDB2403, Mpro-NANPDB2403 duplicate run, M^{pro}-ZINC000095486008, and M^{pro}-ZINC000095486008 duplicate run are represented as black, red, green, blue, yellow and brown, respectively.

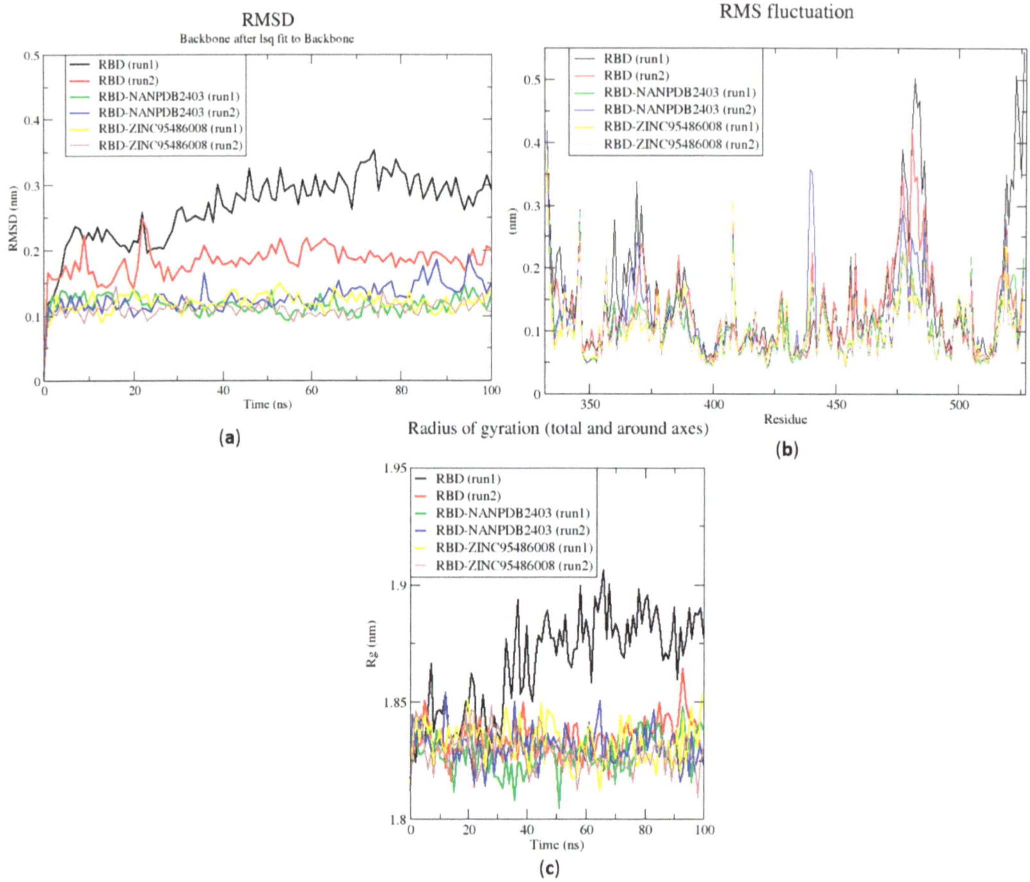

Figure 6. RMSD, RMSF and Rg plots of the extended MD simulations of the RBD-ligand complexes generated over 100 ns using GROMACS. (**a**) RMSD versus time graph of the RBD-ligand complexes, (**b**) analysis of RMSF trajectories of residues of the RBD-ligand complexes, and (**c**) Rg versus time graph of the RBD-ligand complexes. In all the three graphs, the unbound RBD, unbound RBD duplicate run, RBD-NANPDB2403, RBD-NANPDB2403 duplicate run, RBD-ZINC000095486008, and RBD-ZINC000095486008 duplicate run are shown as black, red, green, blue, yellow and brown, respectively.

The RMSD values of the RBD and RBD–ligand complexes were also computed. For the unbound RBD, the RMSD averaged between 0.2 nm and 0.25 nm during the first 20 ns of the simulation (Figure 6a). It then rose to about 0.3 nm, where it was fairly stable during the rest of the simulation. The RMSD of the unbound protein averaged around 0.15 nm (Figure 6a). For both runs, the RBD-ZINC95486008 complex showed the most stable RMSD, with both averaging around 0.1 nm. During the two duplicate runs, the RBD-ZINC000095486008 complex maintained an average RMSD of 0.125 nm throughout the 100 ns simulation period (Figure 6a). Generally, the unbound RBD structures had the highest RMSD values (Figure 6a). A recent molecular dynamics study of RBD and terpene compounds reported a similar RMSD range of 0.08 and 0.25 nm [99].

2.6.2. Root Mean Square Fluctuation of the Complexes for 100 ns MD Simulations

The RMSF was computed to analyze the residual fluctuations over the simulation time. For the M^{pro} and its complexes, fluctuations were observed at regions from residue indexes of 50–55, 150–160, 215–230 and 275–280 (Figure 5b). For the RBD, fluctuations were observed in all the RBD–ligand complexes around regions 420–430, 470–480 and 520–530

(Figure 6b). A few residues in the range 470–480 showed more flexibility in the RBD for both duplicate runs than the RBD-NANPDB2403 and RBD-ZINC95486008 complexes (Figure 6b). The results obtained corroborate with that of a previous study that identified SARS-CoV-2 RBD inhibitors via molecular docking and dynamics simulations studies of terpenes [99].

2.6.3. Radius of Gyration of the Complexes for 100 ns MD Simulations

The radius of gyration was assessed to evaluate the compactness of the structures. A stably folded protein maintains a reasonably steady Rg over the simulation time. Considering M^{pro}, the Rg of the unbound M^{pro} for both duplicate runs was relatively steady and averaged around 2.2 nm (Figure 5c). The duplicate run of the M^{pro}-ZINC000095486008 was observed to have a similar Rg as the other, maintaining an average Rg value of 2.25 nm until about 30 ns, where it spiked and then fluctuated throughout the rest of the simulation time (Figure 5c). For the duplicate runs, M^{pro}-NANPDB2403 had a steady Rg over the simulation and averaged around 2.25 nm. Previous MD simulation studies of M^{pro} in complex with lichen spp. compounds revealed Rg values ranging between 2.175 nm and 2.25 nm [98], consistent with the Rg values reported elsewhere [100] and in this work.

The Rg curves of the unbound RBD and RBD-ligand complexes for 100 ns ranged between 1.8 and 1.91 nm (Figure 5c), consistent with the Rg ranges for the 10 ns MD simulations and in a recent study (1.8 and 1.88 nm) [99]. The unbound RBD had similar Rg with that of the duplicate, with an average of 1.84 nm until 30 ns, where it increased gradually to an average of 1.88 nm until the end of the 100 ns period (Figure 5c).

2.7. *Comparison of Binding Modes Pre-MD and Post-100 ns MD Simulations*

Analyses of the binding modes of potential leads NANPDB2403 and ZINC95486008 in complex with both target structures (M^{pro} and RBD) were undertaken after both docking and 100 ns MD simulation. This was to ascertain if the same binding modes were maintained after undergoing MD simulation. Binding mode superimpositions and visual inspections revealed that the compounds resided well in the active site pocket of each target, with almost the same binding modes for the pre-MD docked complex and post-MD simulations results (Figures 7 and 8). The evaluation of the binding modes based on the superimposition of the complexes gave RMSD values for NANPDB2403 and ZINC95486008 in M^{pro} pre-MD and post-MD as 1.167Å and 0.703Å, respectively. For the RBD target, RMSD values 0.807Å and 1.396Å were obtained for both compounds, which were less than 2Å, considered as the threshold for good alignment [101]. Therefore, the binding poses could be considered similar even after MD simulations.

2.7.1. Binding Modes Interactions Analysis between M^{pro} and Potential Leads

The pre-MD interaction analysis revealed that the NANPDB2403 complex formed hydrogen bond interactions with the Thr199 and Leu287 residues. However, after MD simulations, these hydrogen bond interactions were lost. Instead, the ligand formed hydrophobic interactions with Leu287 and no interactions with Thr199. Likewise, the pre-existing hydrophobic bonds were only maintained for Tyr237, Leu271 and Leu272. The hydrophobic bond interactions for Tyr237 and Tyr239 were lost after the simulation (Table 4, Figure 7a and Figure S2A,B).

ZINC95486008 formed only hydrophobic interactions before and after MD simulations. Surprisingly, the hydrophobic interactions observed after MD simulations were completely different from those observed before. New hydrophobic interactions were found between the ligand and residues Trp31, Ala70, Gly71, Asn72, Val73, Leu75 and Ala94 (Table 4, Figure 7b and Figure S2C,D). Both compounds docked well within the active site after MD simulations.

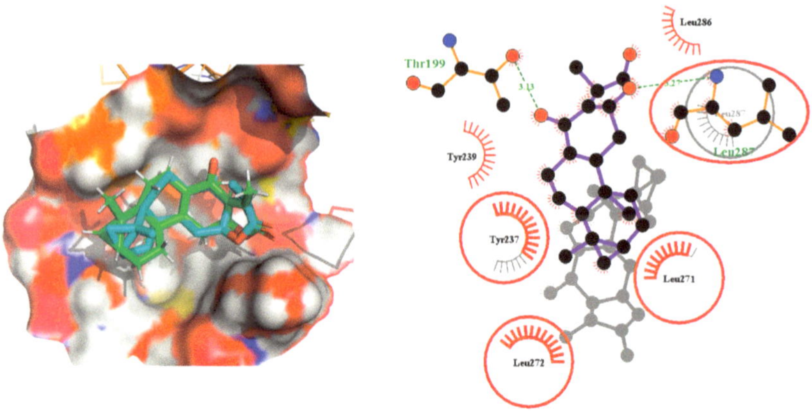

(**a**)

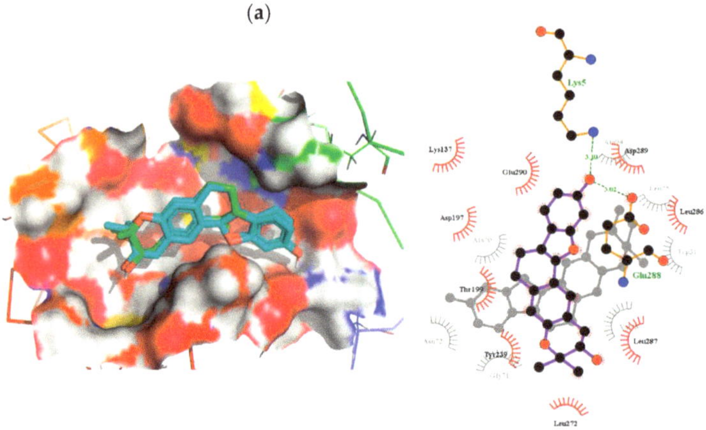

(**b**)

Figure 7. Binding mode characterization using superimposition and LigPlot+ analyses. (**a**) The pre- and post-100 ns MD simulations of M^{pro}-NANPDB2403, and (**b**) M^{pro}-ZINC95486008 complexes. Ligands for the pre- and post-MD simulations are shown in green and cyan colors, respectively. The overlapped interactions between the pre- and post-MD simulations are circled in red color.

Table 4. Molecular interactions between potential lead compounds and the M^{pro} target before and after molecular dynamics (MD) simulations.

M^{pro}				
	Pre-MD Interactions		**Post-MD Interactions (100 ns)**	
Compound Name	**H-Bond Residues**	**Hydrophobic Bond Residues**	**H-Bond Residues**	**Hydrophobic Bond Residues**
NANPDB2403	Thr199, Leu287	Tyr237, Tyr239, Leu271, Leu272, Leu286	-	Tyr237, Leu271, Leu272, Leu287
ZINC95486008	Gly5, Glu288	Lys137, Asp197, Thr199, Tyr239, Leu272, Leu286, Leu287, Asp289, Glu290	-	Trp31, Ala70, Gly71, Asn72, Val73, Leu75, Ala94

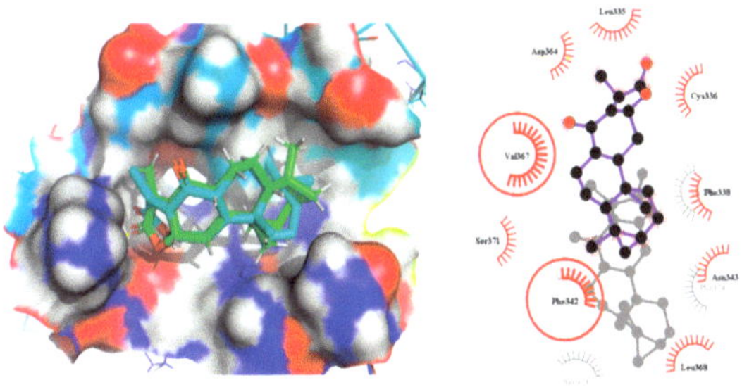

(a)

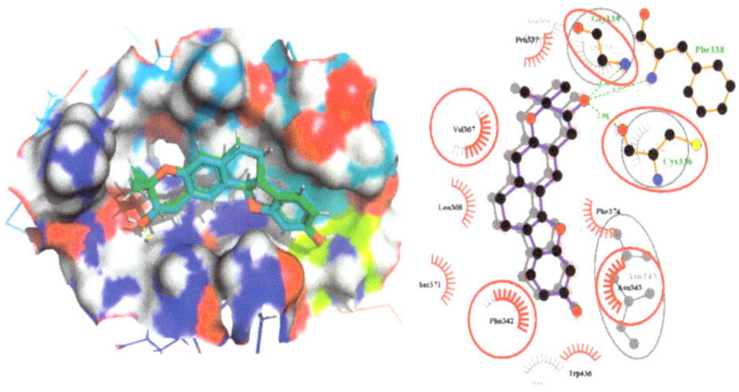

(b)

Figure 8. Binding mode characterization using superimposition and LigPlot+ analyses. (**a**) The pre- and post-100 ns MD simulations of RBD-NANPDB2403, and (**b**) RBD-ZINC95486008 complexes. Ligands for the pre- and post-MD simulations are shown in green and cyan colors, respectively. The overlapped interactions between the pre- and post-MD simulations are circled in red color.

2.7.2. Binding Modes Interactions Analysis between RBD and Potential Leads

For the RBD target, hydrogen bond interactions were not observed for NANPDB2403 ligand before and after MD simulations. However, two residues namely Phe342 and Val367 maintained their hydrophobic interactions after the simulation. New interactions were observed between the ligand and residues Gly339, Ser373 and Phe374 (Table 5, Figure 8a and Figure S2E,F).

Table 5. Molecular interactions between potential lead compounds and RBD targets before and after MD simulations.

RBD				
	Pre-MD Interactions		Post-MD Interactions (100 ns)	
Compound Name	**H-Bond Residues**	**Hydrophobic Bond Residues**	**H-Bond Residues**	**Hydrophobic Bond Residues**
NANPDB2403	-	Leu335, Cys336, Phe338, Phe342, Asn343, Asp364, Val367, Leu368, Ser371	-	Gly339, Phe342, Val367, Ser373, Phe374
ZINC95486008	Cys336, Phe338, Gly339	Pro337, Phe342, Asn343, Val367, Leu368, Ser371, Phe374, Trp436	Asn343	Cys336, Gly339, Phe342, Asp364, Val367, Ser373

The interactions for ZINC95486008 and the target before the simulation showed that hydrogen bond interactions formed between the ligand and residues Cys336, Phe338 and Gly339. Even though, Cys336 has been reported as a critical residue for binding, it did not maintain the hydrogen bond after MD simulations but rather formed a hydrophobic contact. It is worth noting that after MD simulations, ZINC95486008 formed a hydrogen bond with Asn343, a residue predicted as a critical. Residues whose hydrophobic bonds were maintained after MD simulations include Phe342 and Val367. However, ZINC95486008 formed hydrophobic bonds with Gly339, a residue that previously formed a hydrogen bond during pre-MD simulation (Table 5, Figure 8b and Figure S2G,H).

2.8. Summary and Implications of the Results

Six potential anti-SARS-CoV-2 biomolecules were identified as leads comprising NANPDB2245, NANPDB2403, fusidic acid, ZINC000095486008, ZINC0000556656943 and ZINC001645993538 (Table 6). These molecules were obtained by screening libraries made up of ANC, known drugs and machine learning-derived compounds against the SARS-CoV-2 receptors M^{pro} and RBD. The known drugs included antiviral remdesivir, dexamethasone, hydroxychloroquine and chloroquine. The techniques utilized included previously described methods used to identify potential bioactive compounds against the Ebola virus protein VP24 [45].

Table 6. A list of selected compounds with their 2D structures and common/IUPAC names generated using the Marvin suite (http://www.chemaxon.com/).

Ligand ID	Common/IUPAC Name	2D Structure
NANPDB2245	Helioscopinolide B	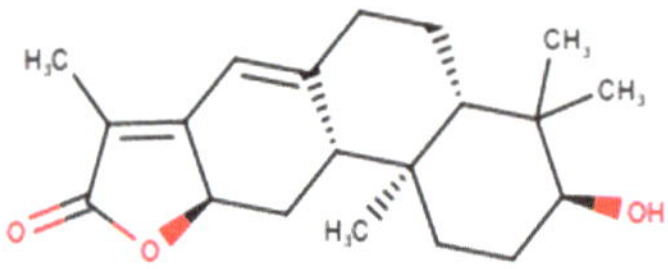
NANPDB2403	Retusolide B	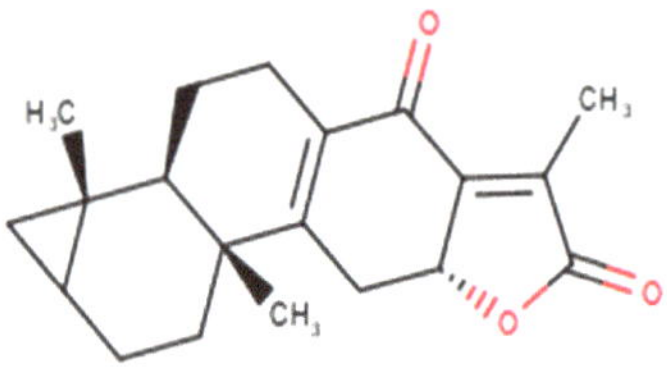
Fusidic acid	Fusidic acid	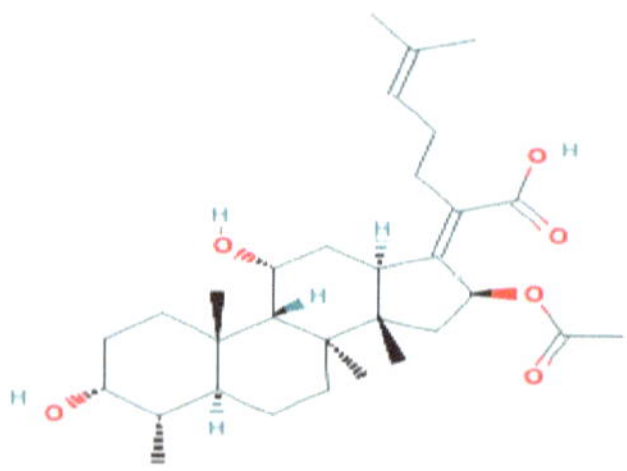
ZINC000095486008	(2R,10R,18S)-17,17-dimethyl-3,16-dioxapentacyclo (1 1.8.0.02,10.04,9.015,20) henicosa-1(13),4,6,8,14,20-hexaene-6,18-diol	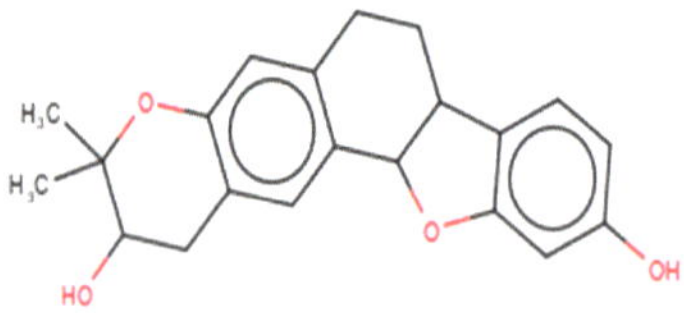
ZINC000055656943	(4S)-7-fluoro-N-((1S)-5-fluoro-2,3-dihydro-1H-inden-1-yl)-2-oxo-1,2,3,4-tetrahydroquinoline-4-carboxamide	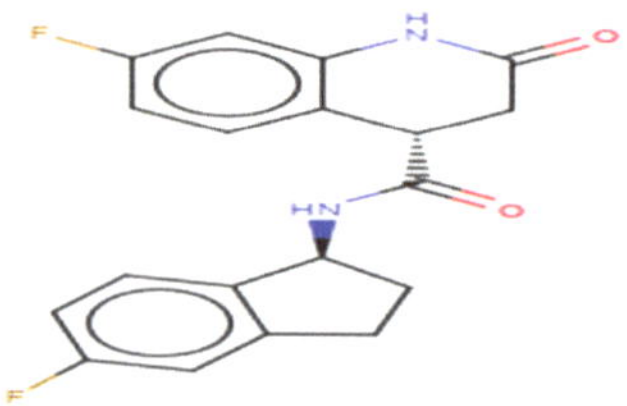

Table 6. *Cont.*

Ligand ID	Common/IUPAC Name	2D Structure
ZINC001645993538	(4S)-N-((1R,3R,6R)-7,7-difluorobicyclo (4.1.0) heptan-3-yl)-7-fluoro-2-oxo-1,2,3,4-tetrahydroquinoline-4-carboxamide	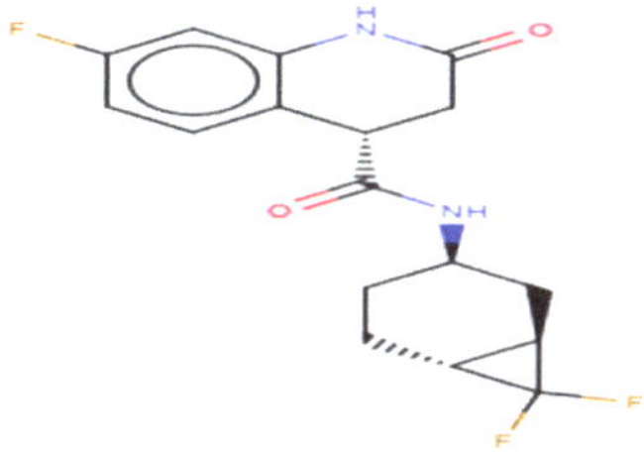

Fusidic acid was shown to be structurally similar to NANPDB2245 and NANPDB2403 via DrugBank. Betulinic acid was also identified as structurally similar to oxymetholone. In our previous study, helioscopinolide C which is a structural analog of NANPDB2245 (helioscopinolide B), and ZINC000095486008 were reported as being plausible anti-Ebola compounds [45]. These compounds warrant further in vitro and in vivo testing to ascertain their anti-SARS-CoV-2 activity. The results also corroborate ongoing research on the inhibitory activities of remdesivir and oxymetholone. Since fusidic acid is already an FDA-approved drug for humans, this offers the opportunity to explore its therapeutic potential against SARS-CoV-2.

The study proposed potential anti-SARS-CoV-2 compounds, which were reinforced with antiviral activity predictions. Additionally, the study highlights the repurposing of existing drugs as potential anti-SARS-CoV-2 molecules. This study complements ongoing efforts geared towards the identification of SARS-CoV-2 inhibitors. Making these predicted compounds accessible to the scientific community could stimulate the pace of searching for effective SARS-CoV-2 drugs.

3. Materials and Methods

A schema showing the step-by-step techniques employed in predicting the potential leads is shown in Figure 9. A library consisting of ANC, FDA-approved drugs and machine learning-derived compounds were screened against the structures of SARS-CoV-2 receptors M^{pro} and RBD of the spike protein. The compounds were prefiltered using molecular weights (MW) between 250 and 350 g/mol, good absorption, distribution, metabolism, excretion and toxicity (ADMET) profiles. The docked complexes were subjected to MD simulations, and the biological activities of the hits were predicted.

3.1. Data Sources for SARS-CoV-2 Targets

The experimentally solved 3D structures of SARS-CoV-2 M^{pro} and RBD of the spike protein (Accession numbers: PDB IDs 5R82 and 6M0J, respectively) were retrieved from the Research Collaboratory for Structural Bioinformatics Protein Data Bank (RCSB PDB) [102]. 5R82 is a monomer of the SARS-CoV-2 M^{pro} with a co-crystallized ligand (6-(ethylamino) pyridine-3-carbonitrile), while 6M0J is also a monomer of the SARS-CoV-2 RBD in complex with the human ACE2.

3.2. The Screening Library

A library of 7675 compounds was created from the North African Natural Product and African Natural Product Databases [42,43]. The natural compounds were filtered using ADMET Predictor™ (V8.0, Simulations Plus, Inc., Lancaster, PA, USA) and those with high toxicity levels and molecular weights greater than 350 g/mol and less than 250 g/mol were eliminated [45,103], which reduced the number to 1470. Additionally, the 1000 top hit compounds generated by a machine learning (ML) study that screened 1.3 billion compounds against the M^{pro} [28] were also screened based on Lipinski's rule of five using OpenBabel [104]. A total of 60 compounds were eliminated and the remaining 940 added to the library. Additionally, a set of 43 FDA-approved antivirals, including those undergoing clinical trials [105], were added to the library. Thus, the library used for molecular docking was composed of 2453 compounds.

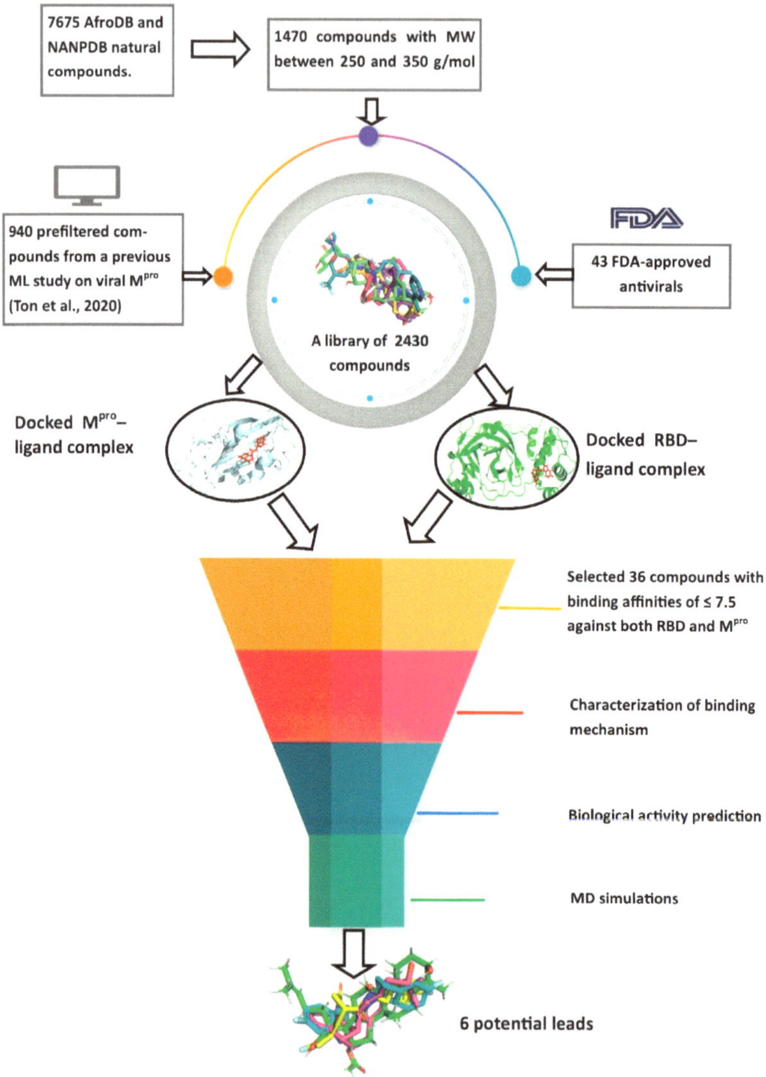

Figure 9. Methodology schema employed in the cheminformatics studies for predicting potential anti-SARS-COV-2 compounds. Natural compounds from the African Natural Product Database (AfroDB), North African Natural Product Database (NANPDB), the Food and Drug Administration (FDA)-approved antivirals and compounds from previous machine learning (ML) studies were docked against SARS-CoV-2 receptors M^{pro} and RBD. Molecular weights (MW) were used to prefilter the molecules. Compounds with binding affinities of ≤ -7.5 kcal/mol against both receptors were selected for the downstream analysis. Methods include the characterizations of the protein-ligand complexes, biological activity predictions and MD simulations.

3.3. Preparation of the Protein Structure and Elucidation-Binding Sites

The 3D protein structures of M^{pro} and RBD were analyzed using PyMOL Version 1.5.0.4 (PyMOL Molecular Graphics System, Schrödinger, LLC) as described [106]. The structures were first cleaned of all available water molecules and ligands before being subjected to energy minimization. A 10 ns molecular dynamics (MD) simulation for each of the structures was performed using the Groningen Machine for Chemical Simulations GROMACS version 2018 [107]. The Optimized Potentials for Liquid Simulations (OPLS)/All Atom (AA) force field was used to generate the protein topologies and position restrain files. Periodic boundary conditions (PBC) were applied to each structure, with the protein centered 1 nm from the edge of a cubic box to monitor the movement of all particles and avoid edge effects on the surface atoms [108]. The system was solvated with SPC water and neutralized, and the steepest descent algorithm used for the energy minimization was at 50,000 steps. A 100 ps equilibration simulation was performed using the NVT and NPT ensembles to ensure that the system was well-equilibrated to an optimal temperature of 300 K and pressure of 1 bar before the MD simulation, which was performed for 10 ns. Xmgrace was used to generate the graphical outputs [109]. The binding sites of the proteins were then predicted using the Computed Atlas of Surface Topography of proteins (CASTp) version 3.0 (available at http://sts.bioe.uic.edu/castp/calculation.html), which utilizes theoretical and algorithmic results of computational geometry to predict pockets and cavities, whiles excluding shallow depressions [49]. The predicted binding cavities were visualized and analyzed using Chimera version 1.12 and PyMOL. Predicted sites with very small volumes and areas such that no ligands could fit were excluded from downstream virtual screening.

3.4. Virtual Screening of Ligand Library

AutoDock Vina [61] was used to screen the integrated library against the energy-minimized M^{pro} and RBD. The library was imported into the OpenBabel workspace [104] in ".sdf" file format and minimized using the Universal Force Field (Uff) for 200 steps and then optimized using the conjugate gradient before finally converting the file into the AutoDock format (".pdbqt"). The file format ".pdbqt" stores the atomic coordinates, partial charges, and describes the rigid and rotational parts of the molecule, as well as serves as the input format for AutoDock Vina. AutoDock Vina employs an empirical and knowledge-based scoring function to predict the binding affinity of compounds [61]. The specified dimensions of the grid box for the M^{pro} were 37.58 * 64.78 * 62.77 $Å^3$ and centered at 49.33, 49.36, and 49.56 Å. The corresponding dimensions for the RBD were 42.54 * 42.73 * 42.44 $Å^3$ and centered at 41.19, 47.74, and 55.37 Å. The grid boxes covered all the predicted binding pockets of the two proteins. The compounds with binding energies of −7.5 kcal/mol or less for both receptors were selected for downstream analysis [62].

3.5. Characterization of the Protein-Ligand Interactions

LigPlot+ [69] was used to generate the 2D protein-ligand interactions which revealed hydrogen and hydrophobic interactions. The best poses of the hits were saved in ".pdb" file format and then visualized in PyMol [106]. The saved protein-ligand complexes served as inputs for LigPlot+ [69]. The hydrogen bonds were denoted as green dashed lines and the arcs with spokes radiating towards the ligands as the hydrophobic interactions. Default parameters were used in generating the interaction profiles.

3.6. Prediction of Antiviral Properties of Hit Compounds

The antiviral activities of the hits were predicted using the Bayesian-based Prediction of Activity Spectra for Substances (PASS) [110] using the Simplified Molecular Input Line-Entry System (SMILES) format of the compounds as inputs. PASS determines the relevant biological activities of compounds based on the structural–activity relationship between the compound of interest and a training set of over 26,000 compounds with known biological activities [110]. For any given compound, PASS predicts the Pa and Pi, with both ranging between 0.000 and 1.000 for a predicted activity. When the Pa is greater than the Pi for a particular compound activity and Pa > 0.3, it is worth exploring the pharmacological activity [111,112]. PASS was used in previous studies to predict the antiviral activity of novel compounds, and the experimental results corroborated the PASS predictions [113].

3.7. Molecular Dynamics Simulation of Protein-Ligand Complexes

MD simulations of the protein-ligand complexes were performed using GROMACS 2018 [107]. Their protein topologies were generated using the CHARMM36 all-atom force field. The ligand topologies were generated using the CHARMM force field via the CHARMM General Force Field (CGenFF) server (available at https://cgenff.umaryland.edu/). Complexes were generated from the ligands and protein topologies for each of the selected cases under study. Each complex was solvated with the transferable intermolecular potential with a 3 points (TIP3P) water model in a cubic box of size 1.0 nm and neutralized with Na and Cl ions. Energy minimization of each complex was conducted for 50,000 steps using the steepest descent algorithm. The ligands were restrained before the NVT and further using the NPT ensemble. Equilibration of each complex was performed for 100 ps apiece and the final MD simulation was conducted for 10 ns with time steps of 2 fs under the PME. Extended MD runs (in duplicates) were also conducted for 100 ns for the unbound proteins and selected protein-ligand complexes of NANPDB2403 and ZINC95486008. Duplicate MD runs were carried out by generating random seeds for the initial velocities of each run.

The root mean square deviation (RMSD), root mean square fluctuation (RMSF), and radius of gyration (Rg) of the unbound proteins and selected protein-ligand complexes were determined. RMSD is a frequently used measure of the differences between the structures sampled during the simulation and the reference structure [114]. MD simulations require systems to be close to their equilibrium (native) conformation. The time trajectory of RMSD shows how a protein structure deviates from a reference structure as a function of time [114].

RMSF measures the movement of a subset of atoms concerning the average structure over the entire simulation. RMSF indicates the flexibility of different regions of a protein, which can be related to crystallographic B factors [114]. Residues contributing to the complex structural fluctuation can be assessed using this stability profile analysis. Higher RMSF values imply greater fluctuations. Greater amounts of structural fluctuations occur in regions known to be involved in ligand binding and catalysis, notably the catalytic loop regions [115]. Adaptive variation in flexibility lies principally in these regions of the sequence that influence the conformational stabilities of the protein-ligand complex [115].

The radius of gyration (Rg) assesses the changes in compactness of a protein-ligand complex over the simulation time. The loss of compactness affects the stability of the complex by introducing weak intermolecular bonds. When the Rg of a complex is relatively steady, the compactness of the protein-ligand complex is high, and the protein is folded well, whereas the Rg value changes over time if the protein unfolds [45,116].

G_mmpbsa [48] was also used to calculate the free binding energies of each complex over the 10 ns simulation period, utilizing frames in steps of 0.1 ns. The binding free energy contribution per residue was calculated using the MM/PBSA and the plots generated using R programming.

3.8. Analysis of Binding Modes

From the aforementioned, two out of six lead compounds comprising NANPDB2403 and ZINC95486008 were selected for an extended 100 ns MD simulation for both targets, and their binding modes were analyzed comparatively for the pre- and post-MD simulations. Binding mode analysis was done based on whether the compounds maintained their poses and interactions with the residues in the binding site pockets after the simulations. This was evaluated using RMSD resulting from superimposition of the complexes, visual inspection and intermolecular interactions via LigPlot+.

4. Conclusions

The study utilized cheminformatics techniques to identify six potential anti-SARS-COV-2 compounds from an integrated compound library made up of natural products from Africa, machine learning-based studies and drugs undergoing clinical trials. The compounds were screened against the binding pockets of two putative drug targets, namely RBD of the spike protein and the M^{pro}. The six potential lead compounds, namely NANPDB2245, NANPDB2403, fusidic acid, ZINC000095486008, ZINC0000556656943 and ZINC001645993538 had binding energies ranging from −6.9 kcal/mol to −8.2 kcal/mol and 7.2 kcal/mol to −8.0 kcal/mol against the M^{pro} and RBD of spike protein, respectively. Additional molecular dynamics simulations coupled with MM/PBSA calculations reinforced the potential inhibition of the two SARS-CoV-2 therapeutic targets. These identified druglike biomolecules have good pharmacological profiles with insignificant toxicity. The compounds were also predicted to have a high propensity in inhibiting viral entry and replication. The predicted scaffolds could form the basis for the de novo design of the next-generation SARS-CoV-2 compounds for clinical evaluations. Potentially novel critical binding residues were identified that could help in the design of new inhibitors. The studies are computational and would therefore need in vitro studies to corroborate the findings.

Supplementary Materials: The following are available online. Figure S1: Cartoon representation of RBD in complex with: (a) NANPDB2245 (helioscopinolide B), (b) ZINC000095486008, (c) ZINC001645993538, and (d) Oxymetholone. Figure S2: Binding mode representation and LigPlot+ characterization of: (a) M^{pro} and NANPDB2403 before MD simulation (pre-MD), (b) M^{pro} and NANPDB2403 after MD simulation (post-MD), (c) M^{pro} and ZINC95486008 before MD simulation (pre-MD), (d) M^{pro} and ZINC95486008 after MD simulation (post-MD), (e) RBD and NANPDB2403 before MD simulation (pre-MD), (f) RBD and NANPDB2403 after MD simulation (post-MD), (g) RBD and ZINC95486008 before MD simulation (pre-MD), and (h) RBD and ZINC95486008 after MD simulation (post-MD). Figure S3: Two-dimensional diagram of the RBD–ligand interaction generated using LigPlot+. (a) Interaction profile of the RBD-NANPDB2245 complex, and (b) Interaction profile of the RBD-ZINC000095486008 complex. Figure S4: Graphs of the RMSD, RMSF and radius of gyration of the RBD–ligand complexes generated over a 10 ns molecular dynamics simulation using GROMACS. (a) RMSD versus time graph of the RBD–ligand complexes, (b) Analysis of the RMSF trajectories of the residues of the RBD–ligand complexes, and (c) Rg versus time graph of the RBD–ligand complexes. Figure S5: MM/PBSA plot of the binding free energy contribution per residue of the protein-ligand complexes (a) M^{pro}-Remdesivir, (b) M^{pro}-NANPDB2245, (c) M^{pro}-NANPDB2403, (d) M^{pro}-fusidic acid, (e) M^{pro}-ZINC000055656943, (f) M^{pro}-ZINC001645993538, (g) M^{pro}-oxymetholone, (h) RBD-Remdesivir, (i) RBD-NANPDB2245, (j) RBD-NANPDB2403, (k) RBD-fusidic acid, (l) RBD-ZINC000095486008, (m) RBD-ZINC000055656943, (n) RBD-ZINC001645993538, and (o) RBD-Oxymetholone. Table S1: The binding energies and intermolecular interactions between the hits and M^{pro} as well as RBD.

Author Contributions: M.D.W., S.K.K., and E.B. conceptualized and designed the study; S.K.K., E.B., and S.O.A. performed the computational analysis with contributions from G.B.K.; B.D., K.S.E., E.K.T., and S.K.K. wrote the first draft, and all authors proofread and agreed to the current version of the manuscript. All authors have read and agreed to the published version of the manuscript.

Funding: This research received no external funding.

Institutional Review Board Statement: Not applicable.

Informed Consent Statement: Not applicable.

Data Availability Statement: All the IDs of the datasets used for the work were cited in the manuscript.

Acknowledgments: The authors are grateful to the West African Centre for Cell Biology of Infectious Pathogens (WACCBIP) at the University of Ghana for making Zuputo, a Dell EMC high-performance computing cluster, available for this study.

Conflicts of Interest: The authors declare no conflict of interest.

Sample Availability: Samples of the compounds are not available from the authors.

References

1. Rabi, A.F.; Al Zoubi, S.M.; Kasasbeh, A.G.; Salameh, M.D.; Al-Nasser, D.A. SARS-CoV-2 and Coronavirus Disease 2019: What We Know So Far. *Pathogens* **2020**, *9*, 231. [CrossRef] [PubMed]
2. World Health Organization Naming the Coronavirus Disease (COVID-19) and the Virus That Causes It. Available online: https://www.who.int/emergencies/diseases/novel-coronavirus-2019/technical-guidance/naming-the-coronavirus-disease-(covid-2019)-and-the-virus-that-causes-it (accessed on 3 July 2020).
3. WHO. WHO Coronavirus Disease. Available online: https://covid19.who.int/ (accessed on 8 December 2020).
4. Aura Vision. Coronavirus (COVID-19) Lockdown Tracker | Aura Vision. Available online: https://auravision.ai/covid19-lockdown-tracker/ (accessed on 18 May 2020).
5. Cascella, M.; Rajnik, M.; Cuomo, A.; Dulebohn, S.C.; Di Napoli, R. *Features, Evaluation and Treatment Coronavirus (COVID-19)*; StatPearls Publishing LLC: Treasure Island, FL, USA, 2020.
6. Hjelmesæth, J.; Skaare, D. Covid-19 med nedsatt lukte-og smakssans som eneste symptom. *Tidsskr. Nor. Laegeforening* **2020**, *140*. [CrossRef] [PubMed]
7. Passarelli, P.C.; Lopez, M.A.; Mastandrea Bonaviri, G.N.; Garcia-Godoy, F.; D'Addona, A. Taste and smell as chemosensory dysfunctions in COVID-19 infection. *Am. J. Dent.* **2020**, *33*, 135–137.
8. Moein, S.T.; Hashemian, S.M.R.; Mansourafshar, B.; Khorram-Tousi, A.; Tabarsi, P.; Doty, R.L. Smell dysfunction: A biomarker for COVID-19. *Int. Forum Allergy Rhinol.* **2020**, *10*, 944–950. [CrossRef]
9. Yan, C.H.; Faraji, F.; Prajapati, D.P.; Boone, C.E.; DeConde, A.S. Association of chemosensory dysfunction and COVID-19 in patients presenting with influenza-like symptoms. *Int. Forum Allergy Rhinol.* **2020**, *10*, 806–813. [CrossRef] [PubMed]
10. Zhou, F.; Yu, T.; Du, R.; Fan, G.; Liu, Y.; Liu, Z.; Xiang, J.; Wang, Y.; Song, B.; Gu, X.; et al. Clinical course and risk factors for mortality of adult inpatients with COVID-19 in Wuhan, China: A retrospective cohort study. *Lancet* **2020**, *395*, 1054–1062. [CrossRef]
11. Woo, P.C.Y.; Huang, Y.; Lau, S.K.P.; Yuen, K.Y. Coronavirus genomics and bioinformatics analysis. *Viruses* **2010**, *2*, 1805–1820. [CrossRef]
12. Sahin, A.R. 2019 Novel Coronavirus (COVID-19) Outbreak: A Review of the Current Literature. *Eurasian J. Med. Oncol.* **2020**, *4*, 1–7. [CrossRef]
13. Luk, H.K.H.; Li, X.; Fung, J.; Lau, S.K.P.; Woo, P.C.Y. Molecular epidemiology, evolution and phylogeny of SARS coronavirus. *Infect. Genet. Evol.* **2019**, *71*, 21–30. [CrossRef]
14. Glowacka, I.; Bertram, S.; Muller, M.A.; Allen, P.; Soilleux, E.; Pfefferle, S.; Steffen, I.; Tsegaye, T.S.; He, Y.; Gnirss, K.; et al. Evidence that TMPRSS2 Activates the Severe Acute Respiratory Syndrome Coronavirus Spike Protein for Membrane Fusion and Reduces Viral Control by the Humoral Immune Response. *J. Virol.* **2011**, *85*, 4122–4134. [CrossRef]
15. Tai, W.; He, L.; Zhang, X.; Pu, J.; Voronin, D.; Jiang, S.; Zhou, Y.; Du, L. Characterization of the receptor-binding domain (RBD) of 2019 novel coronavirus: Implication for development of RBD protein as a viral attachment inhibitor and vaccine. *Cell. Mol. Immunol.* **2020**, *17*, 613–620. [CrossRef] [PubMed]
16. Liu, S.; Xiao, G.; Chen, Y.; He, Y.; Niu, J.; Escalante, C.R.; Xiong, H.; Farmar, J.; Debnath, A.K.; Tien, P.; et al. Interaction between heptad repeat 1 and 2 regions in spike protein of SARS-associated coronavirus: Implications for virus fusogenic mechanism and identification of fusion inhibitors. *Lancet* **2004**, *363*, 938–947. [CrossRef]
17. Hoffmann, M.; Kleine-Weber, H.; Schroeder, S.; Krüger, N.; Herrler, T.; Erichsen, S.; Schiergens, T.S.; Herrler, G.; Wu, N.H.; Nitsche, A.; et al. SARS-CoV-2 Cell Entry Depends on ACE2 and TMPRSS2 and Is Blocked by a Clinically Proven Protease Inhibitor. *Cell* **2020**, *181*, 271–280.e8. [CrossRef] [PubMed]
18. Kuba, K.; Imai, Y.; Rao, S.; Gao, H.; Guo, F.; Guan, B.; Huan, Y.; Yang, P.; Zhang, Y.; Deng, W.; et al. A crucial role of angiotensin converting enzyme 2 (ACE2) in SARS coronavirus-induced lung injury. *Nat. Med.* **2005**, *11*, 875–879. [CrossRef] [PubMed]
19. Prompetchara, E.; Ketloy, C.; Palaga, T. Immune responses in COVID-19 and potential vaccines: Lessons learned from SARS and MERS epidemic. *Asian Pacific J. Allergy Immunol.* **2020**, *38*, 1–9.
20. Smith, T.; Prosser, T.; Information, C.D.; Solutions, C. COVID-19 Drug Therapy—Potential Options. *Elsevier's Nov. Coronavirus Inf. Cent.* **2020**, 19–22. Available online: Repository.phb.ac.id/776/1/COVID-19-Drug-Therapy_5.14.2020.pdf (accessed on 18 May 2020).
21. Food and Drug Administration COVID-19 Frequently Asked Questions. Available online: https://www.fda.gov/emergency-preparedness-and-response/coronavirus-disease-2019-covid-19/covid-19-frequently-asked-questions (accessed on 3 July 2020).
22. Elfiky, A.A. Anti-HCV, nucleotide inhibitors, repurposing against COVID-19. *Life Sci.* **2020**, *248*, 117477. [CrossRef]
23. Liu, J.; Cao, R.; Xu, M.; Wang, X.; Zhang, H.; Hu, H.; Li, Y.; Hu, Z.; Zhong, W.; Wang, M. Hydroxychloroquine, a less toxic derivative of chloroquine, is effective in inhibiting SARS-CoV-2 infection in vitro. *Cell Discov.* **2020**, *6*, 16. [CrossRef]
24. Dong, L.; Hu, S.; Gao, J. Discovering drugs to treat coronavirus disease 2019 (COVID-19). *Drug Discov. Ther.* **2020**, *14*, 58–60. [CrossRef]
25. FDA. Coronavirus (COVID-19) Update: FDA Issues Emergency Use Authorization for Potential COVID-19 Treatment. *Press Announc.* **2020**, 3–5. Available online: https://www.fda.gov/news-events/press-announcements/coronavirus-covid-19-update-fda-issues-emergency-use-authorization-potential-covid-19-treatment (accessed on 3 July 2020).
26. Roberts, M. "Coronavirus: Dexamethasone Proves First Life-Saving Drug—BBC News," BBC News Online. 2020. Available online: https://www.bbc.com/news/health-53061281 (accessed on 3 July 2020).
27. Zhou, Y.; Hou, Y.; Shen, J.; Huang, Y.; Martin, W.; Cheng, F. Network-based drug repurposing for novel coronavirus 2019-nCoV/SARS-CoV-2. *Cell Discov.* **2020**, *6*, 14. [CrossRef] [PubMed]

28. Ton, A.T.; Gentile, F.; Hsing, M.; Ban, F.; Cherkasov, A. Rapid Identification of Potential Inhibitors of SARS-CoV-2 Main Protease by Deep Docking of 1.3 Billion Compounds. *Mol. Inform.* **2020**, *39*. [CrossRef] [PubMed]
29. Jin, Z.; Du, X.; Xu, Y.; Deng, Y.; Liu, M.; Zhao, Y.; Zhang, B.; Li, X.; Zhang, L.; Peng, C.; et al. Structure of M pro from COVID-19 virus and discovery of its inhibitors. *Nature* **2020**, *582*, 289–293. [CrossRef] [PubMed]
30. Luo, H.; Tang, Q.; Shang, Y.; Liang, S.; Yang, M.; Robinson, N.; Liu, J. Can Chinese Medicine Be Used for Prevention of Corona Virus Disease 2019 (COVID-19)? A Review of Historical Classics, Research Evidence and Current Prevention Programs. *Chin. J. Integr. Med.* **2020**, *26*, 243–250. [CrossRef]
31. MacKenzie, D. The hunt for covid-19 drugs. *New Sci.* **2020**, *245*, 10. [CrossRef]
32. Wu, C.; Liu, Y.; Yang, Y.; Zhang, P.; Zhong, W.; Wang, Y.; Wang, Q.; Xu, Y.; Li, M.; Li, X.; et al. Analysis of therapeutic targets for SARS-CoV-2 and discovery of potential drugs by computational methods. *Acta Pharm. Sin. B* **2020**, *10*, 766–788. [CrossRef]
33. Cao, B.; Wang, Y.; Wen, D.; Liu, W.; Wang, J.; Fan, G.; Ruan, L.; Song, B.; Cai, Y.; Wei, M.; et al. A Trial of Lopinavir–Ritonavir in Adults Hospitalized with Severe Covid-19. *N. Engl. J. Med.* **2020**, *382*, 1787–1799. [CrossRef]
34. Talele, T.; Khedkar, S.; Rigby, A. Successful Applications of Computer Aided Drug Discovery: Moving Drugs from Concept to the Clinic. *Curr. Top. Med. Chem.* **2010**, *10*, 127–141. [CrossRef]
35. Hilgenfeld, R. From SARS to MERS: Crystallographic studies on coronaviral proteases enable antiviral drug design. *FEBS J.* **2014**, *281*, 4085–4096. [CrossRef]
36. Zhang, L.; Lin, D.; Sun, X.; Curth, U.; Drosten, C.; Sauerhering, L.; Becker, S.; Rox, K.; Hilgenfeld, R. Crystal structure of SARS-CoV-2 main protease provides a basis for design of improved a-ketoamide inhibitors. *Science* **2020**, *368*, 409–412. [CrossRef]
37. Cragg, G.M.; Newman, D.J. Natural products: A continuing source of novel drug leads. *Biochim. Biophys. Acta Gen. Subj.* **2013**, *1830*, 3670–3695. [CrossRef] [PubMed]
38. Lahlou, M. The Success of Natural Products in Drug Discovery. *Pharmacol. Pharm.* **2013**, *4*, 17–31. [CrossRef]
39. Veeresham, C. Natural products derived from plants as a source of drugs. *J. Adv. Pharm. Technol. Res.* **2012**, *3*, 200–201. [CrossRef]
40. Patridge, E.; Gareiss, P.; Kinch, M.S.; Hoyer, D. An analysis of FDA-approved drugs: Natural products and their derivatives. *Drug Discov. Today* **2016**, *21*, 204–207. [CrossRef] [PubMed]
41. Chen, C.Y.-C. TCM Database@Taiwan: The World's Largest Traditional Chinese Medicine Database for Drug Screening In Silico. *PLoS ONE* **2011**, *6*, e15939. [CrossRef] [PubMed]
42. Ntie-Kang, F.; Zofou, D.; Babiaka, S.B.; Meudom, R.; Scharfe, M.; Lifongo, L.L.; Mbah, J.A.; Mbaze, L.M.; Sippl, W.; Efange, S.M.N. AfroDb: A Select Highly Potent and Diverse Natural Product Library from African Medicinal Plants. *PLoS ONE* **2013**, *8*, e78085. [CrossRef]
43. Ntie-Kang, F.; Telukunta, K.K.; Döring, K.; Simoben, C.V.; Moumbock, A.F.A.; Malange, Y.I.; Njume, L.E.; Yong, J.N.; Sippl, W.; Günther, S. NANPDB: A Resource for Natural Products from Northern African Sources. *J. Nat. Prod.* **2017**, *80*, 2067–2076. [CrossRef] [PubMed]
44. Lin, L.T.; Hsu, W.C.; Lin, C.C. Antiviral natural products and herbal medicines. *J. Tradit. Complement. Med.* **2014**, *4*, 24–35. [CrossRef]
45. Kwofie, S.K.; Broni, E.; Teye, J.; Quansah, E.; Issah, I.; Wilson, M.D.; Miller, W.A.; Tiburu, E.K.; Bonney, J.H.K. Pharmacoinformatics-based identification of potential bioactive compounds against Ebola virus protein VP24. *Comput. Biol. Med.* **2019**, *113*, 103414. [CrossRef]
46. de Castro, S.; Camarasa, M.J. Polypharmacology in HIV inhibition: Can a drug with simultaneous action against two relevant targets be an alternative to combination therapy? *Eur. J. Med. Chem.* **2018**, *150*, 206–227. [CrossRef]
47. Pizzorno, A.; Padey, B.; Terrier, O.; Rosa-Calatrava, M. Drug repurposing approaches for the treatment of influenza viral infection: Reviving old drugs to fight against a long-lived enemy. *Front. Immunol.* **2019**, *10*, 531. [CrossRef]
48. Kumari, R.; Kumar, R.; Lynn, A. g_mmpbsa—A GROMACS Tool for High-Throughput MM-PBSA Calculations. *J. Chem. Inf. Model.* **2014**, *54*, 1951–1962. [CrossRef] [PubMed]
49. Binkowski, T.A. CASTp: Computed Atlas of Surface Topography of proteins. *Nucleic Acids Res.* **2003**, *31*, 3352–3355. [CrossRef] [PubMed]
50. Yoshino, R.; Yasuo, N.; Sekijima, M. Identification of Key Interactions Between SARS-CoV-2 Main Protease and Inhibitor Drug Candidates. *ChemRxiv* **2020**. [CrossRef]
51. Shi, J.; Song, J. The catalysis of the SARS 3C-like protease is under extensive regulation by its extra domain. *FEBS J.* **2006**, *273*, 1035–1045. [CrossRef] [PubMed]
52. Yu, R.; Chen, L.; Lan, R.; Shen, R.; Li, P. Computational screening of antagonists against the SARS-CoV-2 (COVID-19) coronavirus by molecular docking. *Int. J. Antimicrob. Agents* **2020**, *56*, 106012. [CrossRef]
53. Salman, S.; Shah, F.H.; Idrees, J.; Idrees, F.; Velagala, S.; Ali, J.; Khan, A.A. Virtual screening of immunomodulatory medicinal compounds as promising anti-SARS-CoV-2 inhibitors. *Future Virol.* **2020**, *15*, 267–275. [CrossRef]
54. Maurya, S.K.; Maurya, A.K.; Mishra, N.; Siddique, H.R. Virtual screening, ADME/T, and binding free energy analysis of anti-viral, anti-protease, and anti-infectious compounds against NSP10/NSP16 methyltransferase and main protease of SARS CoV-2. *J. Recept. Signal Transduct.* **2020**, *40*, 605–612. [CrossRef]
55. Lan, J.; Ge, J.; Yu, J.; Shan, S.; Zhou, H.; Fan, S.; Zhang, Q.; Shi, X.; Wang, Q.; Zhang, L.; et al. Structure of the SARS-CoV-2 spike receptor-binding domain bound to the ACE2 receptor. *Nature* **2020**, *581*, 215–220. [CrossRef]
56. Tian, X.; Li, C.; Huang, A.; Xia, S.; Lu, S.; Shi, Z.; Lu, L.; Jiang, S.; Yang, Z.; Wu, Y.; et al. Potent binding of 2019 novel coronavirus spike protein by a SARS coronavirus-specific human monoclonal antibody. *Emerg. Microbes Infect.* **2020**, *9*, 382–385. [CrossRef]

57. He, J.; Tao, H.; Yan, Y.; Huang, S.-Y.; Xiao, Y. Molecular Mechanism of Evolution and Human Infection with SARS-CoV-2. *Viruses* **2020**, *12*, 428. [CrossRef] [PubMed]
58. de Oliveira, O.V.; Rocha, G.B.; Paluch, A.S.; Costa, L.T. Repurposing approved drugs as inhibitors of SARS-CoV-2 S-protein from molecular modeling and virtual screening. *J. Biomol. Struct. Dyn.* **2020**, 1–10. [CrossRef] [PubMed]
59. Kiran, G.; Karthik, L.; Shree Devi, M.S.; Sathiyarajeswaran, P.; Kanakavalli, K.; Kumar, K.M.; Ramesh Kumar, D. In Silico computational screening of Kabasura Kudineer—Official Siddha Formulation and JACOM against SARS-CoV-2 spike protein. *J. Ayurveda Integr. Med.* **2020**. [CrossRef] [PubMed]
60. Morris, G.M.; Lim-Wilby, M. Molecular docking. *Methods Mol. Biol.* **2008**, *443*, 365–382. [CrossRef]
61. Trott, O.; Olson, A.J. AutoDock Vina: Improving the speed and accuracy of docking with a new scoring function, efficient optimization, and multithreading. *J. Comput. Chem.* **2010**, *31*, 455–461. [CrossRef]
62. Chang, M.W.; Lindstrom, W.; Olson, A.J.; Belew, R.K. Analysis of HIV wild-type and mutant structures via in silico docking against diverse ligand libraries. *J. Chem. Inf. Model.* **2007**, *47*, 1258–1262. [CrossRef]
63. Ray, A.K.; Gupta, P.S.S.; Panda, S.K.; Biswal, S.; Rana, M.K. Repurposing of FDA Approved Drugs for the Identification of Potential Inhibitors of SARS-CoV-2 Main Protease. *ChemRxiv* **2020**. [CrossRef]
64. Chen, Y.W.; Yiu, C.-P.B.; Wong, K.-Y. Prediction of the SARS-CoV-2 (2019-nCoV) 3C-like protease (3CLpro) structure: Virtual screening reveals velpatasvir, ledipasvir, and other drug repurposing candidates. *F1000Research* **2020**, *9*, 129. [CrossRef]
65. Nejat, R.; Sadr, A.S. Are Losartan and Imatinib Effective Against SARS-CoV2 Pathogenesis? A Pathophysiologic-Based in Silico Study. *ChemRxiv* **2020**. [CrossRef]
66. Cheke, R.S. The Molecular Docking Study of Potential Drug Candidates Showing Anti-COVID-19 Activity by Exploring of Therapeutic Targets of SARS-CoV-2. *Eurasian J. Med. Oncol.* **2020**, *4*, 185–195. [CrossRef]
67. Miroshnychenko, K.; Shestopalova, A.V. Combined Use of Amentoflavone and Ledipasvir Could Interfere with Binding of Spike Glycoprotein of SARS-CoV-2 to ACE2: The Results of Molecular Docking Study. *ChemRxiv* **2020**. [CrossRef]
68. Du, X.; Li, Y.; Xia, Y.L.; Ai, S.M.; Liang, J.; Sang, P.; Ji, X.L.; Liu, S.Q. Insights into protein-ligand interactions: Mechanisms, models, and methods. *Int. J. Mol. Sci.* **2016**, *17*, 144. [CrossRef] [PubMed]
69. Laskowski, R.A.; Swindells, M.B. LigPlot+: Multiple ligand-protein interaction diagrams for drug discovery. *J. Chem. Inf. Model.* **2011**, *51*, 2778–2786. [CrossRef] [PubMed]
70. Kumar, Y.; Singh, H.; Patel, C.N. In silico prediction of potential inhibitors for the main protease of SARS-CoV-2 using molecular docking and dynamics simulation based drug-repurposing. *J. Infect. Public Health* **2020**, *13*, 1210–1223. [CrossRef]
71. Chandel, V.; Raj, S.; Rathi, B.; Kumar, D. In silico identification of potent fda approved drugs against coronavirus covid-19 main protease: A drug repurposing approach. *Chem. Biol. Lett.* **2020**, *7*, 166–175. [CrossRef]
72. Narkhede, R.R.; Pise, A.V.; Cheke, R.S.; Shinde, S.D. Recognition of Natural Products as Potential Inhibitors of COVID-19 Main Protease (Mpro): In-Silico Evidences. *Nat. Prod. Bioprospect.* **2020**, *10*, 297–306. [CrossRef]
73. Shang, J.; Ye, G.; Shi, K.; Wan, Y.; Luo, C.; Aihara, H.; Geng, Q.; Auerbach, A.; Li, F. Structural basis for receptor recognition by the novel coronavirus from Wuhan. *Nature* **2020**. [CrossRef]
74. Wan, Y.; Shang, J.; Graham, R.; Baric, R.S.; Li, F. Receptor Recognition by the Novel Coronavirus from Wuhan: An Analysis Based on Decade-Long Structural Studies of SARS Coronavirus. *J. Virol.* **2020**, *94*. [CrossRef]
75. Choudhary, S.; Malik, Y.S.; Tomar, S. Identification of SARS-CoV-2 Cell Entry Inhibitors by Drug Repurposing using in silico Structure-based Virtual Screening Approach. *Front. Immunol.* **2020**, *11*, 1664. [CrossRef]
76. Mazzon, M.; Mercer, J. Lipid interactions during virus entry and infection. *Cell. Microbiol.* **2014**, *16*, 1493–1502. [CrossRef]
77. Alen, M.M.F.; Schols, D. Dengue virus entry as target for antiviral therapy. *J. Trop. Med.* **2012**, *2012*, 628475. [CrossRef] [PubMed]
78. Schmitz, M.L.; Kracht, M.; Saul, V.V. The intricate interplay between RNA viruses and NF-κB. *Biochim. Biophys. Acta Mol. Cell Res.* **2014**, *1843*, 2754–2764. [CrossRef] [PubMed]
79. Zhang, D.H.; Wu, K.L.; Zhang, X.; Deng, S.; Peng, B. In silico screening of Chinese herbal medicines with the potential to directly inhibit 2019 novel coronavirus. *J. Integr. Med.* **2020**, *18*, 152–158. [CrossRef] [PubMed]
80. Barlough, J.E.; Shacklett, B.L. Antiviral studies of feline infectious peritonitis virus in vitro. *Vet. Rec.* **1994**, *135*, 177–179. [CrossRef]
81. Wen, C.C.; Kuo, Y.H.; Jan, J.T.; Liang, P.H.; Wang, S.Y.; Liu, H.G.; Lee, C.K.; Chang, S.T.; Kuo, C.J.; Lee, S.S.; et al. Specific plant terpenoids and lignoids possess potent antiviral activities against severe acute respiratory syndrome coronavirus. *J. Med. Chem.* **2007**, *50*, 4087–4095. [CrossRef]
82. Fujioka, T.; Kashiwada, Y.; Kilkuskie, R.E.; Cosentino, L.M.; Bailas, L.M.; Jiang, J.B.; Janzen, W.P.; Chen, I.S.; Lee, K.H. Anti-aids agents, 11. betulinic acid and platanic acid as anti-HIV principles from syzigium claviflorum, and the anti-HIV activity of structurally related triterpenoids. *J. Nat. Prod.* **1994**, *57*, 243–247. [CrossRef]
83. Mayaux, J.F.; Bousseau, A.; Pauwels, R.; Huet, T.; Hénin, Y.; Dereu, N.; Evers, M.; Soler, F.; Poujade, C.; De Clercq, E.; et al. Triterpene derivatives that block entry of human immunodeficiency virus type 1 into cells. *Proc. Natl. Acad. Sci. USA* **1994**, *91*, 3564–3568. [CrossRef]
84. Soler, F.; Poujade, C.; Evers, M.; Carry, J.C.; Hénin, Y.; Bousseau, A.; Huet, T.; Pauwels, R.; De Clercq, E.; Mayaux, J.F.; et al. Betulinic acid derivatives: A new class of specific inhibitors of human immunodeficiency virus type 1 entry. *J. Med. Chem.* **1996**, *39*, 1069–1083. [CrossRef]
85. Zhao, H.; Caflisch, A. Molecular dynamics in drug design. *Eur. J. Med. Chem.* **2015**, *91*, 4–14. [CrossRef]

86. Perez, A.; Morrone, J.A.; Simmerling, C.; Dill, K.A. Advances in free-energy-based simulations of protein folding and ligand binding. *Curr. Opin. Struct. Biol.* **2016**, *36*, 25–31. [CrossRef]
87. Kollman, P.A.; Massova, I.; Reyes, C.; Kuhn, B.; Huo, S.; Chong, L.; Lee, M.; Lee, T.; Duan, Y.; Wang, W.; et al. Calculating structures and free energies of complex molecules: Combining molecular mechanics and continuum models. *Acc. Chem. Res.* **2000**, *33*, 889–897. [CrossRef] [PubMed]
88. Ganesan, A.; Coote, M.L.; Barakat, K. Molecular dynamics-driven drug discovery: Leaping forward with confidence. *Drug Discov. Today* **2017**, *22*, 249–269. [CrossRef] [PubMed]
89. Elmezayen, A.D.; Al-Obaidi, A.; Şahin, A.T.; Yelekçi, K. Drug repurposing for coronavirus (COVID-19): In silico screening of known drugs against coronavirus 3CL hydrolase and protease enzymes. *J. Biomol. Struct. Dyn.* **2020**, 1–13. [CrossRef] [PubMed]
90. Gohlke, H.; Kiel, C.; Case, D.A. Insights into protein-protein binding by binding free energy calculation and free energy decomposition for the Ras-Raf and Ras-RalGDS complexes. *J. Mol. Biol.* **2003**, *330*, 891–913. [CrossRef]
91. Deng, N.J.; Zhang, P.; Cieplak, P.; Lai, L. Elucidating the energetics of entropically driven protein-ligand association: Calculations of absolute binding free energy and entropy. *J. Phys. Chem. B* **2011**, *115*, 11902–11910. [CrossRef]
92. Campanera, J.M.; Pouplana, R. MMPBSA decomposition of the binding energy throughout a molecular dynamics simulation of amyloid-beta (Aß10-35) aggregation. *Molecules* **2010**, *15*, 2730–2748. [CrossRef]
93. Gupta, A.; Chaudhary, N.; Aparoy, P. MM-PBSA and per-residue decomposition energy studies on 7-Phenyl-imidazoquinolin-4(5H)-one derivatives: Identification of crucial site points at microsomal prostaglandin E synthase-1 (mPGES-1) active site. *Int. J. Biol. Macromol.* **2018**, *119*, 352–359. [CrossRef]
94. Lee, V.S.; Tue-ngeun, P.; Nangola, S.; Kitidee, K.; Jitonnom, J.; Nimmanpipug, P.; Jiranusornkul, S.; Tayapiwatana, C. Pairwise decomposition of residue interaction energies of single chain Fv with HIV-1 p17 epitope variants. *Mol. Immunol.* **2010**, *47*, 982–990. [CrossRef]
95. Kwofie, S.K.; Dankwa, B.; Enninful, K.S.; Adobor, C.; Broni, E.; Ntiamoah, A.; Wilson, M.D. Molecular docking and dynamics simulation studies predict munc18b as a target of mycolactone: A plausible mechanism for granule exocytosis impairment in Buruli Ulcer Pathogenesis. *Toxins* **2019**, *11*, 181. [CrossRef]
96. Lokhande, K.B.; Doiphode, S.; Vyas, R.; Swamy, K.V. Molecular docking and simulation studies on SARS-CoV-2 Mpro reveals Mitoxantrone, Leucovorin, Birinapant, and Dynasore as potent drugs against COVID-19. *J. Biomol. Struct. Dyn.* **2020**, 1–12. [CrossRef]
97. Cherrak, S.A.; Merzouk, H.; Mokhtari-Soulimane, N. Potential bioactive glycosylated flavonoids as SARS-CoV-2 main protease inhibitors: A molecular docking and simulation studies. *PLoS ONE* **2020**, *15*, e0240653. [CrossRef]
98. Joshi, T.; Sharma, P.; Joshi, T.; Pundir, H.; Mathpal, S.; Chandra, S. Structure-based screening of novel lichen compounds against SARS Coronavirus main protease (Mpro) as potentials inhibitors of COVID-19. *Mol. Divers.* **2020**. [CrossRef] [PubMed]
99. Muhseen, Z.T.; Hameed, A.R.; Al-Hasani, H.M.H.; Tahir ul Qamar, M.; Li, G. Promising terpenes as SARS-CoV-2 spike receptor-binding domain (RBD) attachment inhibitors to the human ACE2 receptor: Integrated computational approach. *J. Mol. Liq.* **2020**, *320*, 114493. [CrossRef] [PubMed]
100. Razzaghi-Asl, N.; Ebadi, A.; Shahabipour, S.; Gholamin, D. Identification of a potential SARS-CoV2 inhibitor via molecular dynamics simulations and amino acid decomposition analysis. *J. Biomol. Struct. Dyn.* **2020**, 1–16. [CrossRef] [PubMed]
101. Alves, M.J.; Froufe, H.J.C.; Costa, A.F.T.; Santos, A.F.; Oliveira, L.G.; Osório, S.R.M.; Abreu, R.M.V.; Pintado, M.; Ferreira, I.C.F.R. Docking studies in target proteins involved in antibacterial action mechanisms: Extending the knowledge on standard antibiotics to antimicrobial mushroom compounds. *Molecules* **2014**, *19*, 1672–1684. [CrossRef]
102. Burley, S.K.; Berman, H.M.; Kleywegt, G.J.; Markley, J.L.; Nakamura, H.; Velankar, S. Protein Data Bank (PDB): The single global macromolecular structure archive. In *Methods in Molecular Biology*; Humana Press Inc.: New York, NY, USA, 2017; Volume 1607, pp. 627–641.
103. Doytchinova, I.; Atanasova, M.; Valkova, I.; Stavrakov, G.; Philipova, I.; Zhivkova, Z.; Zheleva-Dimitrova, D.; Konstantinov, S.; Dimitrov, I. Novel hits for acetylcholinesterase inhibition derived by docking-based screening on ZINC database. *J. Enzyme Inhib. Med. Chem.* **2018**, *33*, 768–776. [CrossRef]
104. O'Boyle, N.M.; Banck, M.; James, C.A.; Morley, C.; Vandermeersch, T.; Hutchison, G.R. Open Babel: An Open chemical toolbox. *J. Cheminform.* **2011**, *3*, 33. [CrossRef]
105. Harrison, C. Coronavirus puts drug repurposing on the fast track. *Nat. Biotechnol.* **2020**, *38*, 379–381. [CrossRef]
106. Yuan, S.; Chan, H.C.S.; Hu, Z. Using PyMOL as a platform for computational drug design. *Wiley Interdiscip. Rev. Comput. Mol. Sci.* **2017**, *7*. [CrossRef]
107. Abraham, M.J.; Murtola, T.; Schulz, R.; Páll, S.; Smith, J.C.; Hess, B.; Lindahl, E. GROMACS: High performance molecular simulations through multi-level parallelism from laptops to supercomputers. *SoftwareX* **2015**, *1–2*, 19–25. [CrossRef]
108. Gajula, M.; Kumar, A.; Ijaq, J. Protocol for Molecular Dynamics Simulations of Proteins. *Bio-Protocol* **2016**, *6*, 1–11. [CrossRef]
109. Vaught, A. Graphing with Gnuplot and Xmgr: Two graphing packages available under Linux. *Linux J.* **1996**, *1996*, 7.
110. Parasuraman, S. Prediction of activity spectra for substances. *J. Pharmacol. Pharmacother.* **2011**, *2*, 52–53. [CrossRef] [PubMed]
111. Kwofie, S.K.; Dankwa, B.; Odame, E.A.; Agamah, F.E.; Doe, L.; Teye, J.; Agyapong, O.; Miller, W.A.; Mosi, L.; Wilson, M.D. In Silico Screening of Isocitrate Lyase for Novel Anti-Buruli Ulcer Natural Products Originating from Africa. *Molecules* **2018**, *23*, 1550. [CrossRef] [PubMed]
112. Jamkhande, P.; Barde, S. Evaluation of anthelmintic activity and in silico PASS assisted prediction of Cordia dichotoma (Forst.) root extract. *Anc. Sci. Life* **2014**, *34*, 39. [CrossRef] [PubMed]

113. Stasevich, M.V.; Zvarich, V.I.; Novikov, V.P.; Zagorodnyaya, S.D.; Povnitsa, O.Y.; Chaika, M.A.; Nesterkina, M.V.; Kravchenko, I.A.; Druzhilovskii, D.S.; Poroikov, V.V. 9,10-Anthraquinone Dithiocarbamates as Potential Pharmaceutical Substances with Pleiotropic Actions: Computerized Prediction of Biological Activity and Experimental Validation. *Pharm. Chem. J.* **2020**, *53*, 905–913. [CrossRef]
114. Cheng, X.; Ivanov, I. Molecular dynamics. *Methods Mol. Biol.* **2012**, *929*, 243–285.
115. Dong, Y.W.; Liao, M.L.; Meng, X.L.; Somero, G.N. Structural flexibility and protein adaptation to temperature: Molecular dynamics analysis of malate dehydrogenases of marine molluscs. *Proc. Natl. Acad. Sci. USA* **2018**, *115*, 1274–1279. [CrossRef]
116. Sinha, S.; Wang, S.M. Classification of VUS and unclassified variants in BRCA1 BRCT repeats by molecular dynamics simulation. *Comput. Struct. Biotechnol. J.* **2020**, *18*, 723–736. [CrossRef]

molecules

MDPI

Review

Andrographolide, an Anti-Inflammatory Multitarget Drug: All Roads Lead to Cellular Metabolism

Rafael Agustín Burgos [1,2,*], Pablo Alarcón [1,2], John Quiroga [1,2,3], Carolina Manosalva [4] and Juan Hancke [1]

1 Laboratory of Inflammation Pharmacology, Faculty of Veterinary Sciences, Institute of Pharmacology and Morphophysiology, Universidad Austral de Chile, Valdivia 5090000, Chile; pabloalarcon.u@gmail.com (P.A.); john.quiroga@uach.cl (J.Q.); juan@hpingredients.com (J.H.)

2 Laboratory of Immunometabolism, Institute of Pharmacology and Morphophysiology, Faculty of Veterinary Sciences, Universidad Austral de Chile, Valdivia 5090000, Chile

3 PhD Program in Veterinary Sciences, Faculty of Veterinary Sciences, Universidad Austral de Chile, Valdivia 5090000, Chile

4 Faculty of Sciences, Institute of Pharmacy, Universidad Austral de Chile, Valdivia 5090000, Chile; carolinamanosalva@uach.cl

* Correspondence: rburgos1@uach.cl; Tel.: +56-63-2293-015

Abstract: Andrographolide is a labdane diterpene and the main active ingredient isolated from the herb *Andrographis paniculata*. Andrographolide possesses diverse biological effects including anti-inflammatory, antioxidant, and antineoplastic properties. Clinical studies have demonstrated that andrographolide could be useful in therapy for a wide range of diseases such as osteoarthritis, upper respiratory diseases, and multiple sclerosis. Several targets are described for andrographolide, including the interference of transcription factors NF-κB, AP-1, and HIF-1 and signaling pathways such as PI3K/Akt, MAPK, and JAK/STAT. In addition, an increase in the Nrf2 (nuclear factor erythroid 2–related factor 2) signaling pathway also supports its antioxidant and anti-inflammatory properties. However, this scenario could be more complex since recent evidence suggests that andrographolide targets can modulate glucose metabolism. The metabolic effect of andrographolide might be the key to explaining the diverse therapeutic effects described in preclinical and clinical studies. This review discusses some of the most recent evidence about the anti-inflammatory and metabolic effects of andrographolide.

Keywords: andrographolide; anti-inflammatory; multitarget; metabolism; glycolysis

check for updates

Citation: Burgos, R.A.; Alarcón, P.; Quiroga, J.; Manosalva, C.; Hancke, J. Andrographolide, an Anti-Inflammatory Multitarget Drug: All Roads Lead to Cellular Metabolism. *Molecules* **2021**, *26*, 5. https://dx.doi.org/10.3390/molecules26010005

Academic Editors: Simona Fabroni, Krystian Marszałek and Aldo Todaro

Received: 3 December 2020
Accepted: 18 December 2020
Published: 22 December 2020

1. Introduction

Andrographis paniculata is a traditional herb from Asian countries that has been used for relieving and reducing the severity and duration of symptoms of common colds and alleviating fever, coughs, and sore throats in uncomplicated respiratory tract infections. Over the last decades, many Asian and European researchers have begun to investigate the composition, activity, safety, and efficacy of this ancient herb, confirming that the traditional medicine has a sound basis. The biological effects of *A. paniculata* are related to the constituents of its aerial parts, a group of diterpene lactones belonging to the ent-labdane class, present in both free and glycosidic forms. The main constituent of *A. paniculata* is andrographolide, a diterpene that contains a γ-lactone ring that is present in leaves in quantities varying from 51.22 ± 0.04 to 68.35 ± 1.50 mg/g [1]. Other diterpenes such as 14-deoxy-11, 12-didehydroandrographolide, and neoandrographolide varied from 5.38 ± 0.30 to 16.01 ± 0.8 mg/g and 5.47 ± 0.03 to 11.72 ± 0.33 mg/g, respectively. Meanwhile, 14-deoxyandrographolide shows high levels in leaves only at the transfer stage in field experiments (30.59 ± 1.39 mg/g) [1].

Methanolic and aqueous extracts of *A. paniculata*, or andrographolide, exhibited an antioxidant and acute anti-inflammatory effect against carrageenan-induced paw edema in rodents [2], suggesting that andrographolide may be responsible for the pharmacological properties described for the herb in traditional medicine [3]. In addition, andrographolide shows the most potent anti-inflammatory [4] and anticancer activities, with stronger effects than 4-deoxy- 11,12-didehydroandrographolide and 14-deoxyandrographolide [5].

Andrographolide's antioxidant properties have been related to its anti-inflammatory mechanism [6]. In fact, together with nitric oxide, reactive oxygen species (ROS) like hydrogen peroxide H_2O_2 or superoxide anion O_2^- are involved in the intracellular killing of microorganisms. In this way, the release of extracellular DNA traps by neutrophils, a process regulated by the mitogen-activated protein kinase (MAPK) pathway that involves the extrusion of DNA decorated with cytoplasmic proteins and with antimicrobial and proteolytic activity, is a defense mechanism of neutrophil and other leukocytes against infectious processes or pro-inflammatory agents [7]. Indeed, in neutrophils, several proinflammatory stimuli induce the ROS-dependent formation of neutrophil extracellular traps (NETs) [7], including microorganisms and proinflammatory factors [8]. However, because of their uncontrolled production by phagocytes during the inflammation process, they may become toxic metabolites, leading to severe tissue damage in several pathologies. For example, excessive ROS production and NETs release by proinflammatory stimuli plays an important pathological factor in the induction of ischemic and reperfusion injury, metabolic impairments, and liver diseases [9,10]. In rat neutrophils, pretreatment with andrographolide prevented phorbol-12-myristate-13-acetate (PMA)- induced ROS production and accumulation, as well as N-formyl-methionyl-leucyl-phenylalanine (fMLP)-induced adhesion and transmigration [11]. Andrographolide inhibited the upregulation of CD11b and CD18 induced by fMLP [12]. The overexpression of Mac-1 (Macrophage-1 antigen or CD11b/Cd18, involved in the adhesion of leukocytes onto the endothelium) by neutrophils is regulated by the accumulation of ROS and intracellular calcium $[Ca^{2+}]_i$ mobilization; andrographolide pretreatment reduced the fMLP-induced production of H_2O_2 and O_2^- but did not affect the $[Ca^{2+}]_i$ mobilization. Andrographolide interferes with the effects of PMA, a direct protein kinase C (PKC) activator, suggesting that an inhibition of ROS production through the modulation of PKC-dependent pathway (Figure 1) could explain, at least in part, the ability of andrographolide to downregulate Mac-1 expression, essential for neutrophil adhesion and transmigration during inflammation [12]. In addition, PMA is a potent NET inducer; in fact, andrographolide alleviates murine arthritis by reduced neutrophil infiltration and NET formation [13].

Andrographolide reduces the expression of several proinflammatory genes, including cyclooxygenase-2 (COX-2), IL-6, IL-8, IL-1β, and inducible nitric oxide synthase (iNOS) in endothelial cells, synoviocytes, colorectal cancer cells, and leukocytes [14]. One milligram per kg b.w. andrographolide administered to rodents by i.v. reduces the pro-inflammatory and hemodynamic effects of lipopolysaccharides (LPS) [15]. In addition, 1 mg/kg i.p. andrographolide protects against LPS-induced acute lung injury, reducing myeloperoxidase (MPO) activity, neutrophils, macrophages, TNF-α, IL-6, and IL-1β in the bronchoalveolar lavage fluid of rodents [16]. In Freund's adjuvant-induced complete arthritis or collagen-induced arthritis models in rodents, andrographolide (3-6 mg/kg i.p.) diminishes the clinical score of arthritis and joint damage and reduces the production of NO and TNF-α [17], indicating that andrographolide may be employed as a natural anti-inflammatory or for the synthesis of more potent derivatives [18].

On the other hand, chronic inflammation of tissue or organ is characterized by the presence of inflammatory cells (immune cells) and the proliferation/regeneration of damaged tissue. Sustained enhanced cellular proliferation in an environment with abundant inflammatory cells, growth factors, and DNA damage and/or mutagenic insult certainly potentiates and/or promotes carcinogenesis risk [19,20]. Andrographolide exhibits antineoplastic activities through COX-2 inhibition and reduced proliferation in human breast cancer [21]. Moreover, andrographolide antagonizes TNF-α-induced IL-8 via inhibition of the NADPH oxidase/ROS/NF-κB and Src/MAPKs/AP-1 axes (Figure 1) in human colorectal cancer HCT116 cells and suppresses angiogenesis in the tumor microenvironment [22].

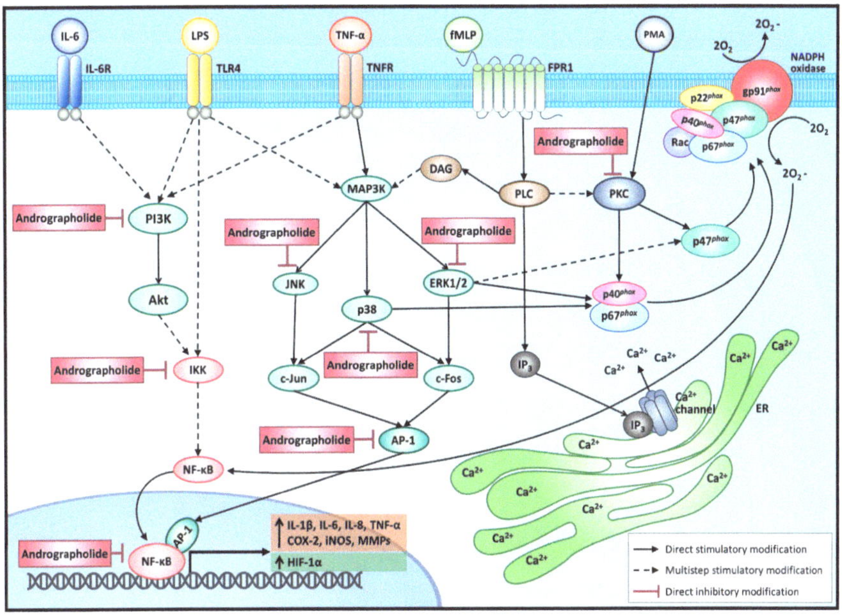

Figure 1. Andrographolide exerts anti-inflammatory effects by inhibiting several intracellular signaling pathways activated by different pro-inflammatory agents. Akt = protein kinase B; AP-1 = activator protein 1; Ca^{+2} = calcium ion; COX-2 = cyclooxygenase 2; DAG = diacylglycerol; ERK1/2 = extracellular signal-regulated kinases 1/2; fMLP = N-Formyl-methionyl-leucyl-phenylalanine; FPR1 = formyl peptide receptor 1; HIF-1α = hypoxia inducible factor 1, alpha subunit; IKK = I-kappa-B kinase; IL-1β = interleukin 1 beta; IL-6 = interleukin 6; IL-6R = IL-6 receptor; IL-8 = interleukin 8; iNOS = inducible nitric oxide synthase; IP3 = inositol triphosphate; JNK = c-Jun N-terminal kinase; LPS = lipopolysaccharide; MAP3K = mitogen activated protein kinase kinase kinase; MMPs = matrix metalloproteinases; NADPH oxidase = nicotinamide adenine dinucleotide phosphate oxidase complex; NF-κB = nuclear factor kapa B; PI3K = phosphatidyl inositol 3 kinase; PKC = protein kinase C; PLC = phospholipase C; PMA = phorbol myristate acetate; TLR-4 = Toll-like receptor 4; TNF-α = tumoral necrosis factor alpha; TNFR = TNF-α receptor.

Hypoxia-inducible factor 1 (HIF-1) controls metabolic reprogramming in cancer cells via gene expression of glucose transporters and glycolytic enzymes [23]. In addition, andrographolide reduced hepatoma cancer cells' growth, inhibiting vascular endothelial growth factor A expression via HIF-1α degradation [24], which also suggests that andrographolide could interfere with cell metabolism in cancer.

It has been demonstrated that an ethanolic extract of *A. paniculata* and andrographolide showed an alpha-glucosidase inhibitory effect in a concentration-dependent manner, supporting a potential use for the management of type 2 diabetes mellitus [25]. Moreover, andrographolide downregulated the expression of sterol regulatory element-binding proteins' (SREBPs) target genes, decreased cellular lipid accumulation in vitro, ameliorated lipid metabolism, and improved glucose use in mice with high-fat-diet-induced obesity [26]. In addition, andrographolide suppressed the TNF-α-induced activation of the NF-κB signaling pathway and its downstream inflammatory factors' expression, ameliorating insulin resistance in 3T3-L1 adipocytes [27], which suggests a close relationship between the anti-inflammatory effects and metabolic modulation by andrographolide.

In addition, andrographolide has been useful in the therapy of inflammatory processes and metabolic disorders [14,28]. Both biological properties have been analyzed separately; however, during inflammation, several disturbances of metabolism occur at once. Moreover, these effects could be part of the common signaling pathways modulated by andrographolide.

2. Andrographolide Target Involved in Anti-inflammatory and Metabolic Effects

2.1. Metabolism and Inflammation

Inflammation involves the coordinated liberation of blood components (plasma and leukocytes) to the injured tissue. The recruitment of leukocytes from the blood, together with tissue-resident immune cells, ensures efficient pathogen killing and contributes to healing. Importantly, the duration and intensity of the inflammatory response is controlled by the secretion of several molecules such as cytokines or lipid mediators and chemokines that recruit innate and adaptive immune cells to the site of injury [29]. A successful inflammatory response is followed by the resolution phase, which gradually reduces the inflammation when the danger signal or the injury has been eliminated; this phase is critical for restoring homoeostasis [30]. Under certain circumstances, the immune cells fail to resolve inflammatory processes, and the inflammation progresses over time to chronic inflammation, which includes several chronic and age-associated human pathologies, such as neurodegenerative, cardiovascular, joint, and muscular diseases [31].

Cells use macromolecules that are degraded to obtain the necessary energy for essential functions. Metabolism is a process fundamental to cellular biology, providing energy and building blocks for macromolecule synthesis. One of the main metabolic pathways that provide energy for all cellular processes is glycolysis (Figure 2), which comprises multiple enzymes that include glucose uptake and its metabolization, generating two net molecules of ATP and NADH. Glycolysis consists of 10 consecutive enzymatic reactions that convert glucose into two molecules of pyruvate, which connect with other metabolic pathways. In general terms, there are three rate-limiting/rate-controlling enzymes of glycolysis: hexokinase, phosphofructokinase, and pyruvate kinase (Figure 2).

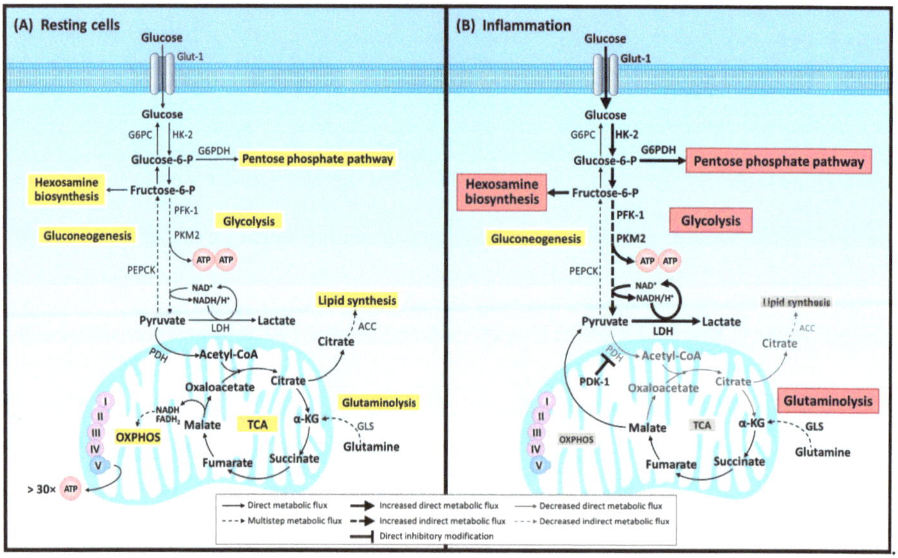

Figure 2. Predominant metabolic pathways activated in resting cells and during inflammation. ACC = acetyl-CoA carboxylase; ATP = adenosine triphosphate; α-KG = alpha-ketoglutarate;FADH$_2$ = reduced flavin adenine dinucleotide; GLS = glutaminase; Glut-1 = glucose transporter 1; G6PC = glucose-6-phosphatase catalytic subunit; G6PDH = glucose-6-phosphate dehydrogenase; HK-2 = hexokinase 2; LDH = lactate dehydrogenase; NAD$^+$ = oxidized nicotinamide adenine dinucleotide; NADH/H$^+$ = reduced nicotinamide adenine dinucleotide; OXPHOS = oxidative phosphorylation; PDH = pyruvate dehydrogenase; PDK-1 = pyruvate dehydrogenase kinase 1; PEPCK = phosphoenolpyruvate carboxykinase; PFK-1 = phosphofructokinase 1; PKM2 = pyruvate kinase M2 isoform; TCA = tricarboxylic acid cycle.

Glycolysis is usually carried out under hypoxic conditions and is only effective as an energetic route during intense and short exercise of less than 2 min. Furthermore, the accumulation of lactic acid, a by-product of anaerobic metabolic, causes tissue pain and fatigue during exercise and inflammation (Figure 2). While ATP is primarily the energy source for metabolic work, NADH provides reducing power in anabolic reactions and can also be oxidized in the respiratory chain. Thus, the generation of glycolytic ATP and NADH serves as an energy source for aerobic respiration and anaerobic fermentation. The pyruvate generated in glycolysis has access to the mitochondria and fully metabolizes to CO_2 through the tricarboxylic acid (TCA) cycle, generating NADH and reducing FADH2 to perform oxidative phosphorylation (OXPHOS) (Figure 2). However, pyruvate can also be fermented to lactate under normoxic conditions, regenerating NAD^+ without generating ATP in a process called aerobic glycolysis. Despite the higher energy efficiency of OXPHOS, highly proliferating cells such as tumor cells prioritize the glycolytic pathway [32]. In fact, cancer cells rewire their metabolism to promote growth, survival, proliferation, and long-term maintenance. The common feature is an increase in glucose uptake and fermentation of glucose to lactate. This metabolic profile is observed even in the presence of oxygen and fully functioning mitochondria; this together is known as the Warburg effect.

The different cell types of the immune system use different metabolic pathways to maintain a good source of energy that allows them a fast and adequate response to an infection or cellular injury. During the inflammatory response, the activation of immune cells also depends on a process of metabolic reprogramming. Thus, while glycolysis is the main energy source at the peak of inflammation, during the resolution phase the immune cells rely mainly on OXPHOS. Therefore, glycolytic metabolism seems to be a new target for the development of anti-inflammatory and antineoplastic drugs [32,33]. The production and influx of neutrophils into the damaged tissue is an earlier event during the acute inflammatory cascade. Mature neutrophils are mainly glycolytic; some examples have shown that mitochondrial respiration and ATP synthesis/release are also important for neutrophil transmigration [34], ROS production [35], and NET formation [36,37]. Moreover, the release of extracellular ROS is mediated by the action of the membrane-bound enzyme NADPH oxidase. The interference of glycolysis led to reduced NADPH oxidase function, NETs formation, and deficient microbial killing [38].

The glycolytic switch is a key for immune cells during inflammation, e.g., pro-inflammatory macrophages (M1 phenotype) or neutrophils. Glycolysis supplies metabolic intermediates for other biosynthetic pathways necessary for cellular growth and differentiation that decrease respiration and a broken Krebs cycle, leading to an accumulation of both citrate and succinate [39]. In addition to glycolysis, the pentose phosphate pathway (PPP), the hexosamine pathway, and glutaminolysis increase upon activation (Figure 2) [38]. On the contrary, M2 macrophages that serve to modulate inflammation, promote tissue repair, and regulate adaptive immunity, the Krebs cycle, and oxidative phosphorylation are intact, and high levels of fatty acid oxidation (FAO) are observed [39].

In T cells, metabolism determinates the quiescent and activated form of naïve T cells. In quiescent T cells, glucose and amino acid catabolism are predominant; however, when T cells exit the naïve state for activation by immunological cues, this induces a metabolic reprogramming (catabolism and anabolism) that allows for increased consumption of extracellular nutrients for the biosynthesis of energy and ATP production by glycolysis or OXPHOS, coupled to the TCA cycle and fatty acid synthesis [39].

The accumulation of several cellular metabolites determines the activation/repression of signaling pathways, the epigenetic and post-transcriptional regulation of inflammatory genes, and the post-translational modification of proteins [39]. Several of these pathways induced in the inflammatory process also control glycolytic metabolism and could contribute to the ability of andrographolide to reduce inflammation and cancer progression [14].

2.2. *Hypoxia-Inducible Factor 1α (HIF-1α)*

HIF-1α is a transcription factor controlled by cellular oxygen concentrations; it is easily degraded in normoxia and stabilized in hypoxia. In fact, proline hydroxylases (PHDs), responsible for HIF-1a hydroxylation, or the von Hippel–Lindau protein (*p*VHL) target HIF-1a for ubiquitination and subsequent proteasomal degradation. However, HIF-1α can be sustained even under normoxic conditions by the products of glycolysis, lactate, and pyruvate [40]. At a transcriptional level, bacterial LPS increase the expression of HIF-1α [41]. Induction of HIF-1α mRNA by LPS under normoxic conditions may be of particular importance in inflammatory processes since it prepares monocytes for survival and function in a hypoxic microenvironment before they extravasate from the vasculature into the tissue [42]. In M1 macrophages stimulated with LPS, HIF-1-mediated metabolic reprogramming is dependent on pyruvate kinase M2 (PKM2), a glycolytic enzyme responsible for converting phosphoenolpyruvate (PEP) to pyruvate. While PKM2 tetramers are highly active at a cytosolic level, supporting the final step of glycolysis, PKM2 monomers and dimers lack enzymatic activity [43]. However, PKM2 dimers can translocate to the nucleus and interact directly with HIF-1α, positively regulating the expression of and promotes the expression of downstream target genes [44]. In fact, PKM2 is highly expressed in tumor cells and LPS-stimulated macrophages, and the stabilization of HIF-1α favors the expression of IL-1β and glycolytic enzymes [44]. In addition, HIF-1α controls the glycolytic machinery via the expression of several proteins including hexokinase-2 (HK-2) that catalyze the first step of glucose metabolism: phosphofructokinase-1 (PFK-1), a rate-limiting enzyme of glycolysis; glucose-6-phosphate-dehydrogenase (G6PDH), which catalyzes the first step in the pentose phosphate pathway; lactate dehydrogenase; pyruvate dehydrogenase kinase-1, which inactivates the TCA cycle enzyme; pyruvate dehydrogenase (PDH), which converts pyruvate to acetyl-CoA; and the glucose transporter-1 (GLUT-1) (Figure 3) [45].

HIF-1α acts as a key regulator during inflammation, increasing pro-inflammatory cytokines (e.g., TNF-α, IL-1β, IL-6, and IL-8), matrix metalloproteinases (e.g., MMP-1, MMP-3, and MMP-9) (Figure 3), and pathways of glucose metabolism in rheumatoid arthritis (RA) [46].

Andrographolide reduces MMP-1, MMP-3, and MMP-9 expression of rheumatoid arthritis fibroblast-like synoviocytes (RA-FLS) in hypoxia, decreasing HIF-1α expression and interfering with HIF-1α binding to DNA (Figure 3) [47]. RA-FLS shows a shift from oxidative phosphorylation to glycolytic ATP production; in addition, synovial tissue is enriched in HIF-1α, key to RA pathogenesis [48,49]. In synovial tissue from RA or chondrocytes from osteoarthritis (OA), patients show an increase in PKM2 expression, suggesting a potential new therapeutical target [50,51]. By using a quantitative chemical proteomic approach, novel potential targets of andrographolide have been proposed in the fields of inflammation and immunity research, including PKM2 and LDH [52]. This suggests that andrographolide could reduce the glycolytic switch induced during the joint inflammatory process.

In human clinical trials, patients with RA or moderate knee osteoarthritis treated with 300 mg daily of a standardized dried extract of *A. paniculata* (ParActin®: ~150 mg daily of andrographolide) for 16 weeks or one year had reduced swelling joint and pain [53,54]. However, whether this also modulates glucose metabolism is uncertain and should be carefully considered in future studies.

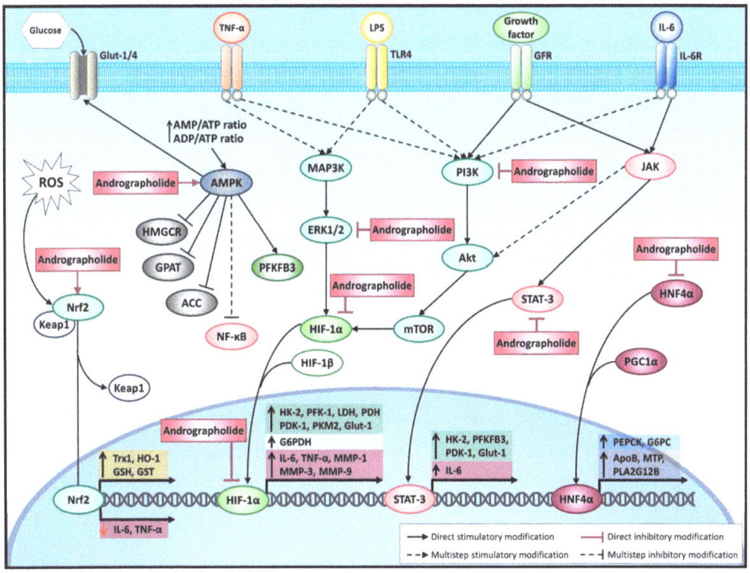

Figure 3. Andrographolide modulates at different levels the metabolic responses induced by proinflammatory agents and growth factors. ACC = acetyl-CoA carboxylase; ADP = adenosine diphosphate; Akt = protein kinase B; AMP = adenosine monophosphate; AMPK = adenosine monophosphate-activated protein kinase; ApoB = apolipoprotein B; ATP = adenosine triphosphate; ERK1/2 = extracellular signal-regulated kinases 1/2; GFR = growth factor transporter; Glut-1/4 = glucose transporter 1/4; GPAT = glycerol-3-phosphate acyltransferase; GSH = Glutathione; GST = Glutathione S-transferase; G6PC = glucose-6-phosphatase catalytic subunit; G6PDH = glucose-6-phosphate dehydrogenase; HIF-1α = hypoxia inducible factor 1, alpha subunit; HIF-1β = hypoxia inducible factor 1, beta subunit; HMGCR = hydroxymethylglutaryl CoA reductase; HK-2 = hexokinase 2; HNF4α = hepatocyte nuclear factor 4 alpha; HO-1 = Heme oxygenase-1; IL-6 = interleukin 6; IL-6R = IL-6 receptor; JAK = Janus kinase; Keap1 = Kelch-like ECH-associated protein 1; LDH = lactate dehydrogenase; LPS = lipopolysaccharide; MAP3K = mitogen activated protein kinase kinase kinase; MMP-1 = matrix metalloproteinase 1; MMP-3 = matrix metalloproteinase 3; MMP-9 = matrix metalloproteinase 9; mTOR = mammalian target of rapamycin; MTP = microsomal triglyceride transfer protein; Nrf2 = nuclear factor erythroid 2–related factor 2; PDH = pyruvate dehydrogenase; PDK-1 = pyruvate dehydrogenase kinase 1; PEPCK = phosphoenolpyruvate carboxykinase; PFK-1 = phosphofructokinase 1; PFKFB3 = 6-phosphofructo-2-kinase/fructose-2,6-biphosphatase; PGC1α = peroxisome proliferator-activated receptor gamma coactivator 1 alpha; PI3K = phosphatidyl inositol 3 kinase; PKM2 = pyruvate kinase M2 isoform; PLA2G12B = phospholipase A2 G12B; ROS = reactive oxygen species; STAT-3 = signal transducer and activator of transcription 3; TLR-4 = Toll-like receptor 4; TNF-α = tumoral necrosis factor alpha; TNFR = TNF-α receptor; Trx1 = Thioredoxin 1.

Andrographolide reduces the expression of HK-2, with GLUT-1 reducing glycolysis in human chondrosarcoma SW 1353 cells [55]. Furthermore, andrographolide has been used to develop new anti-inflammatory derivatives that simultaneously show HK-2 inhibition, reducing glycolysis and interfering with the NF-kB pathway, and iNOS and COX2 expression [56]. It has been reported that andrographolide reduces HIF-1α protein and expression through the inhibition of PI3K/Akt-mTOR in breast cancer cells (Figure 3) [57]. In HCT116 colon cancer cells, the inhibition of PI3K/Akt-mTOR–HIF-1α reduces lactate and ATP concentration and GLUT1, HK-2 and PFK-1 expression, supporting the idea that the inhibition of glycolysis is critical for reducing viability [58]. Andrographolide inhibits the PI3K/Akt pathway in human umbilical vein endothelial cells, reducing IL-6, TNF-α, and IL-1β induced by exposure to high glucose [59]. Furthermore, in endothelial cells, andrographolide abolishes TNF-α-induced Akt phosphorylation (Figure 3), decreasing the expression of adhesion molecules such as ICAM-1 [60]. On the other hand, in 3T3-L1 fibroblasts, andrographolide increases glucose uptake, activating the PI3K pathway and restoring the insulin sensitivity impaired by TNF-α [27]. In spite of this, andrographolide reduces the expression of IL-6, iNOS, SOCS3, and MCP-1 induced by TNF-α in 3T3-L1 cells; this is attributable to the inhibition of NF-κB [27].

2.3. Nuclear Factor Kappa B (NF-κB) Pathway

NF-κB is a key transcription factor for pro-inflammatory genes' expression [61]. Activation of NF-κB involves the phosphorylation of the specific inhibitory factor IκBα, by I-kappa kinase (IKK), and phosphorylated IκBα is rapidly degraded by a proteasome, allowing the resultant free NF-κB heterodimer (*p* 50/*p* 65) to translocate into the nucleus and induce gene transcription (Figure 1) [61]. The analysis of the HIF-1α promoter showed binding sites for NF-κB. Moreover, inhibition of NF-kB reduced the LPS-induced increase in HIF-1α mRNA but also reduced constitutive HIF-1α expression [42], suggesting that proinflammatory stimuli could alter the cellular metabolism.

Andrographolide interferes with NF-κB binding to DNA in neutrophils, thus reducing the expression of proinflammatory proteins. Mechanistically, andrographolide formed a covalent adduct with the reduced cysteine (62) of *p* 50-NF-κB, thus blocking the binding to DNA and reducing transcriptional activity (Figure 1) [14]. Moreover, other authors found that andrographolide inhibits TNF-α-induced IKK activation and IκBα phosphorylation in endothelial cells [62]. This anti-inflammatory property has been associated with its efficacy in ameliorating ulcerative colitis, asthma, hepatic inflammation, and arthritis [14,17].

In addition, NF-κB affects carbohydrate metabolism by increasing HIF-1α expression. Proinflammatory cytokines and short-term hypoxia lead to phosphorylation of IκB, thus releasing NF-κB, which binds to the promoter of the HIF-1α gene, thereby increasing HIF-1α mRNA and protein levels [63]. The inhibition of NF-κB by p65 siRNA or BAY 11–7082 in RA-FLS led to a marked decrease in the HIF-1α, MMP2, and MMP9 expression induced by IL-17A [46]. This supports the interplay between inflammation and metabolism, a hallmark observed during joint inflammation. Moreover, Walmsley et al. [64] demonstrated a close relationship between HIF-1 and NF-κB in neutrophils. In fact, the NF-κB p65 subunit and IκBα regulator IKKα were identified as HIF-1 target genes, indicating that NF-κB is an important downstream regulator of the hypoxic response in neutrophils [64]. These observations further support the notion that HIF-1 and NF-κB work in cooperation in inflammatory conditions and in hypoxia. In fact, there is a relationship between NF-κB and tumor progression-associated inflammation that controls the expression of cancer-associated genes such as cytokines, chemokines, growth factors, and transcription factors such as HIF-1α [65]. Andrographolide suppresses the proliferation of human cancer colon and multiple myeloma cells via TLR4/NF-κB inhibition [66].

2.4. MAPK-Src and AP-1

Src kinases are a large family of nonreceptor protein tyrosine kinases (PTK) that control multiple signaling pathways, are expressed either ubiquitously or predominantly in specific immune-competent cells, and are involved in a variety of immunologic processes, such as immune cell development, proliferation, adhesion, migration, chemotaxis, phagocytosis, and survival. Src kinases have upstream crosstalk activation with MAPK and phosphatidylinositol 3-kinase (PI3K)/Akt pathways [67]. The activation of tyrosine kinase receptor by growth factor induces activation of the PI3K/Akt pathway, which stimulates glucose uptake via GLUTs and increases glycolysis during cell growth and survival [68]. On the other hand, MAPK pathways are signaling modules that transduce extracellular and intracellular cues via the phosphorylation of key protein targets. After the appropriate cellular cue, successive activation of a MAPK kinase kinase (MAPKKK) and a MAPK kinase (MAPKK) results in the phosphorylation/activation of three main families: ERKs (extracellular-signal-regulated kinases), JNKs (Jun amino-terminal kinases), and *p* 38/SAPKs (stress-activated protein kinases) (Figure 1).

The ERK1/2 pathway is involved in HIF-1α activation through p300/CBP [45]. LPS increased HIF-1α via ERK1/2 in human monocytes as well as in nondifferentiated cells and the differentiated human monocytic cell line THP-1 under normoxic conditions [42], with a reducing effect on metabolism and inflammation. Andrographolide reduces the phosphorylation of MAPK such as ERK1/2, p38, and JNK in LPS-stimulated RAW264.7 cells [69] or TNF-α stimulated RA-FLS [70]. Andrographolide suppressed LPS-induced phosphorylation of ERK1/2, JNK, and *p* 38 in mouse RAW264.7 cells (Figure 1) [69]. MAPK (ERK 1/2 and *p* 38 MAPK) activate the transcription factor activator protein-1 (AP-1). In fact, andrographolide inhibits Src and ERK1/2, interfering with the activation of AP-1 (Figure 1) and antagonizing TNF-α-induced IL-8 in HCT116 human colorectal cancer cells [22].

The AP-1 complex (homo- and heterodimer) is a dimeric transcription factor encompassing a group of structurally and functionally related members of the Jun, Fos, ATF, and MAF protein families. In the immune system, a variety of different cytokines and chemokines are predominantly regulated by AP-1. In addition, increased expression of AP-1 is involved in certain types of cancer such as breast cancer, endometrial carcinoma, and colorectal cancer [71]. Moreover, the inhibition of AP-1 has been related to siRNA attenuated cytokine expression (TNF-α, IL-1β, and IL-6) in macrophages and reduces lipid accumulation in hepatocytes induced by palmitic acid [72], suggesting the involvement of AP-1 in inflammation and hepatic lipid metabolism. Andrographolide reduces the activation of AP-1 (Figure 1) in human colorectal cancer [22] and endothelial cells [73], induced by TNF-α. In addition, andrographolide suppresses the nuclear localization of AP-1 and STAT-1 in macrophages stimulated with LPS, evidence of the inhibitory effect of andrographolide in JAK/STAT pathways (Figure 3) [74].

2.5. JAK/STAT Pathway

The Janus kinase (JAK)/signal transducer and activator of transcription (STAT) pathway is activated for several extracellular stimuli such as growth factors (epidermal growth factor receptor (EGFR), granulocyte-macrophage colony-stimulating factors (GM-CSF)), cytokines (IL4, IL6, and IFNγ) [75], and metabolism-relevant hormones (growth hormone, leptin, erythropoietin, and prolactin), leading to critical cellular events such as hematopoiesis, lactation, and the development of the mammary glands and the immune system [75]. Seven mammalian STAT family members have been identified (STAT1, STAT2, STAT3, STAT4, STAT5a, STAT5b, and STAT6) and four JAKs, named JAK1, JAK2, JAK3, and TYK2. The JAK/STAT pathway is associated with autoimmune and inflammatory diseases and cancer [75,76].

Additionally, this pathway is involved in metabolic diseases such as metabolic syndrome, insulin resistance, and obesity-associated metabolic syndrome [75]. The JAK/STAT pathway controls the expression of key glycolytic mediators 6-phosphofructo-2-kinase/fructose-2,6-biphosphatase 3 (PFKFB3), PDK-1, HK-2, glycogen synthase kinase-3α (GSK-3α), and GLUT-1, an effect paralleled by the activation of proinflammatory, proangiogenic, and invasive mechanisms in RA (Figure 3) [77]. Andrographolide reduces IL-6-induced STAT3 phosphorylation and subsequent nuclear translocation in cancer cells (Figure 3) [78]. Moreover, andrographolide suppresses IL-6 cell signaling, including STAT3, ERK, and Akt phosphorylation [79]. In addition, andrographolide reduces the phosphorylation of STAT 1/2, thus interfering with the JAK/STAT pathway in an influenza virus-induced inflammation murine model [80]. In recent years, the JAK/STAT pathway has been implicated in obesity and metabolic syndrome [75]. Indeed, polymorphism of JAK2 is involved in the accumulation of central fat, metabolic syndrome, and lipid metabolic disorder derivates of leptin and insulin activation. Furthermore, polymorphism of STAT5B has been associated with an increase in plasma low-density lipoprotein cholesterol concentrations. Moreover, it has been reported that andrographolide and neoandrographolide reduce the total cholesterol, triglycerides, and low-density lipoprotein cholesterol, suggesting that this molecule has lipidemic effects and could be of help in metabolic syndrome [81]. In addition, STAT3 is essential for normal gluconeogenesis carbohydrate metabolism in the liver [82].

It has been described that andrographolide has a hepatoprotective role. For instance, andrographolide reduces the symptoms of hepatitis [74] and reduces galactosamine and paracetamol-induced liver toxicity [83]. One of the master regulators in the liver, implicated in the metabolism of xenobiotic and metabolic nutrients and waste, is hepatocyte nuclear factor 4 alpha (HNF4α), which has been implicated in drug metabolism, lipid metabolism, and inflammation [84]. In addition, HNF4α possesses a ligand-binding domain that binds with different fatty acids, favoring the interaction with coactivator peroxisome proliferator-activated receptor gamma coactivator 1-alpha (PGC1α) (Figure 3), involved in liver-specific genes related to the metabolism of cholesterol, bile acids, lipids, and glucose [84,85]. In connection with this, andrographolide has been described as a potential HNF4α antagonist, disrupting the interaction with PGC1α (Figure 3). This suppresses the expression of gluconeogenesis enzymes, lowers the plasma glucose level, and ameliorates fatty liver and inflammation in high-fat-diet-fed mice [86]. Moreover, andrographolide reduces RNA replication of the hepatitis C virus in a hepatic cell culture [87] and protects against the acute liver injury induced by LPS and d-galactosamine [88] by activating *p* 38 MAPK pathways and increasing nuclear factor (erythroid-derived-2)-like 2 (Nrf2), which is mediated by the increase in the expression of heme oxygenase-1 (HO-1) (an antioxidant product) [87]. So far, Nrf2 has been implicated in chronic liver damage occurring during viral hepatitis, and in alcoholic and nonalcoholic fatty liver diseases [89].

2.6. Nrf2/keap1 Pathway

The Nrf2 and Kelch-like ECH-associated protein 1 (Nrf2/Keap1) pathway is essential for protection against a plethora of diseases that involve inflammation and oxidative stress [90]. Keap1 is one of the key regulators of Nrf2 protein stability. Under normal conditions, Nrf2 is sequestered by Keap1 in the cytoplasm and then degraded by proteasomes, but under stress or oxidative conditions, Keap1 loses the interaction of Nrf2 and translocates to the nucleus to promote the transcription of antioxidant genes (Figure 3) [91]. In addition, oxidative stress stimulates adipose differentiation, which directly triggers obesity and is considered to feed into this pathway; together with the increase in circulating proinflammatory cytokines, this aggravates the insulin resistance in metabolic disorders [92,93]. Beyond antioxidant control, activated Nrf2 also exerts anti-inflammatory effects through the direct inhibition of pro-inflammatory cytokines' expression, including IL-6 and TNF-α (Figure 3) [94].

Recently, it has been proposed that andrographolide could protect neurons against inflammation-mediated injury via NF-κB inhibition and Nrf2/HO-1 activation. This is attributed to the reduction of NO, TNF-α, and IL-6 release, and decrease in ROS production (Figure 3) in microglia [95]. In addition, andrographolide protects H_2O_2 and 6-OHDA-induced oxidative damage in neurons [95].

It has been reported that andrographolide suppresses HIF-1α in human endothelial cell lines, activating Nrf2/HO-1 [96], and reduces the expression of histone deacetylase 1 (HDAC1) [24]. Andrographolide inhibits oxidative biomarkers (such as 8-isoprostane, 8-OHdG, and 3-nitrotyrosine) and increases antioxidant enzyme activity (such as glutathione peroxidase and glutathione reductase) [97]. In this way, andrographolide increases the expression of antioxidant enzymes (such as superoxide dismutase, catalase, glutathione reductase, glutathione peroxidase, glutathione-S-transferase, and reduced glutathione [GSH]) as well as glutathione disulfite (GSSG) concentrations [98]. Andrographolide protects liver cells from oxidative stress via Nrf2/HO-1, probably activating adenosine A2a receptor and reducing the cell death induced by H_2O_2 [99]. Additionally, andrographolide increases the redox status in liver cells via downregulation of HNF4A, reducing miR-433 and miR-377 and in this way increasing GSH and HO-1 [100]. In spite of this, the role of the Nrf2/keap1 pathway in acute or chronic liver damage, and tumor development is not completely understood [89], and therefore, these findings should be carefully considered.

2.7. AMP-Activated Protein Kinase (AMPK)

Another important pathway involved in the regulation of Nrf2 in the liver is the AMP-activated protein kinase (AMPK) pathway [101,102]. In mammals, AMPK has been described as the main protein that regulates cellular homeostasis, even acting as a metabolic sensor that is activated when the ratios of AMP/ATP and ADP/ATP have been altered. AMPK is a serine/threonine kinase formed by heterotrimeric protein kinase that is composed of α (two isoforms described), β (two isoforms described), and γ subunits (three isoforms described), where the α subunit is the catalytic one that determines the activity of the protein complex, and the β and γ subunits are the regulatory ones; they are also partly involved in modulation of the kinase complex activity. Activation of AMPK occurs when the AMP level increases, and it binds to the γ subunit, resulting in a conformational change that activates AMPK through phosphorylation at Thr172 in the α subunit by upstream kinases (AMPKK) [103]. However, phosphorylation of AMPKα in Ser485/491, mediated by insulin stimulation, has been shown to lead to a decrease in AMPK activity [104]. Increased activation of AMPK is involved in metabolic reprogramming to switch anabolic metabolism to catabolic metabolism and ignite processes aimed at ATP production or involved in carbohydrate, lipid, and protein metabolism, and is also responsible for the regulation of mitochondrial biogenesis (Figure 3) [103]. AMPK has glycolytic activity through the phosphorylation and activation of PFKFB3 (Figure 3) [105], in the absence of fructose-1,6-bisphosphate (FBP) [106]. Moreover, AMPK regulates carbohydrate metabolism via glucose uptake through GLUT [103]. Andrographolide promotes glucose uptake via GLUT-3 and 4 through the AMPK pathway in rat hippocampal neurons [107]. Four decades ago, AMPK activity was first linked to lipid metabolism. For instance, acetyl-CoA carboxylase (ACC) is an enzyme that can catalyze the carboxylation of acetyl-CoA to malonyl-CoA during the synthesis of fatty acids, or allosterically inhibit CPT-1, a key enzyme in β-oxidation, when AMPK is activated, with inactivated ACC phosphorylating Ser79, Ser1200, and Ser1215 to inhibit fatty acid synthesis and promote β-oxidation [108]. In addition, hydroxymethylglutaryl CoA reductase (HMGCR, a key rate-limiting enzyme in the cholesterol biosynthetic pathway) activation was inhibited by phosphorylation at Ser872 when AMPK was activated (Figure 3) [109]. In this way, the activation of AMPK by AICAR also reduces triglyceride synthesis by inhibition of glycerol-3-phosphate acyltransferase (GPAT) [103,109], evidence of the role AMPK plays in lipid metabolism.

Andrographolide increases AMPK phosphorylation, which reduces the degradation of IκBα and translocation of p65-NF-κB to the nucleus in macrophages [110]. In these cells, an increase in AMPK phosphorylation by andrographolide (10 μM) is associated with an inhibition of MAPK, ERK1/2, p38, and JNK, and a reduction in pro-inflammatory protein expression [110]. Since AMPK is a cellular energy sensor that maintains energy homeostasis via stimulation of glucose uptake and fatty acid oxidation, this could suggest that andrographolide, via AMPK activation, increases the cellular carbohydrate metabolism via OXPHOS and reduces inflammatory processes. In support of this, lactate can reduce Thr172 AMPK phosphorylation in L6 myocytes [111]; in a similar fashion urate crystal, a pro-inflammatory agent in gout, reduces Thr172 AMPK phosphorylation in bone marrow-derived macrophages, which is involved in IL-1β and CXCL1 expression [112]. On the contrary, cell stretching-induced AMPK phosphorylation in rabbit FLS and reduces the NF-κB activation and COX-2 and iNOS expression induced by TNF-α [113]. A reduction of AMPK phosphorylation has been observed in FLS isolated from rats with arthritis; conversely, the glycolysis inhibitor 2-deoxyglucose (2-DG) increases AMPK phosphorylation, interfering with the NF-κB pathway, reducing pro-inflammatory cytokine release [114], and supporting the role of glycolysis in joint inflammation. In addition, andrographolide reduces dextran sulfate sodium-induced acute colitis through AMPK activation [110]. In recent years, andrographolide has been demonstrated to reduce the expression of interferon γ, interleukin (IL)-23, and IL-17A from peripheral blood mononuclear cells derived from ulcerative colitis patients. In these patients, andrographolide had inhibitory effects on Th1/Th17 cells and promoted Th2 cells' response [115,116]. So far, there have been three clinical trials using *Andrographis paniculata* extract (HMPL-004) for the treatment for ulcerative colitis (NCT00659802 in phase II and NCT01805791 in phase III) and Crohn's disease (NCT00655733 in phase II). In this respect, an improvement in the clinical severity of ulcerative colitis as assessed by colonoscopy has been observed in patients treated with *Andrographis paniculata* extract [117].

3. Conclusions

Cellular metabolism plays a key role in the development of several inflammatory diseases. Andrographolide has been described as a multitarget drug with a myriad of anti-inflammatory, antioxidant, and antineoplastic effects in different cell types. In recent years, andrographolide has been shown to have an effect on cellular metabolism, with potential use in metabolic-related diseases such as diabetes mellitus, hypercholesterolemia, and metabolic syndrome. Moreover, several andrographolide signaling pathway targets participate in glucose metabolism control and could be involved in its effects on inflammation. In addition, the metabolic effect of andrographolide would help explain the miscellaneous therapeutic effects described in preclinical and clinical studies. Therefore, further studies will be required to clarify the relationship between the andrographolide effects on metabolism and inflammation or cancer, which would contribute to a better understanding of its various therapeutic properties and to the development of new derivatives with more selective effects.

Author Contributions: R.A.B., P.A., and J.Q. wrote the draft manuscript. J.H. and C.M. revised the manuscript. R.A.B. supervised the writing process and edited the manuscript. All authors have read and agreed to the published version of the manuscript.

Funding: This research was funded by INNOVA-CORFO (grant number 19CVC-118713); the APC was funded by HP Ingredients.

Conflicts of Interest: The authors declare no conflict of interest.

References

1. Pholphana, N.; Rangkadilok, N.; Saehun, J.; Ritruechai, S.; Satayavivad, J. Changes in the contents of four active diterpenoids at different growth stages in Andrographis paniculata (Burm.f.) Nees (Chuanxinlian). *Chin. Med.* **2013**, *8*, 2. [CrossRef] [PubMed]
2. Lin, F.L.; Wu, S.J.; Lee, S.C.; Ng, L.T. Antioxidant, antioedema and analgesic activities of Andrographis paniculata extracts and their active constituent andrographolide. *Phytother. Res.* **2009**, *23*, 958–964. [CrossRef] [PubMed]
3. Hossain, M.S.; Urbi, Z.; Sule, A.; Hafizur Rahman, K.M. Andrographis paniculata (Burm. f.) Wall. ex Nees: A review of ethnobotany, phytochemistry, and pharmacology. *Sci. World J.* **2014**, *2014*, 274905. [CrossRef] [PubMed]
4. Mussard, E.; Jousselin, S.; Cesaro, A.; Legrain, B.; Lespessailles, E.; Esteve, E.; Berteina-Raboin, S.; Toumi, H. Andrographis Paniculata and Its Bioactive Diterpenoids Protect Dermal Fibroblasts Against Inflammation and Oxidative Stress. *Antioxidants* **2020**, *9*, 432. [CrossRef] [PubMed]
5. Kumar, R.A.; Sridevi, K.; Kumar, N.V.; Nanduri, S.; Rajagopal, S. Anticancer and immunostimulatory compounds from Andrographis paniculata. *J. Ethnopharmacol.* **2004**, *92*, 291–295. [CrossRef]
6. Mussard, E.; Cesaro, A.; Lespessailles, E.; Legrain, B.; Berteina-Raboin, S.; Toumi, H. Andrographolide, a Natural Antioxidant: An Update. *Antioxidants* **2019**, *8*, 571. [CrossRef]
7. Rosazza, T.; Warner, J.; Sollberger, G. NET formation—Mechanisms and how they relate to other cell death pathways. *FEBS J.* **2020**. [CrossRef]
8. Lu, T.; Kobayashi, S.D.; Quinn, M.T.; Deleo, F.R. A NET Outcome. *Front. Immunol.* **2012**, *3*, 365. [CrossRef]
9. Hilscher, M.B.; Shah, V.H. Neutrophil Extracellular Traps and Liver Disease. *Semin. Liver Dis.* **2020**, *40*, 171–179. [CrossRef]
10. Ravindran, M.; Khan, M.A.; Palaniyar, N. Neutrophil Extracellular Trap Formation: Physiology, Pathology, and Pharmacology. *Biomolecules* **2019**, *9*, 365. [CrossRef]
11. Shen, Y.C.; Chen, C.F.; Chiou, W.F. Suppression of rat neutrophil reactive oxygen species production and adhesion by the diterpenoid lactone andrographolide. *Planta Med.* **2000**, *66*, 314–317. [CrossRef] [PubMed]
12. Shen, Y.C.; Chen, C.F.; Chiou, W.F. Andrographolide prevents oxygen radical production by human neutrophils: Possible mechanism(s) involved in its anti-inflammatory effect. *Br. J. Pharmacol.* **2002**, *135*, 399–406. [CrossRef] [PubMed]
13. Li, X.; Yuan, K.; Zhu, Q.; Lu, Q.; Jiang, H.; Zhu, M.; Huang, G.; Xu, A. Andrographolide Ameliorates Rheumatoid Arthritis by Regulating the Apoptosis-NETosis Balance of Neutrophils. *Int. J. Mol. Sci.* **2019**, *20*, 5035. [CrossRef] [PubMed]
14. Tan, W.S.D.; Liao, W.; Zhou, S.; Wong, W.S.F. Is there a future for andrographolide to be an anti-inflammatory drug? Deciphering its major mechanisms of action. *Biochem. Pharmacol.* **2017**, *139*, 71–81. [CrossRef]
15. Chiou, W.F.; Lin, J.J.; Chen, C.F. Andrographolide suppresses the expression of inducible nitric oxide synthase in macrophage and restores the vasoconstriction in rat aorta treated with lipopolysaccharide. *Br. J. Pharmacol.* **1998**, *125*, 327–334. [CrossRef]

16. Zhu, T.; Wang, D.X.; Zhang, W.; Liao, X.Q.; Guan, X.; Bo, H.; Sun, J.Y.; Huang, N.W.; He, J.; Zhang, Y.K.; et al. Andrographolide protects against LPS-induced acute lung injury by inactivation of NF-kappaB. *PLoS ONE* **2013**, *8*, e56407. [CrossRef]
17. Gupta, S.; Mishra, K.P.; Singh, S.B.; Ganju, L. Inhibitory effects of andrographolide on activated macrophages and adjuvant-induced arthritis. *Inflammopharmacology* **2018**, *26*, 447–456. [CrossRef]
18. Dai, G.F.; Zhao, J.; Jiang, Z.W.; Zhu, L.P.; Xu, H.W.; Ma, W.Y.; Chen, X.R.; Dong, R.J.; Li, W.Y.; Liu, H.M. Anti-inflammatory effect of novel andrographolide derivatives through inhibition of NO and PGE2 production. *Int. Immunopharmacol.* **2011**, *11*, 2144–2149. [CrossRef]
19. Coussens, L.M.; Werb, Z. Inflammation and cancer. *Nature* **2002**, *420*, 860–867. [CrossRef]
20. Singh, N.; Baby, D.; Rajguru, J.P.; Patil, P.B.; Thakkannavar, S.S.; Pujari, V.B. Inflammation and cancer. *Ann. Afr. Med.* **2019**, *18*, 121–126. [CrossRef]
21. Peng, Y.; Wang, Y.; Tang, N.; Sun, D.; Lan, Y.; Yu, Z.; Zhao, X.; Feng, L.; Zhang, B.; Jin, L.; et al. Andrographolide inhibits breast cancer through suppressing COX-2 expression and angiogenesis via inactivation of p300 signaling and VEGF pathway. *J. Exp. Clin. Cancer Res.* **2018**, *37*, 248. [CrossRef] [PubMed]
22. Yuan, M.; Meng, W.; Liao, W.; Lian, S. Andrographolide Antagonizes TNF-alpha-Induced IL-8 via Inhibition of NADPH Oxidase/ROS/NF-kappaB and Src/MAPKs/AP-1 Axis in Human Colorectal Cancer HCT116 Cells. *J. Agric. Food Chem* **2018**, *66*, 5139–5148. [CrossRef] [PubMed]
23. Semenza, G.L. HIF-1: Upstream and downstream of cancer metabolism. *Curr. Opin. Genet. Dev.* **2010**, *20*, 51–56. [CrossRef] [PubMed]
24. Shi, L.; Zhang, G.; Zheng, Z.; Lu, B.; Ji, L. Andrographolide reduced VEGFA expression in hepatoma cancer cells by inactivating HIF-1alpha: The involvement of JNK and MTA1/HDCA. *Chem. Biol. Interact.* **2017**, *273*, 228–236. [CrossRef]
25. Subramanian, R.; Asmawi, M.Z.; Sadikun, A. In vitro alpha-glucosidase and alpha-amylase enzyme inhibitory effects of Andrographis paniculata extract and andrographolide. *Acta Biochim. Pol.* **2008**, *55*, 391–398. [CrossRef]
26. Ding, L.; Li, J.; Song, B.; Xiao, X.; Huang, W.; Zhang, B.; Tang, X.; Qi, M.; Yang, Q.; Yang, Q.; et al. Andrographolide prevents high-fat diet-induced obesity in C57BL/6 mice by suppressing the sterol regulatory element-binding protein pathway. *J. Pharmacol. Exp. Ther.* **2014**, *351*, 474–483. [CrossRef]
27. Jin, L.; Shi, G.; Ning, G.; Li, X.; Zhang, Z. Andrographolide attenuates tumor necrosis factor-alpha-induced insulin resistance in 3T3-L1 adipocytes. *Mol. Cell Endocrinol.* **2011**, *332*, 134–139. [CrossRef]
28. Islam, M.T. Andrographolide, a New Hope in the Prevention and Treatment of Metabolic Syndrome. *Front. Pharmacol.* **2017**, *8*, 571. [CrossRef]
29. Rossaint, J.; Margraf, A.; Zarbock, A. Role of Platelets in Leukocyte Recruitment and Resolution of Inflammation. *Front. Immunol.* **2018**, *9*, 2712. [CrossRef]
30. Wang, F.; Huang, S.; Xia, H.; Yao, S. Specialized pro-resolving mediators: It's anti-oxidant stress role in multiple disease models. *Mol. Immunol.* **2020**, *126*, 40–45. [CrossRef]
31. Barbe-Tuana, F.; Funchal, G.; Schmitz, C.R.R.; Maurmann, R.M.; Bauer, M.E. The interplay between immunosenescence and age-related diseases. *Semin. Immunopathol.* **2020**, *42*, 545–557. [CrossRef] [PubMed]
32. Soto-Heredero, G.; Gomez de Las Heras, M.M.; Gabande-Rodriguez, E.; Oller, J.; Mittelbrunn, M. Glycolysis—A key player in the inflammatory response. *FEBS J.* **2020**, *287*, 3350–3369. [CrossRef] [PubMed]
33. Certo, M.; Tsai, C.H.; Pucino, V.; Ho, P.C.; Mauro, C. Lactate modulation of immune responses in inflammatory versus tumour microenvironments. *Nat. Rev. Immunol.* **2020**. [CrossRef] [PubMed]
34. Bao, Y.; Ledderose, C.; Graf, A.F.; Brix, B.; Birsak, T.; Lee, A.; Zhang, J.; Junger, W.G. mTOR and differential activation of mitochondria orchestrate neutrophil chemotaxis. *J. Cell Biol.* **2015**, *210*, 1153–1164. [CrossRef] [PubMed]
35. Quiroga, J.; Alarcon, P.; Manosalva, C.; Taubert, A.; Hermosilla, C.; Hidalgo, M.A.; Carretta, M.D.; Burgos, R.A. Glycolysis and mitochondrial function regulate the radical oxygen species production induced by platelet-activating factor in bovine polymorphonuclear leukocytes. *Vet. Immunol. Immunopathol.* **2020**, *226*, 110074. [CrossRef] [PubMed]
36. Quiroga, J.; Alarcon, P.; Manosalva, C.; Taubert, A.; Hermosilla, C.; Hidalgo, M.A.; Carretta, M.D.; Burgos, R.A. Mitochondria-derived ATP participates in the formation of neutrophil extracellular traps induced by platelet-activating factor through purinergic signaling in cows. *Dev. Comp. Immunol.* **2020**, *113*, 103768. [CrossRef]
37. Alarcon, P.; Manosalva, C.; Quiroga, J.; Belmar, I.; Alvarez, K.; Diaz, G.; Taubert, A.; Hermosilla, C.; Carretta, M.D.; Burgos, R.A.; et al. Oleic and Linoleic Acids Induce the Release of Neutrophil Extracellular Traps via Pannexin 1-Dependent ATP Release and P2 × 1 Receptor Activation. *Front. Vet. Sci.* **2020**, *7*, 260. [CrossRef]
38. Curi, R.; Levada-Pires, A.C.; Silva, E.B.D.; Poma, S.O.; Zambonatto, R.F.; Domenech, P.; Almeida, M.M.; Gritte, R.B.; Souza-Siqueira, T.; Gorjao, R.; et al. The Critical Role of Cell Metabolism for Essential Neutrophil Functions. *Cell. Physiol. Biochem.* **2020**, *54*, 629–647. [CrossRef]
39. O'Neill, L.A.; Kishton, R.J.; Rathmell, J. A guide to immunometabolism for immunologists. *Nat. Rev. Immunol.* **2016**, *16*, 553–565. [CrossRef]
40. Lu, H.; Forbes, R.A.; Verma, A. Hypoxia-inducible factor 1 activation by aerobic glycolysis implicates the Warburg effect in carcinogenesis. *J. Biol. Chem.* **2002**, *277*, 23111–23115. [CrossRef]

41. Rius, J.; Guma, M.; Schachtrup, C.; Akassoglou, K.; Zinkernagel, A.S.; Nizet, V.; Johnson, R.S.; Haddad, G.G.; Karin, M. NF-kappaB links innate immunity to the hypoxic response through transcriptional regulation of HIF-1alpha. *Nature* **2008**, *453*, 807–811. [CrossRef] [PubMed]
42. Frede, S.; Stockmann, C.; Freitag, P.; Fandrey, J. Bacterial lipopolysaccharide induces HIF-1 activation in human monocytes via p44/42 MAPK and NF-kappaB. *Biochem. J.* **2006**, *396*, 517–527. [CrossRef] [PubMed]
43. Hitosugi, T.; Kang, S.; Vander Heiden, M.G.; Chung, T.W.; Elf, S.; Lythgoe, K.; Dong, S.; Lonial, S.; Wang, X.; Chen, G.Z.; et al. Tyrosine phosphorylation inhibits PKM2 to promote the Warburg effect and tumor growth. *Sci. Signal.* **2009**, *2*, ra73. [CrossRef]
44. Palsson-McDermott, E.M.; Curtis, A.M.; Goel, G.; Lauterbach, M.A.; Sheedy, F.J.; Gleeson, L.E.; van den Bosch, M.W.; Quinn, S.R.; Domingo-Fernandez, R.; Johnston, D.G.; et al. Pyruvate kinase M2 regulates Hif-1alpha activity and IL-1beta induction and is a critical determinant of the warburg effect in LPS-activated macrophages. *Cell Metab.* **2015**, *21*, 65–80. [CrossRef] [PubMed]
45. McGettrick, A.F.; O'Neill, L.A.J. The Role of HIF in Immunity and Inflammation. *Cell Metab.* **2020**, *32*, 524–536. [CrossRef] [PubMed]
46. Hua, S.; Dias, T.H. Hypoxia-Inducible Factor (HIF) as a Target for Novel Therapies in Rheumatoid Arthritis. *Front. Pharmacol.* **2016**, *7*, 184. [CrossRef]
47. Li, G.F.; Qin, Y.H.; Du, P.Q. Andrographolide inhibits the migration, invasion and matrix metalloproteinase expression of rheumatoid arthritis fibroblast-like synoviocytes via inhibition of HIF-1alpha signaling. *Life Sci.* **2015**, *136*, 67–72. [CrossRef]
48. Ahn, J.K.; Koh, E.M.; Cha, H.S.; Lee, Y.S.; Kim, J.; Bae, E.K.; Ahn, K.S. Role of hypoxia-inducible factor-1alpha in hypoxia-induced expressions of IL-8, MMP-1 and MMP-3 in rheumatoid fibroblast-like synoviocytes. *Rheumatology* **2008**, *47*, 834–839. [CrossRef]
49. Garcia-Carbonell, R.; Divakaruni, A.S.; Lodi, A.; Vicente-Suarez, I.; Saha, A.; Cheroutre, H.; Boss, G.R.; Tiziani, S.; Murphy, A.N.; Guma, M. Critical Role of Glucose Metabolism in Rheumatoid Arthritis Fibroblast-like Synoviocytes. *Arthritis Rheumatol.* **2016**, *68*, 1614–1626. [CrossRef]
50. Yang, X.; Chen, W.; Zhao, X.; Chen, L.; Li, W.; Ran, J.; Wu, L. Pyruvate Kinase M2 Modulates the Glycolysis of Chondrocyte and Extracellular Matrix in Osteoarthritis. *DNA Cell Biol.* **2018**, *37*, 271–277. [CrossRef]
51. Li, X.J.; Xu, M.; Zhao, X.Q.; Zhao, J.N.; Chen, F.F.; Yu, W.; Gao, D.Y.; Luo, B. Proteomic analysis of synovial fibroblast-like synoviocytes from rheumatoid arthritis. *Clin. Exp. Rheumatol.* **2013**, *31*, 552–558. [PubMed]
52. Wang, J.; Tan, X.F.; Nguyen, V.S.; Yang, P.; Zhou, J.; Gao, M.; Li, Z.; Lim, T.K.; He, Y.; Ong, C.S.; et al. A quantitative chemical proteomics approach to profile the specific cellular targets of andrographolide, a promising anticancer agent that suppresses tumor metastasis. *Mol. Cell. Proteom.* **2014**, *13*, 876–886. [CrossRef] [PubMed]
53. Burgos, R.A.; Hancke, J.L.; Bertoglio, J.C.; Aguirre, V.; Arriagada, S.; Calvo, M.; Caceres, D.D. Efficacy of an Andrographis paniculata composition for the relief of rheumatoid arthritis symptoms: A prospective randomized placebo-controlled trial. *Clin. Rheumatol.* **2009**, *28*, 931–946. [CrossRef] [PubMed]
54. Hancke, J.L.; Srivastav, S.; Caceres, D.D.; Burgos, R.A. A double-blind, randomized, placebo-controlled study to assess the efficacy of Andrographis paniculata standardized extract (ParActin(R)) on pain reduction in subjects with knee osteoarthritis. *Phytother. Res.* **2019**, *33*, 1469–1479. [CrossRef] [PubMed]
55. Li, Z.Y.; Shi, Y.L.; Liang, G.X.; Yang, J.; Zhuang, S.K.; Lin, J.B.; Ghodbane, A.; Tam, M.S.; Liang, Z.J.; Zha, Z.G.; et al. Visualization of GLUT1 Trafficking in Live Cancer Cells by the Use of a Dual-Fluorescence Reporter. *ACS Omega* **2020**, *5*, 15911–15921. [CrossRef] [PubMed]
56. Wang, W.; Wu, Y.; Yang, K.; Wu, C.; Tang, R.; Li, H.; Chen, L. Synthesis of novel andrographolide beckmann rearrangement derivatives and evaluation of their HK2-related anti-inflammatory activities. *Eur. J. Med. Chem.* **2019**, *173*, 282–293. [CrossRef]
57. Miao, H.; Chen, L.; Hao, L.; Zhang, X.; Chen, Y.; Ruan, Z.; Liang, H. Stearic acid induces proinflammatory cytokine production partly through activation of lactate-HIF1alpha pathway in chondrocytes. *Sci. Rep.* **2015**, *5*, 13092. [CrossRef]
58. Li, X.; Tian, R.; Liu, L.; Wang, L.; He, D.; Cao, K.; Ma, J.K.; Huang, C. Andrographolide enhanced radiosensitivity by downregulating glycolysis via the inhibition of the PI3K-Akt-mTOR signaling pathway in HCT116 colorectal cancer cells. *J. Int. Med. Res.* **2020**, *48*, 300060520946169. [CrossRef]
59. Duan, M.X.; Zhou, H.; Wu, Q.Q.; Liu, C.; Xiao, Y.; Deng, W.; Tang, Q.Z. Andrographolide Protects against HG-Induced Inflammation, Apoptosis, Migration, and Impairment of Angiogenesis via PI3K/AKT-eNOS Signalling in HUVECs. *Mediat. Inflamm.* **2019**, *2019*, 6168340. [CrossRef]
60. Chen, H.W.; Lin, A.H.; Chu, H.C.; Li, C.C.; Tsai, C.W.; Chao, C.Y.; Wang, C.J.; Lii, C.K.; Liu, K.L. Inhibition of TNF-alpha-Induced Inflammation by andrographolide via down-regulation of the PI3K/Akt signaling pathway. *J. Nat. Prod.* **2011**, *74*, 2408–2413. [CrossRef]
61. Taniguchi, K.; Karin, M. NF-kappaB, inflammation, immunity and cancer: Coming of age. *Nat. Rev. Immunol.* **2018**, *18*, 309–324. [CrossRef] [PubMed]
62. Chao, C.Y.; Lii, C.K.; Tsai, I.T.; Li, C.C.; Liu, K.L.; Tsai, C.W.; Chen, H.W. Andrographolide inhibits ICAM-1 expression and NF-kappaB activation in TNF-alpha-treated EA.hy926 cells. *J. Agric. Food Chem.* **2011**, *59*, 5263–5271. [CrossRef] [PubMed]
63. D'Ignazio, L.; Bandarra, D.; Rocha, S. NF-kappaB and HIF crosstalk in immune responses. *FEBS J.* **2016**, *283*, 413–424. [CrossRef] [PubMed]
64. Walmsley, S.R.; Print, C.; Farahi, N.; Peyssonnaux, C.; Johnson, R.S.; Cramer, T.; Sobolewski, A.; Condliffe, A.M.; Cowburn, A.S.; Johnson, N.; et al. Hypoxia-induced neutrophil survival is mediated by HIF-1alpha-dependent NF-kappaB activity. *J. Exp. Med.* **2005**, *201*, 105–115. [CrossRef]

65. Atsumi, T.; Singh, R.; Sabharwal, L.; Bando, H.; Meng, J.; Arima, Y.; Yamada, M.; Harada, M.; Jiang, J.J.; Kamimura, D.; et al. Inflammation amplifier, a new paradigm in cancer biology. *Cancer Res.* **2014**, *74*, 8–14. [CrossRef]
66. Farooqi, A.A.; Attar, R.; Sabitaliyevich, U.Y.; Alaaeddine, N.; de Sousa, D.P.; Xu, B.; Cho, W.C. The Prowess of Andrographolide as a Natural Weapon in the War against Cancer. *Cancers* **2020**, *12*, 2159. [CrossRef]
67. Byeon, S.E.; Yi, Y.S.; Oh, J.; Yoo, B.C.; Hong, S.; Cho, J.Y. The role of Src kinase in macrophage-mediated inflammatory responses. *Mediat. Inflamm.* **2012**, *2012*, 512926. [CrossRef]
68. Papa, S.; Choy, P.M.; Bubici, C. The ERK and JNK pathways in the regulation of metabolic reprogramming. *Oncogene* **2019**, *38*, 2223–2240. [CrossRef]
69. Li, Y.; He, S.; Tang, J.; Ding, N.; Chu, X.; Cheng, L.; Ding, X.; Liang, T.; Feng, S.; Rahman, S.U.; et al. Andrographolide Inhibits Inflammatory Cytokines Secretion in LPS-Stimulated RAW264.7 Cells through Suppression of NF-kappaB/MAPK Signaling Pathway. *Evid. Based Complement. Altern. Med.* **2017**, *2017*, 8248142. [CrossRef]
70. Li, Z.Z.; Tan, J.P.; Wang, L.L.; Li, Q.H. Andrographolide Benefits Rheumatoid Arthritis via Inhibiting MAPK Pathways. *Inflammation* **2017**, *40*, 1599–1605. [CrossRef]
71. Trop-Steinberg, S.; Azar, Y. AP-1 Expression and its Clinical Relevance in Immune Disorders and Cancer. *Am. J. Med. Sci.* **2017**, *353*, 474–483. [CrossRef] [PubMed]
72. Hu, X.; Zhou, J.; Song, S.S.; Kong, W.; Shi, Y.C.; Chen, L.L.; Zeng, T.S. TLR4/AP-1-Targeted Anti-Inflammatory Intervention Attenuates Insulin Sensitivity and Liver Steatosis. *Mediat. Inflamm.* **2020**, *2020*, 2960517. [CrossRef] [PubMed]
73. Lu, C.Y.; Yang, Y.C.; Li, C.C.; Liu, K.L.; Lii, C.K.; Chen, H.W. Andrographolide inhibits TNFalpha-induced ICAM-1 expression via suppression of NADPH oxidase activation and induction of HO-1 and GCLM expression through the PI3K/Akt/Nrf2 and PI3K/Akt/AP-1 pathways in human endothelial cells. *Biochem. Pharmacol.* **2014**, *91*, 40–50. [CrossRef] [PubMed]
74. Shen, T.; Yang, W.S.; Yi, Y.S.; Sung, G.H.; Rhee, M.H.; Poo, H.; Kim, M.Y.; Kim, K.W.; Kim, J.H.; Cho, J.Y. AP-1/IRF-3 Targeted Anti-Inflammatory Activity of Andrographolide Isolated from Andrographis paniculata. *Evid. Based Complement. Altern. Med.* **2013**, *2013*, 210736. [CrossRef] [PubMed]
75. Dodington, D.W.; Desai, H.R.; Woo, M. JAK/STAT—Emerging Players in Metabolism. *Trends Endocrinol. Metab.* **2018**, *29*, 55–65. [CrossRef]
76. Owen, K.L.; Brockwell, N.K.; Parker, B.S. JAK-STAT Signaling: A Double-Edged Sword of Immune Regulation and Cancer Progression. *Cancers* **2019**, *11*, 2002. [CrossRef]
77. McGarry, T.; Orr, C.; Wade, S.; Biniecka, M.; Wade, S.; Gallagher, L.; Low, C.; Veale, D.J.; Fearon, U. JAK/STAT Blockade Alters Synovial Bioenergetics, Mitochondrial Function, and Proinflammatory Mediators in Rheumatoid Arthritis. *Arthritis Rheumatol.* **2018**, *70*, 1959–1970. [CrossRef]
78. Zhou, J.; Ong, C.N.; Hur, G.M.; Shen, H.M. Inhibition of the JAK-STAT3 pathway by andrographolide enhances chemosensitivity of cancer cells to doxorubicin. *Biochem. Pharmacol.* **2010**, *79*, 1242–1250. [CrossRef]
79. Chun, J.Y.; Tummala, R.; Nadiminty, N.; Lou, W.; Liu, C.; Yang, J.; Evans, C.P.; Zhou, Q.; Gao, A.C. Andrographolide, an herbal medicine, inhibits interleukin-6 expression and suppresses prostate cancer cell growth. *Genes Cancer* **2010**, *1*, 868–876. [CrossRef]
80. Ding, Y.; Chen, L.; Wu, W.; Yang, J.; Yang, Z.; Liu, S. Andrographolide inhibits influenza A virus-induced inflammation in a murine model through NF-kappaB and JAK-STAT signaling pathway. *Microbes Infect.* **2017**, *19*, 605–615. [CrossRef]
81. Yang, T.; Shi, H.X.; Wang, Z.T.; Wang, C.H. Hypolipidemic effects of andrographolide and neoandrographolide in mice and rats. *Phytother. Res.* **2013**, *27*, 618–623. [CrossRef] [PubMed]
82. Inoue, H. Central insulin-mediated regulation of hepatic glucose production [Review]. *Endocr. J.* **2016**, *63*, 1–7. [CrossRef] [PubMed]
83. Handa, S.S.; Sharma, A. Hepatoprotective activity of andrographolide against galactosamine & paracetamol intoxication in rats. *Indian J. Med. Res.* **1990**, *92*, 284–292. [PubMed]
84. Lu, H. Crosstalk of HNF4alpha with extracellular and intracellular signaling pathways in the regulation of hepatic metabolism of drugs and lipids. *Acta Pharm. Sin. B* **2016**, *6*, 393–408. [CrossRef]
85. Dubois, V.; Staels, B.; Lefebvre, P.; Verzi, M.P.; Eeckhoute, J. Control of Cell Identity by the Nuclear Receptor HNF4 in Organ Pathophysiology. *Cells* **2020**, *9*, 2185. [CrossRef]
86. Zhang, M.; Yang, M.; Wang, N.; Liu, Q.; Wang, B.; Huang, T.; Tong, Y.; Ming, Y.; Wong, C.W.; Liu, J.; et al. Andrographolide modulates HNF4alpha activity imparting on hepatic metabolism. *Mol. Cell Endocrinol.* **2020**, *513*, 110867. [CrossRef]
87. Lee, J.C.; Tseng, C.K.; Young, K.C.; Sun, H.Y.; Wang, S.W.; Chen, W.C.; Lin, C.K.; Wu, Y.H. Andrographolide exerts anti-hepatitis C virus activity by up-regulating haeme oxygenase-1 via the p38 MAPK/Nrf2 pathway in human hepatoma cells. *Br. J. Pharmacol.* **2014**, *171*, 237–252. [CrossRef]
88. Pan, C.W.; Yang, S.X.; Pan, Z.Z.; Zheng, B.; Wang, J.Z.; Lu, G.R.; Xue, Z.X.; Xu, C.L. Andrographolide ameliorates d-galactosamine/lipopolysaccharide-induced acute liver injury by activating Nrf2 signaling pathway. *Oncotarget* **2017**, *8*, 41202–41210. [CrossRef]
89. Orru, C.; Giordano, S.; Columbano, A. Nrf2 in Neoplastic and Non-Neoplastic Liver Diseases. *Cancers* **2020**, *12*, 2932. [CrossRef]
90. Cuadrado, A.; Rojo, A.I.; Wells, G.; Hayes, J.D.; Cousin, S.P.; Rumsey, W.L.; Attucks, O.C.; Franklin, S.; Levonen, A.L.; Kensler, T.W.; et al. Therapeutic targeting of the NRF2 and KEAP1 partnership in chronic diseases. *Nat. Rev. Drug Discov.* **2019**, *18*, 295–317. [CrossRef]

91. Habeos, I.G.; Ziros, P.G.; Chartoumpekis, D.; Psyrogiannis, A.; Kyriazopoulou, V.; Papavassiliou, A.G. Simvastatin activates Keap1/Nrf2 signaling in rat liver. *J. Mol. Med.* **2008**, *86*, 1279–1285. [CrossRef] [PubMed]
92. Lee, H.; Lee, Y.J.; Choi, H.; Ko, E.H.; Kim, J.W. Reactive oxygen species facilitate adipocyte differentiation by accelerating mitotic clonal expansion. *J. Biol. Chem.* **2009**, *284*, 10601–10609. [CrossRef] [PubMed]
93. Stepien, M.; Stepien, A.; Wlazel, R.N.; Paradowski, M.; Banach, M.; Rysz, J. Obesity indices and inflammatory markers in obese non-diabetic normo- and hypertensive patients: A comparative pilot study. *Lipids Health Dis.* **2014**, *13*, 29. [CrossRef] [PubMed]
94. Boyanapalli, S.S.; Paredes-Gonzalez, X.; Fuentes, F.; Zhang, C.; Guo, Y.; Pung, D.; Saw, C.L.; Kong, A.N. Nrf2 knockout attenuates the anti-inflammatory effects of phenethyl isothiocyanate and curcumin. *Chem. Res. Toxicol.* **2014**, *27*, 2036–2043. [CrossRef]
95. Xu, Y.; Tang, D.; Wang, J.; Wei, H.; Gao, J. Neuroprotection of Andrographolide Against Microglia-Mediated Inflammatory Injury and Oxidative Damage in PC12 Neurons. *Neurochem. Res.* **2019**, *44*, 2619–2630. [CrossRef]
96. Lin, H.C.; Su, S.L.; Lu, C.Y.; Lin, A.H.; Lin, W.C.; Liu, C.S.; Yang, Y.C.; Wang, H.M.; Lii, C.K.; Chen, H.W. Andrographolide inhibits hypoxia-induced HIF-1alpha-driven endothelin 1 secretion by activating Nrf2/HO-1 and promoting the expression of prolyl hydroxylases 2/3 in human endothelial cells. *Environ. Toxicol.* **2017**, *32*, 918–930. [CrossRef]
97. Guan, S.P.; Tee, W.; Ng, D.S.; Chan, T.K.; Peh, H.Y.; Ho, W.E.; Cheng, C.; Mak, J.C.; Wong, W.S. Andrographolide protects against cigarette smoke-induced oxidative lung injury via augmentation of Nrf2 activity. *Br. J. Pharmacol.* **2013**, *168*, 1707–1718. [CrossRef]
98. Das, S.; Gautam, N.; Dey, S.K.; Maiti, T.; Roy, S. Oxidative stress in the brain of nicotine-induced toxicity: Protective role of Andrographis paniculata Nees and vitamin E. *Appl. Physiol. Nutr. Metab.* **2009**, *34*, 124–135. [CrossRef]
99. Mittal, S.P.K.; Khole, S.; Jagadish, N.; Ghosh, D.; Gadgil, V.; Sinkar, V.; Ghaskadbi, S.S. Andrographolide protects liver cells from H2O2 induced cell death by upregulation of Nrf-2/HO-1 mediated via adenosine A2a receptor signalling. *Biochim. Biophys. Acta* **2016**, *1860*, 2377–2390. [CrossRef]
100. Khole, S.; Mittal, S.; Jagadish, N.; Ghosh, D.; Gadgil, V.; Sinkar, V.; Ghaskadbi, S. Andrographolide enhances redox status of liver cells by regulating microRNA expression. *Free Radic. Biol. Med.* **2019**, *130*, 397–407. [CrossRef]
101. Lv, H.; An, B.; Yu, Q.; Cao, Y.; Liu, Y.; Li, S. The hepatoprotective effect of myricetin against lipopolysaccharide and D-galactosamine-induced fulminant hepatitis. *Int. J. Biol. Macromol.* **2020**, *155*, 1092–1104. [CrossRef] [PubMed]
102. Zhao, P.; Saltiel, A.R. From overnutrition to liver injury: AMP-activated protein kinase in nonalcoholic fatty liver diseases. *J. Biol. Chem.* **2020**, *295*, 12279–12289. [CrossRef] [PubMed]
103. Novikova, D.S.; Garabadzhiu, A.V.; Melino, G.; Barlev, N.A.; Tribulovich, V.G. AMP-activated protein kinase: Structure, function, and role in pathological processes. *Biochemistry* **2015**, *80*, 127–144. [CrossRef] [PubMed]
104. Valentine, R.J.; Coughlan, K.A.; Ruderman, N.B.; Saha, A.K. Insulin inhibits AMPK activity and phosphorylates AMPK $Ser^{485/491}$ through Akt in hepatocytes, myotubes and incubated rat skeletal muscle. *Arch. Biochem. Biophys.* **2014**, *562*, 62–69. [CrossRef] [PubMed]
105. Domenech, E.; Maestre, C.; Esteban-Martinez, L.; Partida, D.; Pascual, R.; Fernandez-Miranda, G.; Seco, E.; Campos-Olivas, R.; Perez, M.; Megias, D.; et al. AMPK and PFKFB3 mediate glycolysis and survival in response to mitophagy during mitotic arrest. *Nat. Cell Biol.* **2015**, *17*, 1304–1316. [CrossRef] [PubMed]
106. Zhang, C.S.; Hawley, S.A.; Zong, Y.; Li, M.; Wang, Z.; Gray, A.; Ma, T.; Cui, J.; Feng, J.W.; Zhu, M.; et al. Fructose-1,6-bisphosphate and aldolase mediate glucose sensing by AMPK. *Nature* **2017**, *548*, 112–116. [CrossRef] [PubMed]
107. Gherardelli, C.; Cisternas, P.; Gutierrez, J.; Martinez, M.; Inestrosa, N.C. Andrographolide restores glucose uptake in rat hippocampal neurons. *J. Neurochem.* **2020**. [CrossRef]
108. Davies, S.P.; Carling, D.; Munday, M.R.; Hardie, D.G. Diurnal rhythm of phosphorylation of rat liver acetyl-CoA carboxylase by the AMP-activated protein kinase, demonstrated using freeze-clamping. Effects of high fat diets. *Eur. J. Biochem.* **1992**, *203*, 615–623. [CrossRef]
109. Loh, K.; Tam, S.; Murray-Segal, L.; Huynh, K.; Meikle, P.J.; Scott, J.W.; van Denderen, B.; Chen, Z.; Steel, R.; LeBlond, N.D.; et al. Inhibition of Adenosine Monophosphate-Activated Protein Kinase-3-Hydroxy-3-Methylglutaryl Coenzyme A Reductase Signaling Leads to Hypercholesterolemia and Promotes Hepatic Steatosis and Insulin Resistance. *Hepatol. Commun.* **2019**, *3*, 84–98. [CrossRef]
110. Kim, N.; Lertnimitphun, P.; Jiang, Y.; Tan, H.; Zhou, H.; Lu, Y.; Xu, H. Andrographolide inhibits inflammatory responses in LPS-stimulated macrophages and murine acute colitis through activating AMPK. *Biochem. Pharmacol.* **2019**, *170*, 113646. [CrossRef]
111. Cerda-Kohler, H.; Henriquez-Olguin, C.; Casas, M.; Jensen, T.E.; Llanos, P.; Jaimovich, E. Lactate administration activates the ERK1/2, mTORC1, and AMPK pathways differentially according to skeletal muscle type in mouse. *Physiol. Rep.* **2018**, *6*, e13800. [CrossRef] [PubMed]
112. Wang, Y.; Viollet, B.; Terkeltaub, R.; Liu-Bryan, R. AMP-activated protein kinase suppresses urate crystal-induced inflammation and transduces colchicine effects in macrophages. *Ann. Rheum. Dis.* **2016**, *75*, 286–294. [CrossRef] [PubMed]
113. Kunanusornchai, W.; Muanprasat, C.; Chatsudthipong, V. Adenosine monophosphate-activated protein kinase activation and suppression of inflammatory response by cell stretching in rabbit synovial fibroblasts. *Mol. Cell. Biochem.* **2016**, *423*, 175–185. [CrossRef] [PubMed]
114. Wang, Y.; Xian, H.; Qi, J.; Wei, F.; Cheng, X.; Li, S.; Wang, Q.; Liu, Z.; Yu, Y.; Zhou, J.; et al. Inhibition of glycolysis ameliorate arthritis in adjuvant arthritis rats by inhibiting synoviocyte activation through AMPK/NF-small ka, CyrillicB pathway. *Inflamm. Res.* **2020**, *69*, 569–578. [CrossRef]

115. Zhu, Q.; Zheng, P.; Zhou, J.; Chen, X.; Feng, Y.; Wang, W.; Zhou, F.; He, Q. Andrographolide affects Th1/Th2/Th17 responses of peripheral blood mononuclear cells from ulcerative colitis patients. *Mol. Med. Rep.* **2018**, *18*, 622–626. [CrossRef]
116. Zhu, Q.; Zheng, P.; Chen, X.; Zhou, F.; He, Q.; Yang, Y. Andrographolide presents therapeutic effect on ulcerative colitis through the inhibition of IL-23/IL-17 axis. *Am. J. Transl. Res.* **2018**, *10*, 465–473.
117. Sandborn, W.J.; Targan, S.R.; Byers, V.S.; Rutty, D.A.; Mu, H.; Zhang, X.; Tang, T. Andrographis paniculata extract (HMPL-004) for active ulcerative colitis. *Am. J. Gastroenterol.* **2013**, *108*, 90–98. [CrossRef]

molecules

Brief Report

Preconceptional Resveratrol Supplementation Partially Counteracts Age-Related Reproductive Complications in C57BL/6J Female Mice

Marta Ziętek [1,2,3], Katarzyna Barłowska [1], Barbara Wijas [1], Ewa Szablisty [1], Atanas G. Atanasov [4,5,6,7], Jacek A. Modliński [1], Artur H. Świergiel [2,3,*] and Silvestre Sampino [1,*]

1 Department of Experimental Embryology, Institute of Genetics and Animal Biotechnology of the Polish Academy of Sciences, ul. Postępu 36A, Jastrzębiec, 05-552 Magdalenka, Poland; m.zietek@igbzpan.pl (M.Z.); k.barlowska@igbzpan.pl (K.B.); b.wijas@igbzpan.pl (B.W.); e.szablisty@igbzpan.pl (E.S.); j.a.modlinski@igbzpan.pl (J.A.M.)
2 Department of Animal and Human Physiology, Faculty of Biology, University of Gdańsk, ul. Wita Stwosza 59, 80-308 Gdańsk, Poland
3 Prof. Wacław Dąbrowski Institute of Agricultural and Food Biotechnology, ul. Rakowiecka 36, 02-532 Warsaw, Poland
4 Ludwig Boltzmann Institute for Digital Health and Patient Safety, Medical University of Vienna, 1090 Vienna, Austria; atanas.atanasov@univie.ac.at
5 Department of Biotechnology and Nutrigenomics, Institute of Genetics and Animal Biotechnology of the Polish Academy of Sciences, ul. Postepu 36A, Jastrzębiec, 05-552 Magdalenka, Poland
6 Institute of Neurobiology, Bulgarian Academy of Sciences, Acad. G Bonchev Str. bl. 23, 1113 Sofia, Bulgaria
7 Department of Pharmacognosy, University of Vienna, Althanstrasse 14, 1090 Vienna, Austria
* Correspondence: swiergiel@yahoo.com (A.H.Ś.); s.sampino@igbzpan.pl (S.S.); Tel.: +48-50-393-4080 (A.H.Ś.); +48-22-736-7038 (S.S.)

Citation: Ziętek, M.; Barłowska, K.; Wijas, B.; Szablisty, E.; Atanasov, A.G.; Modliński, J.A.; Świergiel, A.H.; Sampino, S. Preconceptional Resveratrol Supplementation Partially Counteracts Age-Related Reproductive Complications in C57BL/6J Female Mice. *Molecules* **2021**, *26*, 1934. https://doi.org/10.3390/molecules26071934

Academic Editors: Simona Fabroni, Krystian Marszałek and Aldo Todaro

Received: 3 February 2021
Accepted: 25 March 2021
Published: 30 March 2021

Abstract: Aging is associated with a drastic decline in fertility/fecundity and with an increased risk of pregnancy complications. Resveratrol (RES), a natural polyphenolic compound, has shown anti-oxidant and anti-inflammatory activities in both human and animal models, thus representing a potential therapeutic and prophylactic anti-aging supplement. Here, we investigated whether preconceptional resveratrol supplementation improved reproductive outcomes in mid-aged (8-month-old) and old (12-month-old) C57BL/6J female mice. Female siblings were cohoused and assigned to either RES or vehicle supplementation to drinking water for 10 consecutive weeks. Subsequently, females were mated with non-supplemented males and their pregnancy outcomes were monitored. RES improved mating success in old, but not in mid-aged females, and prevented the occurrence of delivery complications in the latter. These results indicate that preconceptional RES supplementation could partially improve age-related reproductive complications, but it was not sufficient to restore fecundity in female mice at a very advanced age.

Keywords: aging; natural compounds; resveratrol; reproduction; mice

1. Introduction

The current trend toward postponing childbearing is increasing in all developed countries in both men and women, who prefer to pursue social and career achievements rather than establish a family [1]. In women, aging is accompanied by a physiologic decline of fertility leading to menopause. Nowadays, improved welfare and assisted reproductive technologies allow women at advanced reproductive age to conceive children naturally or artificially [2,3]. Nevertheless, epidemiological and animal studies have shown that advanced maternal age (AMA) is associated with an increased risk of pregnancy complications, as well as with negative health outcomes in the offspring, including miscarriages and abortions, multisystem malformations, and neurodevelopmental disorders [4–7].

Recently, growing interest has been devoted to natural compounds as anti-aging dietetic supplements, including the flavonoid, resveratrol (RES) [8]. RES is a polyphenol produced by a vast variety of plants, including grapes, peanuts, and a variety of berries, among others [9]. A growing number of studies have indicated that RES has anti-inflammatory, anti-tumorigenic, antioxidant, and anti-aging properties, and improves general health in mammals [9–14]. Previous studies have evaluated the safety, pharmacokinetics, and metabolism of RES, and have reported it to be well tolerated in humans, even at high doses [15]. RES may protect against age-associated infertility in mice [16,17] and can decrease inflammatory responses in the placenta and embryonic oxidative stress [18], as well as improve fetal weight in compromised pregnancies [19], with no evidence of teratogenesis [20]. There is limited knowledge about the eventual beneficial or adverse effects of preconceptional RES supplementation on female age-related fertility and pregnancy complications. Therefore, the aim of this study was to test whether 10 weeks preconceptional RES supplementation may rescue reproductive decline in aged female mice. Offspring and pregnancy outcomes were evaluated in 8- and 12-month-old female siblings supplemented with either RES or vehicle (ethanol). The younger cohort were left to term delivery, whereas, in the 1-year-old cohort, pregnancies were interrupted at 12.5 days post coitum (dpc) to examine fetal and placental development.

2. Results

First, we evaluated whether RES (or its vehicle ethanol) influenced the amount of liquid drunk by the mice. Although there was no effect of RES on liquid intake in the 8-month-old cohort, older females supplemented with RES drunk significantly more than their VEH-supplemented siblings at all timepoints analyzed (1st measurement, $p = 0.02$; 2nd measurement, $p = 0.0009$; 3rd measurement, $p = 0.02$; 4th measurement, $p = 0.002$; 5th measurement, $p = 0.0006$; with paired t-test). Treatment did not influence body weight during the supplementation period ($F(3,35) = 1.89$, $p = 0.1498$; with 2-way ANOVA). The cumulative percentage of females being successfully mated after 10 weeks of supplementation, thus presenting a vaginal plug after overnight male encountering, was similar between RES- and VEH-supplemented 8-month-old sibling females (Figure 1A).

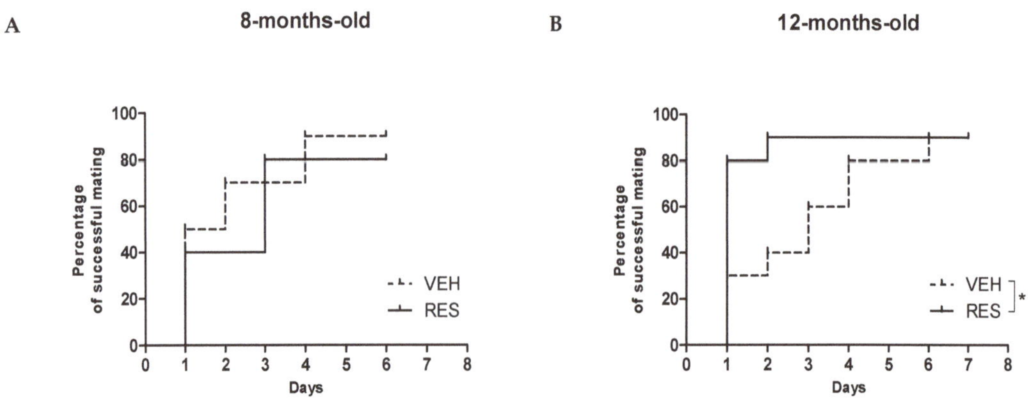

Figure 1. Percentage of females successfully mating (presence of the vaginal plug) over 7 days of daily encounters with males. Differences in time-to-mating between resveratrol (RES)- and vehicle (VEH)-supplemented females were analyzed with Gehan–Breslow–Wilcoxon survival test. No differences were observed in the 8-month-old cohort (**A**), whereas 12-month-old females (**B**) supplemented with RES mated faster than controls (* indicates $p = 0.035$). $n = 10$ females per experimental groups per age group.

Conversely, 12-month-old females supplemented with RES mated significantly faster than age-matched controls ($p = 0.035$ with Gehan–Breslow–Wilcoxon test) (Figure 1B). As showed in Table 1, no significant differences were observed in fertility rate, neither at 0.5 dpc, nor at 18.5 dpc. In the youngest cohort, 50% of control females showed delivery complications and died during labor, whereas all pregnant RES-supplemented mothers delivered within 22 dpc, with no complications ($p < 0.03$ with Chi square test, Figure 2A). Considering only the females delivering at least one pup, litter size was decreased in 8-month-old females supplemented with RES, as compared with VEH-supplemented controls, although this difference was not statistically significant (RES, 2.8 ± 1.4; VEH, 5.2 ± 2.8; $p = 0.342$, with paired t-test).

Table 1. Reproductive outcomes of 8-month-old C57BL/6J females supplemented with resveratrol and vehicle for 10 weeks. dpc: days post coitum.

	Evidence of Pregnancy at 0.5 dpc No. (%)	Evidence of Pregnancy at 18 dpc No. (%)	Litter Size at Delivery [b] *Mean ± SEM*	Offspring at 21 Days Postnatal No.
Resveratrol	9/10 (90%)	7/9 (78%)	2.8 ± 1.4	5
Vehicle	9/9 (100%) [a]	6/9 (67%)	5.2 ± 2.8	9

[a] One of the VEH-supplemented females died during the supplementation period. [b] Average litter size calculated considering females delivering at least one pup.

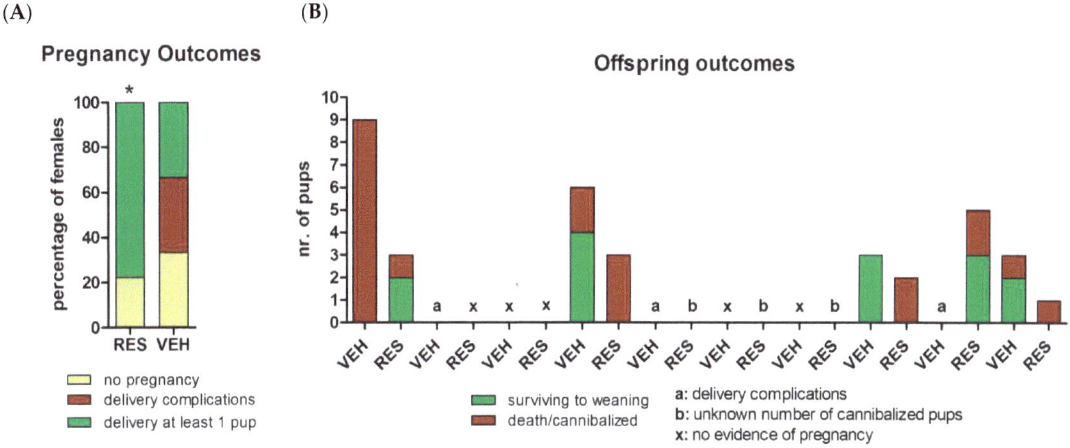

Figure 2. Pregnancy and offspring outcomes in 8-month-old RES- and VEH-supplemented female mice. (**A**) Percentage of pregnancies, delivery complications, and succsessful deliveries. Differences in delivery complications between RES- and VEH-supplemented females were analyzed with Chi square test (* indicates $p = 0.329$). (**B**) General view of the offspring outcomes observed in each female enrolled, including the number of pups surviving to weaning or death.

No statistically significant effects of treatment were observed in cannibalization events, with 57% of RES-supplemented females cannibalizing at least one of their pups, while only 33.3% VEH control females displayed such behavior. Considering only the females which did not cannibalized any of their pups, there was no effect of treatment on pups' survival to weaning, with a total of five pups obtained from RES-supplemented females, and nine pups from controls. Figure 2B depicts the delivery and offspring outcomes of each enrolled female.

In experiment 2, 12-month-old putative pregnant females were sacrificed 12.5 days after mating to collect conceptuses, however none of them were found in any of the RES- or VEH-supplemented females enrolled.

3. Discussion

Considering the increasing trends toward delaying childbearing, there is a growing interest in establishing prophylactic and therapeutic strategies to counteract age-related reproductive failure. RES has been indicated as an anti-aging dietetic supplement with potential beneficial effects on age-related reproductive decline [21,22]. The present study shows that female mice supplemented with RES over 10 consecutive weeks displayed slightly better reproductive performances compared to not-supplemented sibling controls. On one hand, RES significantly reduced the occurrence of delivery complications in mid-aged females, while on the other it was not sufficient to counteract age-related reproductive failure in the older females' cohort. Moreover, the effects of RES were also partial, in terms of the outcome considered, as there were not significant differences between groups concerning fertility and fecundity, whereas the occurrence of pregnancy complications was observed significantly more frequently in the control group compared to the RES-supplemented one. Moreover, RES was not sufficient to rescue fertility/fecundity in very old females under these experimental conditions.

Advanced maternal age is reported to be associated with a range of maternal illness (such as hypertension and gestational diabetes), as well as with several pregnancy complications including fetal growth restriction, preeclampsia, placental defects, pre-term birth, and stillbirth [4–7]. Age-related failure in female reproductive ability is not exclusive to humans, and it is also evident in other mammalian species, such as the rat [23], the mouse [24], and large domestic species [25,26]. Previous studies indicated that C57 female mice display a reproductive decline starting at 6–8 months, which results in decreased pregnancy rate, increased fetal resorption and diminished litter size [27]. Such effects of ageing were also observed by us in an outbred strain [28] and in inbred C57 and BTBR T+ Itpr3tf/J mice [29]. In the present study, one-year-old females supplemented with RES were mated faster than controls. Considering that co-housed female mice tend to synchronize their oestrous cycles [30], this result indicates that the siblings supplemented with RES may have been more prone to mate compared to controls, or that RES may increase female attractiveness toward male breeders. Moreover, a beneficial effect of RES was observed concerning the occurrence of delivery complications in the younger cohort, which were not observed in RES-supplemented females, but occurred at a high rate in controls. Conversely, although 12-month-old RES-supplemented females had an increased chance of being mated, they were still not able to generate detectable offspring at 12.5 dpc. This suggests that a 10 weeks preconceptional supplementation may not be sufficient for RES to rescue all aspects of age-related reproductive decline, or that at an older age reproductive functions may be at a not-rescuable point. Overall, the present study retains some limitations related to the small sample size employed, combined with the low fertility of aged animals, which do not allow coming to consistent conclusions concerning the offspring outcomes. Nonetheless, the experiments were designed to control genetic and environmental variability among supplemented subjects, since pairs of sibling females were cohoused and assigned to either RES or VEH treatment. Therefore, although further animal and human studies will be needed, the present results suggest a beneficial effect of RES on age-related delivery complications.

4. Materials and Methods

4.1. Mice

C57BL/6J (C57) mice used in this study were purchased from Jackson Laboratory (Bar Harbor, ME, USA) and maintained at the Institute of Genetics and Animal Biotechnology PAS (Jastrzębiec, Poland) for the last 4 years, following a sibling mating breeding scheme. Mice were housed under a 12 h light-dark cycle, 19–23 °C, and 40–68% humidity, with standard pellet food and a source of water ad libitum. Sister females were co-housed in $26.5 \times 20.7 \times 14.0$ cm^3 cages, divided by a transparent and perforated PVC (polyvinyl) partition to overcome isolation-induced social stress. Bedding materials and environmental enrichment (including a nest) were provided to each subject on each side of the partition, together with food and one drinking bottle (two in total: one containing resveratrol (RES) and the other containing 0.9% ethanol (VEH), assigned to either sister respectively). This strategy allowed minimizing genetic and environmental confounding factors between the RES and VEH experimental groups. All experiments were conducted with the approval of the local institutional committee for animal welfare, and in line with Polish and European regulations.

4.2. Experimental Design

Two experiments were performed, each including 20 female mice ($n = 10$ females per treatment in each age cohort). In experiment 1, mice aged 22 weeks, were supplemented with RES or vehicle (VEH: ethanol) for 10 weeks, then mated and allowed to deliver spontaneously. In experiment 2, mice aged 38 weeks were supplemented in the same way, mated, and sacrificed 12.5 days after mating in order to collect conceptuses. During the first seven days of the experiments, mice were habituated to the new environment and to the 50 mL bottles. For the subsequent seven days, mice were habituated to 0.9% solution of ethanol (VEH) in drinking water, before the beginning of the supplementation period. The amount of liquid consumed by each mouse was measured every day to accommodate RES doses.

4.3. RES Supplementation to Drinking Water

RES (trans-Resveratrol; catalogue number 72862-72864, Stemcell Technologies Inc., Vancouver, Canada) was first dissolved in ethanol (50 mg/mL) and then diluted in drinking tap water (0.1 mg/mL RES – 0.9% *v/v* ethanol) and provided to mice based on a water intake measurement taken during the habituation periods, to obtain an average 16 mg/kg/day intake of RES. During the supplementation periods volumes of RES or VEH solutions consumed by each mouse were scored every 3–4 days, when mice received a fresh supply. Dosage of RES was selected based on previous studies [31].

4.4. Breeding

After 10 weeks of supplementation, the females were marked by ear cut and the cage separator was removed. Females were then introduced overnight to two 2–4-month-old stud males of the same strain, daily, for 6 consecutive days. The presence of a vaginal plug was checked each morning after overnight encountering and considered as day 0.5 dpc. Pregnancy at 18.5 dpc was scored by the presence of an evident belly swell. Gestational age >22 dpc and unsuccessful labor were considered as delivery complications.

4.5. Outcomes and Statistics

Liquid intake was measured at five timepoints during the supplementation period, and differences between groups were analyzed by paired t-test. The effects of treatment and time on body weight before and after supplementation were analyzed by 2-way ANOVA. For assessment of time-to-mating success at 0.5 dpc, Gehan–Breslow–Wilcoxon survival tests was employed to compare experimental groups. Pregnancies at 0.5 and at 18.5 dpc are expressed as percentage of females and were analyzed by Chi square test. Delivery was classified as complicated when the female was not able to deliver the pups and died soon after. Delivery complications are expressed in percentage of females and analyses by Chi square test. Differences in litter size were analyzed by paired t-test between sister pairs. Differences in postnatal outcomes, such as pups' survival and cannibalization, were expressed as percentage and analyzed by Fisher exact test. $p < 0.05$ was considered as statistically significant. All analyses were performed using GraphPad Prism version 5.0 (GraphPad Software Inc., San Diego, CA, USA) and R (available online: https://www.R-project.org/, accessed on 10 November 2020).

5. Conclusions

RES is a promising therapeutic/prophylactic compound in gerontology and reproductive medicine. The present study indicates that preconceptional RES supplementation could partially improve age-related reproductive complications, but it is not sufficient to rescue fecundity in female mice at a very advanced age. Long-term RES supplementation is known to improve ovarian function, increasing the ovarian follicular reserve and extending the ovarian life span in rats [32], as well as to improve the number of follicles in aged mice ovaries [33]. However, RES supplementation may also induce adverse effects on implantation and decidualization in both humans and animals [34]. Therefore, more research is needed to establish optimal doses and periods of RES supplementation to obtain a beneficial impact on reproduction, while preventing adverse effects on implantation and subsequent pregnancy.

Author Contributions: M.Z. designed the study, conducted the experiments, acquired and analyzed the data, and drafted the article; K.B., B.W. and E.S. conducted the experiments; A.G.A., J.A.M. and A.H.Ś. critically revised the manuscript for important intellectual content; S.S. conceived and designed the study, analyzed and interpreted the data, and prepared the final manuscript. All authors have read and agreed to the published version of the manuscript.

Funding: This work was supported by the Polish National Science Centre under grant number 2014/15/D/NZ4/04274, to S.S.

Institutional Review Board Statement: The study was conducted according to the guidelines of the Declaration of Helsinki, complies with the current Polish laws and European directives, and was approved by the Institutional Review Board of the Institute of Genetics and Animal Biotechnology, PAN.

Informed Consent Statement: Not applicable.

Data Availability Statement: The authors confirm that the data supporting the findings of this study are available within the article, and that raw data are available from the corresponding authors on request.

Conflicts of Interest: The authors report no conflict of interest.

Sample Availability: Samples of this work are available from the authors.

References

1. Hamilton, B.E.; Martin, J.A.; Osterman, M.J.K.; Curtin, S.C.; Matthews, T.J. Births: Final data for 2014. *Natl. Vital Stat. Rep.* **2015**, *64*, 1–64. [PubMed]
2. Haebe, J.; Martin, J.; Tekepety, F.; Tummon, I.; Shepherd, K. Success of intrauterine insemination in women aged 40–42 years. *Fertil. Steril.* **2002**, *78*, 29–33. [CrossRef]
3. Dal Prato, L.; Borini, A.; Cattoli, M.; Preti, M.S.; Serrao, L.; Flamigni, C. Live birth after IVF in a 46-year-old woman. *Reprod. Biomed. Online* **2005**, *1*, 452–454. [CrossRef]
4. Nassar, A.; Usta, I. Advanced maternal age. Part II: Long-term consequences. *Am. J. Perinatol.* **2009**, *26*, 107–112. [CrossRef]
5. Sauer, M.V. Reproduction at an advanced maternal age and maternal health. *Fertil. Steril.* **2015**, *103*, 1136–1143. [CrossRef] [PubMed]
6. Tearne, J.E. Older maternal age and child behavioral and cognitive outcomes: A review of the literature. *Fertil. Steril.* **2015**, *103*, 1381–1391. [CrossRef] [PubMed]
7. Lean, S.C.; Derricott, H.; Jones, R.L.; Heazell, A.E.P. Advanced maternal age and adverse pregnancy outcomes: A systematic review and meta-analysis. *PLoS ONE* **2017**, *12*, e0186287. [CrossRef]
8. McCubrey, J.A.; Lertpiriyapong, K.; Steelman, L.S.; Abrams, S.L.; Yang, L.V.; Murata, R.M.; Rosalen, P.L.; Scalisi, A.; Neri, L.M.; Cocco, L.; et al. Effects of resveratrol, curcumin, berberine and other nutraceuticals on aging, cancer development, cancer stem cells and microRNAs. *Aging* **2017**, *9*, 1477–1536. [CrossRef]
9. Smoliga, J.M.; Baur, J.A.; Hausenblas, H.A. Resveratrol and health–A comprehensive review of human clinical trials. *Mol. Nutr. Food Res.* **2011**, *55*, 1129–1141. [CrossRef]
10. Wung, B.S.; Hsu, M.C.; Wu, C.C.; Hsieh, C.W. Resveratrol suppresses IL-6-induced ICAM-1 gene expression in endothelial cells: Effects on the inhibition of STAT3 phosphorylation. *Life Sci.* **2005**, *78*, 389–397. [CrossRef]
11. Csiszar, A.; Smith, K.; Labinskyy, N.; Orosz, Z.; Rivera, A.; Ungvari, Z. Resveratrol attenuates TNF-α-induced activation of coronary arterial endothelial cells: Role of NF-κB inhibition. *Am. J. Physiol. Heart Circ. Physiol.* **2006**, *291*, H1694–H1699. [CrossRef]
12. Baur, J.A.; Pearson, K.J.; Price, N.L.; Jamieson, H.A.; Lerin, C.; Kalra, A.; Prabhu, V.V.; Allard, J.S.; Lopez-Lluch, G.; Lewis, K.; et al. Resveratrol improves health and survival of mice on a high-calorie diet. *Nature* **2006**, *444*, 337–342. [CrossRef]
13. Breuss, J.; Atanasov, A.; Uhrin, P. Resveratrol and its effects on the vascular system. *Int. J. Mol. Sci.* **2019**, *20*, 1523. [CrossRef]
14. Yeung, A.W.K.; Bhushan Aggarwal, B.; Erdogan Orhan, I.; Horbańczuk, O.K.; Barreca, D.; Battino, M.; Belwal, T.; Bishayee, A.; Daglia, M.; Devkota, H.A.; et al. Resveratrol, a popular dietary supplement for human and animal health: Quantitative research literature analysis—A review. *Anim. Sci. Pap. Rep.* **2019**, *37*, 103–118.

15. Patel, K.R.; Scott, E.; Brown, V.A.; Gescher, A.J.; Steward, W.P.; Brown, K. Clinical trials of resveratrol: Clinical trials. *Ann. N. Y. Acad. Sci.* **2011**, *1215*, 161–169. [CrossRef]
16. Barger, J.L.; Kayo, T.; Vann, J.M.; Arias, E.B.; Wang, J.; Hacker, T.A.; Wang, Y.; Raederstorff, D.; Morrow, J.D.; Leeuwenburgh, C.; et al. A low dose of dietary resveratrol partially mimics caloric restriction and retards aging parameters in mice. *PLoS ONE* **2008**, *3*, e2264. [CrossRef]
17. Pearson, K.J.; Baur, J.A.; Lewis, K.N.; Peshkin, L.; Price, N.L.; Labinskyy, N.; Swindell, W.R.; Kamara, D.; Minor, R.K.; Perez, E.; et al. Resveratrol delays age-related deterioration and mimics transcriptional aspects of dietary restriction without extending life span. *Cell Metab.* **2008**, *8*, 157–168. [CrossRef] [PubMed]
18. Zheng, S.; Feng, Q.; Cheng, J.; Zheng, J. Maternal resveratrol consumption and its programming effects on metabolic health in offspring mechanisms and potential implications. *Biosci. Rep.* **2018**, *38*, BSR20171741. [CrossRef]
19. Singh, C.K.; Kumar, A.; Hitchcock, D.B.; Fan, D.; Goodwin, R.; LaVoie, H.A.; Nagarkatti, P.; DiPette, D.J.; Singh, U.S. Resveratrol prevents embryonic oxidative stress and apoptosis associated with diabetic embryopathy and improves glucose and lipid profile of diabetic dam. *Mol. Nutr. Food Res.* **2011**, *55*, 1186–1196. [CrossRef]
20. Williams, L.D.; Burdock, G.A.; Edwards, J.A.; Beck, M.; Bausch, J. Safety studies conducted on high-purity trans-resveratrol in experimental animals. *Food Chem. Toxicol.* **2009**, *47*, 2170–2182. [CrossRef]
21. Bhullar, K.S.; Hubbard, B.P. Lifespan and healthspan extension by resveratrol. *Biochim. Biophys. Acta* **2015**, *1852*, 1209–1218. [CrossRef]
22. Pasquariello, R.; Verdile, N.; Brevini, T.A.L.; Gandolfi, F.; Boiti, C.; Zerani, M.; Maranesi, M. The role of resveratrol in mammalian reproduction. *Molecules* **2020**, *25*, 4554. [CrossRef]
23. Day, J.R.; La Polt, P.S.; Morales, T.H.; Lu, J.K. An abnormal pattern of embryonic development during early pregnancy in aging rats. *Biol. Reprod.* **1989**, *41*, 933–939. [CrossRef]
24. Finn, C.A. Reproductive capacity and litter size in mice: Effect of age and environment. *Reproduction* **1963**, *6*, 205–214. [CrossRef]
25. Scoggin, C.F. Not just a number: Effect of age on fertility, pregnancy and offspring vigour in thoroughbred brood-mares. *Reprod. Fertil. Dev.* **2015**, *27*, 872. [CrossRef]
26. Shorten, P.R.; Morris, C.A.; Cullen, N.G. The effects of age, weight, and sire on pregnancy rate in cattle. *J. Anim. Sci.* **2015**, *93*, 1535–1545. [CrossRef]
27. Woods, L.; Perez-Garcia, V.; Kieckbusch, J.; Wang, X.; DeMayo, F.; Colucci, F.; Hemberger, M. Decidualisation and placentation defects are a major cause of age-related reproductive decline. *Nat. Commun.* **2017**, *8*. [CrossRef]
28. Sampino, S.; Stankiewicz, A.M.; Zacchini, F.; Goscik, J.; Szostak, A.; Swiergiel, A.H.; Drago, G.; Modlinski, J.A.; Ptak, G.E. Pregnancy at advanced maternal age affects behavior and hippocampal gene expression in mouse offspring. *J. Gerontol. A Biol. Sci. Med. Sci.* **2017**, *72*, 1465–1473. [CrossRef]
29. Ziętek, M.; Sampino, S. Pregnancy-related gene-environment interactions modulate the risk for autism-like behavioral disorders in mouse offspring. Unpublished (manuscript currently in preparation).
30. Meziane, H.; Ouagazzal, A.-M.; Aubert, L.; Wietrzych, M.; Krezel, W. Estrous cycle effects on behavior of C57BL/6J and BALB/cByJ female mice: Implications for phenotyping strategies. *Genes Brain Behav.* **2007**, *6*, 192–200. [CrossRef]
31. Zordoky, B.N.M.; Robertson, I.M.; Dyck, J.R.B. Preclinical and clinical evidence for the role of resveratrol in the treatment of cardiovascular diseases. *Biochim. Biophys. Acta* **2015**, *1852*, 1155–1177. [CrossRef]
32. Ortega, I.; Duleba, A.J. Ovarian actions of resveratrol: Ovarian actions of resveratrol. *Ann. N. Y. Acad. Sci.* **2015**, *1348*, 86–96. [CrossRef]
33. Liu, M.; Yin, Y.; Ye, X.; Zeng, M.; Zhao, Q.; Keefe, D.L.; Liu, L. Resveratrol protects against age-associated infertility in mice. *Hum. Reprod.* **2013**, *28*, 707–717. [CrossRef] [PubMed]
34. Ochiai, A.; Kuroda, K.; Ozaki, R.; Ikemoto, Y.; Murakami, K.; Muter, J.; Matsumoto, A.; Itakura, A.; Brosens, J.J.; Takeda, S. Resveratrol inhibits decidualization by accelerating downregulation of the CRABP2-RAR pathway in differentiating human endometrial stromal cells. *Cell Death Dis.* **2019**, *10*, 276. [CrossRef]

molecules

Article

The Ameliorative Effects of Fucoidan in Thioacetaide-Induced Liver Injury in Mice

Ming-Yang Tsai [1,2], Wei-Cheng Yang [3], Chuen-Fu Lin [4], Chao-Min Wang [5], Hsien-Yueh Liu [6], Chen-Si Lin [3], Jen-Wei Lin [6], Wei-Li Lin [6,7], Tzu-Chun Lin [5], Pei-Shan Fan [5], Kuo-Hsiang Hung [2,*], Yu-Wen Lu [8,9,*] and Geng-Ruei Chang [5,*]

1 Animal Industry Division, Livestock Research Institute, Council of Agriculture, Executive Yuan, 112 Muchang, Xinhua Dist, Tainan 71246, Taiwan; mytsai@mail.tlri.gov.tw
2 Graduate Institute of Bioresources, National Pingtung University of Science and Technology, 1 Shuefu Road, Neipu, Pingtung 91201, Taiwan
3 School of Veterinary Medicine, National Taiwan University, 4 Section, 1 Roosevelt Road, Taipei 10617, Taiwan; yangweicheng@ntu.edu.tw (W.-C.Y.); cslin100@ntu.edu.tw (C.-S.L.)
4 Department of Veterinary Medicine, College of Veterinary Medicine, National Pingtung University of Science and Technology, Shuefu Road, Neipu, Pingtung 912301, Taiwan; cflin2283@mail.npust.edu.tw
5 Department of Veterinary Medicine, National Chiayi University, 580 Xinmin Road, Chiayi 60054, Taiwan; leowang@mail.ncyu.edu.tw (C.-M.W.); lin890090@gmail.com (T.-C.L.); babybelle349@gmail.com (P.-S.F.)
6 Bachelor Degree Program in Animal Healthcare, Hungkuang University, 6 Section, 1018 Taiwan Boulevard, Shalu District, Taichung 433304, Taiwan; lhy_vet@hk.edu.tw (H.-Y.L.); jenweilin@hk.edu.tw (J.-W.L.); ivorylily99@gmail.com (W.-L.L.)
7 General Education Center, Chaoyang University of Technology, 168 Jifeng Eastern Road, Taichung 413310, Taiwan
8 Department of Chinese Medicine, Show Chwan Memorial Hospital, 1 Section, 542 Chung-Shan Road, Changhua 50008, Taiwan
9 Department of Chinese Medicine, Chang Bing Show Chwan Memorial Hospital, 6 Lugong Road, Changhua 50544, Taiwan
* Correspondence: khhung424@mail.npust.edu.tw (K.-H.H.); luffy1964@gmail.com (Y.-W.L.); grchang@mail.ncyu.edu.tw (G.-R.C.)

Citation: Tsai, M.-Y.; Yang, W.-C.; Lin, C.-F.; Wang, C.-M.; Liu, H.-Y.; Lin, C.-S.; Lin, J.-W.; Lin, W.-L.; Lin, T.-C.; Fan, P.-S.; et al. The Ameliorative Effects of Fucoidan in Thioacetaide-Induced Liver Injury in Mice. *Molecules* **2021**, *26*, 1937. https://doi.org/10.3390/molecules26071937

Academic Editor: Simona Fabroni

Received: 27 February 2021
Accepted: 23 March 2021
Published: 30 March 2021

Abstract: Liver disorders have been recognized as one major health concern. Fucoidan, a sulfated polysaccharide extracted from the brown seaweed Fucus serratus, has previously been reported as an anti-inflammatory and antioxidant. However, the discovery and validation of its hepatoprotective properties and elucidation of its mechanisms of action are still unknown. The objective of the current study was to investigate the effect and possible modes of action of a treatment of fucoidan against thioacetamide (TAA)-induced liver injury in male C57BL/6 mice by serum biochemical and histological analyses. The mouse model for liver damage was developed by the administration of TAA thrice a week for six weeks. The mice with TAA-induced liver injury were orally administered fucoidan once a day for 42 days. The treated mice showed significantly higher body weights; food intakes; hepatic antioxidative enzymes (catalase, glutathione peroxidase (GPx), and superoxide dismutase (SOD)); and a lower serum alanine aminotransferase (ALT), aspartate aminotransferase (AST), tumor necrosis factor-α (TNF-α), interleukin-1β (IL-1β), and C-reactive protein (CRP) levels. Additionally, a reduced hepatic IL-6 level and a decreased expression of inflammatory-related genes, such as cyclooxygenase-2 (COX-2), and inducible nitric oxide synthase (iNOS) mRNA was observed. These results demonstrated that fucoidan had a hepatoprotective effect on liver injury through the suppression of the inflammatory responses and acting as an antioxidant. In addition, here, we validated the use of fucoidan against liver disorders with supporting molecular data.

Keywords: fucoidan; inflammation; liver; mice; thioacetamide

1. Introduction

Liver is a vital organ that coordinates several major functions to maintain the homeostasis of the body. It plays the crucial role in the metabolism of sugars, proteins, lipids, vitamins, and hormones. The liver is also an important detoxification organ. When substances are absorbed into the body, blood can flow directly from the surface of the small intestine to the hepatic portal vein. Since the liver's cell membrane presents an optimal permeability and contacts directly with the blood [1], toxic substances brought by the hepatic portal vein can be eliminated after being metabolized by the liver [2]. In addition, inflammation and wound healing are interrelated processes, since inflammatory signals stimulate immune cells toward injury sites [3]. The repair and regeneration of the injured tissues is caused by apoptosis and regenerative cellular processes [4]. Additionally, the accumulation extracellular matrix and scarring tissues are the endpoints of inflammation. However, in the case of chronic injury, the wound-healing process becomes maladaptive, causing the loss of functional hepatic parenchyma. This can further develop into liver fibrosis, which may be a percussor for hepatic cirrhosis and liver cancer [5,6]. Possible causes of abnormal liver function include viral infections, alcohol intake, autoimmunity, genetics, hepatobiliary tumors or infections, chemical toxins, and drugs. Around 50% of liver damage is caused by drugs [7].

Hepatotoxic molecules have the ability to react with cell components and induce different types of lesions in the liver. The chemical toxins such as acetaminophen, carbon tetrachloride (CCl4), galactosamine, and thioacetamide have been efficiently used in the past in vitro, as well as in vivo, to create models of experimental hepatocyte injury [8,9]. The process of hepatotoxicity caused by thioacetamide (TAA) is recognized to be similar to that in humans [10]. TAA-induced hepatic inflammation, fibrosis, and liver damage in mice, in terms of clinical features, are very close to chronic human liver disease [11]. After TAA exposure, the levels of plasma alanine aminotransferase (ALT) and aspartate aminotransferase (AST) are upregulated due to hepatocyte membrane damage, which causes a leakage of transaminase enzymes into the systemic circulation [12]. The advantage of using the TAA model for the hepatotoxin animal model is its unique specificity for targeting the liver [12,13]. The toxicity is because of its bioactivation led by a mixed-function oxidase system—mainly, the cytochrome P2E1 and flavin adenine dinucleotide monooxygenases [13,14]. Furthermore, the activation of TAA causes the formation of reactive metabolites such as thioacetamide-S-oxide radicals and reactive oxygen species (ROS) intermediates [13]. Hence, the biochemical processing of TAA within the cellular milieu causes liver oxidative damage, which can further degenerate and cause necrosis in liver tissues [15].

Fucoidan belongs to a class of fucose-rich sulfated carbohydrates usually present in a brown, marine algae, echinoderms [16], and seagrasses [17]. Fucoidan-rich brown algae is commercially available as a dietary supplement or nutraceutical. With the advantages of having low toxicity and high oral bioavailability, it has been studied for its potential pharmacological effects. Pharmacologically, fucoidan mediates the pathogenesis of numerous cellular processes, such as inflammation, carcinogenesis, and oxidative damage [18,19]. Fucoidan has also been reported to cause cytotoxicity and induce apoptosis in various cancer cells [20], including MCF-7 breast cancer cells, in vitro [21] and in Lewis lung adenocarcinoma cells transplanted into mice [22]. In addition, pronounced antitumor and antimetastatic effects of fucoidan were observed in mice with transplanted lung adenocarcinoma [23]. T-cell-mediated natural killer (NK) cells were also efficiently upregulated in mice treated with fucoidan, and NK cell activation was found to be associated with an increased production of interferon (IFN)-γ and interleukin (IL)-12 by splenic T cells [24]. Fucoidan has also been observed to suppress angiogenic activity while downregulating the vascular endothelial growth factor (VEGF) receptor expression and inhibiting the proliferation of VEGF-induced human umbilical vein endothelial cells [25]. Additional studies have reported the inhibitory activity of fucoidan against matrix metalloproteinases and nuclear factor-kappa β (NF-$\kappa\beta$), preventing metastasis in tumor mice models [26]. Overall, fucoidan is considered to provide anticancer immunity via the activation of an immune cell influx and by upregulating the production of anticancer cytokines.

In other reports, concanavalin A-induced liver injury and the concomitant increase of plasma tumor necrosis factor-α (TNF-α) and IFN-γ levels were inhibited by fucoidan; however, the plasma IL-10 levels were enhanced [27]. It has also been proven that fucoidan has a protective effect against liver ischemia injury via suppression of the inflammatory activation pathway, several inflammatory mediators, and cell infiltration in the inflamed sites [28]. The mechanisms of action of fucoidan against liver injury caused by TAA in male C57BL/6J mice are not completely understood. Therefore, in this study, we analyzed the hepatic function levels, inflammation cytokines, histopathology, and antioxidative enzymes in mice with liver injuries undergoing treatment with or without fucoidan to better understand the mode of action of fucoidan.

2. Results

2.1. Fucoidan Affected Body Weight and Food Intake in Mice with TAA-Induced Liver Damage

Significant differences in body weight, daily food intake, daily body weight gain, and daily food efficiency were observed between the control (saline-treated TAA-induced liver injury mice) and fucoidan group (fucoidan-treated TAA-induced liver injury mice) (Figure 1). TAA was fed to male C57BL6/J mice for eight weeks to induce hepatotoxicity. After six weeks of fucoidan treatment, the fucoidan-treated mice exhibited greater body weights, food intakes, weekly body weight gains, and daily food efficiency than the controls. The body weights and food intakes after six weeks, for all experimental groups, are represented in Figure 1a,b, respectively. Mice with TAA-induced liver damage treated with fucoidan presented 1.07-fold higher body weights than the control mice; this difference was significant (Figure 1a). Moreover, in comparison to the control mice, it was observed that the food intake of the treated mice increased by 1.28-fold (Figure 1b). The fucoidan-treated mice presented significantly higher weekly body weight gains by 1.61-fold (Figure 1c) than that of the control mice. In addition, these changes were demonstrated, along with the significant changes in the daily food efficiency in the group treated with curcumin, as 1.27-fold higher than the control mice (Figure 1d). However, the body weight, daily food intake, daily gain in body weight, and daily efficiency in food intake were all significantly lower in the fucoidan-treated group than in the control group.

2.2. Fucoidan Affected Serum ALT, AST, and Liver Weight in Mice with TAA-Induced Liver Damage

There were significant differences in the serum ALT, AST, and liver weights among the three groups. In fucoidan-treated mice, the serum ALT and AST were downregulated by 23% (Figure 2a) and by 45% (Figure 2b), respectively. This represents a significant downregulation of the hepatic function markers when in comparison to the TAA-induced liver injury control mice. In addition, the fucoidan-treated group exhibited a 27% elevation in liver weights (Figure 2c) and a 21% elevation in body weight-normalized liver weights (Figure 2d). In the normal group, the serum ALT and AST, liver weights, and the relative liver/body weight ratios were downregulated when compared to those in fucoidan-treated mice.

2.3. Fucoidan Affected the Serum Alkaline Phosphatase (ALP), Bilirubin, Globulin, and γ-Glutamyl Transferase (γ-GT)

Additionally, we also determined the effects of fucoidan on the serum enzymes, ALP, bilirubin, globulin, and γ-GT levels in the TAA-induced hepatotoxicity mice models. The results are shown in Figure 3. The levels of ALP, bilirubin, globulin, and γ-GT of the TAA-treated group after an intraperitoneal (ip) injection dose of 100 mg/kg were significantly elevated after six weeks of treatment when compared to the control mice. The increased levels of these biochemical biomarkers match with the hepatic cell damage [29]. In the fucoidan-treated mice, the levels of serum ALP (15%), bilirubin (12%), globulin (21%), and γ-GT (20%) were significantly decreased, demonstrating the overall successful optimization of the hepatic injuries. Overall, TAA mice treated with fucoidan showed a significant downregulation of ALT, AST, and ALP in the serum, confirming the role of fucoidan in the recovery of the liver function. However, these biomarker levels in fucoidan-treated mice did not decrease to the normal baseline observed in healthy mice.

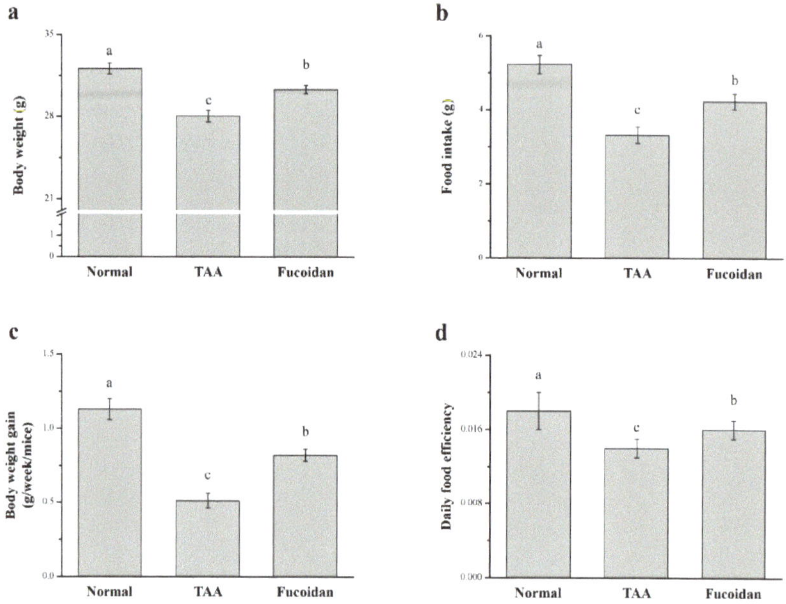

Figure 1. Changes of the (**a**) body weight, (**b**) food intake, (**c**) weekly body weight gain, and (**d**) daily food efficiency in normal mice, mice with thioacetamide (TAA)-induced liver injury, and fucoidan-treated mice with TAA-induced liver injury within a time period of 42 days. All data are presented as the means ± standard error (SEMs) (n = 10). [a–c] Data with different letters in the columns are significantly different with one-way analysis of variance (ANOVA), and the means of different groups were compared by Duncan's test at $p < 0.05$.

2.4. *Fucoidan Affected Liver Damage*

The histopathological changes, observed with hematoxylin and eosin (H&E) staining for all the mice groups, are shown in Figure 4. In the histological profile, the normal control group presented regular hepatocytes with well-preserved cytoplasms, prominent nuclei, and nucleoles. No sign of inflammation, fatty change, or necrosis was noted. However, in TAA-treated mice, the liver tissue showed extensive injuries, with a multifocal area of the necrosis of hepatocytes, cell swelling, disruption of the membranes, and contraction of the nucleus. Fucoidan-treated mice presented markedly diminished histological alterations (Figure 4a). The regular hepatocyte structure was observed with much reduced ballooning and tissue degeneration. The necrotic activity was also decreased in comparison to the TAA-treated group. Additionally, the liver damage score, using Suzuki scoring (0–4), decreased markedly for fucoidan-treated mice (Figure 4b) [30,31].

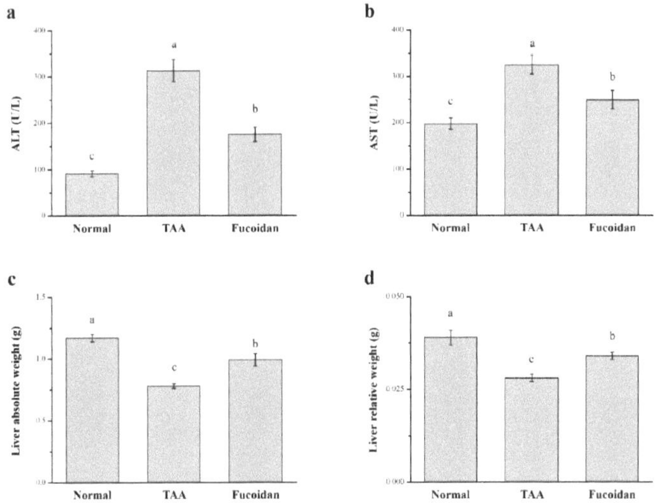

Figure 2. Changes in the (**a**) serum alanine aminotransferase (ALT) levels, (**b**) serum aspartate aminotransferase (AST) levels, (**c**) absolute weights, and (**d**) body weight-normalized weights (%) of the livers in normal mice, mice with TAA-induced liver injury, and fucoidan-treated mice with TAA-induced liver injury over a period of 42 days. All data are presented as the means ± SEMs (n = 10). [a–c] Data with different letters in the columns are significantly different with one-way ANOVA, and the means of different groups were compared by Duncan's test at $p < 0.05$.

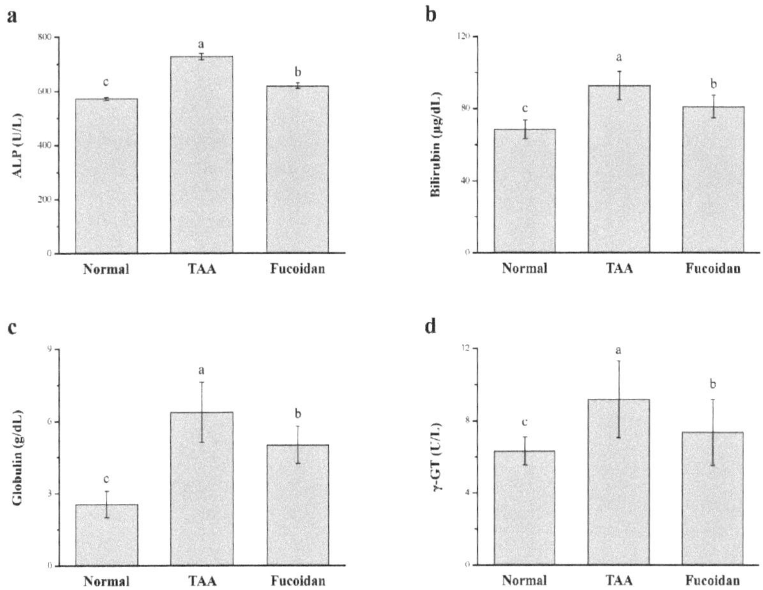

Figure 3. Changes of the serum (**a**) alkaline phosphatase (ALP), (**b**) bilirubin, (**c**) globulin, and (**d**) γ-glutamyl transferase (γ-GT) levels in normal mice, in mice with TAA-induced liver injury, and fucoidan-treated mice with TAA-induced liver injury over a period of 42 days. All data are presented as the means ± SEMs (n = 10). [a–c] Data with different letters in the columns are significantly different with one-way ANOVA, and the means of the different groups were compared by Duncan's test at $p < 0.05$.

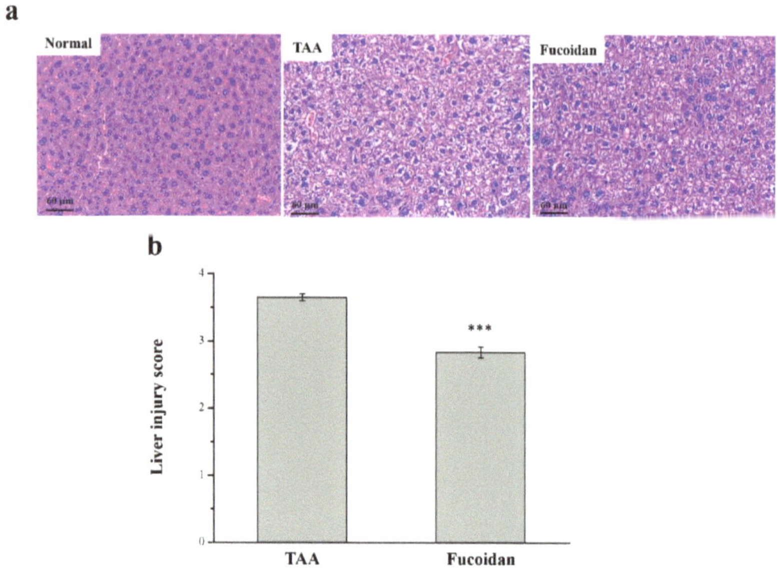

Figure 4. Changes of the (**a**) liver histology by hematoxylin and eosin (H&E) staining (magnification 200×), and (**b**) the scoring index of liver injuries in normal mice, mice with TAA-induced liver injury, and fucoidan-treated mice with TAA-induced liver injury over a period of 42 days. All data are presented as the means ± SEMs (n = 10). Statistical significance at *** $p < 0.001$.

2.5. Fucoidan Affected Tumor Necrosis Factor-α(TNF-α), Interleukin-1β (IL-1β), Fibroblast Growth Factor-21 (FGF21), C-Reactive Protein (CRP), and Cytokine Levels in the Serum

The levels of serum inflammatory cytokines were shown to be significantly altered among the three groups analyzed (Figure 5). The serum TNF-αlevels were 18% lower in the fucoidan treatment group than that of the TAA-induced liver damage group (Figure 5a). Compared to the TAA group, the serum IL-1β levels were 39% lower in the fucoidan-treated TAA-induced liver damage group (Figure 5b). Additionally, we also analyzed serum FGF21, a regulator of hepatic metabolic pathways, to improve the steatosis and inflammation [32]. The serum FGF21 levels in fucoidan-treated-mice were 1.18-fold higher than in TAA-induced liver-damaged mice (Figure 5c). In contrast, the serum CRP levels of fucoidan-treated mice were 25% lower than in the mice with TAA-induced liver damage (Figure 5d). Compared to normal mice, the liver damage model treated with fucoidan revealed a 1.25-fold upregulation in the TNF-αlevels, 1.50-fold upregulation in IL-1β, and 47% decrease in FGF21, with a 1.87-fold increase in the serum CRP levels.

2.6. Fucoidan Affected Interleukin-6 (IL-6), Patatin-Like Phospholipid Domain Containing Protein 3 (PNPLA3), the mRNA of Liver Fatty Acid-Binding Protein (L-FABP), Cyclooxygenase-2 (COX-2), and Inducible Nitric Oxide Synthase (iNOS)

IL-6 is known to induce hepatic inflammatory cell infiltration [33], while PNPLA3 represents a modifier of progression of hepatocellular injury and liver-associated disorders such as steatohepatitis, chronic hepatitis, and hepatocellular carcinoma by promoting the release of inflammatory cytokines [34,35]. Here, we found that the expression of IL-6 and PNPLA3 were significantly different between the three analyzed groups by using a Western blotting analysis (Figure 6a). The expression of hepatic IL-6 was 39% lower in fucoidan-treated mice than the mice with TAA-induced liver injury (Figure 6b). Compared to the TAA group, the hepatic PNPLA3 expression was 36% lower in fucoidan-treated mice (Figure 6c). Contrarily, the mRNA levels of L-FABP in the liver were 31% lower in the latter group (Figure 6d). Moreover, the hepatic mRNA levels of COX-2 (Figure 6e) and iNOS

(Figure 6f) were 29% and 17% lower, respectively, in mice under fucoidan treatment than that of mice with TAA-induced liver damage.

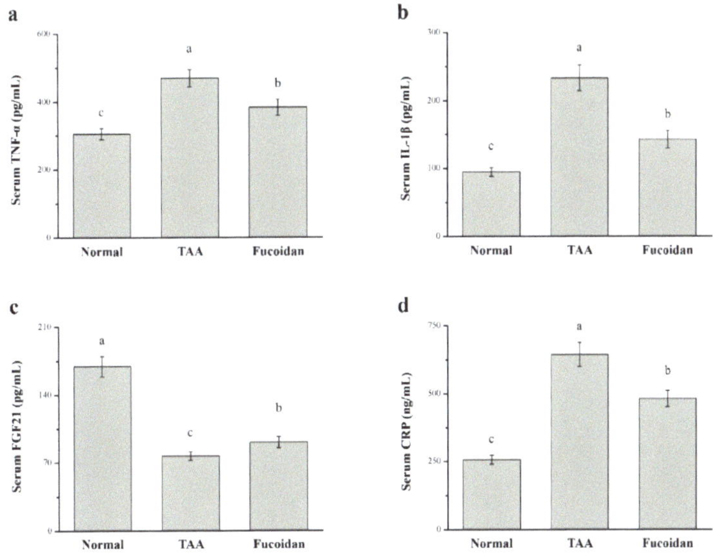

Figure 5. Changes of the serum (**a**) tumor necrosis factor-α (TNF-α), (**b**) interleukin-1β (IL-1β), (**c**) fibroblast growth factor-21 (FGF21), and (**d**) C-reactive protein (CRP) levels in normal mice, mice with TAA-induced liver injury, and fucoidan-treated mice with TAA-induced liver injury after 42 days. All data are presented as the means ± SEMs (n = 10). [a–c] Data with different letters in the columns are significantly different with one-way ANOVA, and the means of different groups were compared by Duncan's test at $p < 0.05$.

In contrast, the hepatic IL-6 and PNPLA3 expression of fucoidan-treated TAA-induced liver-damaged mice were 1.85-fold and 3.78-fold higher, respectively, compared to normal mice treated with saline. Additionally, the hepatic L-FABP, COX-2, and iNOS mRNA levels of mice treated with fucoidan in TAA-induced liver damage revealed increases of 1.69-fold, 4.58-fold, and 2.23-fold, respectively, when compared to normal mice.

2.7. *Fucoidan Affected Antioxidant Enzymes and ROS in the Liver*

We next evaluated whether fucoidan improved liver damage, as the evidence indicated from the TAA treatment results in the development of liver injury by inducing oxidative damage [13,14]. A significant difference in the liver antioxidant enzymes, such as superoxide dismutase (SOD), catalase, and glutathione peroxidase (GPx), was observed in our study (Figure 7). The progress of a liver injury is closely related to the downregulation of antioxidant enzymes in the liver; when the activity of these enzymes is increased, the liver function is enhanced, counteracting the hepatotoxicity issues [36]. Indeed, mice with TAA-induced liver damage treated with fucoidan exhibited significant increases of 1.59-fold, 4.12-fold, and 1.41-fold of catalase (Figure 7a), GPx (Figure 7b), and SOD (Figure 7c), respectively. The overproduction of hepatic ROS was predominantly accelerated in liver inflammatory injuries, affecting the hepatic structure and subsequently leading to severe hepatocellular dysfunction with a poor prognosis [37,38]. Here, we also found that fucoidan-treated mice expressed a significant decrease in hepatic ROS by 29% compared to the liver damage control mice (Figure 7d). The increase in hepatic antioxidant enzymes was lowered by 21%, 22%, and 18% for catalase, GPx, and SOD, respectively, in fucoidan-treated liver-damaged mice compared to normal mice. In contrast, the hepatic ROS production was increased by 1.20-fold in fucoidan-treated mice.

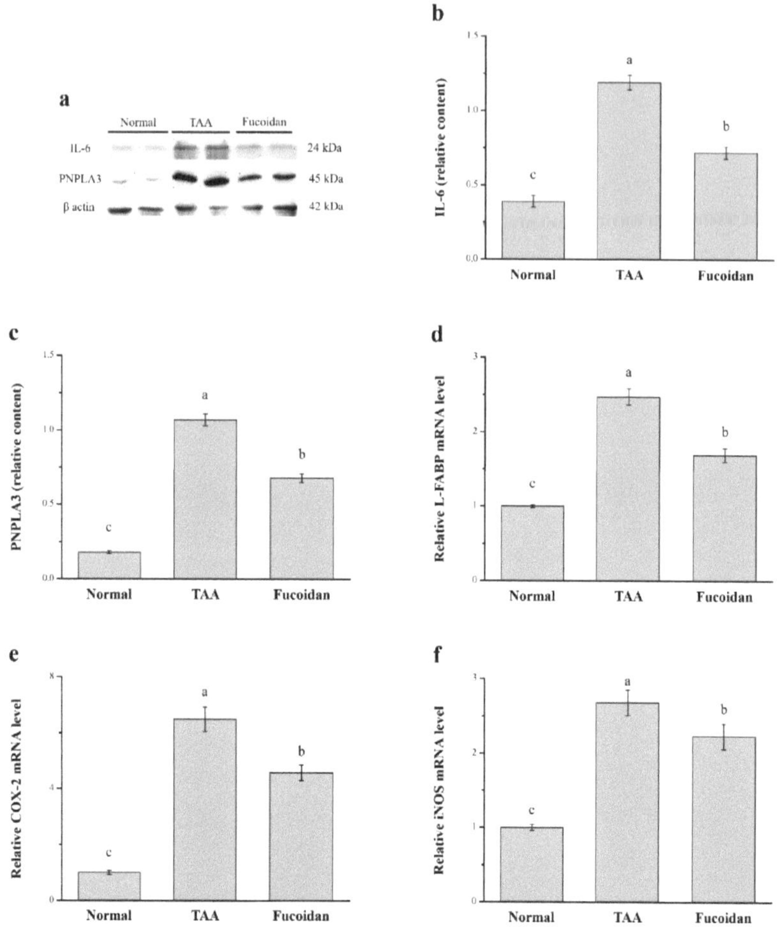

Figure 6. (**a**) Representative Western blot of the liver extracts for interleukin-6 (IL-6) and patatin-like phospholipid domain containing protein 3 (PNPLA3) expression, (**b**) IL-6, and (**c**) PNPLA3 expression, as well as changes in the (**d**) liver fatty acid-binding protein (L-FABP), (**e**) cyclooxygenase-2 (COX-2), and (**f**) inducible nitric oxide synthase (iNOS) mRNA levels in the livers of normal mice, TAA-induced liver injury mice, and fucoidan-treated TAA-induced liver injury mice over a time period of 42 days. All data are presented as the means ± SEMs (n = 10). [a–c] Data with different letters in the columns are significantly different with one-way ANOVA, and the means of different groups were compared by Duncan's test at $p < 0.05$.

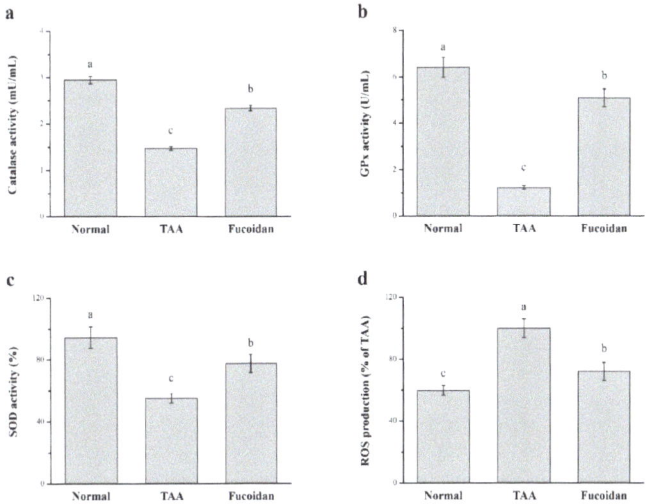

Figure 7. Changes of the hepatic (**a**) catalase activity, (**b**) glutathione peroxidase (GPx) activity, (**c**) superoxide dismutase (SOD) activity, and (**d**) reactive oxygen species (ROS) levels in normal mice, mice with TAA-induced liver injury, and fucoidan-treated mice with TAA-induced liver injury over a period of 42 days. All data are presented as the means ± SEMs (n = 10). $^{a–c}$ Data with different letters in the columns are significantly different with one-way ANOVA, and the means of different groups were compared by Duncan's test at $p < 0.05$.

3. Discussion

Liver diseases, whether caused by viral infections or drugs, are considered severe health issues and require immediate treatment with minimal toxic side effects [39]. Hepatic inflammation is common to all forms of liver injury, including cirrhosis, fibrosis, and liver cancer. Therefore, beneficial anti-inflammatory molecules may be considered as potential therapeutic agents for liver diseases [40]. Natural products present fewer side effects and are often associated with greater biocompatibility [41]. The main impediments to employing natural products involve establishing a comprehensive knowledge of medicinal plants, purifying their components, and identifying their mechanisms of action. In the current study, we successfully established a TAA-induced hepatic injury model to determine the benefits of fucoidan, a plant-based medicinal bioactive compound. In the TAA and fucoidan group, mice were administered fucoidan through oral gavage once per day for 42 days. We then measured the body weight; food intake; and levels of liver function enzymes, antioxidant enzymes, inflammatory cytokines, and molecular proteins of the mice and performed a histological examination, RNA extraction, and quantitative polymerase chain reaction (PCR) analysis. Finally, we demonstrated that fucoidan induced the regulation of the hepatic inflammatory and related regulatory pathways.

Following a TAA treatment for six weeks, this molecule contributed significantly to the decrease in body weight and body weight gain when compared to the control mice. A decrease in weight gain occurs when the food intake decreases as a result of a decrease in appetite [42]. This supports the idea that rats with TAA-induced liver injuries undergo changes in the brain's tryptophan, resulting in a decrease in food intake associated with anorexia [43]. Fucoidan can retard body weight loss, change anorexia, and enhance the quality of life in cancer patients under chemotherapy [44]. Additionally, the expression of insulin-like growth factor 1 (IGF-1) and the mammalian target of rapamycin (mTOR)/S6 ribosomal protein kinase 1 protein are elevated when under the effect of chemotherapeutic drugs such as cisplatin and gemcitabine [45]. Moreover, reports have indicated that the IGF-1 and mTOR pathways are linked to an increased caloric intake and elevate their expression toward an enhanced growth/weight [46,47]. Taken together, fucoidan is known to act as a potential agent to improve liver damage-related anorexia.

The ALT and AST levels in the serum are the vital indicators of liver damage [41]. The treatment of TAA causes major changes in the histology of liver and, also, significantly upregulates the ALT and AST levels in the serum [48]. Fucoidan ameliorated the damage and attenuated the liver histological damage with the significant downregulation of the ALT and AST levels in comparison to the TAA model group. On the other hand, the serum levels of liver enzymes such as ALP, bilirubin, globulin, and γ-GT increase [29]. These enzymes are biomarkers for liver injury. In our findings, considerable elevations in ALT, AST, ALP, the total proteins, and globulin in the serum were also observed [41,49]. This was in line with previous findings. Mohamed et al. showed that fucoidan ameliorated diazinon-induced injuries to hepatic tissues by changing the outflow of hepatic transaminases to the bloodstream [50]. Additionally, increased serum globulin resulted from inflammation, infection, tissue necrosis, and stress [51]. Here, TAA-induced liver damage treated with fucoidan could decrease the serum globulin levels, which was attributed to fucoidan improving the hepatic inflammatory response in high-fat high-sucrose diet-fed mice [52]. In addition, after the treatment with fucoidan, a significant recovery of hepatic damage was observed by the downregulated plasma levels of the hepatic enzymes. These results validated the hepatoprotective activity of fucoidan.

Apart from the biochemistry, similar effects were validated from our histopathological results. Overall, 100-mg/kg TAA applied thrice a week successfully induced chronic liver damage in mice [49,53], while mice treated with TAA had lower body weights compared with the untreated mice [54]. The lower body weights in the TAA-treated mice were different from most nonalcoholic fatty liver disease models, since the liver damage was not caused by dieting or factors related to obesity [42,47]. Additionally, some studies have indicated that animals treated with TAA for over eight weeks have higher liver weights due to the accumulation of collagen and extracellular matrix protein in the liver after TAA exposure [54,55]. We found, using Sirius red staining, that a liver injury induced with a treatment of TAA for six weeks did not develop liver fibrosis (Figure S1). The discrepancies of our results with other studies are probably indicative of the effects of the length of the TAA treatment. Whether liver damage affects fibrogenesis, many studies have suggested a hepatoprotective role of medicinal plants in TAA injury models [53]. In the present animal model, the fucoidan treatment offered protection to mice for substantial lethality and, also, because it was able to lower the ALT, AST, and ALP levels in the serum. After 42 days of fucoidan treatment, mice markedly improved their histological changes, with less degeneration and necrosis. In chronic liver injury experiments done in combination with other studies, fucoidan was shown to promote the restoration of hepatic physiology, further confirmed with the ALT, AST, ALP, bilirubin, globulin, and γ-GT levels [56].

Inflammation is a common stimulus for all liver diseases, triggering the progression of liver damage to severe fibrogenesis or culminating in hepatic carcinoma [57]. Therefore, we investigated the proteins such as TNF-α, IL-1β, FGF21, and CRP that have an effect on the inflammatory response. TNF-α is a pleiotropic proinflammatory cytokine that can induce multiple mechanisms to initiate apoptosis in hepatocytes and, hence, cause liver injury by either directly or indirectly targeting the mitochondria [58]. The research indicates that the TNF receptor one knockout mice showed reduced CCl4-induced liver fibrosis [59]. There is no doubt that TNF-α can promote fibrosis and lead to chronic liver injury and inflammation. Interleukin-1β is associated with the activation of NF-κβ signaling, the upregulation of proinflammatory cytokines, and liver damage [60]. Another study showed that mice with an IL-1β deficiency had less diet-induced inflammation and liver fibrosis. In addition, a decreased expression of IL-6, TNF-α, and TGF-β, in comparison to wild-type mice, was observed [61]. Next, the elevated FGF21 was able to lower the content of AST, ALT, and γ-GT [62]. For example, acetaminophen-induced hepatotoxicity and mortality was exacerbated in FGF21 null mice. The opposite effect was observed when the treatment with recombinant FGF21 produced significant protective effects on the overall liver function [63].

Finally, the CRP levels correlated closely with the changes in inflammation/disease activity, radiological damage, and functional disability [64]. CRP is primarily synthesized in the liver and is involved in many chronic diseases. Moreover, the elevated CRP level reveals that chronic ethanol administration is responsible for severe hepatic endothelial injury [65]. Our results demonstrated that TNF-α, IL-1β, and CRP in fucoidan-treated mice were lower than of TAA-treated mice. On the other hand, FGF21 was enhanced after the treatment with fucoidan. This means that fucoidan is able to reduce the inflammation and severity of liver damage. In light of all our findings, fucoidan demonstrated a significant hepatoprotective effect against TAA-induced liver injuries by regulating the inflammatory process.

The other important factors related to liver damage are IL-6, PNPLA3, L-FABP, COX-2, and iNOS. IL-6, upregulated in chronic liver inflammation [66], is produced by activated Kupffer cells exacerbating the local inflammatory response and inducing hepatocyte proliferation, with a compensatory transformation of hepatocytes into malignant hepatocytes [67]. PNPLA3 increases the release of proinflammatory and profibrogenic cytokines, known to be active key players in the liver injury process [68,69]. Human hepatic stellate cells with the PNPLA3 variant have a higher expression of inflammatory cytokines and chemokines [70]. L-FABP can be used for the diagnosis of acute hepatitis, chronic hepatitis, and cirrhosis [71]. In another study, it was demonstrated that the upregulated serum L-FABP levels are related to a degree of fibrosis and inflammation in the liver, which indicates that serum L-FABP can be a noninvasive biomarker used to assess the severity of fibrosis and inflammation in nonalcoholic steatohepatitis patients [72]. The COX-2 expression was related to the early phases of different chronic liver diseases and found to be responsible for the induction of hepatic cancer [73]. A real-time quantitative PCR analysis of total RNA demonstrated that the hepatic COX-2 gene expression is upregulated in both acutely and chronically damaged livers [74]. Finally, iNOS has been also reported to be involved in the excessive production of proinflammatory mediators [75]. Chronic liver diseases begin with an inflammatory phase, followed by fibrosis and continuous oxidative stress. In this state, the iNOS production is increased, causing an overproduction of nitric oxide [76]. By using Western blotting, we can find that the expression of IL-6 and PNPLA3 are reduced by fucoidan. Moreover, the gene expression of L-FABP, COX-2, and iNOS are also lower in fucoidan-treated mice than in TAA-treated mice. The application of fucoidan was able to reduce the inflammation and liver damage-related factors, reducing the effect of the liver damage caused by TAA.

Oxidative stress is a major factor in the progression of liver damage and its pathogenesis. ROS cause the induction of hepatic inflammation, necrosis, and cholestasis [77]. Additionally, the antioxidant enzymes, including catalase, GPx, and SOD, offer protection against the harmful effects of ROS [78]. GPx is a family of enzymes that constitutes a main antioxidant defense system in mammals [79]. The GPx and SOD levels are regularly downregulated, with an accompanied increase in the ALT and AST levels [80]. CAT is present in peroxisomes, where it decomposes two hydrogen peroxide (H_2O_2) molecules into two H_2O molecules and O_2 [81]. While the concentration of H_2O_2 increases, CAT shows a greater contribution to H_2O_2 degradation [82]. The serum CAT activity was moderately increased in fatty liver and acute alcoholic hepatitis [83]. Our results indicate that fucoidan decreases oxidative stress and the contents of ROS by increasing the GPx, SOD, and CAT activity. Therefore, the treatment with fucoidan was able to reduce the ROS to prevent liver cells from developing into a more severe liver disease.

Together, our findings demonstrated and validated that fucoidan showed hepatoprotective activity that was mediated via the inhibition of the inflammatory pathways and upregulation of the antioxidant enzymes. Hence, we emphasized the importance of fucoidan as a new potential treatment against liver damage.

4. Materials and Methods

4.1. Animal, Liver Injury Mouse, and Fucoidan

Male C57BL/6J mice, five-weeks-old, 18–20 g, were purchased from Education Research Resource, National Laboratory Animal Center, Taipei, Taiwan. In accordance with the Taiwan government's recommendations, all animal housing and experimentation were performed under the Guidelines for the Care and Use of Laboratory Animals. The review of our experimental protocol was conducted by the National Chiayi University's Institutional Animal Care and Use Committee, who approved it under the archive no. 108017. Animals were housed in normal cages at 22 ± 1 °C under 55% ± 5% humidity, with a 12:12-h light–dark schedule. The mice were fed a standard diet (SD, 3.3 kcal/g of metabolizable energy; diet 5008; PMI Nutrition International, Brentwood, MO, USA) for 4 weeks ad libitum until the mice weighed approximately 25 g. They were subsequently divided into three groups.

Mice fed the SD were randomly divided into the normal control, TAA model, and TAA and fucoidan groups. Within each housing group, the mice were randomly assigned to three groups consisting of 10 mice. The mice in the TAA model and fucoidan groups were injected with 100-mg/kg TAA ip (Sigma, St. Louis, MO, USA) thrice a week for six weeks, and the mice in the fucoidan group were administered fucoidan from *Fucus vesiculosus* (20 mg/kg; Sigma) by oral gavage once a day for 42 days. The dose of TAA was based on the available literature [49,53,54], and the dose of fucoidan was based on previous articles that revealed fucoidan as a choice candidate for chronic treatment in blood glucose homeostasis, lung injury, hepatic ischemia–reperfusion injury, and endometriotic studies in mouse models [84–86]. In our preliminary studies, the serum ALT, AST, globulin, and γ-GT levels in the fucoidan-treated (10 mg/kg) TAA-induced liver injury mice did not differ significantly from those of the TAA-induced liver injury mice (Figure S2). Consequently, we initiated this study by using a dose of fucoidan at 20 mg/kg in our mice model. All mice were weighed weekly. On termination of our experiment, mice were euthanized and their livers harvested, as well as their serum, for the subsequent analysis of the liver function enzymes, antioxidant enzymes, histological examination, inflammatory cytokine, molecular proteins, RNA extraction, and quantitative PCR analysis.

4.2. Body Weight and Food Intake

Weekly measurements of the food intake, as well as body weight, throughout the study period were taken. For the food intake measurements, the leftover food within each cage dispenser, in addition to what was spilled, was measured.

4.3. Liver Function Enzymes Tests

The levels of ALT, AST, ALP, bilirubin, globulin, and γ-GT in the serum were measured from the blood samples using an automated chemistry analyzer (Catalyst One Chemistry Analyzer, IDEXX Laboratories, Westbrook, ME, USA) from a commercial kit. Our method followed the manufacturer's recommendations. The coefficient of variation within and between the analysis runs was <2%.

4.4. Liver Histopathological Evaluation

We measured the liver weights and determined their percentages in relation to the total body weight. The isolated liver tissues were treated with formalin, entrenched in paraffin, and further processed for the histological analysis. The sections (3 μm) were stained with H&E staining, and the liver damage was quantified. The injuries were classified using the mentioned four-point scale (Suzuki scoring 0–4). Scoring was assigned based on the degree of liver damage: 0 = normal, 1 = the development of sinusoidal congestion space, 2 or 3 = the presence and/or severity of sinusoidal congestion and cytoplasmic vacuolization, and 4 = tthe necrosis of parenchymal cells and hemorrhage [30,31].

All the sections were reassessed for liver damage changes by Taiwan-certified veterinary pathologists blinded to the animals' treatments.

4.5. Determination of Serum Inflammatory Cytokines, FGF21, and CRP

The levels of TNF-αand IL-1β in the plasma were determined using mouse TNF-α and IL-1β ELISA kits (Invitrogen, Carlsbad, CA, USA) according to the manufacturer's instructions. Serum mouse FGF-21 ELISA kits were obtained from Zgenebio Biotech (Taipei, Taiwan). Serum concentrations of CRP were measured using a mouse CRP quantitative ELISA kit (Abcam, Cambridge, UK).

4.6. Western Blotting Assay

Animals were anesthetized with an ip injection at the termination of the experiment. Livers were quickly removed, minced coarsely, and immediately homogenized prior to the Western blot [47,49] using the following antibodies: anti-IL-6 (Cell Signaling Technology, Beverly, MA, USA), anti-PNPLA3 (Sigma), and anti-actin antibodies (Sigma). We used enhanced chemiluminescence reagents (Thermo Scientific, Rockford, MA, USA) to produce immunoreactive signals and UVP ChemStudio (Analytik Jena, Upland, CA, USA) to detect these signals. Protein expression and phosphorylation studies were quantified using ImageJ by the National Institutes of Health (Bethesda, MA, USA).

4.7. RNA Extraction and Real-Time Quantitative PCR

The mRNA levels of L-FABP, COX-2, and iNOS in the livers were determined by the CFX Connect Quantitative Real-Time PCR System (Bio-Rad, Hercules, CA, USA). Briefly, the total RNA was isolated from the liver samples using TRI Reagent (Sigma), and the concentration was assessed based on the absorbance at 260–280 and 230–260 nm on a Qubit fluorometer (Invitrogen, Carlsbad, CA, USA). Next, we performed real-time PCR using this complementary DNA and iTaq universal SYBR Green supermix (Bio-Rad), in accordance with the manufacturer's protocols. The PCR was performed as follows: 95 °C for 5 min and then 45 cycles at 95 °C for 15 s, followed by 60 °C for 25 s. The L-FABP sequence primers employed in this study were forward 5′-GCAGAGCCAGGAGAACTTTG-3′ and reverse 5′-GGGTCCATAGGTGATGGTGAG-3′ [87]; those for COX-2 were forward 5′- AAAACCGTGGGGAATGTATGAGC-3′ and reverse 5′-GATGGGTGAAGTGCTGGGGAAAG-3′ [88]; those for the iNOS sequence primers employed in this study were forward 5′- TCTTGGGTCTCCGCTTCTCGTC-3′ and reverse 5′-TGGCTGGTACATGGGCAC AGAG-3′ [89]. Each target gene expression level was calculated in relation to the Actb level and expressed using the $2-\Delta\Delta Ct$ method.

4.8. Evaluation of Hepatic Catalase, GPx, SOD, and ROS

An evaluation of the functional activity of the antioxidant system was analyzed for the enzymatic activity of hepatic catalase, GPx, and SOD. Livers were perfused with ice-cold saline (0.9% sodium chloride) and homogenized in chilled potassium chloride (1.17%) [90]. After, the obtained homogenates underwent centrifugation at 800× *g* for 5 min at 4 °C. The supernatant was recentrifuged at 10,500× *g* for 20 min at 4 °C to finally obtain the post-mitochondrial supernatant. This was used for measuring the catalase, GPx, and SOD. The enzymes were quantified using a commercially available colorimetrical kit (#K773-100 for catalase, #K762-100 for GPx, and #K335-100 for SOD; BioVision, Milpitas, CA, USA). In addition, small portions of fresh liver tissue were collected and homogenized in an ice-cold lysis solution (10-mM Tris, 150-mM NaCl, 0.1-mM EDTA, and 0.5% Triton X 100, pH 7.5) [91] and centrifuged at 1000× *g* for 10 min. The supernatant was collected, and the level of ROS was measured using a commercially available colorimetrical kit (#:KTE71621; Abbkine, Redlands, CA, USA), according to the manufacturer's instructions.

4.9. Statistical Analyses

All values were expressed as the mean ± SEMs. Any other experimental data, except for the pathological findings, were analyzed by one-way analysis of variance (ANOVA) followed by Duncan's multiple range test. The injury scores of the liver pathological examination were analyzed using a *t*-test. A p-value < 0.05 was considered statistically significant.

5. Conclusions

The study demonstrated that fucoidan was effective in preventing TAA-induced hepatic damage in mice. The protective effect of fucoidan was due to its anti-inflammatory, antioxidant, and free radical scavenger effects. The underlying mechanisms included an increase in the expression of catalase, GPx, and SOD and a decrease in ROS production. The reduced inflammatory conditions were revealed by a decrease in the TNF-α, IL-1β, and CRP levels and the inhibition of hepatic IL-6 and PNPLA3 expression, as well as an under-expression of the L-FABP, COX-2, and iNOS mRNA levels, which ultimately prevented the progression of hepatic damage. Hence, this study not only validated the use of fucoidan as a hepatoprotective agent but also supported its future exploration as a therapeutic drug for liver disorders.

Supplementary Materials: The following are available online: Figure S1: Changes of the liver histology by Sirius red staining (magnification, 200×). Figure S2: Changes of the serum ALT, AST, globulin, and γ-GT levels in TAA-induced liver injury mice and fucoidan (10 mg/kg)-treated TAA-induced liver injury mice over a period of 42 days. To perform Sirius red staining of the liver collagen, we used a commercially available kit (Abcam, Cambridge, MA, USA) and followed the manufacturer's instructions.

Author Contributions: M.-Y.T. conceived the idea and performed experiments. W.-C.Y., C.-F.L., C.-M.W., H.-Y.L., C.-S.L. and J.-W.L. assisted in recombinant construction. W.-L.L., T.-C.L. and P.-S.F. analyzed the data. K.-H.H., Y.-W.L. and G.-R.C. wrote, reviewed, and edited the manuscript. All authors have read and agreed to the published version of the manuscript.

Funding: This study was supported in part by grant SRD109043 from the Show Chwan Memorial Hospital (Changhua, Taiwan) and National Chiayi University (Chiayi, Taiwan; 108A3-093, 108A3-094, 108A3-160, 110A3-038, 109D1-100D6, and 110D1-100D6).

Institutional Review Board Statement: The review of our experimental protocol was conducted by the National Chiayi University's Institutional Animal Care and Use Committee, who approved it under approval no. 108017.

Informed Consent Statement: Not applicable.

Data Availability Statement: The data presented in this study are available on request from the corresponding author.

Acknowledgments: The authors would like to thank Li-Tzung Biotechnology, Kaohsiung, Taiwan, for providing pathological assistance for this study.

Conflicts of Interest: The authors declare no conflict of interest.

References

1. Treyer, A.; Müsch, A. Hepatocyte polarity. *Compr. Physiol.* **2013**, *3*, 243–287. [PubMed]
2. Blouin, A.; Bolender, R.P.; Weibel, E.R. Distribution of organelles and memebranes between hepatocytes and nonhepatocytes in the rat liver parenchyma. A stereological study. *Cell Biol.* **1977**, *72*, 441–455. [CrossRef]
3. De Mingo Pulido, Á.; de Gregorio, E.; Chandra, S.; Colell, A.; Morales, A.; Kronenberg, M.; Marí, M. Differential role of cathepsins S and B in hepatic APC-mediated NKT cell activation and cytokine secretion. *Front. Immunol.* **2018**, *9*, 391. [CrossRef] [PubMed]
4. Hirsova, P.; Gores, G.J. Death receptor-mediated cell death and proinflammatory signaling in nonalcoholic steatohepatitis. *Cell. Mol. Gastroenterol. Hepatol.* **2015**, *1*, 17–27. [CrossRef] [PubMed]
5. Yang, L.; Seki, E. Toll-like receptors in liver fibrosis: Cellular crosstalk and mechanisms. *Front. Physiol.* **2012**, *3*, 138. [CrossRef]
6. Fisher, C.P.; Kierzek, A.M.; Plant, N.J.; Moore, J.B. Systems biology approaches for studying the pathogenesis of non-alcoholic fatty liver disease. *World J. Gastroenterol.* **2014**, *20*, 15070. [CrossRef]
7. Mermelstein, C.S.; Guma, F.C.; Mello, T.G.; Fortuna, V.A.; Guaragna, R.M.; Costa, M.L.; Borojevic, R. Induction of the lipocyte phenotype in murine hepatic stellate cells: Reorganization of the actin cytoskeleton. *Cell Tissue Res.* **2001**, *306*, 75–83. [CrossRef]
8. Domenicali, M.; Caraceni, P.; Giannone, F.; Baldassarre, M.; Lucchetti, G.; Quarta, C.; Bernardi, M. A novel model of CCl4-induced cirrhosis with ascites in the mouse. *J. Hepatol.* **2009**, *51*, 991–999. [CrossRef] [PubMed]
9. Kučera, O.; Červinková, Z.; Lotkova, H.; Křiváková, P.; Roušar, T.; Mužáková, V.; Rudolf, E. Protective effect of S-adenosylmethionine against galactosamine-induced injury of rat hepatocytes in primary culture. *Physiol. Res.* **2006**, *55*, 551–560. [PubMed]

10. Zhan, F.; Zhao, G.; Li, X.; Yang, S.; Yang, W.; Zhou, S.; Zhang, F. Inositol-requiring enzyme 1 alpha endoribonuclease specific inhibitor STF-083010 protects the liver from thioacetamide-induced oxidative stress, inflammation and injury by triggering hepatocyte. *Int. Immunopharmacol.* **2019**, *73*, 261–269. [CrossRef] [PubMed]
11. Loh, Z.; Fitzsimmons, R.L.; Reid, R.C.; Ramnath, D.; Clouston, A.; Gupta, P.K.; Iyer, A. Inhibitors of class I histone deacetylases attenuate thioacetamide-induced liver fibrosis in mice by suppressing hepatic type 2 inflammation. *Br. J. Pharmacol.* **2019**, *176*, 3775–3790. [CrossRef] [PubMed]
12. Lin, X.; Wei, J.; Nie, J.; Bai, F.; Zhu, X.; Zhuo, L.; Lu, Z.; Huang, Q. Inhibition of RKIP aggravates thioacetamide-induced acute liver failure in mice. *Exp. Ther. Med.* **2018**, *16*, 2992–2998. [CrossRef] [PubMed]
13. Chilakapati, J.; Shankar, K.; Korrapati, M.C.; Hill, R.A.; Mahendale, H.M. Saturation toxicokinetics of thioacetamide: Role in initiation of liver injury. *Drug Metab. Dispos.* **2005**, *33*, 1877–1885. [CrossRef] [PubMed]
14. Salama, S.M.; AlRashdi, A.S.; Abdulla, M.A.; Hassandarvish, P.; Bilgen, M. Protective activity of Panduratin A against thioacetamide-induced oxidative damage: Demonstration with in vitro experiments using WRL-68 liver cell line. *BMC Complement. Altern. Med.* **2013**, *13*, 279. [CrossRef]
15. Fennema, D.; Phillips, I.R.; Shephard, E.A. Trimethylamine and trimethylamine N-oxide, a flavin-containing monooxygenase 3 (FMO3)-mediated host-microbiome metabolic axis implicated in health and disease. *Drug Metab. Dispos.* **2016**, *44*, 1839–1850. [CrossRef]
16. Berteau, O.; Mulloy, B. Sulfated fucans, fresh perspectives: Structures, functions, and biological properties of sulfated fucans and an overview of enzymes active toward this class of polysaccharide. *Glycobiology* **2003**, *13*, 29R–40R. [CrossRef]
17. Kannan, R.R.R.; Arumugam, R.; Anantharaman, P. Pharmaceutical potential of a fucoidan-like sulphated polysaccharide isolated from Halodule pinifolia. *Int. J. Biol. Macromol.* **2013**, *62*, 30–34. [CrossRef]
18. Fitton, J.H. Therapies from fucoidan; multifunctional marine polymers. *Mar. Drugs* **2011**, *9*, 1731–1760. [CrossRef]
19. Pomin, V.H. Eucanomics and galactanomics: Current status in drug discovery, mechanisms of action and role of the well-defined structures. *Biochim. Biophys. Acta. Gen. Subj.* **2012**, *1820*, 1971–1979. [CrossRef] [PubMed]
20. Senthilkumar, K.; Manivasagan, P.; Venkatesan, J.; Kim, S.K. Brown seaweed fucoidan: Biological activity and apoptosis, growth signaling mechanism in cancer. *Int. J. Biol. Macromol.* **2013**, *60*, 366–374. [CrossRef] [PubMed]
21. Zhang, Z.; Teruya, K.; Eto, H.; Shirahata, S. Fucoidan extract induces apoptosis in MCF-7 cells via a mechanism involving the ROS-dependent JNK activation and mitochondria-mediated pathways. *PLoS ONE* **2011**, *6*, e27441. [CrossRef]
22. Alekseyenko, T.V.; Zhanayeva, S.Y.; Venediktova, A.A.; Zvyagintseva, T.N.; Kuz-netsova, T.A.; Besednova, N.N.; Korolenko, T.A. Antitumor and antimetastatic activity of fucoidan, a sulfated polysaccharide isolated from the Okhotsk Sea Fucus evanescens brown alga. *Bull. Exp. Biol. Med.* **2007**, *143*, 730–732. [CrossRef] [PubMed]
23. Atashrazm, F.; Lowenthal, R.M.; Woods, G.M.; Holloway, A.F.; Dickinson, J.L. Fucoidan and cancer: A multifunctional molecule with anti-tumor potential. *Mar. Drugs* **2015**, *13*, 2327–2346. [CrossRef] [PubMed]
24. Maruyama, H.; Tamauchi, H.; Iizuka, M.; Nakano, T. The role of NK cells in antitumor activity of dietary fucoidan from Undaria pinnatifida Sporophylls. *Planta Med.* **2006**, *72*, 1415–1417. [CrossRef] [PubMed]
25. Liu, F.; Wang, J.; Chang, A.K.; Liu, B.; Yang, L.; Li, Q.; Zou, X. Fucoidan extract derived from Undaria pinnatifida inhibits angiogenesis by human umbilical vein endothelial cells. *Phytomedicine* **2012**, *19*, 797–803. [CrossRef] [PubMed]
26. Huang, T.H.; Chiu, Y.H.; Chan, Y.L.; Chiu, Y.H.; Wang, H.; Huang, K.C.; Wu, C.J. Prophylactic administration of fucoidan represses cancer metastasis by inhibiting vascular endothelial growth factor (VEGF) and matrix metalloproteinases (MMPs) in Lewis tumor-bearing mice. *Mar. Drugs* **2015**, *13*, 1882–1900. [CrossRef]
27. Saito, A.; Yoneda, M.; Yokohama, S.; Okada, M.; Haneda, M.; Nakamura, K. Fucoidan prevents concanavalin A-induced liver injury through induction of endogenous IL-10 in mice. *Hepatol. Res.* **2006**, *35*, 190–198. [CrossRef]
28. Li, X.J.; Ye, Q.F. Fucoidan reduces inflammatory response in a rat model of hepatic ischemia–reperfusion injury. *Can. J. Physiol. Pharmacol.* **2015**, *93*, 999–1005. [CrossRef]
29. Zheng, J.; Yan, Q.; Zhang, K.; Zheng, Y.; Zhao, S. Protective effects of different extracts of Eucommia ulmoides Oliv. against thioacetamide-induced hepatotoxicity in mice. *Indian J. Exp. Biol.* **2012**, *50*, 875–882. [PubMed]
30. Tak, E.; Jung, D.H.; Kim, S.H.; Park, G.C.; Jun, D.Y.; Lee, J.; Jung, B.H.; Kirchner, V.A.; Hwang, S.; Song, G.W.; et al. Protective role of hypoxia-inducible factor-1alpha-dependent CD39 and CD73 in fulminant acute liver failure. *Toxicol. Appl. Pharmacol.* **2017**, *314*, 72–81. [CrossRef] [PubMed]
31. Hart, M.L.; Much, C.; Gorzolla, I.C.; Schittenhelm, J.; Kloor, D.; Stahl, G.L.; Eltzschig, H.K. Extracellular adenosine production by ecto-5′-nucleotidase protects during murine hepatic ischemic. *Gastroenterology* **2008**, *135*, 1739–1750. [CrossRef]
32. Keinicke, H.; Sun, G.; Mentzel, C.M.J.; Fredholm, M.; John, L.M.; Andersen, B.; Raun, K.; Kjaergaard, M. FGF21 regulates hepatic metabolic pathways to improve steatosis and inflammation. *Endocr. Connect.* **2020**, *9*, 755–768. [CrossRef]
33. Kang, J.M.; Lee, W.J.; Kim, W.B.; Kim, T.Y.; Koh, J.M.; Hong, S.J.; Huh, J.; Ro, J.Y.; Chi, H.S.; Kim, M.S. Systemic inflammatory syndrome and hepatic inflammatory cell infiltration caused by an interleukin-6 producing pheochromocytoma. *Endocr. J.* **2005**, *52*, 193–198. [CrossRef] [PubMed]
34. Bruschi, F.V.; Claudel, T.; Tardelli, M.; Starlinger, P.; Marra, F.; Trauner, M. PNPLA3 I148M variant impairs liver x receptor signaling and cholesterol homeostasis in human hepatic stellate cells. *Hepatol. Commun.* **2019**, *3*, 1191–1204. [CrossRef]
35. Chang, M.L.; Yang, S.S. Metabolic signature of hepatic fibrosis: From individual pathways to systems biology. *Cells* **2019**, *8*, 1423. [CrossRef] [PubMed]

36. Uma, N.J.; Fakurazi, S.; Hairuszah, I. Moringa oleifera enhances liver antioxidant status via elevation of antioxidant enzymes activity and counteracts paracetamol-induced hepatotoxicity. *Malays. J. Nutr.* **2010**, *16*, 293–307. [PubMed]
37. Sakai, N.; Van Sweringen, H.L.; Belizaire, R.M.; Quillin, R.C.; Schuster, R.; Blanchard, J.; Burns, J.M.; Tevar, A.D.; Edwards, M.J.; Lentsch, A.B. Interleukin-37 reduces liver inflammatory injury via effects on hepatocytes and non-parenchymal cells. *J. Gastroenterol. Hepatol.* **2012**, *27*, 1609–1616. [CrossRef] [PubMed]
38. Jaeschke, H. Reactive oxygen and mechanisms of inflammatory liver injury: Present concepts. *J. Gastroenterol. Hepatol.* **2011**, *26* (Suppl. 1), 173–179. [CrossRef] [PubMed]
39. Xin, Y.; Wei, J.; Chunhua, M.; Danhong, Y.; Jianguo, Z.; Zongqi, C.; Jian-An, B. Protective effects of Ginsenoside Rg1 against carbon tetrachloride-induced liver injury in mice through suppression of inflammation. *Phytomedicine* **2016**, *23*, 583–588. [CrossRef] [PubMed]
40. Chang, G.R.; Hou, P.H.; Yang, W.C.; Wang, C.M.; Fan, P.S.; Liao, H.J.; Chen, T.P. Doxepin exacerbates renal damage, glucose intolerance, nonalcoholic fatty liver disease and urinary chromium loss in obese mice. *Pharmaceuticals* **2021**, *14*, 267. [CrossRef]
41. Azam, F.; Sheikh, N.; Ali, G.; Tayyeb, A. Fagonia indica repairs hepatic damage through expression regulation of toll-like receptors in a liver injury model. *J. Immunol. Res.* **2018**, *2018*, 7967135. [CrossRef]
42. Chang, G.R.; Wu, Y.Y.; Chiu, Y.S.; Chen, W.Y.; Liao, J.W.; Hsu, H.M.; Chao, T.H.; Hung, S.W.; Mao, F.C. Long-term administration of rapamycin reduces adiposity, but impairs glucose tolerance in high-fat diet-fed KK/HlJ mice. *Basic Clin. Pharmacol. Toxicol.* **2009**, *105*, 188–198. [CrossRef] [PubMed]
43. Haider, S.; Saleem, S.; Shameem, S.; Ahmed, S.P.; Parveen, T.; Haleem, D.J. Is anorexia in thioacetamide-induced cirrhosis related to an altered brain serotonin concentration? *Pol. J. Pharmacol.* **2004**, *56*, 73–78. [PubMed]
44. Hsu, H.Y.; Hwang, P.A. Clinical applications of fucoidan in translational medicine for adjuvant cancer therapy. *Clin. Transl. Med.* **2019**, *8*, 15. [CrossRef] [PubMed]
45. Chen, M.C.; Hsu, W.L.; Hwang, P.A.; Chen, Y.L.; Chou, T.C. Combined administration of fucoidan ameliorates tumor and chemotherapy-induced skeletal muscle atrophy in bladder cancer-bearing mice. *Oncotarget* **2016**, *7*, 51608–51618. [CrossRef] [PubMed]
46. Mahachoklertwattana, P.; Wanasuwankul, S.; Poomthavorn, P.; Choubtum, L.; Sriphrapradang, A. Short-term cyproheptadine therapy in underweight children: Effects on growth and serum insulin-like growth factor-I. *J. Pediatr. Endocrinol. Metab.* **2009**, *22*, 425–432. [CrossRef] [PubMed]
47. Chang, G.R.; Chiu, Y.S.; Wu, Y.Y.; Lin, Y.C.; Hou, P.H.; Mao, F.C. Rapamycin impairs HPD-induced beneficial effects on glucose homeostasis. *Br. J. Pharmacol.* **2015**, *172*, 3793–3804. [CrossRef] [PubMed]
48. Alkiyumi, S.S.; Abdullah, M.A.; Alrashdi, A.S.; Salama, S.M.; Abdelwahab, S.I.; Hadi, A.H.A. Ipomoea aquatica extract shows protective action against thioacetamide-induced hepatotoxicity. *Molecules* **2012**, *17*, 6146–6155. [CrossRef]
49. Wu, H.T.; Chuang, Y.W.; Huang, C.P.; Chang, M.H. Loss of angiotensin converting enzyme II (ACE2) accelerates the development of liver injury induced by thioacetamide. *Exp. Anim.* **2018**, *67*, 41–49. [CrossRef] [PubMed]
50. Abdel-Daim, M.M.; Abushouk, A.I.; Bahbah, E.I.; Bungau, S.G.; Alyousif, M.S.; Aleya, L.; Alkahtani, S. Fucoidan protects against subacute diazinon-induced oxidative damage in cardiac, hepatic, and renal tissues. *Environ. Sci. Pollut. Res.* **2020**, *27*, 11554–11564. [CrossRef]
51. Schuppan, D.; Kim, Y.O. Evolving therapies for liver fibrosis. *J. Clin. Investig.* **2013**, *123*, 1887–1901. [CrossRef] [PubMed]
52. Wang, J.; Hu, S.; Jiang, W.; Song, W.; Cai, L.; Wang, J. Fucoidan from sea cucumber may improve hepatic inflammatory response and insulin resistance in mice. *Int. Immunopharmacol.* **2016**, *31*, 15–23. [CrossRef] [PubMed]
53. Wang, H.; Zhang, H.; Zhang, Y.; Wang, D.; Cheng, X.; Yang, F.; Zhang, Q.; Xue, Z.; Li, Y.; Zhang, L.; et al. Plumbagin protects liver against fulminant hepatic failure and chronic liver fibrosis via inhibiting inflammation and collagen production. *Oncotarget* **2016**, *7*, 82864–82875. [CrossRef] [PubMed]
54. Chen, I.S.; Chen, Y.C.; Chou, C.H.; Chuang, R.F.; Sheen, L.Y.; Chiu, C.H. Hepatoprotection of silymarin against thioacetamide-induced chronic liver fibrosis. *J. Sci. Food. Agric.* **2012**, *92*, 1441–1447. [CrossRef] [PubMed]
55. Dwivedi, D.K.; Jena, G.; Kumar, V. Dimethyl fumarate protects thioacetamide-induced liver damage in rats: Studies on Nrf2, NLRP3, and NF-κB. *J. Biochem. Mol. Toxicol.* **2020**, *34*, e22476. [CrossRef] [PubMed]
56. Hong, S.W.; Jung, K.H.; Lee, H.S.; Zheng, H.M.; Choi, M.J.; Lee, C.; Hong, S.S. Suppression by Fucoidan of liver fibrogenesis via the TGF-beta/Smad pathway in protecting against oxidative stress. *Biosci. Biotechnol. Biochem.* **2011**, *75*, 833–840. [CrossRef] [PubMed]
57. Del Campo, J.A.; Gallego, P.; Grande, L. Role of inflammatory response in liver diseases: Therapeutic strategies. *World J. Hepatol.* **2018**, *10*, 1–7. [CrossRef]
58. Ding, W.X.; Yin, X.M. Dissection of the multiple mechanisms of TNF-α-induced apoptosis in liver injury. *J. Cell. Mol. Med.* **2004**, *8*, 445–454. [CrossRef]
59. Sudo, K.; Yamada, Y.; Moriwaki, H.; Saito, K.; Seishima, M. Lack of tumor necrosis factor receptor type 1 inhibits liver fibrosis induced by carbon tetrachloride in mice. *Cytokine* **2005**, *29*, 236–244. [CrossRef]
60. Sultan, M.; Ben-Ari, Z.; Masoud, R.; Pappo, O.; Harats, D.; Kamari, Y.; Safran, M. Interleukin-1α and Interleukin-1β play a central role in the pathogenesis of fulminant hepatic failure in mice. *PLoS ONE* **2017**, *12*, e0184084. [CrossRef]

61. Kamari, Y.; Shaish, A.; Vax, E.; Shemesh, S., Kandel-Kfir, M.; Arbel, Y.; Harats, D. Lack of interleukin-1α or interleukin-1β inhibits transformation of steatosis to steatohepatitis and liver fibrosis in hypercholesterolemic mice. *J. Hepatol.* **2011**, *55*, 1086–1094. [CrossRef]
62. Markova, M.; Pivovarova, O.; Hornemann, S.; Sucher, S.; Frahnow, T.; Wegner, K.; Pfeiffer, A.F. Isocaloric diets high in animal or plant protein reduce liver fat and inflammation in individuals with type 2 diabetes. *Gastroenterology* **2017**, *152*, 571–585. [CrossRef] [PubMed]
63. Ye, D.; Wang, Y.; Li, H.; Jia, W.; Man, K.; Lo, C.M.; Xu, A. Fibroblast growth factor 21 protects against acetaminophen-induced hepatotoxicity by potentiating peroxisome proliferator-activated receptor coactivator protein-1α-mediated antioxidant capacity in mice. *Hepatology* **2014**, *60*, 977–989. [CrossRef]
64. Zhao, J.; Liu, J.; Pang, X.; Zhang, X.; Wang, S.; Wu, D. Rosiglitazone attenuates angiotensin II-induced c-reactive protein expression in hepatocytes via inhibiting AT1/ROS/MAPK signal pathway. *Int. Immunopharmacol.* **2016**, *31*, 178–185. [CrossRef] [PubMed]
65. Liu, W.H.; Liu, T.C.; Yin, M.C. Beneficial effects of histidine and carnosine on ethanol-induced chronic liver injury. *Food Chem. Toxicol.* **2008**, *46*, 1503–1509. [CrossRef]
66. Devière, J.; Content, J.; Denys, C.; Vandenbussche, P.; Schandené, L.; Wybran, J.; Dupont, E. High interleukin-6 serum levels and increased production by leucocytes in alcoholic liver cirrhosis. Correlation with IgA serum levels and lymphokines production. *Clin. Exp. Immunol.* **1989**, *77*, 221. [PubMed]
67. Oyanagi, Y.; Takahashi, T.; Matsui, S.; Takahashi, S.; Boku, S.; Takahashi, K.; Asakura, H. Ehanced expression of interleukin-6 in chronic hepatitis C. *Liver* **1999**, *19*, 464–472. [CrossRef] [PubMed]
68. Bruschi, F.V.; Claudel, T.; Tardelli, M.; Caligiuri, A.; Stulnig, T.M.; Marra, F.; Trauner, M. The PNPLA3 I148M variant modulates the fibrogenic phenotype of human hepatic stellate cells. *Hepatology* **2017**, *65*, 1875–1890. [CrossRef] [PubMed]
69. Valenti, L.; Alisi, A.; Galmozzi, E.; Bartuli, A.; Del Menico, B.; Alterio, A.; Nobili, V. I148M patatin-like phospholipase domain-containing 3 gene variant and severity of pediatric nonalcoholic fatty liver disease. *Hepatology* **2010**, *52*, 1274–1280. [CrossRef]
70. Bruschi, F.V.; Tardelli, M.; Einwallner, E.; Claudel, T.; Trauner, M. PNPLA3 I148M up-regulates hedgehog and yap signaling in human hepatic stellate cells. *Int. J. Mol. Sci.* **2020**, *21*, 8711. [CrossRef] [PubMed]
71. Cakir, O.O.; Toker, A.; Ataseven, H.; Demir, A.; Polat, H. The importance of liver-fatty acid binding protein in diagnosis of liver damage in patients with acute hepatitis. J. Clin. *Diagn. Res.* **2017**, *11*, OC17–OC21.
72. Özenirler, S.; Degertekin, C.K.; Erkan, G.; Elbeğ, Ş.; Tuncer, C.; Kandilc, U.; Akyol, G. Serum liver fatty acid binding protein shows good correlation with liver histology in NASH. *Hepatogastroenterology* **2013**, *60*, 1095–1100. [PubMed]
73. Giannitrapani, L.; Ingrao, S.; Soresi, M.; Florena, A.M.; Spada, E.L.; Sandonato, L.; Montalto, G. Cyclooxygenase-2 expression in chronic liver diseases and hepatocellular carcinoma: An immunohistochemical study. *Ann. N. Y. Acad. Sci.* **2009**, *1155*, 293–299. [CrossRef] [PubMed]
74. Wójcik, M.; Ramadori, P.; Blaschke, M.; Sultan, S.; Khan, S.; Malik, I.A.; Schultze, F.C. Immunodetection of cyclooxygenase-2 (COX-2) is restricted to tissue macrophages in normal rat liver and to recruited mononuclear phagocytes in liver injury and cholangiocarcinoma. *Histochem. Cell Biol.* **2012**, *137*, 217–233. [CrossRef] [PubMed]
75. Raso, G.M.; Meli, R.; Di Carlo, G.; Pacilio, M.; Di Carlo, R. Inhibition of inducible nitric oxide synthase and cyclooxygenase-2 expression by flavonoids in macrophage J774A. 1. *Life Sci.* **2001**, *68*, 921–931. [CrossRef]
76. Diesen, D.L.; Kuo, P.C. Nitric oxide and redox regulation in the liver: Part II. Redox biology in pathologic hepatocytes and implications for intervention. *J. Surg. Res.* **2011**, *167*, 96–112. [CrossRef] [PubMed]
77. Cressman, D.E.; Greenbaum, L.E.; DeAngelis, R.A.; Ciliberto, G.; Furth, E.E.; Poli, V.; Taub, R. Liver failure and defective hepatocyte regeneration in interleukin-6-deficient mice. *Science* **1996**, *274*, 1379–1383. [CrossRef] [PubMed]
78. Halliwell, B.; Gutteridge, J.M. Role of free radicals and catalytic metal ions in human disease: An overview. *Meth. Enzymol.* **1990**, *186*, 1–85.
79. Espinoza, S.E.; Guo, H.; Fedarko, N.; DeZern, A.; Fried, L.P.; Xue, Q.L.; Walston, J.D. Glutathione peroxidase enzyme activity in aging. *J. Gerontol. A Biol. Sci. Med. Sci.* **2008**, *63*, 505–509. [CrossRef] [PubMed]
80. Lin, Y.; Li, Y.; Hu, X.; Liu, Z.; Chen, J.; Lu, Y.; Liao, X. The hepatoprotective role of reduced glutathione and its underlying mechanism in oxaliplatin-induced acute liver injury. *Oncol. Lett.* **2018**, *15*, 2266–2272. [CrossRef]
81. Jones, D.P.; Eklöw, L.; Thor, H.; Orrenius, S. Metabolism of hydrogen peroxide in isolated hepatocytes: Relative contributions of catalase and glutathione peroxidase in decomposition of endogenously generated H_2O_2. *Arch. Biochem. Biophys.* **1981**, *210*, 505–516. [CrossRef]
82. Verkerk, A.; Jongkind, J.F. Vascular cells under peroxide induced oxidative stress: A balance study on in vitro peroxide handling by vascular endothelial and smooth muscle cells. *Arch. Biochem. Biophys.* **1992**, *17*, 121–132. [CrossRef] [PubMed]
83. Goth, L.; Meszaros, I.; Nemeth, H. Serum catalase enzyme activity in liver diseases. *Acta. Biol. Hung.* **1987**, *38*, 287–290.
84. Kim, K.J.; Yoon, K.Y.; Lee, B.Y. Fucoidan regulate blood glucose homeostasis in C57BL/KSJ m+/+db and C57BL/KSJ db/db mice. *Fitoterapia* **2012**, *83*, 1105–1109. [CrossRef]
85. Li, J.; Zhang, Q.; Li, S.; Dai, W.; Feng, J.; Wu, L.; Liu, T.; Chen, K.; Xia, Y.; Lu, J.; et al. The natural product fucoidan ameliorates hepatic ischemia-reperfusion injury in mice. *Biomed. Pharmacother.* **2017**, *94*, 687–696. [CrossRef] [PubMed]
86. Zhu, D.Z.; Wang, Y.T.; Zhuo, Y.L.; Zhu, K.J.; Wang, X.Z.; Liu, A.J. Fucoidan inhibits LPS-induced acute lung injury in mice through regulating GSK-3 beta-Nrf2 signaling pathway. *Arch. Pharm. Res.* **2020**, *43*, 646–654. [CrossRef] [PubMed]

87. Smathers, R.L.; Fritz, K.S.; Galligan, J.J.; Shearn, C.T.; Reigan, P.; Marks, M.J.; Petersen, D.R. Characterization of 4-HNE modified L-FABP reveals alterations in structural and functional dynamics. *PLoS ONE* **2012**, *7*, 38459. [CrossRef] [PubMed]
88. Lin, C.C.; Hsieh, H.L.; Shih, R.H.; Chi, P.L.; Cheng, S.E.; Yang, C.M. Up-regulation of COX-2/PGE2 by endothelin-1 via MAPK-dependent NF-kappaB pathway in mouse brain microvascular endothelial cells. *Cell Commun. Signal.* **2013**, *11*, 8. [CrossRef]
89. Mungrue, I.N.; Gros, R.; You, X.; Pirani, A.; Azad, A.; Csont, T.; Schulz, R.; Butany, J.; Stewart, D.J.; Husain, M. Cardiomyocyte overexpression of iNOS in mice results in peroxynitrite generation, heart block, and sudden death. *J. Clin. Investig.* **2002**, *109*, 735–743. [CrossRef] [PubMed]
90. Tirkey, N.; Pilkhwal, S.; Kuhad, A.; Chopra, K. Hesperidin, a citrus bioflavonoid, decreases the oxidative stress produced by carbon tetrachloride in rat liver and kidney. *BMC Pharmacol.* **2005**, *5*, 2. [CrossRef]
91. Xu, J.; Wang, X.; Cao, K.; Dong, Z.; Feng, Z.; Liu, J. Combination of β-glucan and morus alba l. leaf extract promotes metabolic benefits in mice fed a high-fat diet. *Nutrients* **2017**, *9*, 1110. [CrossRef] [PubMed]

molecules

Article

Enhanced Pharmaceutically Active Compounds Productivity from *Streptomyces* SUK 25: Optimization, Characterization, Mechanism and Techno-Economic Analysis

Muhanna Mohammed Al-Shaibani [1,2], Radin Maya Saphira Radin Mohamed [1,*], Noraziah Mohamad Zin [2,*], Adel Al-Gheethi [1,*], Mohammed Al-Sahari [1] and Hesham Ali El Enshasy [3,4]

1 Micro-Pollutant Research Centre (MPRC), Faculty of Civil Engineering & Built Environment, Universiti Tun Hussein Onn Malaysia, Parit Raja, 86400 Batu Pahat, Malaysia; muhanan@uthm.edu.my (M.M.A.-S.); mohammedalsahari@gmail.com (M.A.-S.)

2 Center for Diagnostic, Therapeutic and Investigative Studies, Faculty of Health Sciences, Universiti Kebangsaan Malaysia, Jalan Raja Muda Abdul Aziz, 50300 Kuala Lumpur, Malaysia

3 Institute of Bioproducts Development (IBD), Universiti Teknologi Malaysia (UTM), 81310 Skudai, Malaysia; henshasy@ibd.utm.my

4 City of Scientific Research and Technology Applications (SRTA), New Burg Al Arab, 21934 Alexandria, Egypt

* Correspondence: maya@uthm.edu.my (R.M.S.R.M.); noraziah.zin@ukm.edu.my (N.M.Z.); adel@uthm.edu.my (A.A.-G.); Tel.: +607-4564236 (R.M.S.R.M.); +603-92897373 (N.M.Z.); +607-4564346 (A.A.-G.); Fax: +607-4536588(R.M.S.R.M.); +601-93924639 (N.M.Z.); +607-4536588 (A.A.-G.)

Citation: Al-Shaibani, M.M.; Radin Mohamed, R.M.S.; Zin, N.M.; Al-Gheethi, A.; Al-Sahari, M.; El Enshasy, H.A. Enhanced Pharmaceutically Active Compounds Productivity from *Streptomyces* SUK 25: Optimization, Characterization, Mechanism and Techno-Economic Analysis. *Molecules* **2021**, *26*, 2510. https://doi.org/10.3390/molecules26092510

Academic Editors: Simona Fabroni, Krystian Marszałek and Aldo Todaro

Received: 13 March 2021
Accepted: 20 April 2021
Published: 25 April 2021

Abstract: The present research aimed to enhance the pharmaceutically active compounds' (PhACs') productivity from *Streptomyces* SUK 25 in submerged fermentation using response surface methodology (RSM) as a tool for optimization. Besides, the characteristics and mechanism of PhACs against methicillin-resistant *Staphylococcus aureus* were determined. Further, the techno-economic analysis of PhACs production was estimated. The independent factors include the following: incubation time, pH, temperature, shaker rotation speed, the concentration of glucose, mannitol, and asparagine, although the responses were the dry weight of crude extracts, minimum inhibitory concentration, and inhibition zone and were determined by RSM. The PhACs were characterized using GC-MS and FTIR, while the mechanism of action was determined using gene ontology extracted from DNA microarray data. The results revealed that the best operating parameters for the dry mass crude extracts production were 8.20 mg/L, the minimum inhibitory concentrations (MIC) value was 8.00 μg/mL, and an inhibition zone of 17.60 mm was determined after 12 days, pH 7, temperature 28 °C, shaker rotation speed 120 rpm, 1 g glucose /L, 3 g mannitol/L, and 0.5 g asparagine/L with R^2 coefficient value of 0.70. The GC-MS and FTIR spectra confirmed the presence of 21 PhACs, and several functional groups were detected. The gene ontology revealed that 485 genes were upregulated and nine genes were downregulated. The specific and annual operation cost of the production of PhACs was U.S. Dollar (U.S.D) 48.61 per 100 mg compared to U.S.D 164.3/100 mg of the market price, indicating that it is economically cheaper than that at the market price.

Keywords: pharmaceutical active compounds; fermentation; optimization; *Streptomyces* SUK 25; response surface methodology; techno-economic analysis

1. Introduction

Methicillin-resistant *Staphylococcus aureus* (MRSA) strain is a group of strains belonging to *S. aureus* that has acquired resistance to a class of beta-lactam antibiotics. This strain represents a worldwide health problem due to its capability to become resistant to the currently available antibiotics. Moreover, MRSA infections can become severe and cause sepsis and life-threatening disease [1]. *Streptomyces* species have been isolated from a different environment with high potential to produce more than 10,000 of the pharmaceutically active compounds (PhACs), including alkaloid, polyene macrolide, saccharide, and antibiotics [2]. However, the PhACs produced depend on the source of *Streptomyces* species. Therefore, more studies are required to explore the ability of *Streptomyces* sp. to produce PhACs with high antimicrobial activity against pathogenic microorganisms. Moreover, production efficiency relies on external factors including temperature, pH, incubation time, and shaker rotation speed, and several internal factors such as the sources of nitrogen and carbon, which should be optimized.

The one-variable-at-a-time approach design is time-consuming and complex to conduct for each separated factor, since the optimization method entails the use of a single parameter for every trial [3,4]. For the optimum levels to be ascertained, multiple numbers of experimental trials need to be conducted. One of the investigational techniques is the response surface methodology (RSM), which represents an important tool that employs mathematical models and statistics for the optimization of parameters of fermentation processes. Besides, RSM consists of multivariable polynomial models, which are used to optimize a response based on a given set of variables. In addition, there is a wide application of the RSM approach in the optimization of microbial fermentation processes, as well as in the determination of the effect of various factors. A different approach used to increase engineered strain's expression is the application of biostatistics to optimize culture conditions, with the RSM optimization with central composite design (CCD) being the widely accepted design [5].

The gas chromatography mass spectroscopy (GC-MS) analysis involves the combination of technologies for the analysis of chemical compounds. The GC executes the separation of the compounds, whereas the MS produces the specific mass profiles of each of the detected compounds. The analysis of emissions from biogenic volatile organic compounds (VOCs) is commonly executed using GC-MS [6].

VOCs are typically small and odorous compounds manufactured as secondary metabolites produced by some microorganisms and microorganisms associated with plants and soil [7].

The chloramphenicol (CAP) and seven diketopiperazines (DKP) were isolated, purified, and identified from SUK 25 as a bacteriostatic antibiotic [3,4]. The mode of action of CAP and DKP involves the inhibition of protein synthesis, which is achieved by preventing the elongation of the protein chain through the inactivation of the activity of peptidyl transferase present in the bacterial ribosomes as reported by Zin et al. [8].

Gene ontology (GO) is a bioinformatics technique that combines intra- and interspecies representation of gene and gene product attributes. GO consists of three main aspects: the biological process, its molecular function, and the cellular function, as the technique allows operators to describe and define a gene/gene product in detail [9].

The techno-economic analysis (TEA) is an ingrained process, which is developed in performance with technology, to confirm that market-driven prices can be realized [10]. TEA is a part of the stage-gate process in product development and related research. It is a technically and economically interrelated invention and encouragement in most industries [11]. TEA is a study of a production process in an industry sector to determine roughly how effectively the economy or something within it is operating. In addition, it assesses the details of how you intend to deliver a product or service to customers and has shown the costs involved in the production of products [12,13].

The present study aimed to optimize the PhACs productivity from *Streptomyces* SUK 25 in submerged fermentation (SmF) using RSM. Furthermore, the characteristics and mechanism of action of the PhACs against MRSA were determined. Besides, the techno-economic analysis of PhACs production was estimated to studying the applicability of the generated compounds to be used as an alternative for the commercial antimicrobial products.

2. Results

2.1. RSM-Based Optimization of PHACs Production

The optimization of the production using SmF process was conducted with seven independent factors and three dependent variables as represented in Table S1. The maximum crude extracts production (y_1) was 8.20 vs. 8.58 mg/L of the actual and the predicted results, respectively. The maximum MIC (y_2) activity obtained 10.40 vs. 8.00 μg/mL for the predicted and actual results, respectively. For the maximum I.Z (y_3) values, the predicted and actual results were 18.03 vs. 17.60 mm, respectively. These findings were recorded after 12 days, at pH 7, temperature 28 °C, speed 120 rpm, 1 g glucose /L, 3 g mannitol/L, and 0.5 g asparagine/L. These results specified that the explored factors represent an essential starring role in the production of crude extracts, increasing inhibition zone (I.Z), and improving the MIC activity. The productivity of the amount of crude extracts (y_1) was associated positively and significantly ($p < 0.05$) with the incubation time (days) x_1 and pH x_2 factors, while having a non-significant correlation with other factors $x_3 - x_7$, as described in Table 1 and Table S2. In contrast, the incubation time x_1, pH x_2, and temperature x_3 revealed a significant negative quadratic effect on amount of crude extract (y_1), while the other factors $x_4 - x_7$ had a non-significant quadratic effect on the amount of crude extracts. The good activity and low MIC (y_2) was related positively and statistically significant ($p < 0.05$) with factors x_3, x_4, x_5, and x_7, while having a non-significant correlation with factors x_1 and x_2. In contrast, all responses exhibited a significant negative quadratic effect on the MIC values. The best and wide I.Z (y_3) was related positively and statistically significant ($p < 0.05$) with factors x_1, and x_2, , while having a non-significant correlation with factors temperature x_3, speed x_4, glucose x_5, mannitol x_6, and asparagine x_7. In contrast, x_1, x_2, x_3, and x_4 displayed a significant negative quadratic effect on the size of the I.Z, while there was a non-significant correlation for x_5, x_6 , and x_7 factors, as described in Table 1. The ANOVA analysis for the quadratic model, as clarified in Table 2 and Table S2, indicated that independent factors ($x_1 - x_7$) contributed significantly ($p < 0.05$) with $R^2 = 0.70$ of the coefficient for amount of crude extracts y_1, $R^2 = 0.63$ for MIC y_2, and $R^2 = 0.75$ for the I.Z y_3. These findings verify the appropriateness of the model.

Table 1. Analysis of the variance (ANOVA) of the response surface quadratic model for pharmaceutically active compounds production from *Streptomyces* SUK 25.

Source	DF	Sum of Squares			Mean Square			F Value			*p*-Value		
		y_1	y_2	y_3	y_1	y_2	y_3	y_1	y_2	y_3	y_1	y_2	y_3
Model	14	6.98	636.49	23.16	6.98	636.49	23.16	3.80	2.70	4.83	0.0026 Significant	0.0180 Significant	0.0005 Significant
Residual error	22	1.84	235.51	4.80	1.84	235.51	4.80						
Lack-of-fit	16	2.40	211.33	5.82	2.40	211.33	5.82	7.20	0.7044	2.79	0.0112 Significant	0.7327 Not significant	0.1053 Not significant
Pure error	6	0.3333	300.00	2.08	0.3333	300.00	2.08						
Total	36												

x_1 (time) (day), x_2 (pH), x_3 (temperature) (°C), x_4 (speed) (rpm), x_5 (glucose) (g/L), x_6 (mannitol) (g/L), x_7 (asparagine) (g/L), y_1 (crude extracts) (mg/L), y_2((MIC) (μg/mL), y_3((I.Z) (mm). y_1 R^2=0.70; R^2 (adj.) = 0.52,y_2 R^2 = 0.63; R^2 (adj.) = 0.40, y_3 R^2 = 0.75; R^2 (adj.) = 0.6.

The standard error of regression was used to detect the fitting of the experimental and predicted results. The value of the standard error of regression recorded in this study ranged from 0.49 to 2.17 for amount of crude extracts (y_1) and from 0.79 to 3.52 for I.Z y_3, which indicate the accuracy of the experimental results. In contrast, the standard error of y_2 was between 8.22 and 24.63; this error was high and might be related to the nature of the response and measurement methods where the MIC represented good response when it was in low value.

Table 2. Antimicrobial compounds identified in the ethyl acetate extract by GC-MS from fraction no. 2.

Peak no	R. Time	Name of the Compound	Molecular Formula	Molecular Weight (g/mol)	Area %	Quality (%)	Activity	References
1	5.799	n-dodecane	$C_{12}H_{26}$	170	0.66	76	Antioxidants, antimicro-bial	[14,15]
2	9.908	Eicosane	$C_{20}H_{42}$	283	1.38	74	Antibacterial, antifungal	[16]
3	10.200	Phenol, 2,5-bis (1,1-dimethyl ethyl)	$C_{14}H_{22}O$	206	2.24	87	Antimicrobial	[17,18]
4	21.893	Cetene	$C_{16}H_{32}$	29	0.82	75	Antioxidants	[19]
5	23.482	Diethylphthalate	$C_{12}H_{14}O4$	222	3.92	88	Antimicrobial	[16]
6	25.037	2-methyloctacosane	$C_{29}H_{60}$	409	0.81	94	Antifungal	[20]
7	25.330	1-octadecane	$C_{18}H_{36}$	252	0.86	75	Antifungal	[21]
8	28.775	Phthalic acid, isobutyl nonyl ester	$C_{21}H_{32}O_4$	348	1.11	79	Antimicrobial, antioxidants	[22]
9	29.896	Heneicosane	$C_{21}H_{44}$	297	1.86	74	Antibacterial	[23]
10	31.735	n-hexa-decanoic acid	$C_{16}H_{32}O_2$	256	1.02	73	Cosmetics, antioxidants	[24,25]

The equation for the linear and quadratic influence of the independent factors on the dependent variables were presented in Equations (1)–(3).

$$y_1 = 1.86 + 2.49\, x_{1*} + 3.07x_{2*} + 0.22x_3 - 1.13x_4 - 0.39 - 0.099x_6 - 0.27x_7 - 4.42x_{1*}^2 - 2.19x_3^2 + 0.22x_4^2 + 1.36x_5^2 + 1.70x_6^2 + 1.31x_7^2 \quad (1)$$

$$y_2 = 10.03 + 11.90\, x_1 + 1.79x_2 - 17.01x_{3*} - 25.60x_{4*} - -26.50x_{5*} + 13.41x_{6*} + 8.03x_7 - 8.48\, x_1^2 - 0.015x_2^2 - 33.44x_{3*}^2 - 24.00x_{4*}^2 - 26.15\, x_{5*}^2 - 10.75x_6^2\ 19.03x_7^2 \quad (2)$$

$$y_3 = 9.07 + 6.48x_{1*} + 4.18x_{2*} - 0.75x_3 - 1.93x_4 - 1.45x_5 + 0.63x_6 - 0.25x_7 - 7.23x_{1*}^2 - 1.68x_2^2 - 3.37x_3^2 - 1.61x_4^2 + 1.33x_5^2 + 2.30x_6^2 + 1.59x_7^2 \quad (3)$$

where * represents the factors that have a significant role; x_1 (time) (day), x_2 (pH), x_3 (temperature) (°C), x_4 (speed) (rpm), x_5 (glucose) (g/L), x_6 (mannitol) (g/L), x_7 (asparagine) (g/L), y_1 (crude extracts) (mg/L), y_2 (MIC) (µg/mL), y_3 (I.Z) (mm).

2.2. Validation of the Optimal Parameters

The best operating parameters for producing PhACs were determined after 12 days at pH 7, temperature 28 °C, shaker rotation speed 120 rpm, (1 g) glucose /L, (3 g) mannitol /L, (0.5 g) asparagine/L with 95% of the confidence level and 0.5% of significance. The autonomous considerations correlated significantly ($p < 0.05$) just at perfect circumstances of the SmF process. This evidently shows that interaction effects occur, as one factor's influence depends mostly on another factor as represented in (Figure 1a–l). The results revealed that the internal factors glucose, mannitol, and asparagine (x_5, x_6, x_7) have independent effects on the productivity of amount of crude extract y_1 without a significant interaction (Figure 1a–c). However, the external factors x_1 (incubation time), x_2 (pH), x_3 (temperature), x_4 (speed) exhibited a significant synergistic effect on the productivity of the amount of the crude extract as represented in (Figure 1d,g,k,l). Moreover, the external factors showed more influence on the amount of crude extract y_1 compared to the internal factors. Nonetheless, the external factors exhibited a synergistic interaction with the internal factors (Figure 1e,f,h–j)). In contrast, the data analysis of the interactions of the factors revealed that the internal factors (x_5, x_6, x_7) have a synergistic interaction to MIC activity as shown in (Figure S1b,e,f). The internal factors exhibited significant synergistic effects on MIC activity, as revealed in (Figure S1a,d,e), more than the external factors (Figure S1j–l)). The internal factors have no significant interaction and effects on y_3 (Figure S2i), while the external factors exhibited more efficiency in improving y_3 (Figure S2a–c,g,h).

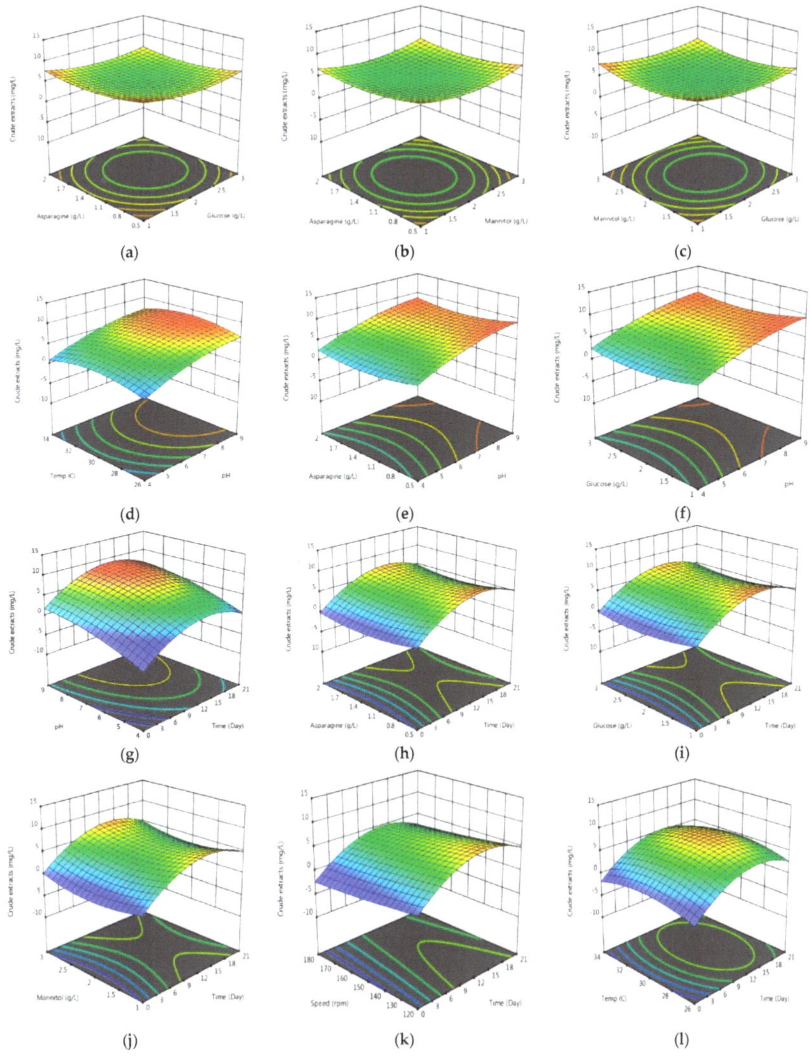

Figure 1. Three-dimensional response surface plot for interactions between x_1 (time) (day), x_2 (pH), x_3 (temperature) (°C), x_4 (speed) (rpm), x_5 (glucose) (g/L), x_6 (mannitol) (g/L), x_7 (asparagine) (g/L), and their effects on y_1 (crude extracts) (mg/L). (**a**) Effects of asparagine and glucose on weight of crude extract; (**b**) Effects of asparagine and mannitol on weight of crude extract; (**c**) Effects of mannitol and glucose on weight of crude extract; (**d**) Effects of temperature and pH on the weight of crude extract; (**e**) Effects of asparagine and pH on weight of crude extract; (**f**) Effects of glucose and pH on the weight of crude extract; (**g**) Effects of pH and time on the weight of crude extract; (**h**) Effects of asparagine and time on the weight of crude extract; (**i**) Effects of glucose and time on the weight of crude extract; (**j**) Effects of mannitol and time on the weight of crude extract; (**k**) Effects of speed and time on the weight of crude extract; (**l**) Effects of temperature and time on the weight of crude extract.

The Plackett–Burman design was used to screen and select the components of the compelling media. The resulting data indicated that the internal factors such as glucose, asparagine, and mannitol had positive impacts in contrast to the other components in the medium. Additionally, other external factors including the pH, duration of incubation, and shaker rotation speed had positive effects. In this study, the obtained R^2 value (0.70) indicated that the model explained 70% of the overall variation. When R^2 is closer to 1, the strength of the model is improved and makes a better prediction of the response. Following the RSM-based optimization, the increment in the antibacterial activity of PHACs was affirmed after comparing the results to the one-at-a-time strategy design medium. With deference to the weight of the crude extract, the antibacterial activities against MRSA ATCC 43300 and MIC were improved from 5.6 to 8.2 g/L (42.9%) and from 16 to 8 and 4 µg/mL (50%), respectively.

The increment in the zone of inhibition was from 11.4 ± 1.5 to 17.2 ± 1.5 mm. Based on the (50.9%) increase in the modified media, the finding suggests that the antibacterial metabolite production by SUK 25 was affected by the quantity of the media components. Therefore, the present experimental design enhanced the optimization of significant media components with a high level of accuracy. Accordingly, this is the first study to report a 50.9% increase in antibacterial activity against MRSA ATCC 43300 from an endophytic *Streptomyces* SUK 25.

2.3. Column Chromatography and Thin-Layer Chromatography

The present study showed that 13 fractions were collected from the crude ethyl acetate extract of SUK 25 according to the gradient elution solvent by using column chromatography. The fractions with similar R_f in TLC were evenly mixed. Only seven pure compounds were separated from fraction number 2 and fraction number 7. These compounds were purified, identified, and characterized by using HPLC, LS-MS, and NMR as reported by our previous studies carried by [6,7].

2.4. Determination of Antibacterial Activity by Using MIC and Disc Diffusion Method

The mean MICs of the triplicate samples for fraction number 2 and fraction number 7 at the best fermentation conditions against MRSA ATCC 43300 were 4 and 8 µg, respectively. In addition, the mean diameters of the triplicate samples for I.Z against the same bacteria were 17.2 ± 1.2 mm and 17 ± 1.3 mm.

2.5. Fourier Transform Infrared Spectroscopy

The presence of different compounds from various functional groups was detected. In this study, the FTIR spectroscopy demonstrated its reliability and sensitivity for the detection of bimolecular composition. The dominant bands in the case of crude extract were observed at 3348, 3319, 2945, 2833, 1558, 1449, 1394 cm^{-1}. The bands at 1121, 1092, and 972 cm^{-1} are due to the presence of C–O stretch (primary alcohol) and =C–H bend alkenes, O–H bending as carboxylic acids. The absorption spectra of the column compounds' extract samples are depicted in (Figure 2) and (Table S3).

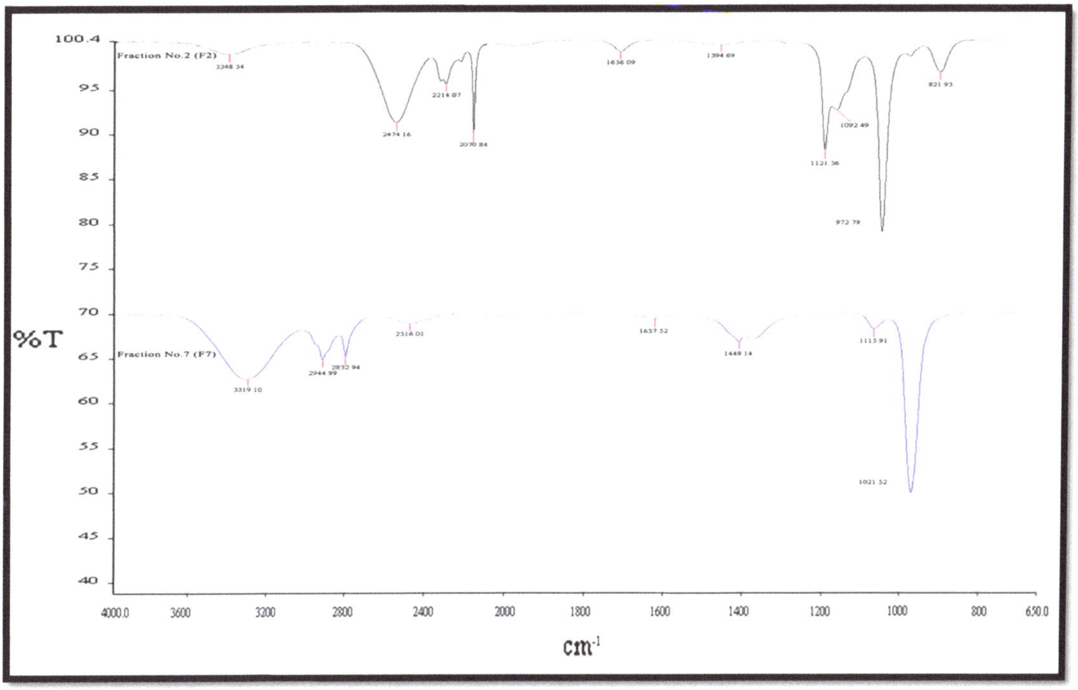

Figure 2. The Fourier transform infrared spectroscopy for fraction number 2 and 7.

2.6. GC-MS Analysis of Volatile Components

The GC-MS analysis of the volatile components of *Streptomyces* SUK 25 using the NCBI PubChem bioassay database. The peaks were prudently identified based by comparing the features of the mass spectral and NIST database. The analysis revealed that among the 38 peaks, as demonstrated in (Figure 3), the 21 compounds that demonstrated the presence of bioactive constituents have been previously documented for their antioxidant, antimicrobial, antifungal, and anti-adherence activities. Other functions include neurotropic action and presence of anti-inflammatory activates such as n-dodecane (1), eicosane (2), phenol, 2,5-bis (1,1-dimethyl ethyl) (3), cetene (4), diethylphthalate (5), 2-methyloctacosane (6), 1-octadecane (7), phthalic acid, isobutyl nonyl ester (8), heneicosane (9), and n-hexa decanoic acid (10) were isolated from fraction no. 2 as represented in Table 2. From fraction no. 7, eleven compounds were identified as di-butyl phthalate (11), 1-nonadecene, thieno [3,2-e] benzofuran(12), (1-decene) (13), di-isooctyl phthalate (14), bis (2-ethyl hexyl) phthalate (15), dodecane (16), eicosane (17), heneicosane (18), 1,2-benzenedi carboxylic acid (19), lauric acid (20), and dodecanoic acid (21). According to the previous literature, these compounds were isolated from other aforementioned studies.

Tables 2 and 3 represents the name of the compound, molecular formula, molecular weight, area, quality, and activity. In addition, Figure 4 illustrates the compounds and their chemical structures.

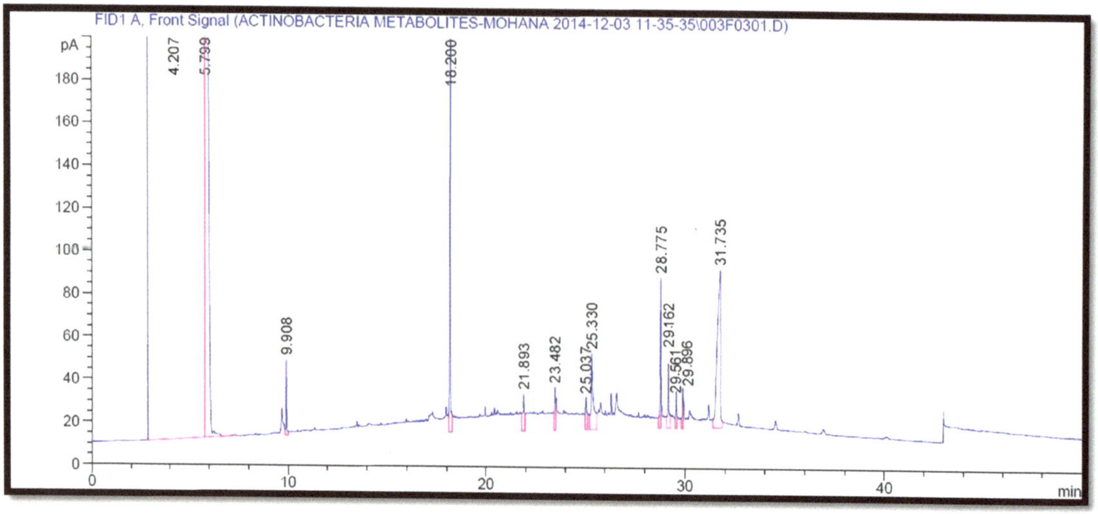

Figure 3. The GC-MS chromatograms analysis of volatile organic compounds of the ethanolic crude extract in fraction number 2 at different retention times.

Table 3. Antimicrobial compounds identified in the ethyl acetate extract by GC-MS from fraction no. 7.

Peak no	R. Time	Name of the Compound	Molecular Formula	Molecular Weight (g/mol)	Area %	Quality (%)	Activity	References
1	17.671	Di-butyl phthalate	$C_{16}H_{22}O_4$	278	2.64	86	Antifungal	[26]
2	12.301	1-nonadecene	$C_{19}H_{38}$	267	0.85	79	Antioxidants, antimicro-bial	[27]
3	12.734	Thieno[3,2-e] benzofuran	$C_{10}H_6OS$	174	1.20	92	Antibacterial	[28]
4	13.279	(1-decene)	$C_{10}H_{20}$	140	0.46	85	Antifungal	[29]
5	13.509	Diisooctyl phthalate	$C_{24}H_{38}O_4$	391	1.11	98	Anticancer, antibacterial	[30]
6	13.590	Bis (2-ethyl hexyl) phthalate	$C_{24}H_{38}O_4$	391.56	79.06	94	Antimicrobial	[31]
7	14.084	Dodecane	$C_{12}H_{26}$	170	19.41	69	neurotropic action	[14,15]
8	17.762	Eicosane	$C_{20}H_{42}$	283	5.31	71	Antimicrobial, Antifungal	[16]
9	18.532	Heneicosane	$C_{21}H_{44}$	296	3.70	72	Antifungal	[23]
10	19.648	1,2-benzenedi carboxylic acid	$C_8H_6O_4$	166.14	3.1	93	Antimicrobial and anti-inflammatory Activities	[32]
11	25.431	Lauric acid, dodecanoic acid	$C_{12}H_{24}O_2$	200.32	2.1	74	Antimicrobial and anti-inflammatory Activities	[33]

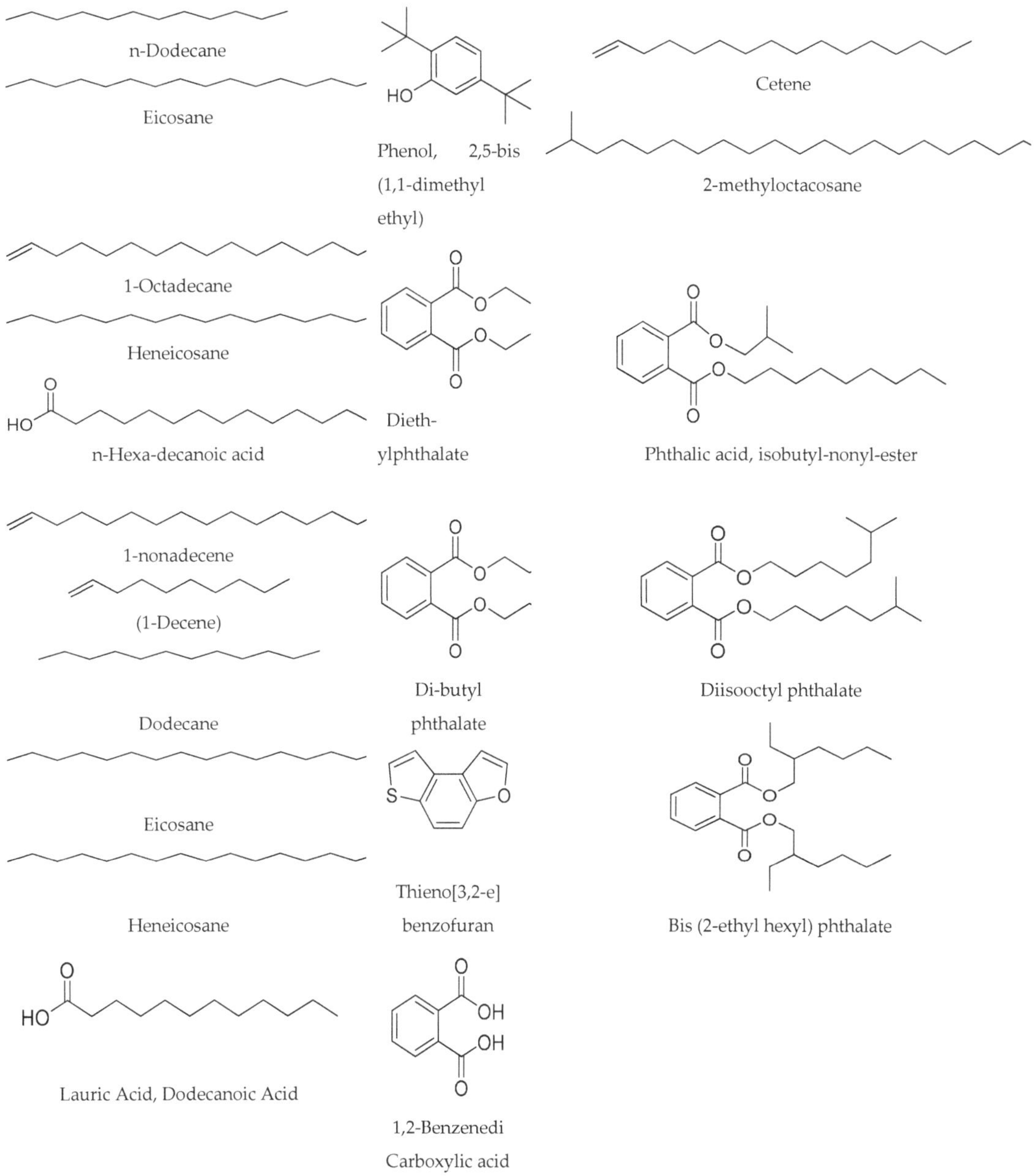

Figure 4. The chemical structure of the antimicrobial compounds identified by GC-MS from crude extract of SUK 25.

2.7. Gene Ontology and Pathway Analysis

Gene ontology analyses of hits with DAVID for the DNA microarray experiments using CAP and *cyclo*-(L-Val-L-Pro), as a representative for DKP revealed 485 genes upregulated and nine genes that were downregulated. A total of 74 genes were considered to be upregulated in the biological function group. Most categories were under the adenosine monophosphate (AMP) salvage pathways (25 gens), translation category (24 genes), metabolic process (13 genes), followed by an oxidation-reduction process (12 genes). Besides, nine genes were downregulated in the biological function under categories of the cell cycle, cell division, and cellular amino acid metabolic process. In addition, 153 genes were considered upregulated in the cellular function. The furthermost categories were under ribonucleoprotein complex (16 genes), ribosome (25 genes), and cytoplasm (19 genes). Moreover, 258 genes upregulated in the molecular function level.

Scattered under RNA binding (28 genes), transferase activity (20 genes), and structural constituent of ribosome (19 genes), oxidoreductase activity (15 genes), and catalytic activity (14 genes). The other genes, which were not documented, were under hypothetical proteins The mode of action of PhACs against MRSA ATCC 43300 was an inhibition of translation process, as shown in (Figure 5).

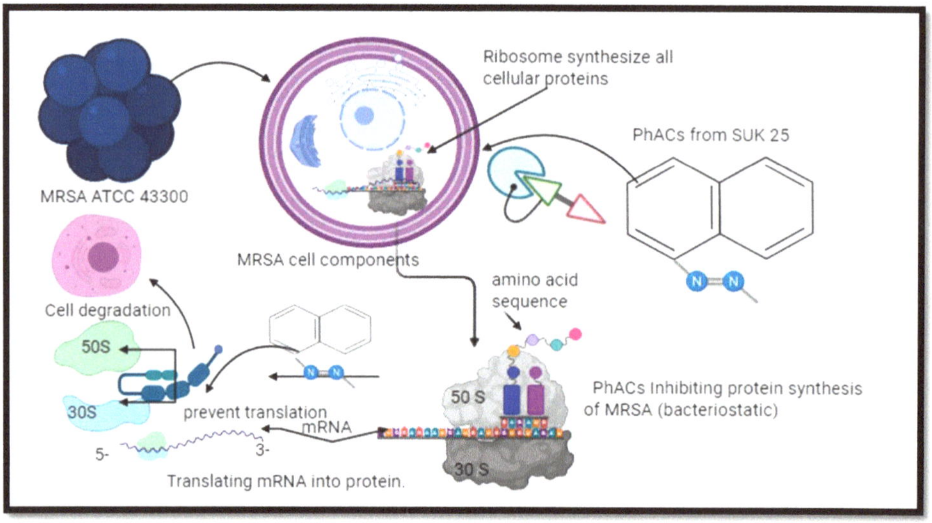

Figure 5. The mode of action of PhACs from SUK 25 against MRSA ATCC 43300. PhACs from SUK 25 stop the translation process and prevent the cell to converts genetic information carried in an mRNA molecule.

2.8. Techno-Economic Analysis of the PhACs Productivity from SUK 25

The annual report functioning period for production PhACs in the SmF process is 3300 h/year (interchangeable for 330 handling days). The complete capital expenditure (TCI) for a recommended plant together with the secure capital guesstimate (FCE) and functioning working capital cost (WCC) (Equation (4)).

$$TCI = FCE + WCC \quad (4)$$

The FCE involves the cost of purchasing equipment, installation of the system, process piping, electronic systems, percussion and sensors, yard upgrades, buildings, and perhaps even the cost of WCC, which, as indicated by Herrera-Rodriguez et al. [34], may constitute 6.5% of the FCE. Consequently, the FCE to plan a production process with 1000 m^3/day of aptitude reaches U.S. Dollar (U.S.D) 507,000.00, as described in Table 4.

Table 4. The fixed capital estimate (FCE) for production of PhACs by the SUK 25 using the submerged fermentation process.

Item Code *	Quantity	Item	Percentage of FCE	Cost
P-1/BR-101	1	Cell culture bio-reactor	33.33%	10,000.00
P-1/DS-101	2	Centrifuge		30,000.00
P-1/MSX-101	1	Purification system		17,000.00
P-1/FDR-101	1	Freeze drying		5000.00
P-1/RBS-101	1	Inoculum preparation		5000.00
P-2/V-101	2	Storage tank		25,000.00
P-1/MF-101	1	Microfiltration		14,000.00
P-1/AB-101	1	Aerobic bio-oxidation		30,000.00
P-2/DE-101	1	Dead end filtration		20,000.00
P-1/DB-101	1	Biomass storage tank		13,000.00
		Total equipment purchase cost		169,000.00
		Equipment installation	9.86%	50,000.00
		Process piping	9.86%	50,000.00
		Instrumentation and controls	9.47%	48,000.00
		Electrical systems	9.86%	50,000.00
		Buildings	7.89%	40,000.00
		Yard improvements	3.94%	20,000.00
		Construction	15.78%	80,000.00
		TOTAL		507,000.00

* Refer the item code to (Figure S3).

The surplus treatment practice costs may reach 50% of the equipment cost, which is equivalent to U.S.D 84,500.00. Therefore, the TCI of the production unit could outstretch U.S.D 591,500.00. WCC is considered to be 6.5% of the FCE (U.S.D 38,447.50). As a result, permitting Equation (4), the TCI of U.S.D 629,947.50 could be premeditated. The TEA is considered one of the superlative methods to assess the efficacy of any eccentric scheme for PhACs productivity.

2.9. Annual Operation Cost

The annual operating cost (AOC) for the production, purification, and applications of PhACs consists of the cost of underdone materials (C_{-RM}), production process waste (C_{-WG}), utilities (C_{-U}), and extra costs (C_{-E}) such as extra maintenance or any other emergency things such as downing tools due to any technical defect in the equipment during the production of bioactive compounds as assessed on an anniversary basis in accordance with Equation (5).

$$AOC = C_{RM} + C_{WG} + C_U + C_E \tag{5}$$

The C_{RM} calculates the portion of raw materials used as a production medium and the chemicals needed for the production and purification manufactured by a domestic supplier. The C_U comprises electricity and water that are mandatory for the operation progression and predictable grounded on the price for each component in the native exchange. The C_{WG} characterizes the ultimate biomass yield produced in the production process of the PhACs. The AOC is for the production, purification, and applications of pharmaceutical active compounds. The raw supplies cost (C_{RM}) was probable at U.S.D 438,700.00/year, as described in Table 5. The estimated cost of the entire charge of utilities was U.S.D 100,000/year, as conveyed in China. This approximation is on the contrary reported in Malaysia. The operating labor entails 10 hands with a mediocre salary of U.S.D 120,000 year. The preservation and insurance were projected as 2 and 1% of the FCE, respectively, as reported Gunukula et al. [35].

Table 5. Annual operation cost of the production of PhACs in submerged fermentation unit using Thornton's medium after optimization via RSM.

	Component	Price (U.S.D)	Unit	Quantity	Total Cost (U.S.D)
Raw material	K_2HPO_4	80	kg	1000 kg	80,000.00
(chemicals)	KNO_3	100	kg	500 kg	50,000.00
	$MgSO_4.2H_2O$	300	kg	200 kg	60,000.00
	$CaCl_2.H_2O$	100	kg	100 kg	10,000.00
	NaCl	100	kg	100 kg	10,000.00
	$FeCl_3$	350	kg	10 kg	3500
	Glucose	415	kg	1500 kg	62,200.00
	Asparagine	1500	kg	50 kg	75,000.00
	HCl	100/L	1 L	100 L	10,000.00
	NaOH	160/L	1 L	50 L	8000
	Reagents required for the purification				20,000.00
	Reagents required for the application test				50,000.00
Utilities	Electricity	0.04	kWh	1,000,000	40,000.00
	Water	0.01	U.S.D/m^3	6,000,000	60,000.00
Other costs	Labor	12000	U.S.D/ employee	10	120,000.00
	Maintenance	2	% of FCE		5070
	Insurance	1	% of FCE		10,140.00
		Total			673,910.00

2.10. *PhACs Profitability and Annual Revenue*

The present rate of seven PhACs that were produced and purified in the current work are *cyclo*-(tryptophanyl-prolyl) (U.S.D 40/100 mg), chloramphenicol (U.S.D 6/1000 mg), *cyclo*-(L-Val-L-Pro) (U.S.D 200/100 mg), *cyclo*-(L-Leu-L-Pro) (U.S.D 220/100 mg), *cyclo*-(L-Phe-L-Pro) (U.S.D 240/100 mg), *cyclo*-(L-Val-L-Phe) (U.S.D 250/100 mg), N-(7-hydroxy-6-methyl-octyl)-acetamide (U.S.D 200/100 mg).

The specific cost of these products C_Z (kg/U.S.D) is the deliberated, which is dependent on the utility cost C_U, annual costs of capital C_C, raw material cost C_{RM}, extra cost C_E, and the annual PhACs production, as presented by [36] (Equation (6)).

$$C_Z = (C_C + C_U + C_{RM} + C_E)/E_P \quad (6)$$

The techno-economic analysis of the PhACs productivity from SUK 25 has revealed that at the optimum productivity of PhACs with 0.1 m^3 was 17.142 g of *cyclo*-(tryptophanyl-prolyl), 14.27 g of chloramphenicol, 5 g of *cyclo*-(L-Val-L-Pro), 3.57 g of cyclo-(L-Leu-L-Pro), 2.857 g of *cyclo*-(L-Phe-L-Pro), 2.857 g of *cyclo*-(L-Val-L-Phe), and 4.285 g N-(7-hydroxy-6-methyl-octyl)-acetamide was generated. Thus, the 27.75 runs throughout the year in SmF with 100 m^3 for each run, the total quantity generated consists of 475 kg of *cyclo*-(tryptophanyl-prolyl), 396 kg of chloramphenicol, 138.75 kg of *cyclo*-(L-Val-L-Pro), 99.1 kg of *cyclo*-(L-Leu-L-Pro), 79.2 kg of *cyclo*-(L-Phe-L-Pro), 79.2 kg of *cyclo*-(L-Val-L-Phe), and 118.9 kg N-(7-hydroxy-6-methyl-octyl)-acetamide.

The total income for manufacturing these quantities offers U.S.D 1,313,776.00. An approximation of the local tax in Malaysia is 15% (U.S.D 197,066.40) and AOC with FCE (U.S.D 673,910.00). The annual turnover (after the tax) of PhACs was U.S.D 442,799.60/year, as shown in Table 5. The average precise cost of 100 mg of PhACs was evaluated to be U.S.D 48.61 per 100 mg, which is low-priced compared with the market price of U.S.D 164.3/100 mg for the average. In this investigation, the 10-year records of the internal rate of return (IRR), payback period (PBP), and net present value (NPV) were gleaned for the assessment of the economics of PhACs production. The IRR designated the competence of the investment was more than 45% in this investigation, as shown in (Figure 6), which specifies that green PhACs are economically practicable. In addition, the cost of waste production will be calculated when this techno-economic analysis is applied for a future study.

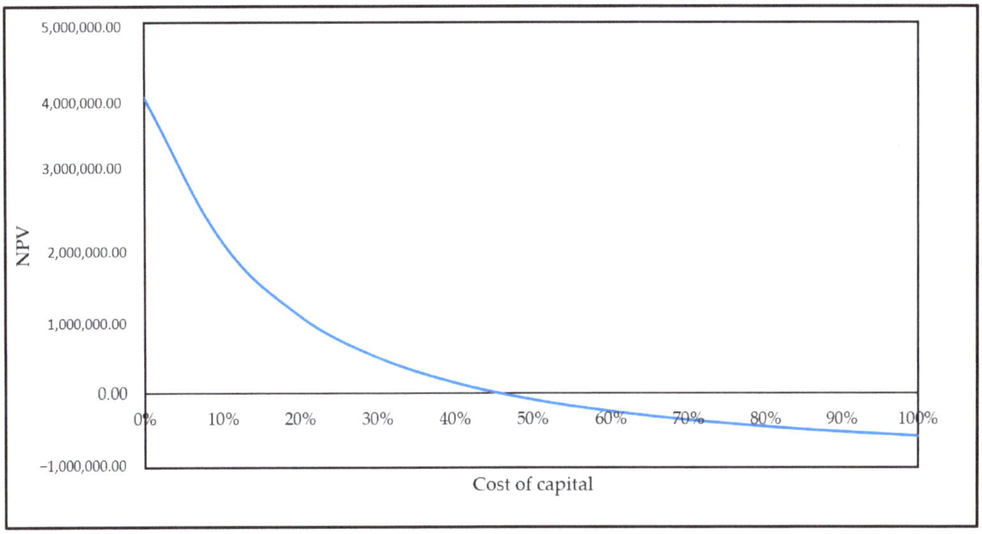

Figure 6. The IRR of the investment was more than 50% in this investigation.

3. Discussion

Different stages of optimization of fermentation have been analyzed using RSM. The key finding in this study was the statistical optimization of the significant media components that enhanced antibacterial activities against MRSA ATCC 43300 by using RSM. The model suggested that the glucose, asparagine, and mannitol affected the metabolite production from secondary metabolites produced by SUK 25, as reported with several earlier studies [37,38]. If the steepest descent experiments could be designed according to the findings of the dual-level factorial experiments, it will facilitate more insight into the optimal area of each significant factor. Moreover, the minor changes in media components, such as sources of carbon and nitrogen, or physical factors including agitation, aeration, temperature, fermentation incubation periods, and pH can considerably impact the quantity of the bacterial growth curve as in the level of $\log_{10}$ Colony-Forming Unit (C.F.U)/mL [39].

In this study, we determined the value of the secondary metabolites and metabolic profiles of associated microorganisms, their activity based on the size of inhibition zone diameter, and the values of MIC. There is a need to develop viable options, which involve the application of statistical methods for optimization with the capacity to correct existing problems related to conventional optimization [23]. The fact that any component required for the growth of microorganisms is a potential substrate adds to the complexity in culture media optimization [40]. Therefore, the improvement of antibiotic yield will entail a design of the correct medium and determining the necessary conditions for cultivation [41]. In a previous study, the one-at-a-time strategy design was applied to assess the impact of carbon and nitrogen sources, pH, and culture temperature on the production of bioactive compounds produced by SUK 25 strain as reported by Ahmad et al. [39]. On the contrary, the current study optimized the bioprocess using a combination of two methods. In the first method, the best nitrogen and carbon sources were selected for the growth of SUK 25 and precise measurement of the maximum weight of the crude extract, MIC, and maximum inhibition zone using the one-at-a-time strategy design. The findings from the present study are consistent with previous reports. The MIC values showed an improvement in the MIC values. Comparison with previous studies using *Nocardia* and *Streptosporangium* against *S. aureus* and MRSA showed that MIC against *S. aureus* were 30 μg and MRSA 40 μg [29]. The inhibition zone results are similar to a previous study carried by Mangzira Kemung et al. [42] where MUSC 125 towards MRSA ATCC 43300 showed the I.Z of 19 ± 0 mm, while the I.Z was 19.33 ± 0.58 mm against MRSA ATCC 33591.

The representations of the bonds of N–H and O–H stretching functional group represents carboxylic acids [43], and 2516, 2070 cm^{-1} bond H–C=O: C–H stretch represents the aldehydes functional group and were also attributed to C=C stretching of alkynes [44]. The peaks at 1558 cm−1 show the bond C–C stretch (in–ring) as the functional group for aromatics [45]. The peaks at 1449, 1394 cm^{-1} represent C–H bond, C–H rock as the functional group for alkanes. The peaks at 1121, 1092, and 972 cm^{-1} are due to the presence of C–O stretch (primary alcohol) and =C–H bend alkenes, O–H bending as carboxylic acids, and revealed the presence of C=O stretching of acid anhydrides [27]. The band at 821 cm^{-1} shows aromatic compounds [16].

The GC-MS analysis revealed that among the 38 peaks, 21 compounds that demonstrate the presence of bioactive constituents have been previously reported for their antimicrobial [33], antioxidant [14], antifungal [15], anti-inflammatory activates [18], and neurotropic action [18,19].

According to the gene ontology analyses of the biological function group, the most categories were under the adenosine monophosphate (AMP) salvage pathways, translation category, and metabolic process, followed by an oxidation-reduction process. Our findings are consistent with previous reports by Cui et al. [46]. The downregulated genes were under categories of the cell cycle, cell division, and cellular amino acid metabolic process, which corroborates the results from an earlier study by Guihua et al. [47]. The cellular function groups with the furthermost categories were under ribosome, cytoplasm, and ribonucleoprotein complex. These findings were published with the previous study that was conducted by Desaulniers et al. [48].

The most upregulated genes for the molecular function level were under categories of RNA binding, transferase activity, structural constituent of ribosome, oxidoreductase activity, and catalytic activity. These results are similar to the previous research conducted by Alam et al. [49].

The techno-economic analysis of the PhACs productivity from SUK 25 has revealed that the optimum productivity of PhACs with 0.1 m^3 was cheaper than that in the market price. The average of the specific cost of 100 mg of PhACs was evaluated to be U.S.D 48.61 per 100 mg, the market price was U.S.D 164.3/100 mg for the average. These results are similar to the previous researches conducted by [35,36].

In this study, the techno-economic viability and limitations that need more attention, while seeing the development-associated technologies with the major steps of the production of PhACs such as cultivation, harvesting, collection, logistics, pretreatment, fermentation, and separation will be improved during the production of the PhACs at the industrial level. In conclusion, the PhACs production activity against MRSA ATCC 43300 was improved by 42.9% of crude extract, from 16 to 4 and 8 μg/mL for MIC, while the I.Z increased by 5.8 mm. The mechanism of CAP and DKP against MRSA acted by inhibiting protein synthesis, via preventing protein chain elongation by preventing the peptidyl transferase activity of the bacterial ribosome. The average specific cost of 100 mg of PhACs was evaluated to be cheaper than that at market price.

This study represents a preliminary study in the TEA, which is an initial exploration of the TEA technology by using SuperPro Designer software. The complete information will be in the next future study, which will calculate the amount of hot water, steam, heat, and waste production. In addition, it will calculate the flowrates of individual substances.

4. Materials and Methods

4.1. Strain and Culture Condition

The antibacterial compound producing strain, *Streptomyces* SUK 25, was previously isolated from the root of the *Zingiber spectabile* plant and extensively studied for their antibacterial secondary metabolites. The SUK 25 strain was maintained over the surface of ISP2 agar for 14 days before used [3].

4.2. Basal Production Medium

The SUK 25 was cultured on ISP2 agar media, followed by incubation at 28°C for 12 days. Then, a few blocks of ISP2 media containing pure and good growth of SUK 25 were transferred into 250 mL of modified Thornton's medium, which is composed of (g/L) K_2HPO_4 1.0, KNO_3 0.5, $MgSO_4{\cdot}2H_20$ 0.2, $CaCl_2{\cdot}H_2O$ 0.1, NaCl 0.1, $FeCl_3$ 0.01, asparagine 0.5, and glucose 1.5 and pH 7.4 [50].

4.3. RSM-Based Optimization of PHACs Production

The experimental setup in the present research included optimizing the internal and external factors of the submerged fermentation (SmF) with endophytic *Streptomyces* SUK 25, production, purification, and application of PhACs against MRSA ATCC 43300. The best operating parameters for PhACs production were assessed by response surface methodology (RSM) via central composite design (CCD), designed using the Design-Expert version 11 software. Factorial complete randomized design (CRD) (7 × 3 × 3) in triplicate was employed to examine the independent variables affecting the PhACs production. Seven independent factors including incubation time (x_1) (3–21 days), pH (x_2) (4–9), (x_3) temperature (26–32 °C), (x_4) shaker rotation speed (120–180 rpm), (x_5) glucose concentration (1–3 g/L), (x_6) mannitol concentration (1–3 g/L), (x_7) asparagine concentration (0.5–2.5 g/L), and three points for each factor as well as three dependent variables were studied, and it included y_1 (crude extracts), y_2 minimum inhibitory concentration (MIC), and y_3 inhibition zone (I.Z), as described in Table 6. The selection of the range of each independent factor was as described by Kang et al. [13]. Three-stage testing, high (+), medium, and low (−), was conducted for each independent factor as presented in Table 6.

Table 6. The min and max range for each independent factor used for pharmaceutically active compounds production from *Streptomyces* SUK 25.

Medium Components	Codes	High Level (+)	Medium Level	Low Level (–)
Incubation Time (day)	x_1	21	12	0
pH	x_2	9	6	4
Temperature °C	x_3	26	30	34
Shaker rotation speed (rpm)	x_4	180	160	120
Glucose (g/L)	x_5	3	2	1
Mannitol (g/L)	x_6	3	2	1
Asparagine (g/L)	x_7	2	1.0	0.5

A total of thirty-seven experimental runs were carried out simultaneously to study the efficiency of the selected factors on the production. For the outcomes to be optimized in terms of the independent variables, the quadratic model was implemented as the following Equation (7) reported by Noman et al. [51].

$$y = \beta 0 + \sum_{i=1}^{k} \beta_i x_i + \sum_{i=1}^{k} \beta_{ii} {x_i}^2 + \sum_{i=1}^{n} \sum_{i<j}^{n} \beta_{ij} x_i x_j \quad (7)$$

where y represents the expected responses for the production process; coefficients of the linear model are represented as β_0, βi, βii, and βij; coded independent factors = x_i; while k represents the results of the responses.

The affiliation between the coded variable and experimental data is denoted as the following Equation (8):

$$x_i = e_i \frac{[\mathrm{HL} + \mathrm{LL}]/2}{[\mathrm{HL} - \mathrm{LL}]/2} \quad (8)$$

where x_i denotes the coded variable, e_i is the experiment results, LL and HL denote the lowest and the highest level of the independent factors.

4.4. Submerged Fermentation and Extraction of Secondary Metabolites Methods

The submerged fermentation (SmF) process was carried out in a 250 mL Erlenmeyer flask containing Thornton's medium (g/L), according to the composition presented by Alshaibani et al., 2016 [6]. The SmF pH medium was set to pH 7.4 using 0.1 M each of HCl and NaOH. The flasks were plugged with foam caps and enclosed with aluminum foil, before being autoclaved at 121 °C for 15 min. The autoclaved flasks were left to cool, then 100 microliter (µL) of SUK 25 spore suspension were inoculated, as described by Thakur et al. [49]. The 37 runs of the fermentation process were carried out according to the RSM design, as presented in Table S1.

A Büchner funnel incorporated with Whatman® Grade 1 filter paper was used to filter the production medium by using the vacuum filtration. Thereafter, 100 mL of the supernatant was mixed with half volume (50 mL) of ethyl acetate (EtOAC) in 3 separated times, and homogenized at 125 rotation/minutes (rpm) for 1 min. The separation and extraction of the PhACs from the supernatant was completed using 1 L separating funnels. The mixture of PhACs with EtOAC was separated by evaporation using a rotary evaporator RV 10 at 40 °C for 30 min and vacuum at 240 mbar [3].

All 37 runs were analyzed in triplicates, and the response was computed based on the mean antibacterial activity against MRSA ATCC 43300.

4.5. Determination of Dependent Variables and Determination of Antibacterial Activity

Three dependent variables were selected to be studied in this research according to the previous study reported by [25]. The first variable was (y_1), which represents the weight of the crude extracts in milligram for each run. The weight of the crude extract obtained as ethanolic extract was determined using high-precision milligram balance (Intelligent-Lab™ Milligram Balance, PM series with 0.001 g accuracy); the means of the three runs were calculated. The second variable was (y_2), which represents the MIC assay in µg/mL, and a 96-well plate was used for the MIC, which allowed a wider range of sample concentration and lower volume of samples to be tested. The third variable was (y_3), which represents that the inhibition zone in mm was performed against MRSA ATCC 43300, according to the Clinical and Laboratory Standards Institute (CLSI) M 07- A10 (2016) [52].

4.6. Purification of PhACs by Column Chromatography and Thin-Layer Chromatography

To purify the PhACs, the production process was conducted using SmF in a 1 L Erlenmeyer flask containing 800 mL Thornton's medium based on the working parameters presented in Section 2.1. The procedure was performed at incubation time (day 12), pH 7, temperature 28 °C, shaker rotation speed (140 rpm), glucose (2 g/L), mannitol (1 g/L), asparagine (2 g/L), as recommended by RSM. The crude containing the PhACs were separated, as described in Section 2.1. Thereafter, air-dried powder samples of ethyl acetate crude extracts were processed by weighing and mixing them with silica powder (2 g) and dried at room temperature. This mixture was subjected to separation through column chromatography (CC) size exclusion of 2 × 40 cm containing 150 mg of the Sephadex LH_{20}. A linear gradient of 100% chloroform was used for the starting gradient elution, and the polarity of methanol was gradually increased (100:0, 95:5, 90:10, 85:15, 80:20, 70:30, 60:40, 50:50, and 0:100 *v*:*v*) to 100% in 30 min. All the separates were dried upon column elution under lower pressure before using thin-layer chromatography (TLC) for another round of purification. Aluminum plates pre-coated with silica gel plates 60 F_{254} (20 × 20 cm, 0.25 thickness, MERCK) were used. The TLC plates were developed using a solvent system and a mobile phase consisting of ethyl acetate, hexane, and methanol (4:4:2 volume/volume/volume (*v*/*v*/*v*)). To visualize the chromatograms, a short ultraviolet (UV) (λ = 254 nm) and long-wavelength UV (λ365 nm) were used for the absorbance and fluorescence, respectively. Then, the chromatograms were sprayed with 10% H_2SO_4 in ethanol and heated at 100 °C for 10 min. The compounds with comparable R_f in TLC were mixed homogeneously, as reported by [53].

4.7. Fourier Transform Infrared Spectroscopy Spectrometer

The Fourier transform infrared spectroscopy (FT-IR) spectra was collected using a Perkin-Elmer-2000 spectrophotometer under a resolution of 4 cm^{-1} at 64 scans. A fixed weight (2 mg) of the crude extract and pure compounds were deposited in the KBr disks to create a thin film. The records of all spectra were conducted from the range 4000 to 400 cm^{-1} and analyzed with the computer software program, Spectrum for Windows (Perkin-Elmer), as reported by [54]. For every sample, triplicate spectra were processed and analyzed.

4.8. GC-MS Analysis of Volatile Components of *Streptomyces* SUK 25

The Shimadzu GC-MS QF2010 EI/NCI system was used to analyze the incomplete purified active fractions from SUK 25. Other parts incorporated with the system include a film thickness of 0.25 μm, ZB-5MS capillary column I.D. 30 m × 0.25 mm, and 5%-phenyl-arylene stationary phase, as previously reported by Sharma et al. [55]. Thereafter, 1.0 microliter (μL) of crude extract sample injected into the GC column for analysis, while maintaining the column and injector temperature at 70 and 200 °C, respectively. A split mode ratio of 40 and a flow rate of 1.51 mL/min were used. A similar flow rate (pressure of 105 kPa) was maintained for the carrier gas containing helium. The MS with ion 200 °C, interface temp: 240 °C, scan range: 40–1000 *m*/*z*, event time 0.5 s, solvent cut time: 5 min, MS start time: 5 min, MS end time: 35 min. In addition, the ionization voltage was EI (−70 ev). A comparison between the mass spectra and the data from the National Institute of Standards and Technology, US (NIST05) was employed to identify the chemical compounds in the extract [56]. The software used in observing the chromatogram was again implemented for the final analysis, by determining and confirming the molecular weight, name, and peak area percentage of unknown compounds.

4.9. Gene Ontology and Pathway Analysis

The gene ontology, classified genes, gene products, and pathway analysis were conducted by using the web-based Kyoto Encyclopedia of Genes and Genomes (KEGG) databases and using the Database software for Annotation, Visualization, and Integrated Discovery (DAVID) (https://david.ncifcrf.gov, Accessed on: 13 November 2020).

4.10. Techno-Economic Analysis of the PhACs Productivity from SUK 25

4.10.1. Process Simulation

The TEA was conducted for a facility that handles 1000 m^3/day of dry crude extract production in SmF process, isolated, and purified from SUK 25.

The facility was assumed operating 3300 h/year (interchangeable for 330 handling days). The TEA was developed based on experimental data from the previous studies, as indicated by [57]. The methods proposed have three stages intended as the production (A), purification (B), and applications (C) as illustrated in (Figure 7).

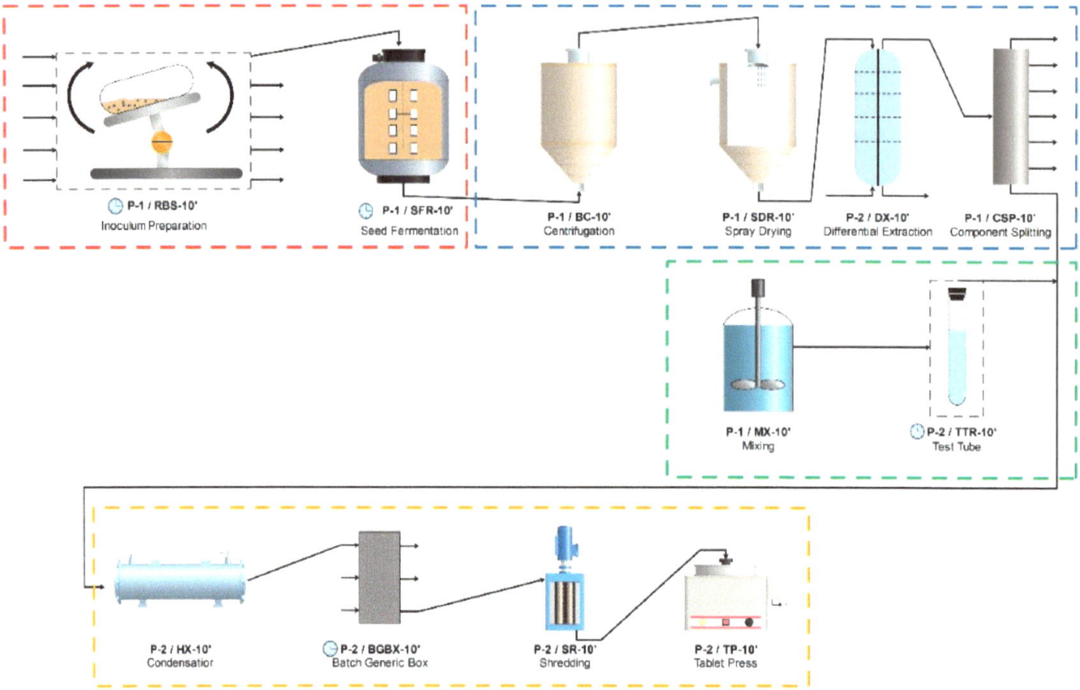

Figure 7. The proposed flowchart of PhACs production in SmF process.

4.10.2. Production and Extraction of PhACs in SmF Media

In phase A, the production of PhACs in SmF process was conducted. To improve the PhACs productivity, the SmF process was performed under the proper operating parameters as presented previously in the Section 4.4.

4.10.3. Column Chromatography and Thin-Layer Chromatography

In stage B, column chromatography and TLC were applied for the purification of PhACs, as presented in Section 4.6.

4.10.4. Effectiveness of the PhACs and Determination of Antibacterial Activity

In stage C, the effectiveness of the PhACs against the MRSA ATCC 43300 is carried out, as presented in Section 4.5.

4.10.5. Economic Analysis

The models built in the techno-economic analysis was applied by using the SuperPro Designer for the production of seven pure active compounds isolated and identified in our previous research from SUK 25 against MRSA ATCC 43300, as represents the first attempt of a TEA of production of PhACs from SUK 25. The proposed flowchart of PhACs was then established to estimate the capital and operating costs of each process. Table 4 details the fixed capital estimate (FCE) for production of PhACs by the SUK 25 using SmF and the other economic assumptions made in the study.

The cell culture bio-reactor, centrifuge, purification system, freeze drying, inoculum preparation, storage tank, microfiltration, aerobic bio-oxidation, dead end filtration, and biomass storage tank. In addition, the equipment installation, process piping, instrumentation and controls, electrical systems, buildings, yard improvements, and construction were considered in the analysis. This approach assumes that the technology is mature and that several of facilities employing the same process are established and are fully operational. Vendor quotes were used when the costs were not available for the row material [58]. The total direct costs were estimated as the sum of all the installed equipment costs, plus the costs for buildings, piping, and site development, as shown in Table 4.

4.10.6. Annual Operation Cost

The annual operating cost (AOC) for the production, purification, and applications of PhACs consists of the cost of underdone materials ($C\text{-}_{RM}$), production process waste ($C\text{-}_{WG}$), utilities ($C_{\text{-}U}$), and extra costs ($C_{\text{-}E}$). The operating costs that include variable costs (feedstocks, utilities, etc.), fixed costs (salaries, maintenance, taxes, etc.), and general costs (e.g., marketing expenses) were estimated based on the principles, as reported by [58]. The $C\text{-}_{U}$ comprises electricity and water that are mandatory for the operation progression and predictable grounded on the price for each component in the native exchange. The $C\_{_{WG}}$ characterizes the ultimate biomass yield produced in the production process of the PhACs. Techno-economic problems derive into demonstration when a procedure is scaled-up or needs to be scaled-up to a marketable level. The main problem to be kept in mind is choosing the right optimized parameters, which are measured for cost economics. Reagents required for the purification and reagents required for the application test, electricity, and water as well as equipment used in the bioprocess are inclusive in the calculation. The sensitivity analysis was laboring to evaluate the influence that different conditions may have on the price of the purified PhACs.

4.11. *Statistical Analysis*

The linear effect of the independent factors and interactions between these factors were examined using analysis of variance (ANOVA), and p-values ≤ 0.05 were considered significant. The efficiency of SmF for producing PhACs was investigated as a function for the quadratic model and coefficient adjusted R (R^2 adj.). All experiments were conducted in triplicate. Three-dimensional (3D) graphical illustrations were used to depict the association between factors and the degradation process. To show the association between the experimental phases of each explanatory variable, 3D response surface plots were produced.

5. Conclusions

The most recognizable finding to appear from this study is that the PhACs' production activity against MRSA ATCC 43300 was improved by 42.9% of dry weight of the crude extract; MIC also enhanced from 16 to 8 µg/mL, while the I.Z increased by 5.8 mm. The GC-MS analysis revealed that 21 compounds that demonstrate the presence of bioactive constituents were identified. The mechanism of CAP and DKP against MRSA acted by inhibiting protein synthesis, via preventing protein chain elongation by preventing the peptidyl transferase activity of the bacterial ribosome. The average specific cost of 100 mg of PhACs was evaluated to be cheaper than that at the market price.

Supplementary Materials: The following are available online, Figure S1: Three-dimensional response surface plot for interactions between x_1 (time) (day), x_2 (pH), x_3 (temperature) (°C), x_4 (speed) (rpm), x_5 (glucose) (g/L), x_6 (mannitol) (g/L), x_7 (asparagine) (g/L) and their effects on y_2 ((MIC) (µg/mL). Figure S2: Three-dimensional response surface plot for interactions between x_1 (time) (day), x_2 (pH), x_3 (temperature) (°C), x_4 (speed) (rpm), x_5 (glucose) (g/L), x_6 (mannitol) (g/L), x_7 (asparagine) (g/L) and their effects on y_3 (I.Z) mm. Table S1: The central composite design arrangement and responses for pharmaceutically active compounds production from *Streptomyces* SUK 25. Table S2; The regression coefficient and their significance of the quadratic model for pharmaceutically active compounds production from *Streptomyces* SUK 25. Table S3: Peaks absorbance values of FT-IR spectra of PHACs.

Author Contributions: M.M.A.-S., N.M.Z., R.M.S.R.M. developed the theoretical formalism; A.A.-G., M.A.-S., H.A.E.E. performed the analytic calculations, and both authors A.A.-G. and M.M.A.-S. contributed to the final version of the manuscript. N.M.Z. and R.M.S.R.M. supervised the project. All authors have read and agreed to the published version of the manuscript.

Funding: FRGS (UKM-NN-03-FRGS0042-2009 and FRGS/1/2020/WAB02/UTHM/03/1.

Institutional Review Board Statement: Not applicable.

Informed Consent Statement: Not applicable.

Data Availability Statement: The data presented in this study are available in the article and in the Supplementary Data online at www.mdpi.com/xxx/s1 and https://drive.google.com/file/d/1-DxbaSBpAia0q38SVc78LltY3WddPoNa/view?usp=sharing.

Acknowledgments: The authors appreciate the support and funding provided by the Ministry of Higher Education Malaysia (KPT) for the FRGS grant with reference code: FRGS UKM-NN-03-FRGS0042-2009 and FRGS/1/2020/WAB02/UTHM/03/1-Photocatalysis of Trisiloxane and Pathogenic Bacteria in Greywater using Eco-Friendly Green Nanoparticles (ECO-GNPs) for Safe Disposal for this research project.

Conflicts of Interest: The authors declare no conflict of interest.

Sample Availability: Samples of the crude extract and secondary metabolites are available from the authors for a limited time.

References

1. Butini, M.E.; Abbandonato, G.; Di Rienzo, C.; Trampuz, A.; Di Luca, M. Isothermal Microcalorimetry Detects the Presence of Persister Cells in a *Staphylococcus aureus* Biofilm After Vancomycin Treatment. *Front. Microbiol.* **2019**, *10*, 332. [CrossRef]
2. Qi, D.; Zou, L.; Zhou, D.; Chen, Y.; Gao, Z.; Feng, R.; Zhang, M.; Li, K.; Xie, J.; Wang, W. Taxonomy and Broad-Spectrum Antifungal Activity of *Streptomyces* sp. SCA3-4 Isolated from Rhizosphere Soil of Opuntia stricta. *Front. Microbiol.* **2019**, *10*, 1390. [CrossRef]
3. Alshaibani, M.M.; Jalil, J.; Sidik, N.M.; Edrada-Ebel, R.; Zin, N.M. Isolation and characterization of cyclo-(tryptophanyl-prolyl) and chloramphenicol from *Streptomyces* sp. SUK 25 with antimethicillin-resistant *Staphylococcus aureus* activity. *Drug Des. Devel. Ther.* **2016**, *10*, 1817–1827.
4. AlShaibani, M.M.; Jalil, J.; Sidik, N.M.; Ahmad, S.J.; Kamal, N.; Edrada-Ebel, R.; Zin, N.M. Isolation, purification and characterization of five active diketopiperazine derivatives from endophytic *Streptomyces* SUK 25 with antimicrobial and cytotoxic activities. *J. Microbiol. Biotechnol.* **2017**, *27*, 1249–1256. [CrossRef]
5. Pathak, L.; Singh, V.; Niwas, R.; Osama, K.; Khan, S.; Haque, S.; Tripathi, C.K.M.; Mishra, B.N. Artificial Intelligence versus Statistical Modeling and Optimization of Cholesterol Oxidase Production by using *Streptomyces* Sp. *PLoS ONE* **2015**, *10*, e0137268. [CrossRef]
6. Konappa, N.; Udayashankar, A.C.; Krishnamurthy, S.; Pradeep, C.K.; Chowdappa, S.; Jogaiah, S. GC–MS analysis of phytoconstituents from Amomum nilgiricum and molecular docking interactions of bioactive serverogenin acetate with target proteins. *Sci. Rep.* **2020**, *10*, 1–23. [CrossRef]
7. Schulz-Bohm, K.; Martín-Sánchez, L.; Garbeva, P. Microbial Volatiles: Small Molecules with an Important Role in Intra- and Inter-Kingdom Interactions. *Front. Microbiol.* **2017**, *8*, 2484. [CrossRef]
8. Zin, N.M.; Al-Shaibani, M.M.; Jalil, J.; Sukri, A.; Al-Maleki, A.R.; Sidik, N.M. Profiling of gene expression in methicillin-resistant *Staphylococcus aureus* in response to cyclo-(l-Val-l-Pro) and chloramphenicol isolated from *Streptomyces* sp., SUK 25 reveals gene downregulation in multiple biological targets. *Arch. Microbiol.* **2020**, *202*, 1–10. [CrossRef]
9. Zhao, Y.; Wang, J.; Chen, J.; Zhang, X.; Guo, M.; Yu, G. A Literature Review of Gene Function Prediction by Modeling Gene Ontology. *Front. Genet.* **2020**, *11*, 400. [CrossRef]
10. Mikhelkis, L.; Govindarajan, V. Techno-Economic and Partial Environmental Analysis of Carbon Capture and Storage (CCS) and Carbon Capture, Utilization, and Storage (CCU/S): Case Study from Proposed Waste-Fed District-Heating Incinerator in Sweden. *Sustainability* **2020**, *12*, 5922. [CrossRef]
11. Kumar, V.; Al-Gheethi, A.; Asharuddin, S.M.; Othman, N. Potential of cassava peels as a sustainable coagulant aid for institutional wastewater treatment: Characterisation, optimisation and techno-economic analysis. *Chem. Eng. J.* **2020**. [CrossRef]
12. Teymouri, A.; Adams, K.J.; Dong, T.; Kumar, S. Evaluation of lipid extractability after flash hydrolysis of algae. *Fuel* **2018**, *224*, 23–31. [CrossRef]
13. Kang, C.; Wen, T.-C.; Kang, J.-C.; Meng, Z.-B.; Li, G.-R.; Hyde, K.D. Optimization of Large-Scale Culture Conditions for the Production of Cordycepin with Cordyceps militarisby Liquid Static Culture. *Sci. World J.* **2014**, *2014*, 510627. [CrossRef] [PubMed]
14. Kang, E.B.; Mazrad, Z.A.I.; Robby, A.I.; In, I.; Park, S.Y. Alkaline phosphatase-responsive fluorescent polymer probe coated surface for colorimetric bacteria detection. *Eur. Polym. J.* **2018**, *105*, 217–225. [CrossRef]

15. Jayakumar, V.; Sundar, A.R.; Viswanathan, R. Biocontrol of *Colletotrichum falcatum* with volatile metabolites produced by endophytic bacteria and profiling VOCs by headspace SPME coupled with GC–MS. *Sugar Tech.* **2021**, *23*, 94–107. [CrossRef]
16. Ahsan, T.; Chen, J.; Zhao, X.; Irfan, M.; Wu, Y. Extraction and identification of bioactive compounds (eicosane and dibutyl phthalate) produced by *Streptomyces* strain KX852460 for the biological control of *Rhizoctonia solani* AG-3 strain KX852461 to control target spot disease in tobacco leaf. *AMB Express* **2017**, *7*, 54. [CrossRef]
17. Padmavathi, A.R.; Abinaya, B.; Pandian, S.K. Phenol, 2,4-bis(1,1-dimethylethyl) of marine bacterial origin inhibits quorum sensing mediated biofilm formation in the uropathogen *Serratia marcescens*. *Biofouling* **2014**, *30*, 1111–1122. [CrossRef] [PubMed]
18. Ren, J.; Wang, J.; Karthikeyan, S.; Liu, H.; Cai, J. Natural anti-phytopathogenic fungi compound phenol, 2, 4-bis (1, 1-dimethylethyl) from *Pseudomonas fluorescens* TL-1. *Indian J. Biochem. Biophys.* **2019**, *56*, 162–168.
19. Bouaziz, F.; Koubaa, M.; Chaabene, M.; Barba, F.J.; Ghorbel, R.E.; Chaabouni, S.E. High Throughput Screening for Bioactive Volatile Compounds and Polyphenols from Almond (Prunus amygdalus) Gum: Assessment of Their Antioxidant and Antibacterial Activities. *J. Food Process. Preserv.* **2016**, *41*, e12996. [CrossRef]
20. Barretto, D.; Vootla, S. GC-MS Analysis of Bioactive Compounds and Antimicrobial Activity of *Cryptococcus rajasthanensis* Ky627764 Isolated from Bombyx Mori Gut Microflora. *Int. J. Adv. Res.* **2018**, *6*, 525–538. [CrossRef]
21. Girija, S.; Duraipandiyan, V.; Kuppusamy, P.S.; Gajendran, H.; Rajagopal, R. Chromatographic Characterization and GC-MS Evaluation of the Bioactive Constituents with Antimicrobial Potential from the Pigmented Ink of *Loligo duvauceli*. *Int. Sch. Res. Not.* **2014**, *2014*, 820745. [CrossRef]
22. Gogoi, D.; Bora, G.; Borgohain, R.; Handique, J.G. Antioxidant Capacity and GC-MS Analysis of Hexane, Ethylacetate and Methanol Extracts of *Ficus bhotanica*–A Potential Folklore Medicinal Plant. *Int. J. Pharmacogn. Phytochem. Res.* **2018**, *10*, 201–212.
23. Vanitha, V.; Vijayakumar, S.; Nilavukkarasi, M.; Punitha, V.; Vidhya, E.; Praseetha, P. Heneicosane—A novel microbicidal bioactive alkane identified from *Plumbago zeylanica* L. *Ind. Crop. Prod.* **2020**, *154*, 112748. [CrossRef]
24. Aparna, V.; Dileep, K.V.; Mandal, P.K.; Karthe, P.; Sadasivan, C.; Haridas, M. Anti-inflammatory property of n-hexadecanoic acid: Structural evidence and kinetic assessment. *Chem. Biol. Drug Des.* **2012**, *80*, 434–439. [CrossRef]
25. Manikandan, S.; Lakshmanan, G.M.A.; Ansarali, S. Identification of Bioactive Compounds from Selected *Plectranthus* Species by Gas Chromatography-Mass Spectroscopy and Fourier Transform Infra-Red Spectroscopy. *J. Biol. Act. Prod. Nat.* **2017**, *7*, 438–451. [CrossRef]
26. Rahnama, K.; Jahanshahi, M.; Nasrollanejad, S.; Fatemi, M.H.; Tabarestani, M.S. Identification of Volatile Organic Compounds from Trichoderma virens (6011) by GC-MS and Separation of a Bioactive Compound via Nanotechnology. *Int. J. Eng.* **2016**, *29*, 1347–1353. [CrossRef]
27. Tonisi, S.; Okaiyeto, K.; Hoppe, H.; Mabinya, L.V.; Nwodo, U.U.; Okoh, A.I. Chemical constituents, antioxidant and cytotoxicity properties of *Leonotisleonurus* used in the folklore management of neurological disorders in the Eastern Cape, South Africa. *3 Biotech* **2020**, *10*, 1–14. [CrossRef]
28. Karabay-Yavasoglu, N.U.; Sukatar, A.; Ozdemir, G.; Horzum, Z. Antimicrobial activity of volatile components and various extracts of the red *AlgaJania rubens*. *Phytother. Res.* **2007**, *21*, 153–156. [CrossRef]
29. Palanna, K.; Narendrappa, T.; Basavaraj, S.; Shreenivasa, K. Efficacy of Fungal and Bacterial Bio-control Agents on *Ganoderma* Spp. Causing Foot Rot of Arecanut. *Int. J. Agric. Innov. Res.* **2017**, *6*, 299–304.
30. Nguyen, N.H.; Nguyen, T.T.; Ma, P.C.; Ta, Q.T.H.; Duong, T.-H.; Vo, V.G. Potential Antimicrobial and Anticancer Activities of an Ethanol Extract from *Bouea macrophylla*. *Molecules* **2020**, *25*, 1996. [CrossRef]
31. Perveen, I.; Raza, M.A.; Iqbal, T.; Naz, I.; Sehar, S.; Ahmed, S. Isolation of anticancer and antimicrobial metabolites from *Epicoccum nigrum*; endophyte of Ferula sumbul. *Microb. Pathog.* **2017**, *110*, 214–224. [CrossRef]
32. Salem, M.Z.M.; Zayed, M.Z.; Ali, H.M.; El-Kareem, M.S.M.A. Chemical composition, antioxidant and antibacterial activities of extracts from *Schinus molle* wood branch growing in Egypt. *J. Wood Sci.* **2016**, *62*, 548–561. [CrossRef]
33. Yang, H.-T.; Chen, J.-W.; Rathod, J.; Jiang, Y.-Z.; Tsai, P.-J.; Hung, Y.-P.; Ko, W.-C.; Paredes-Sabja, D.; Huang, I.-H. Lauric Acid Is an Inhibitor of *Clostridium difficile* Growth in Vitro and Reduces Inflammation in a Mouse Infection Model. *Front. Microbiol.* **2018**, *8*, 2635. [CrossRef]
34. Herrera-Rodriguez, T.; Parejo-Palacio, V.; Gonzalez-Delgado, A.D. Technoeconomic sensibility analysis of industrial agar production from red Algae. *Chem. Eng. Trans.* **2018**, *70*, 2029–2034.
35. Gunukula, S.; Klein, S.J.; Pendse, H.P.; DeSisto, W.J.; Wheeler, M.C. Techno-economic analysis of thermal deoxygenation based biorefineries for the coproduction of fuels and chemicals. *Appl. Energy* **2018**, *214*, 16–23. [CrossRef]
36. Han, W.; Hu, Y.; Li, S.; Huang, J.; Nie, Q.; Zhao, H.; Tang, J. Simultaneous dark fermentative hydrogen and ethanol production from waste bread in a mixed packed tank reactor. *J. Clean. Prod.* **2017**, *141*, 608–611. [CrossRef]
37. Bhatia, S.K.; Lee, B.-R.; Sathiyanarayanan, G.; Song, H.-S.; Kim, J.; Jeon, J.-M.; Kim, J.-H.; Park, S.-H.; Yu, J.-H.; Park, K.; et al. Medium engineering for enhanced production of undecylprodigiosin antibiotic in *Streptomyces coelicolor* using oil palm biomass hydrolysate as a carbon source. *Bioresour. Technol.* **2016**, *217*, 141–149. [CrossRef]
38. Navarrete-Bolaños, J.L.; Téllez-Martínez, M.G.; Miranda-López, R.; Jiménez-Islas, H. An experimental strategy validated to design cost-effective culture media based on response surface methodology. *Prep. Biochem. Biotechnol.* **2017**, *47*, 578–588. [CrossRef]
39. Ahmad, S.J.; Suhaini, S.; Sidek, H.M.; Basri, D.F.; Zin, N.M. Anti-methicillin resistant *Staphylococcus aureus* activity and optimal culture condition of *Streptomyces* sp. SUK 25. *Jundishapur J. Microbiol.* **2015**, *8*, e16784. [CrossRef]

40. Das, R.; Romi, W.; Das, R.; Sharma, H.K.; Thakur, D. Antimicrobial potentiality of actinobacteria isolated from two microbiologically unexplored forest ecosystems of Northeast India. *BMC Microbiol.* **2018**, *18*, 1–16. [CrossRef]
41. Yun, T.Y.; Feng, R.J.; Zhou, D.B.; Pan, Y.Y.; Chen, Y.F.; Wang, F.; Yin, L.Y.; Zhang, Y.D.; Xie, J.H. Optimization of fermentation conditions through response surface methodology for enhanced antibacterial metabolite production by *Streptomyces* sp. 1-14 from cassava rhizosphere. *PLoS ONE* **2018**, *13*, e0206497. [CrossRef]
42. Kemung, H.M.; Tan, L.T.-H.; Chan, K.-G.; Ser, H.-L.; Law, J.W.-F.; Lee, L.-H.; Goh, B.-H. *Streptomyces* sp. Strain MUSC 125 from Mangrove Soil in Malaysia with Anti-MRSA, Anti-Biofilm and Antioxidant Activities. *Molecules* **2020**, *25*, 3545. [CrossRef] [PubMed]
43. Song, H.; Liu, G.; Zhang, J.; Wu, J. Pyrolysis characteristics and kinetics of low rank coals by TG-FTIR method. *Fuel Process. Technol.* **2017**, *156*, 454–460. [CrossRef]
44. Lindenmaier, R.; Williams, S.D.; Sams, R.L.; Johnson, T.J. Quantitative Infrared Absorption Spectra and Vibrational Assignments of Crotonaldehyde and Methyl Vinyl Ketone Using Gas-Phase Mid-Infrared, Far-Infrared, and Liquid Raman Spectra: S-cis vs s-trans Composition Confirmed via Temperature Studies and ab Initio Methods. *J. Phys. Chem. A* **2017**, *121*, 1195–1212. [CrossRef]
45. Dong, C.-D.; Chen, C.-W.; Hung, C.-M. Synthesis of magnetic biochar from bamboo biomass to activate persulfate for the removal of polycyclic aromatic hydrocarbons in marine sediments. *Bioresour. Technol.* **2017**, *245*, 188–195. [CrossRef]
46. Cui, L.; Wang, X.; Huang, D.; Zhao, Y.; Feng, J.; Lu, Q.; Pu, Q.; Wang, Y.; Cheng, G.; Wu, M.; et al. CRISPR-cas3 of *Salmonella* Upregulates Bacterial Biofilm Formation and Virulence to Host Cells by Targeting Quorum-Sensing Systems. *Pathogens* **2020**, *9*, 53. [CrossRef]
47. Guihua, Z.; Cao, S.; Zhang, G.; Xiao, Y.; Liu, S.; Shang, Y. Florfenicol-induced Mitochondrial Dysfunction Suppresses Cell Proliferation and Autophagy in Fibroblasts. *Sci. Rep.* **2017**, *7*, 1–13. [CrossRef]
48. Desaulniers, A.B.; Kishore, N.; Adames, K.; Nargang, F.E. Characterization of Single Gene Deletion Mutants Affecting Alternative Oxidase Production in *Neurospora crassa*: Role of the *yvh1* Gene. *Microorganisms* **2020**, *8*, 1186. [CrossRef]
49. Alam, A.; Imam, N.; Ahmed, M.M.; Tazyeen, S.; Tamkeen, N.; Farooqui, A.; Malik, Z.; Ishrat, R. Identification and Classification of Differentially Expressed Genes and Network Meta-Analysis Reveals Potential Molecular Signatures Associated with Tuberculosis. *Front. Genet.* **2019**, *10*, 2029–2034. [CrossRef]
50. Thakur, D.; Bora, T.; Bordoloi, G.; Mazumdar, S. Influence of nutrition and culturing conditions for optimum growth and antimicrobial metabolite production by *Streptomyces* sp. 201. *J. Med Mycol.* **2009**, *19*, 161–167. [CrossRef]
51. Noman, E.; Al-Gheethi, A.; Talip, B.A.; Mohamed, R.; Kassim, A.H. Oxidative enzymes from newly local strain Aspergillus iizukae EAN605 using pumpkin peels as a production substrate: Optimized production, characterization, application and techno-economic analysis. *J. Hazard. Mater.* **2020**, *386*, 121954. [CrossRef]
52. Clinical and Laboratory Standards Institute (CLSI). *Performance Standards for Antimicrobial Susceptibility Testing*, 26th ed.; CLSI document M100-S26; Clinical and Laboratory Standards Institute: Wayne, PA, USA, 2016.
53. Gerlach, A.D.C.L.; Gadea, A.; da Silveira, R.M.B.; Clerc, P.; Lohézic-le Dévéhat, F. The use of anisaldehyde sulfuric acid as an alternative spray reagent in TLC analysis reveals three classes of compounds in the genus *Usnea adans.(*Parmeliaceae, li-chenized Ascomycota). *Preprints* **2018**. [CrossRef]
54. She, W.; Ye, W.; Shi, Y.; Zhou, L.; Zhang, Z.; Chen, F.; Qian, P.-Y. A novel chresdihydrochalcone from *Streptomyces chrestomyceticus* exhibiting activity against Gram-positive bacteria. *J. Antibiot.* **2020**, *73*, 429–434. [CrossRef]
55. Sharma, P.; Kalita, M.C.; Thakur, D. Broad Spectrum Antimicrobial Activity of Forest-Derived Soil Actinomycete, *Nocardia* sp. PB-52. *Front. Microbiol.* **2016**, *7*, 347. [CrossRef]
56. Al-Dhabi, N.A.; Esmail, G.A.; Duraipandiyan, V.; Arasu, M.V.; Salem-Bekhit, M.M. Isolation, identification and screening of antimicrobial thermophilic *Streptomyces* sp. Al-Dhabi-1 isolated from Tharban hot spring, Saudi Arabia. *Extremophiles* **2015**, *20*, 79–90. [CrossRef]
57. Gopalakrishnan, Y.; Al-Gheethi, A.; Malek, M.A.; Azlan, M.M.; Al-Sahari, M.; Mohamed, R.R.; Alkhadher, S.; Noman, E. Removal of Basic Brown 16 from Aqueous Solution Using Durian Shell Adsorbent, Optimisation and Techno-Economic Analysis. *Sustainability* **2020**, *12*, 8928. [CrossRef]
58. Jones, S.B.; Zhu, Y.; Anderson, D.B.; Hallen, R.T.; Elliott, D.C.; Schmidt, A.J.; Drennan, C. *Process Design and Economics for the Conversion of Algal Biomass to Hydrocarbons: Whole Algae Hydrothermal Liquefaction and Upgrading. (No. PNNL-23227)*; Pacific Northwest National Lab.(PNNL): Richland, WA, USA, 2014.

molecules

MDPI

Article

Microbial Application to Improve Olive Mill Wastewater Phenolic Extracts

Flora V. Romeo [1,*], Gina Granuzzo [1], Paola Foti [2], Gabriele Ballistreri [1], Cinzia Caggia [2] and Paolo Rapisarda [1]

[1] Council for Agricultural Research and Economics (CREA), Research Centre for Olive, Fruit and Citrus Crops, 95024 Acireale, Italy; gina.granuzzo@gmail.com (G.G.); gabriele.ballistreri@crea.gov.it (G.B.); paolo.rapisarda@crea.gov.it (P.R.)

[2] Department of Agricultural, Food and Environment, University of Catania, 95123 Catania, Italy; paola.foti@phd.unict.it (P.F.); ccaggia@unict.it (C.C.)

* Correspondence: floravaleria.romeo@crea.gov.it; Tel.: +39-095-765-3136

Abstract: Olive mill wastewater (OMW) contains valuable and interesting bioactive compounds, among which is hydroxytyrosol, which is characterized by a remarkable antioxidant activity. Due to the health claims related to olive polyphenols, the aim of this study was to obtain an extract from OMW with an increased level of hydroxytyrosol by means of microbial enzymatic activity. For this purpose, four commercial adsorbent resins were selected and tested. The beta-glucosidase and esterase activity of strains of *Wickerhamomyces anomalus*, *Lactiplantibacillus plantarum*, and *Saccharomyces cerevisiae* were also investigated and compared to those of a commercial enzyme and an *Aspergillus niger* strain. The *W. anomalus* strain showed the best enzymatic performances. The SP207 resin showed the best efficiency in selective recovery of hydroxytyrosol, tyrosol, oleuropein, and total phenols. The bioconversion test of the OMW extract was assessed by using both culture broths and pellets of the tested strains. The results demonstrated that the pellets of *W. anomalus* and *L. plantarum* were the most effective in hydroxytyrosol increasing in phenolic extract. The interesting results suggest the possibility to study new formulations of OMW phenolic extracts with multifunctional microorganisms.

Keywords: adsorbent resins; beta-glucosidase; esterase; hydroxytyrosol; *Lactiplantibacillus plantarum*; oleuropein; tyrosol; *Wickerhamomyces anomalus*

Citation: Romeo, F.V.; Granuzzo, G.; Foti, P.; Ballistreri, G.; Caggia, C.; Rapisarda, P. Microbial Application to Improve Olive Mill Wastewater Phenolic Extracts. *Molecules* **2021**, *26*, 1944. https://doi.org/10.3390/molecules26071944

Academic Editor: Simona Fabroni

Received: 25 February 2021
Accepted: 26 March 2021
Published: 30 March 2021

1. Introduction

The Mediterranean diet is becoming more and more popular due to its nutritional and health benefits, which are associated with lower incidences of atherosclerosis, certain cancers, and cardiovascular and neurodegenerative diseases [1–4]. These health benefits can be strongly related to the high content of antioxidant molecules present in food products widely consumed within the Mediterranean diet, such as extra virgin olive oil and table olives. Olives and their industrial derived products are rich in unsaturated fatty acids, tocopherols, and phenols. However, the olive oil industry generates large amounts of high-polluting by-products, such as a solid residue and an effluent commonly named olive mill wastewater (OMW) [5], which is a mix of olive vegetation water and water added during oil extraction, above all with the three-phase centrifugation system [6].

Olive waste still contains most of the valuable and interesting compounds. During the olive oil processing, most of the phenolic compounds are found in the aqueous phase, with a hydroxytyrosol concentration up to 100-fold higher than that found in olive oil [7]. OMW is rich in phenols, such as hydroxytyrosol, tyrosol, oleuropein, flavonoids, and other compounds. These molecules have great potential in the pharmaceutical and nutraceutical fields [8]. In particular, hydroxytyrosol is characterized by remarkable antioxidant activity, which is similar to that of the main synthetic antioxidants [7,9]. In 2011 [10], the scientific opinion on the substantiation of health claims related to olive polyphenols, standardized by their content of hydroxytyrosol and its derivatives (e.g., oleuropein complex), was published by European Food Safety Authority (EFSA).

Substantial interest has been given to the production of hydroxytyrosol both from natural sources and by synthetic procedures [11]. Several techniques have been reported in literature aiming at recovering phenolic compounds from OMW, such as liquid–liquid extraction [12], membrane filtration [13], adsorption on selective polymeric resins [14,15], and enzymatic reactions [11,16,17]. Compared to chemical treatments, enzymatic treatment is advantageous because it requires mild operating conditions, such as pH, temperature, and the absence of toxic organic solvents [18]. To overcome the high costs of enzymes and their poor operational stabilities, new research has focused on enzyme immobilization strategies [19].

Among microorganisms, *Aspergillus* spp. has been extensively studied for its ability to produce extracellular enzymes that can be easily applied to food processing, because they are recognized as Generally Regarded as Safe (GRAS) by the Food and Drug Administration (FDA) [20]. Within the *Aspergillus* genus, *A. niger* is the most efficient producer of β-glucosidase [21]. However, few reports focused on the effect of fungal enzyme addition on OMW phenols have been published, but even less on the direct application of live microorganisms into an OMW extract to improve antioxidant activity. In particular, yeasts and lactic acid bacteria (LAB) appear particularly interesting as they are commonly used as starter cultures in fermented food and widely isolated by OMW or spontaneous fermentation of table olives. During the last two decades, several *Lactiplantibacillus plantarum* strains have been extensively described for their ability to degrade oleuropein and for their probiotic potential [22,23]. Aponte et al. [24] for the first time tested the in vivo bioconversion of oleuropein into hydroxytyrosol by oral granules containing probiotic *L. plantarum* and an olive leaves standardized extract. Their results showed that co-administration of live *L. plantarum* bacteria with the extract provides for higher amounts of bioavailable hydroxytyrosol, compared to the extract alone. Several yeasts isolated from table olives have been found able to produce noticeable enzymes, such as β-glucosidase, and to exhibit probiotic traits [25,26]. Among yeasts, *Saccharomyces cerevisiae* and *Wickerhamomyces anomalus* have been described to produce enzymes useful in the wine making process and their application has been extensively investigated. In addition, *W. anomalus* (formerly known as *Pichia anomala* or *Hansenula anomala*) shows several probiotic traits including cholesterol removal capability [27], and it is an interesting source of different enzymes and a biotechnologically relevant microorganism [28].

In light of this knowledge, this study aimed to investigate the use of live microorganisms together with OMW extracts. In fact, a selection of different types of resins was performed to obtain a phenol-rich extract with the main aim of evaluating the bioconversion activity of different microbial species. The hydroxytyrosol increase in the OMW extract provided by cultures already known as starters is the first step to design innovative formulations of new nutraceuticals supplemented with probiotics.

2. Results

2.1. Extraction of Phenols from OMW

2.1.1. OMW Analyses

Raw OMW was centrifuged and filtered to remove solid soil residues, olive oil and fragments of olive fruit and stone. The obtained sample was analyzed for the values of pH, dry matter, and total phenols, while hydroxytyrosol (HT), tyrosol (TY), and oleuropein (OLE) concentrations were detected by high performance liquid chromatography (HPLC) (Table 1). These results confirmed the acidic trait of the sample, which showed 6.24 g/100 mL of dry matter value, although subjected to centrifugation, and a high content of total phenolic compounds, for which it is considered a polluting effluent. As expected, among the three quantified peaks, HT showed the highest concentration.

Table 1. Chemical parameters of olive mill wastewater (OMW) after pretreatment (centrifugation and filtration).

Parameters	Means ± SD
pH	4.80 ± 0.06
Dry matter (g/100 mL)	6.24 ± 0.02
Hydroxytyrosol (HT) (mg/L)	218.29 ± 4.94
Tyrosol (TY) (mg/L)	67.24 ± 5.57
Oleuropein (OLE) (mg/L)	207.49 ± 56.37
Total phenols (mg/L)	1343.26 ± 0.54

Moreover, HT was the most abundant phenol in OMW, as shown in Figure 1. OLE was quantified as gallic acid too, but this was done by calculating the response factor of the instrument in respect to the Internal Standard (I.S.) under the same HPLC operating conditions. HT, TY, and OLE were analyzed as principal markers of OLE enzymatic degradation. They were identified by comparing their retention times and absorption spectra with those of pure commercial standards.

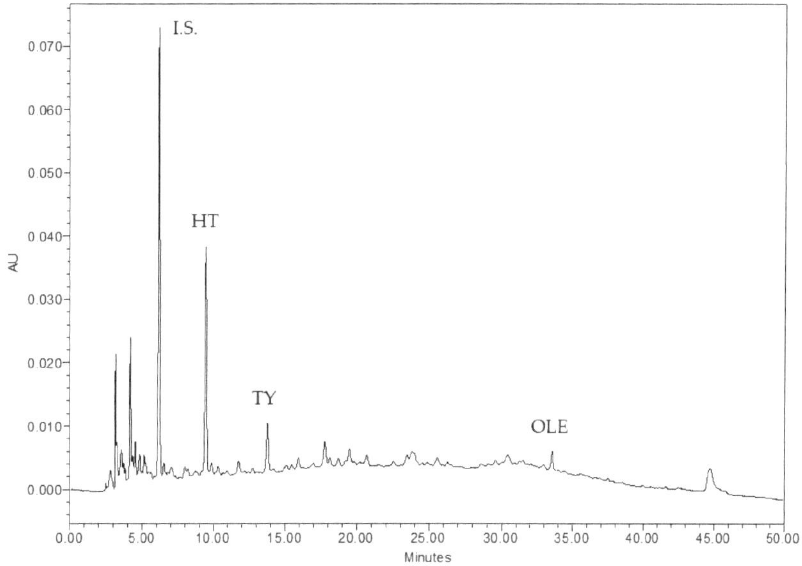

Figure 1. HPLC phenolic profile (at 280 nm) of OMW sample (I.S. = Internal Standard).

2.1.2. Resins Adsorption

The results of OMW adsorption on resins are shown in Table 2. The results confirmed a different adsorbing resins' capacity for both polyphenols and individual bioactive compounds. The four tested resins can be divided into three groups. The most efficient were the S-DVB copolymers resins (namely SP207 and XAD16), which adsorbed the highest concentration of total phenolic compounds, followed by the C18 and the PAD900C resins, respectively. Concerning the capacity to adsorb the individual bioactive compounds, the SP207 resin showed the highest efficiency in concentrating HT, TY, and OLE from OMW, whereas the three others (XAD16, PAD900C, and C18 resins) showed a lower efficiency. The extract obtained by the most effective resin (SP207) was then selected for bioconversion experiments.

Table 2. Phenols detected in OMW extracts obtained with the different tested resins.

Adsorbent Resin	Hydroxytyrosol (HT)	Tyrosol (TY)	Oleuropein (OLE)	Total Phenols
SP207	3239.7 ± 57.34 a	2016.8 ± 49.36 a	11,336.6 ± 199.65 a	16,447.9 ± 16.31 a
XAD16	1381.7 ± 66.45 b	1000.3 ± 120.75 b	10,047.8 ± 120.29 ab	13,144.6 ± 21.75 b
PAD900C	294.3 ± 2.67 c	90.0 ± 10.59 c	4443.8 ± 176.32 c	4880.3 ± 10.87 d
C18	191.1 ± 16.59 c	50.4 ± 10.34 c	8092.3 ± 218.15 b	8763.6 ± 43.50 c

Data are expressed as mg/L of means ± standard deviations. Different letters indicate statistical differences within the same column (Significance at $p < 0.001$).

2.2. Enzymatic Tests

The enzymatic activity of three strains was compared to that obtained by the commercial enzyme (Lallzyme beta), and to the activity of germinated conidia of *A. niger* strain. The *A. niger* DSM 2466 was selected among other commercial strains for its proven enzymatic activity (data not shown) and used as positive control. As shown in Table 3, after germination in culture broth, the strain produced the highest value (IU/mL) of both beta-glucosidase and esterase activities. The *A. niger* activity was also higher than that highlighted by the commercial tested enzyme, which is widely used in oenological applications for its known poly-enzymatic activity. Among *W. anomalus, L. plantarum,* and *S. cerevisiae* tested strains, *W. anomalus* was found as the most effective for beta-glucosidase, together with the commercial enzyme and followed by *L. plantarum,* while, for esterase activity, results obtained from *W. anomalus* and *S. cerevisiae* were not statistically different.

Table 3. Beta-glucosidase and esterase activities detected for the commercial enzyme and the different microorganisms tested under their specific growth culture conditions.

Samples	Beta-Glucosidase	Esterase
Lallzyme beta enzyme	6979.9 ± 3.22 b	316,713.9 ± 1032.48 c
W. anomalus	7066.4 ± 5.36 b	405,417.6 ± 516.24 b
L. plantarum	2387.6 ± 3.22 c	316,348.9 ± 1540.62 c
S. cerevisiae	1437.9 ± 78.31 d	399,577.0 ± 1548.72 b
A. niger	314,313.8 ± 107.28 a	4,506,229.2 ± 4646.15 a

Data are expressed as IU/mL of broth sample (means ± standard deviations). Different letters indicate statistical differences within the same column (Significance at $p < 0.001$).

2.3. Bioconversion Efficiency by *W. anomalus* and *L. plantarum*

In the bioconversion test, the two microorganisms showing the best performances at the enzymatic tests (Table 3) were used to evaluate their ability to increase the HT and TY concentrations in the OMW extract. The SP207 extract was used as substrate for both pellets and filtered supernatants. The kinetic analysis of bioconversion was done through HPLC quantification of single phenols up to 6 h. The treatment with the commercial enzyme highlighted a slow and constant increasing of both HT and TY, whereas the treatment with the two tested microorganisms showed the highest value of phenols, reaching the plateau after 2 h of reaction. The only exception was detected after 6 h in the sample treated with pellets, when a further increase of HT value was registered (Table S1, on-line Supplementary Materials). Overall, the highest phenolic values were obtained using the microbial pellets rather than the filtered broths. The mean values of HT, TY, and OLE obtained after 2 h are shown in Figure 2. The highest values of HT were reached using *W. anomalus* and *L. plantarum* pellets (+35.5% and +33.7%, respectively), whereas no statistical difference between the control sample and the *W. anomalus* broth was observed. The TY content increased significantly only after contact with both pellets (+32%). Regarding the OLE amount, although a decrease was registered after 2 h by using the commercial enzyme, its concentration remained almost constant up to 6 h (Table S1). On the contrary, the OLE concentration increased in the presence of both strains' pellet (+30.2% and +29.5%) and to a lesser degree with *W. anomalus* broth (+7%).

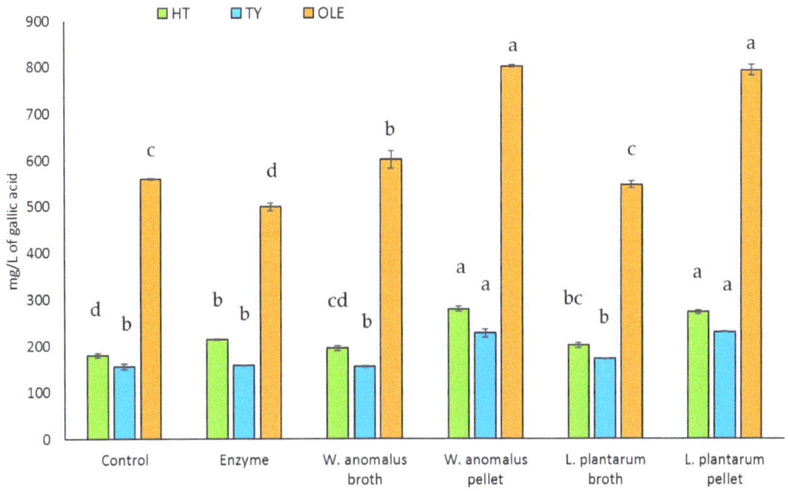

Figure 2. Concentration of hydroxytyrosol, tyrosol, and oleuropein after bioconversion of diluted OMW extract by Lallzyme, extracellular broths and pellets of the two tested strains after 2 h at 50 °C. Different letters indicate statistical differences among the columns of the same compound (Significance at $p < 0.001$).

3. Discussion

Crude OMW was sampled from a collection tank during a continuous oil extraction from different olive cultivars. The sample showed a pH value of 4.8, within the range of 2.2–5.9 as reported by Dermeche et al. [5], and close to those reported by Aggoun et al. [6] and Dammak et al. [18] who found values between 4.5 and 5.16 for OMW from the three-phase extracting system. Here, the dry matter value was slightly lower than those reported by the previous authors, as a consequence of sample pre-treatment, namely centrifugation and filtration. The total phenols were also lower than those discussed by Belaqziz et al. [29] who found value of 6.46 g/L in effluent generated from the three-phase extraction system. The amount of 1.34 g/L detected in the present work was comparable to 1.8 g/L reported by Dammak [18] and within the range 0.03–6.13 g/L reported by Davies et al. [30]. The variability in total phenol concentration may be attributed to several factors, such as the olive cultivars and their degree of ripening. Moreover, the agronomic conditions and treatments applied to extract the olive oil and to treat the OMW, may also significantly impact the quantitative and qualitative phenolic content and solid residues of OMW [5,31]. The major compounds detected in OMW are HT and TY, with HT known for its high antioxidant activity [32]. Dermeche et al. [5] in their review reported that in OMW, HT is found in the range of 20–1224 mg/L, while TY is between 145–208 mg/L. Regarding OLE, it has not been detected in OMW deriving from late-harvest olives [31], and its content has been described as highly variable [11,29,33].

The solid phase recovery of polyphenols from OMW has been successfully proposed using adsorbent resins, particularly with the S-DVB-based resins [14,15,34]. In the present work, in order to optimize the process, different resins, including the SP207, never employed before in the recovery of phenolic compounds from OMW, were tested. All the tested resins were capable of successfully adsorbing polyphenols [34,35], with the SP207 resin showing a high and selective affinity for OMW polyphenols. This non-polar resin, generally used for the recovery of phenolic compounds from citrus, apple, and winery by-products and wastes [36–38], showed the best efficiency in terms of total phenols concentration and selective recovery of the main OMW bioactive compounds. This type of approach to the recovery of bioactive compounds from OMW is desirable for the scale-up from laboratory to industrial production thanks to the simple and reproducible operating conditions and to the low costs of the process.

The enzymatic activity, as expected, highlighted the absolute supremacy of the *A. niger* strain, confirming its successful employment at an industrial level to produce purified enzymes [11]. The hyphal mode of fungal growth and its robustness to a wide range of pH and water activity make fungi extremely efficient in the bioconversion of solid substrates [39]. In fact, live *A. niger* culture showed an enzymatic activity up to 45-fold higher than the commercial enzyme. The *W. anomalus* and *L. plantarum* tested strains gave interesting results. In particular, *W. anomalus* showed a beta-glucosidase activity statistically equal to that of the commercial enzyme, and an esterase value higher than that of the commercial enzyme. Despite its potential to produce enzymes of technological importance, *W. anomalus* has been scarcely explored in food and byproduct applications.

The interest in microbial enzymes for biotechnology applications has grown above all in the bioconversion of industrial by-products [40]. The enzymatic test assessed in the present work highlighted the importance of testing microbial cultures already known for their technological or probiotic properties. A considerable amount of research has been conducted to determine the enzymatic properties of LAB [41], while other studies reported a strong β-glucosidase activity for *W. anomalus* yeast [42–46]. The bioconversion test showed that *W. anomalus* and *L. plantarum* were able to produce more HT than the commercial enzyme after 2 h testing. Moreover, the bioconversion test highlighted that microbial live cells were more effective in increasing the HT and TY content than their filtered broths, probably because the phenol extract, generating stressful conditions, could induce a stimulating effect on microbial metabolism. The statistically significant increase in TY (the TY recently highlighted and quantified to apply the EFSA health claim on olive oil polyphenols) obtained using the single pellet of both *W. anomalus* and *L. plantarum* cultures, is also valuable [47]. In the present work, the OLE value was detected as increased, but this may be due to a depolymerizing effect that microbial enzymes may have had on OMW polyphenols.

4. Materials and Methods

4.1. Raw Material Treatment

Fresh OMW was sampled in October 2018 from an olive oil producing plant of a factory located in Catania, Italy, which used a three-phase continuous oil extraction system. Samples were centrifuged at 10,000 rpm for 5 min at +4 °C, then filtered (Miracloth paper, Calbiochem, Canada) and stored at −20 °C until further analyses.

4.2. Extraction of Phenols from OMW

Four commercial adsorbent resins, Amberlite XAD-16 (Sigma Aldrich, Milan, Italy), Sepabeads SP-207 (Mitsubishi Chem. Co., Tokyo, Japan), Purosorb PAD900C (Purolite, Milan, Italy), and Hamilton C18 (Thermo Fischer Scientific, San Jose, CA, USA) were selected and tested. XAD-16 and SP-207 are styrene-divinylbenzene (S-DVB) copolymers; PAD900C is polydivinylbenzene polymer; and finally, C18 having Poly(S-DVB) matrix was used as analytical reference.

Before their employment, the adsorbent resins were pretreated with 95% ethanol (food grade, Alcoolita, Italy), washed with water (HPLC grade, Carlo Erba, Italy), and then oven-dried at 70 °C up to constant weight. All adsorption experiments were conducted with determined quantities of activated dried resins. Then, 1 bed volume (BV) of each dry resin (about 20 mL) was loaded on a glass preparative column (length, 30 cm; i.d., 0.5 cm) connected with a peristaltic pump. After rinsing the resins with water, the OMW, pretreated as described in Section 4.1, was passed on resins until their saturation. Loading was stopped when the outflow OMW reached the same absorbance values and color of the loaded OMW. Before the desorption phase, each saturated resin was washed with 4 BV of water to remove water-soluble compounds. The adsorbed phenolic fraction of each resin was then recovered with 2 BV of a 95% ethanol/water solution (60:40, *v/v*). The collected desorbed fractions were finally concentrated after vacuum distillation of ethanol at 40 °C by using a rotary evaporator (Rotavapor RE111, Büchi, Cornaredo, Italy).

4.3. Physicochemical Characterisation

The pH value of OMW samples was measured using a Mettler DL25 pH meter (Mettler-Toledo International Inc., Columbus, OH, USA). Dry weight was determined by weighing the samples before and after drying in an oven at 105 °C up to constant weight.

The total phenolic content of OMW and extracts was determined according to the Folin–Ciocalteu's (FC) colorimetric method. The OMW samples were mixed with 5 mL of FC commercial reagent (Labochimica, Italy) diluted with water 1:10, *v/v*, and 4 mL of a 7.5% sodium carbonate solution. After stirring for 2 h at room temperature away from light, the absorbance of the blue solution was measured spectrophotometrically at 765 nm (Cary 100 Scan UV-Visibile, Agilent, CA, USA). The total phenolic content was expressed as mg of gallic acid equivalents (GAE)/L of the sample.

4.4. HPLC Analysis

HPLC analyses of phenol fraction of OMW and extracts were obtained by directly injecting the filtered samples (0.45 μm Millipore filters) in the chromatographic HPLC system. The system consisted of a liquid chromatography Waters Alliance 2695 HPLC equipped with a Waters 996 photodiode array detector (PDA) set at 280 nm and with Waters Empower software (Waters Corporation, MA, USA). The column was a Luna C18 (250 mm × 4.6 mm i.d., 5 μm, 100 Å; Phenomenex, Torrence, CA, USA) maintained in an oven at 40 °C. Chromatographic separation was achieved by elution gradient using an initial composition of 95% of A solution (water acidified with 2% acetic acid) and 5% of B solution (methanol). Solvents were HPLC grade (Merck KGaA, Darmstadt, Germany). The B solution increased to 30% in 15 min and to 70% in 25 min and then, after 2 min in isocratic, the mobile phase was set at the initial conditions for 8 min. A flow of 1 mL/min was used [48]. The internal standard (I.S.) of 5 mM pure gallic acid (Fluka, Switzerland) was used to quantify the phenolic compounds. The identification of phenolic compounds was obtained by comparing retention time with pure tyrosol, oleuropein, and hydroxytyrosol standards (Extrasynthese, Genay, France). All the analyses were carried out in triplicate for each sample.

4.5. Microorganisms

Lactiplantibacillus plantarum DSM 20205, *Saccharomyces cerevisiae* DSM 1333, and *Aspergillus niger* DSM 2466, were all purchased from Leibniz-Institute DSMZ, German collection. Moreover, *Wickerhamomyces anomalus* F6.05, a strain isolated from brine samples of naturally fermented table olives, was from the Culture Collection of the Department of Agricultural, Food, and Environment (Di3A), University of Catania, Italy. All the used cultures were maintained as stock solution in 20% (*v/v*) glycerol at −80 °C before use. *A. niger* stock suspension was obtained by growing the strain on Potato Dextrose Agar (PDA, Oxoid, UK) at 30 °C for 5 days and by harvesting the spores with 5 mL Tween 80, 2% (*v/v*) in sterile water. The spore solution was stored in 20% (*v/v*) glycerol at -20 °C before use.

4.6. Enzyme Production Test

The *L. plantarum* strain was cultured at 32 °C for 24 h in De Man, Rogosa, and Sharpe broth (MRS broth, Oxoid, Milan, Italy) while *S. cerevisiae* and *W. anomalus* strains were cultured at 30 °C for 24 h in Yeast Malt broth (YM broth, Merk, Italy). A cell density of 10^8 CFU/mL of each microorganism was used to inoculate test tubes containing media without a carbon source, MRS broth without Dextrose (Likson, Italy) for *L. plantarum*, and yeast nitrogen base (Likson, Italy) for *S. cerevisiae* and *W. anomalus* strains. After an overnight culture at the same temperature shown above, the samples were filtered with

0.45 µm paper filters (Minisart, Sartorius Stedim, France) to collect the filtrate containing the extracellular enzymes for subsequent enzymatic tests.

The *A. niger* strain was used as a positive control by using the method reported by Hamza et al. [11]. Flasks containing 100 mL of yeast nitrogen base (YNB, Likson, Italy) with 6.7 g of wheat bran (Ki, Italy) as carbon source were inoculated at 10^7 spores/mL concentration. The flasks were left in agitation at 120 rpm for 10 days at 25 °C. The culture broth was filtered through a muslin cloth to collect the filtrate containing the extracellular enzymes.

4.7. Beta-Glucosidase Activity

Beta-glucosidase activity was measured determining the quantity of p-nitrophenol released from the p-nitrophenyl-beta-D-glucopyranoside (p-NPG) by the enzymatic activity of the tested strains. The activity was determined according to the method of Khoufi et al. [16], slightly modified. For the enzymatic assay, the incubation mixture contained 0.9 mL of 5 mM p-NPG (Sigma-Aldrich, Milan, Italy) in 50 mM citrate buffer (pH 4.8), and 0.1 mL of centrifuged broth from each overnight culture. The reaction was maintained at 50 °C for 2 h and stopped by the addition of 2 mL of Na_2CO_3 buffer (1.0 M). The amount of p-nitrophenol (p-NP) released was spectrophotometrically determined at 400 nm and quantified by using p-NP (Glentham, Life Science, UK) to obtain a standard curve. One unit (IU) of enzyme activity was defined as the amount of enzyme that produced 1 µmol of p-NP/min under the same conditions. The enzymatic results were expressed as IU/mL of sample by means of the following equation: $IU/mL = Q/T \times C/V$, where Q is the quantity (µmole) of p-NP generated at the time T (min), V is the volume (mL) of the sample, and C is the total reaction volume (mL).

4.8. Esterase Activity

Esterase activity was determined according to the method of Mackness et al. [49]. The reaction mixture contained 50 µL of centrifuged broth from each overnight culture, 50 µL of 150 mM p-nitrophenyl-acetate (p-NPA, Acros Organics, Fair Lawn, NJ, USA) in ethanol, and 2.9 mL of Tris-HCl buffer (pH 7.5). The solution was incubated at 25 °C for 4 min. The amount of p-NP released was spectrophotometrically determined at 400 nm and quantified by using p-NP to obtain a standard curve. The enzymatic results were expressed as described in Section 4.7.

4.9. Bioconversion of Phenolic Compounds

The OMW extract obtained by the SP207 resin was used for the proposed enzymatic hydrolysis aimed at hydroxytyrosol increase. The procedure described in the enzyme production test above mentioned (Section 4.6) was repeated. In this case, only *L. plantarum* and *W. anomalus*, the two strains that showed the highest enzymatic activity, were tested. The two overnight cultures were centrifuged to collect separately the pellets and supernatants. The pellets were centrifuged twice at 5000 rpm for 5 min at +4 °C in sterile water and standardized at the same cell density of 10^8 CFU/mL in sterile water.

Bioconversion reactions were assessed according to Hamza et al. [11], with modifications. The reactions were conducted in Erlenmeyer flasks by adding 5 mL of cultured samples (both pellet and filtered supernatant were tested separately) and 15 mL of OMW extract (diluted 1:10 in water). The enzymatic reaction was performed at 50 °C for 6 h under static conditions. Samples were collected every hour up to 6 h, and then filtered for subsequent analyses for phenols by HPLC.

Moreover, the enzymatic activity of all the samples was compared to the activity of Lallzyme Beta (Lallemand, Blagnac, France) a commercial enzyme produced by *A. niger*, with β-glucosidase and polygalacturonase activity. The enzyme was used as a positive control (5 mL) at a dose of 5 mg/100 mL sterile water, in accordance with the manufacturer's instructions. The negative control was incubated with 5 mL of sterile water.

4.10. Statistical Analysis

All analyses carried out were performed in triplicate. SPSS software (version 21.0, IBM Statistics, Armonk, NY, USA) was used for data processing. Statistical analysis of the results was performed using one-way analysis of variance (ANOVA), and Tukey's HSD post hoc test for means separation at a significance level of $P \leq 0.05$.

5. Conclusions

The effects of different adsorbent resins were tested to obtain an OMW extract rich in bioactive compounds. The OMW extract obtained from the SP207 resin showed the best efficiency in selective recovery of the main OMW bioactive compounds. The beta-glucosidase and esterase activity of different microorganisms, already known for their technological or probiotic properties, were investigated and compared to those of a commercial enzyme and an *A. niger* strain. Remarkably, the results demonstrated that the *W. anomalus* and *L. plantarum* tested strains were the most effective in increasing HT and TY, especially when they were used as pellets in contact with the OMW extract. The results of the present study are promising and suggest the possibility to further explore liquid or solid formulations for antioxidant/nutraceutical supplements containing health promoting compounds and microorganisms.

Supplementary Materials: The following are available online. Table S1: Kinetics of bioconversion of single phenols analyzed through HPLC.

Author Contributions: Conceptualization, F.V.R., G.G. and G.B.; methodology, F.V.R., G.G. and G.B.; formal analysis, G.G. and P.F.; resources, F.V.R. and P.R.; data curation, F.V.R. and G.B.; writing—original draft preparation, F.V.R. and G.B.; writing-review and editing, F.V.R., G.B., P.R. and C.C.; project administration, P.R.; funding acquisition, P.R. All authors have read and agreed to the published version of the manuscript.

Funding: This research was funded by the Italian Ministry of Agriculture, Food and Forestry (MiPAAF), grant number D.M. n. 12479 of Project 'Informed traceability and processing/product innovations in the olives for oil and table olives supply chain'—INFOLIVA.

Informed Consent Statement: Not applicable.

Conflicts of Interest: The authors declare no conflict of interest.

Sample Availability: Samples of the OMW extracts as obtained and described in this paper are available from the authors.

References

1. Owen, R.W.; Giacosa, A.; Hull, W.E.; Haubner, R.; Spiegelhalder, B.; Bartsch, H. The antioxidant/anticancer potential of phenolic compounds isolated from olive oil. *Eur. J. Cancer.* **2000**, *36*, 1235–1247. [CrossRef]
2. Perona, J.S.; Cabello-Moruno, R.; Ruiz-Gutierrez, V. The role of virgin olive oil components in the modulation of endothelial function. *J. Nutr. Biochem.* **2006**, *17*, 429–445. [CrossRef]
3. Visioli, F.; Caruso, D.; Grande, S.; Bosisio, R.; Villa, M.; Galli, G.; Sirtori, C.; Galli, C. Virgin olive oil study (VOLOS): Vasoprotective potential of extra virgin olive oil in mildly dyslipidemic patients. *Eur. J. Nutr.* **2005**, *44*, 121–127. [CrossRef] [PubMed]
4. Visioli, F.; Davalos, A.; López de las Hazas, M.C.; Crespo, M.C.; Tomé-Carneiro, J. An overview of the pharmacology of olive oil and its active ingredients. *Br. J. Pharmacol.* **2020**, *177*, 1316–1330. [CrossRef]
5. Dermeche, S.; Nadour, M.; Larroche, C.; Moulti-Mati, F.; Michaud, P. Olive mill wastes: Biochemical characterizations and valorization strategies. *Process Biochem.* **2013**, *48*, 1532–1552. [CrossRef]
6. Aggoun, M.; Arhab, R.; Cornu, A.; Portelli, J.; Barkat, M.; Graulet, B. Olive mill wastewater microconstituents composition according to olive variety and extraction process. *Food Chem.* **2016**, *209*, 72–80. [CrossRef]
7. Fernández-Bolaños, J.; Rodríguez, G.; Rodríguez, R.; Guillén, R.; Jiménez, A. Extraction of interesting organic compounds from olive oil waste. *Grasas Aceites* **2006**, *57*, 95–106. [CrossRef]
8. Alfano, A.; Corsuto, L.; Finamore, R.; Savarese, M.; Ferrara, F.; Falco, S.; Santabarbara, G.; De Rosa, M.; Schiraldi, C. Valorization of olive millwastewater by membrane processes to recover natural antioxidant compounds for cosmeceutical and nutraceutical applications or functional foods. *Antioxidants* **2018**, *7*, 72. [CrossRef]

9. Fki, I.; Allouche, N.; Sayadi, S. The use of polyphenolic extract, purified hydroxytyrosol and 3,4-dihydroxyphenyl acetic acid from olive mill wastewater for the stabilization of refined oils: A potential alternative to synthetic antioxidants. *Food Chem.* **2005**, *93*, 197–204. [CrossRef]
10. EFSA Panel on Dietetic Products, Nutrition and Allergies (NDA). Scientific Opinion on the substantiation of health claims related to polyphenols in olive and protection of LDL particles from oxidative damage (ID 1333, *1638*, 1639, 1696, 2865), maintenance of normal blood HDL-cholesterol concentrations (ID 1639), maintenance of normal blood pressure (ID 3781), anti-inflammatory properties (ID 1882), contributes to the upper respiratory tract health (ID 3468), can help to maintain a normal function of gastrointestinal tract (3779), and contributes to body defences against external agents (ID 3467) pursuant to Article 13(1) of Regulation (EC) No 1924/2006. *EFSA J.* **2011**, *9*, 2033. Available online: www.efsa.europa.eu/efsajournal (accessed on 25 February 2021). [CrossRef]
11. Hamza, M.; Khoufi, S.; Sayadi, S. Fungal enzymes as a powerful tool to release antioxidants from olive mill wastewater. *Food Chem.* **2012**, *131*, 1430–1436. [CrossRef]
12. De Leonardis, A.; Aretini, A.; Alfano, G.; Macciola, V.; Ranalli, G. Isolation of a hydroxytyrosol-rich extract from olive leaves (Olea europaea L.) and evaluation of its antioxidant properties and bioactivity. *Eur. Food Res. Technol.* **2008**, *226*, 653–659. [CrossRef]
13. Zagklis, D.P.; Vavouraki, A.I.; Kornaros, M.E.; Paraskeva, C.A. Purification of olive mill wastewater phenols through membrane filtration and resin adsorption/desorption. *J. Hazard. Mater.* **2015**, *285*, 69–76. [CrossRef]
14. Bertin, L.; Ferri, F.; Scoma, A.; Marchetti, L.; Fava, F. Recovery of high added value natural polyphenols from actual olive mill wastewater through solid phase extraction. *Chem. Eng. J.* **2011**, *171*, 1287–1293. [CrossRef]
15. Scoma, A.; Bertin, L.; Zanaroli, G.; Fraraccio, S.; Fava, F. A physicochemical–biotechnological approach for an integrated valorization of olive mill wastewater. *Bioresource Technol.* **2011**, *102*, 10273–10279. [CrossRef]
16. Khoufi, S.; Hamza, M.; Sayadi, S. Enzymatic hydrolysis of olive wastewater for hydroxytyrosol enrichment. *Bioresource Technol.* **2011**, *102*, 9050–9058. [CrossRef] [PubMed]
17. Liu, M.; Yong, Q.; Yu, S. Efficient bioconversion of oleuropein from olive leaf extract to antioxidant hydroxytyrosol by enzymatic hydrolysis and high temperature degradation. *Biotechnol. Appl. Bioc.* **2018**, *65*, 680–689. [CrossRef]
18. Dammak, I.; Khoufi, S.; Sayadi, S. A performance comparison of olive oil mill wastewater enzymatic treatments. *Food Bioprod. Process.* **2016**, *100*, 61–71. [CrossRef]
19. Liu, M.; Yong, Q.; Lian, Z.; Huang, C.; Yu, S. Continuous bioconversion of oleuropein from olive leaf extract to produce the bioactive product hydroxytyrosol using carrier-immobilized enzyme. *Appl. Biochem. Biotech.* **2020**, *190*, 148–165. [CrossRef] [PubMed]
20. Mazzei, R.; Giorno, L.; Piacentini, E.; Mazzuca, S.; Drioli, E. Kinetic study of a biocatalytic membrane reactor containing immobilized b-glucosidase for the hydrolysis of oleuropein. *J. Membr. Sci.* **2009**, *339*, 215–223. [CrossRef]
21. Kang, S.W.; Ko, E.H.; Lee, J.S.; Kim, S.W. Over-production of β-glucosidase by Aspergillus niger mutant from lignocellulosic biomass. *Biotechnol. Lett.* **1999**, *21*, 647–650. [CrossRef]
22. Benítez-Cabello, A.; Calero-Delgado, B.; Rodríguez-Gómez, F.; Garrido-Fernández, A.; Jiménez-Díaz, R.; Arroyo-López, F.N. Biodiversity and multifunctional features of lactic acid bacteria isolated from table olive biofilms. *Front. Microbiol.* **2019**, *10*, 836. [CrossRef] [PubMed]
23. Vaccalluzzo, A.; Pino, A.; De Angelis, M.; Bautista-Gallego, J.; Romeo, F.V.; Foti, P.; Caggia, C.; Randazzo, C.L. Effects of different stress parameters on growth and on oleuropein-degrading abilities of Lactiplantibacillus plantarum strains selected as tailored starter cultures for naturally table olives. *Microorganisms* **2020**, *8*, 1607. [CrossRef]
24. Aponte, M.; Ungaro, F.; d'Angelo, I.; De Caro, C.; Russo, R.; Blaiotta, G.; Dal Piaz, F.; Calignano, A.; Miro, A. Improving in vivo conversion of oleuropein into hydroxytyrosol by oral granules containing probiotic Lactobacillus plantarum 299v and an Olea europaea standardized extract. *Int. J. Pharmaceut.* **2018**, *543*, 73–82. [CrossRef]
25. Porru, C.; Rodríguez-Gómez, F.; Benítez-Cabello, A.; Jiménez-Díaz, R.; Zara, G.; Budroni, M.; Mannazzu, I.; Arroyo-López, F.N. Genotyping, identification and multifunctional features of yeasts associated to Bosana naturally black table olive fermentations. *Food Microbiol.* **2018**, *69*, 33–42. [CrossRef]
26. Bonatsou, S.; Karamouza, M.; Zoumpopoulou, G.; Mavrogonatou, E.; Kletsas, D.; Papadimitriou, K.; Tsakalidou, E.; Nychas, G.-J.E.; Panagou, E.Z. Evaluating the probiotic potential and technological characteristics of yeasts implicated in cv. Kalamata natural black olive fermentation. *Int. J. Food Microbiol.* **2018**, *271*, 48–59. [CrossRef]
27. Zullo, B.A.; Ciafardini, G. Evaluation of physiological properties of yeast strains isolated from olive oil and their in vitro probiotic trait. *Food Microbiol.* **2019**, *78*, 179–187. [CrossRef] [PubMed]
28. Padilla, B.; Gil, J.V.; Manzanares, P. Challenges of the non-conventional yeast Wickerhamomyces anomalus in winemaking. *Fermentation* **2018**, *4*, 68. [CrossRef]
29. Belaqziz, M.; Tan, S.P.; El-Abbassi, A.; Kiai, H.; Hafidi, A.; O'Donovan, O.; McLoughlin, P. Assessment of the antioxidant and antibacterial activities of different olive processing wastewaters. *PLoS ONE* **2017**, *12*, e0182622. [CrossRef]
30. Davies, L.C.; Vilhena, A.M.; Novais, J.M.; Martins-Dias, S. Olive mill wastewater characteristics: Modelling and statistical analysis. *Grasas Aceites* **2004**, *55*, 233–241. [CrossRef]
31. Allouche, N.; Fki, I.; Sayadi, S. Toward a high yield recovery of antioxidants and purified hydroxytyrosol from olive mill wastewaters. *J. Agric. Food Chem.* **2004**, *52*, 267–273. [CrossRef]

32. Obied, H.K.; Allen, M.S.; Bedgood, D.R.; Prenzler, P.D.; Robards, K.; Stockmann, R. Bioactivity and analysis of biophenols recovered from olive mill waste. *J. Agric. Food Chem.* **2005**, *53*, 823–837. [CrossRef]
33. D'Antuono, I.; Kontogianni, V.G.; Kotsiou, K.; Linsalata, V.; Logrieco, A.F.; Tasioula-Margari, M.; Cardinali, A. Polyphenolic characterization of olive mill wastewaters, coming from Italian and Greek olive cultivars, after membrane technology. *Food Res. Int.* **2014**, *65*, 301–310. [CrossRef]
34. Agalias, A.; Magiatis, P.; Skaltsounis, A.-L.; Mikros, E.; Tsar-Bopoulos, A.; Gikas, E.; Spanos, I.; Manios, T. A new process for the management of olive oil mill wastewater and recovery of natural antioxidants. *J. Agric. Food Chem.* **2007**, *55*, 2671–2676. [CrossRef] [PubMed]
35. Kammerer, J.; Carle, R.; Kammerer, D.R. Adsorption and ion exchange: Basic principles and their application in food processing. *J. Agric. Food Chem.* **2011**, *59*, 22–42. [CrossRef]
36. Grohmann, K.; Manthey, J.A.; Cameron, R.G.; Buslig, B.S. Purification of citrus peel juice and molasses. *J. Agric. Food Chem.* **1999**, *47*, 4859–4860. [CrossRef]
37. Kammerer, J.; Boschet, J.; Kammerer, D.R.; Carle, R. Enrichment and fractionation of major apple flavonoids, phenolic acids and dihydrochalcones using anion exchange resins. *LWT-Food Sci. Technol.* **2011**, *44*, 1079–1087. [CrossRef]
38. Soto, M.L.; Conde, E.; González-López, N.; Conde, M.J.; Moure, A.; Sineiro, J.; Falqué, E.; Domínguez, H.; Núñez, M.J.; Parajó, J.C. Recovery and concentration of antioxidants from winery wastes. *Molecules* **2012**, *17*, 3008–3024. [CrossRef] [PubMed]
39. Oliveira, F.; Moreira, C.; Salgado, J.M.; Abrunhosa, L.; Venâncio, A.; Belo, I. Olive pomace valorization by Aspergillus species: Lipase production using solid-state fermentation. *J. Sci. Food Agr.* **2016**, *96*, 3583–3589. [CrossRef]
40. Bhanja, T.; Kumari, A.; Banerjee, R. Enrichment of phenolics and free radical scavenging property of wheat koji prepared with two filamentous fungi. *Bioresource Technol.* **2009**, *100*, 2861–2866. [CrossRef]
41. Matthews, A.; Grimaldi, A.; Walker, M.; Bartowsky, E.; Grbin, P.; Jiranek, V. Lactic Acid Bacteria as a potential source of enzymes for use in vinification. *Appl. Environ. Microbiol.* **2004**, 5715–5731. [CrossRef] [PubMed]
42. Bautista-Gallego, J.; Rodriguez-Gomez, F.; Barrio, E.; Querol, A.; Garrido-Fernandez, A.; Arroyo-López, F.N. Exploring the yeast biodiversity of green table olive industrial fermentations for technological applications. *Int. J. Food Microbiol.* **2011**, *147*, 89–96. [CrossRef] [PubMed]
43. Arroyo-López, F.N.; Romero-Gil, V.; Bautista-Gallego, J.; Rodríguez-Gomez, F.; Jiménez-Díaz, R.; García-García, P.; Querol, A.; Garrido-Fernández, A. Yeasts in table olive processing: Desirable or spoilage microorganisms? *Int. J. Food Microbiol.* **2012**, *160*, 42–49. [CrossRef] [PubMed]
44. Romero-Gil, V.; Bautista-Gallego, J.; Rodríguez-Gómez, F.; García-García, P.; Jiménez-Díaz, R.; Garrido-Fernández, A.; Arroyo-López, F.N. Evaluating the individual effects of temperature and salt on table olive related microorganisms. *Food Microbiol.* **2013**, *33*, 178–184. [CrossRef]
45. Bonatsou, S.; Benitez, A.; Rodriguez-Gómez, F.; Panagou, E.Z.; Arroyo-Lopez, F.N. Selection of yeasts with multifunctional features for application as starters in natural black table olive processing. *Food Microbiol.* **2015**, *46*, 66–73. [CrossRef]
46. Pino, A.; Vaccalluzzo, A.; Solieri, L.; Romeo, F.V.; Todaro, A.; Caggia, C.; Arroyo-López, F.N.; Bautista-Gallego, J.; Randazzo, C.L. Effect of Sequential Inoculum of Beta-Glucosidase Positive and Probiotic Strains on Brine Fermentation to Obtain Low Salt Sicilian Table Olives. *Front. Microbiol.* **2019**, *10*, 174. [CrossRef]
47. Bellumori, M.; Cecchi, L.; Innocenti, M.; Clodoveo, M.L.; Corbo, F.; Mulinacci, N. The EFSA Health Claim on Olive Oil Polyphenols: Acid Hydrolysis Validation and Total Hydroxytyrosol and Tyrosol Determination in Italian Virgin Olive Oils. *Molecules* **2019**, *24*, 2179. [CrossRef] [PubMed]
48. Sorrentino, G.; Muzzalupo, I.; Muccilli, S.; Timpanaro, N.; Russo, M.P.; Guardo, M.; Rapisarda, P.; Romeo, F.V. New accessions of Italian table olives (Olea europaea): Characterization of genotypes and quality of brined products. *Sci. Hortic.* **2016**, *213*, 34–41. [CrossRef]
49. Mackness, M.I.; Walker, C.H.; Rowlands, D.C.; Price, N.R. Esterase activity in homogenates of three strains of the rust red flour beetle Tribolium castaneum (Herbst). *Comp. Biochem. Physiol.* **1983**, *74*, 65–68. [CrossRef]

molecules

MDPI

Review

A Review on Antistaphylococcal Secondary Metabolites from Basidiomycetes

Vimalah Vallavan [1], Getha Krishnasamy [2,*], Noraziah Mohamad Zin [1] and Mazlyzam Abdul Latif [3]

[1] Center for Diagnostic, Therapeutics & Investigative Studies, Faculty of Health Sciences, Universiti Kebangsaan Malaysia, Jalan Raja Muda Abdul Aziz, Kuala Lumpur 50300, Malaysia; p99918@siswa.ukm.edu.my (V.V.); noraziah.zin@ukm.edu.my (N.M.Z.)
[2] Bioactivity Program, Natural Products Division, Forest Research Institute Malaysia (FRIM), Kepong 52109, Selangor, Malaysia
[3] Center for Toxicology and Health Risk Studies, Faculty of Health Sciences, Universiti Kebangsaan Malaysia, Jalan Raja Muda Abdul Aziz, Kuala Lumpur 50300, Malaysia; mazlyzam@ukm.edu.my
* Correspondence: getha@frim.gov.my; Tel.: +603-627-97-652

Received: 30 September 2020; Accepted: 3 December 2020; Published: 11 December 2020

check for updates

Abstract: Fungi are a rich source of secondary metabolites with several pharmacological activities such as antifungal, antioxidant, antibacterial and anticancer to name a few. Due to the large number of diverse structured chemical compounds they produce, fungi from the phyla Ascomycota, Basidiomycota and Muccoromycota have been intensively studied for isolation of bioactive compounds. Basidiomycetes-derived secondary metabolites are known as a promising source of antibacterial compounds with activity against Gram-positive bacteria. The continued emergence of antimicrobial resistance (AMR) poses a major challenge to patient health as it leads to higher morbidity and mortality, higher hospital-stay duration and substantial economic burden in global healthcare sector. One of the key culprits for AMR crisis is *Staphylococcus aureus* causing community-acquired infections as the pathogen develops resistance towards multiple antibiotics. The recent emergence of community strains of *S. aureus* harbouring methicillin-resistant (MRSA), vancomycin-intermediate (VISA) and vancomycin-resistant (VRSA) genes associated with increased virulence is challenging. Despite the few significant developments in antibiotic research, successful MRSA therapeutic options are still needed to reduce the use of scanty and expensive second-line treatments. This paper provides an overview of findings from various studies on antibacterial secondary metabolites from basidiomycetes, with a special focus on antistaphylococcal activity.

Keywords: Basidiomycota; bioactive natural products; antibacterial; antimicrobial resistance; methicillin-resistant *Staphylococcus aureus* (MRSA)

1. Introduction

Antimicrobial resistance (AMR) crisis is associated with more than 2 million hard-to-treat infectious diseases. The Center for Disease Control and Prevention (CDC) reported that increasing mortality rate at an average of 23,000 deaths per year was recorded in developing countries [1]. Major pathogen that contributes to the AMR incidence is *Staphylococcus aureus* with the emergence of multidrug-resistant strains such as methicilin-resistant (MRSA), vancomycin-intermediate (VISA) and vancomycin-resistant (VRSA) *S. aureus* [2–4]. The rising incidence of these resistant pathogens leads to inadequate antimicrobial therapeutic effects that are related to poor healthcare outcome in patients. Community-acquired methicillin-resistant *S. aureus* (CA-MRSA) strains also account for an increasing proportion of hospital-acquired MRSA (HA-MRSA) infections [3,5]. These pathogenic strains of *Staphylococci* with their intrinsic virulence factor can cause a diverse array of life-threatening infections [4]. The high antibiotic selective pressure in crowded populations, like in Asia, creates an environment that allows rapid development and successful spread of multidrug-resistant pathogens such as HA-MRSA and CA-MRSA [3,6]. The last resort treatment for MRSA infections is vancomycin [2]. However, current loss in sensitivity toward vancomycin limits the conventional therapeutic choice for *Staphylococcal* infections [2,7].

Fungal secondary metabolites have been reported as a potential source of bioactive compounds with antibacterial activity. The accidental discovery of penicillin from fungi in 1929 by Fleming drew attention of scientific community to the possible role of fungi as antibiotics and this has contributed to the isolation and development of other antibiotics [8]. Fungi are rich sources of secondary metabolites with diverse bioactivities, many that have been developed into important pharmaceutical products. With more than 15,000 secondary metabolites discovered to date, fungi stand out as an important group of microbes in bioactive natural products research [9]. Advances in analytical chemistry, computational tools, and drug discovery research have enabled the development of some fungal-derived antimicrobial compounds with potential therapeutic effects to be used individually or in adjunctive therapies to control difficult-to-treat pathogens [10]. Secondary metabolites from saprotrophic and easily cultivable fungi of the phyla Ascomycota, Muccoromycota and Basidiomycota have been studied intensively [11].

Previous studies have shown that secondary metabolites from basidiomycetes have a wide range of pharmacological activities including antimicrobials [12]. Basidiomycetes, from the phylum Basidiomycota, are a group of higher fungi with distinctive fruiting bodies and reproductive structures with edible and non-edible properties. Mushroom-forming fungi, mostly from the basidiomycete group, have been used as remedies for various diseases owing to their ability to produce compounds with high structural diversity, including terpenes, anthraquinone, derivatives of benzoic acid, quinolines, cyclic peptides, steroids, sesquiterpenes, oxalic acid, epipolythiopiperazine-2,5 diones and polysaccharides [13,14]. Traditionally, bioactive components have been extracted from fruiting bodies or mycelial extracts of mushrooms [15]. They are known to produce secondary metabolites with a range of pharmacological activities including antimicrobial, antioxidant, anti-angiogenesis, anticancer, immunomodulatory and anti-inflammatory [16].

In many studies, however, antimicrobial activities of different extracts of mushroom were reported without identifying the active compound/s responsible for the observed high activity against Gram-positive bacteria [13,17,18]. Despite the challenges faced in explorative studies to access the bioactive metabolites originating from fruiting bodies of mushrooms as they occur temporarily in the environment, their importance has been significant in recent decades [13]. With regards to this, more studies have been focusing on metabolites produced from submerged fermentation of mycelial culture of mushrooms where frequently these metabolites differ from those of fruiting bodies [19]. This work is a brief review on antistaphylococcal activities of Basidiomycetes that have been reported.

1.1. Antimicrobial Resistance (AMR)

Antimicrobial resistance is described as lowered efficiency or loss of antibiotics' effectiveness against pathogens and this is a major problem in the medical sector globally. Antimicrobial resistance is correlated with high medical costs because of a longer period of disease, additional testing and needless usage of second-line treatments [1,5]. As mentioned by the Organization for Economic Co-operation and Development (OECD), the key risk factor for development of resistance is excessive usage or intake of antibiotics [6,20,21]. The high emergence of AMR has led to a shift change in therapeutic practices towards use of newer wide-spectrum drugs and increased usage (42%) of last resort classes of antibiotics such as vancomycin [6]. Many reports have indicated that the resistance epidemiology is global and spreads through nations and across borders [20].

In vitro antibacterial activity of antibiotics is typically determined by biological assays. The most popular methods include agar well diffusion, disc diffusion, minimum inhibitory concentration (MIC), minimum bactericidal concentration (MBC) and time kill assays [22]. Clinical breakpoints used as interpretive criteria to consider susceptibility of a bacterial isolate to an antimicrobial agent are provided by the Clinical and Laboratory Standards Institute (CLSI) [23,24], the European Committee on Antimicrobial Susceptibility Testing (EUCAST) and/or the US Food and Drug Administration (FDA).

In agar well diffusion assay, a hole was punched into agar inoculated with the test organism and filled with the antibiotic solution. Alternatively, a filter paper disc containing antibiotic was placed on inoculated agar. In both methods, zone of inhibition produced by the diffusion of antibiotic compound into the agar was measured. Due to the agar being an aqueous preparation, non-polar compounds do not diffuse as well as polar compounds, thus producing smaller diameter of inhibition zones despite their higher activity. This could be a limitation in the agar diffusion method [25,26]. The MIC is defined as the lowest concentration of a drug that inhibits the growth of bacteria after incubation. The MBC of a drug is determined upon reading of MIC by streaking the broth dilutions onto general or selective agar with 24–48 h of incubation. Absence of growth of the viable organisms on agar indicated the lowest broth dilution of drug, which caused a 99.9% suppression of the bacterial growth. The most appropriate in vitro approach to study bactericidal activity of a vast variety of antimicrobial agents is the time-kill assay [22]. The outcome of this assay indicates if an antimicrobial effect is dependent on exposure time or concentration of the drug. The assay is often used as initial descriptive analysis in pharmacodynamic analysis of a drug [22,27].

Preventing the development of resistant bacterial strains is important to ensure the effectiveness of current drugs in managing dangerous and life-threatening infections as an attempt to reduce the severity of AMR crisis [28,29]. Thus, there is an urgent need to carry out continuous research and development of new antibacterial drugs to counter the loss in efficacy of current antibiotics [28,30,31].

1.2. Multidrug Resistance in *Staphylococcus aureus*

Staphylococcus aureus infections produce wide spectrum of pyogenic lesions involving several organs, and it can cause hospital outbreaks and community acquired infections. Selective pressure on the bacteria due to high consumption of wide-spectrum antibiotics could stimulate the emergence of antibacterial resistant strains. Burden of infections in low-income countries is high since the solution to overcome this crisis is by replacing ineffective first line antibiotics to more costly second line or third line antibiotics [6]. Thus, the development of new antibiotics, combination drugs, bioprospecting for potential antibacterial natural compounds and improved drug delivery systems are some of the current strategies to control the antimicrobial resistance threat [6,32].

Emergence of HA-MRSA strains are associated with profligate use of antibiotics in healthcare settings [33]. The MRSA strains have demonstrated resistance to a range of antibiotics belonging to isoxazoyl penicillin group (methicillin, oxacillin, flucloxacillin), cephalosporins and carbapenems [34–36]. The first MRSA variant strain was isolated in United Kingdom in 1961 after methicillin was introduced in 1959 [37–40]. Thereafter, the changing epidemiology of variants of the strains found in many other countries like Europe, Australia, Japan and United States eventually makes MRSA as a major threat in nosocomial infections worldwide [20]. MRSA has more propensity to develop resistance to macrolides, quinolones and aminogylycosides, and this led to reduced therapeutic options [34–36,41,42]. In hospitals worldwide, a high prevalence of MRSA with rates above 50% has been documented [43,44]. A new variant strain of CA-MRSA was reported to be prevalent in Asian healthcare settings. This was documented by several studies which showed an occurrence rate of 2.5% in Thailand and 38.8% in Sri Lanka [45].

Emergence of antimicrobial resistance in *S. aureus* to glycopeptide group of antibiotics which is the last resort of staphylococcal treatment, became a global concern in managing staphylococcal infections [29]. Three classes of limited vancomycin susceptibility strains of *S. aureus* that have emerged in different locations around the world are VISA, heterogeneous VISA (hVISA) and VRSA [37,46]. Owing to the dynamic re-organisation of cell wall metabolism, VISA and hVISA strains have thickened cell walls with decreased glycopeptide cross-linking [44]. The first report of VISA and hVISA was detected in Japan in 1996 and 1997, respectively, while VRSA from a hospital in the United States was reported in 2002 [47]. The resistance phenotypic of VISA (Minimum Inhibitory Concentration:

8 μg/mL) has the ability of reverting back to the susceptibility phenotype towards vancomycin when the selective pressure is removed (MIC at 2 μg/mL) [48].

Prevalence of VRSA strains have been documented in South Nigeria (0–6%), Zaria, North Nigeria (57.7%), South India (1.4%), Australia, South Africa, Scotland, Hong Kong, Thailand and Korea (0–74%) [39,48–50]. No reports of vancomycin-resistant *S. aureus* (VRSA) have been documented in Malaysia [51]. The emergence of antibiotic resistance globally could lead to serious problems of limited therapeutic options available [52]. The emergence of VISA and VRSA strains causes more life-threatening infections in the healthcare sector [53,54]. Scanty and expensive drugs like teicoplanin, daptomycin and linezolid are also being used as next therapeutic options for MRSA infection due to the limited sensitivity of vancomycin [15,35,54–56].

2. Anti-MRSA Drug Discovery from Fungi

Research on the development of novel drugs from natural products is faced with limitations and obstacles since pharmaceutical companies earn low profits from this sector. However, screening of natural products for new bioactive compounds in discovery programmes hold high prospects to slow down the development of antimicrobial resistance crisis. Penicillin, which was discovered from *Penicillium notatum*, had been used to derive some commercial antibiotics that acts against Gram-positive bacteria deep infections such as Piperacillin, Amoxicillin and Ampicillin [9,57–59].

Cephalosporin C, which belongs to the class of β-lactam antibiotics, was developed from *Cephalosporium acremonium* and the commercialised products are Cephalexin, Cephradine and Cefadroxil [9]. The β-lactam antibiotics are active against Gram-positive bacteria including *S. aureus*. Fusidic acid, the steroid-like topical antibiotics that was produced from *Fusidium coccineum* or *Acremonium fusidioides*, showed antistaphylococcal activity against penicillin and methicillin-resistant strains of *S. aureus* [60,61]. The commercialised products from tripene structured fusidic acid metabolite are Usidin, Fucidin, Fucicort, Fucibet and Taksta [9]. Retapamulin (Altabax) is a derivative from the secondary metabolite pleuromutilin produced from *Pleurotus mutilis* (Fr.) Sacc. and *Pleurotus passeckerianus* Pilat (Pleurotaceae). Pleuromutilin which was discovered from Basidiomycete origin, has been used to treat infections induced by Gram-positive bacteria including MRSA [62,63].

Glycopeptides such as the semi-synthetic oritavancin are broad spectrum antibiotics against Gram-positive bacteria including MRSA and VRSA. The bacteria-derived compound teixobactin, has potential antistaphyloccocal activity with MIC value of 0.25 μg/mL and is comparatively superior to vancomycin in killing late exponential phase populations of the pathogen and bactericidal activity against VISA [58].

3. Antistaphylococcal Natural Products from Basidiomycetes

Basidiomycetes derived secondary metabolites are known as a promising source of antibacterial compounds with activity against Gram-positive bacteria in natural product discovery. Crude extracts of natural products were reported to target on cell wall biosynthesis and cell membrane permeability as their mechanism of action to exhibit antibacterial activity [64]. Many species have been studied for their potential to produce bioactive secondary metabolites with antibacterial activity against *S.aureus* and other drug resistant strains of bacteria as shown in Tables S1–S3.

The fruiting body extracts of *Ganoderma* species such as *G. applanatum* have shown positive antistaphylococcal activity [65,66]. Quereshi et al. [67] showed that metabolites derived from *Ganoderma* mushroom possessed bacteriolytic enzyme, lysozyme, acid protease and polysaccharides as bioactive principles that play a role in antibacterial activity [67]. The most common species that has been explored for pharmaceutical activity is *G. lucidum* [68,69]. *Ganoderma lucidum* has been regarded as a potential producer of broad-spectrum antibacterial compounds highly potent against Gram-positive bacteria, as indicated by preclinical (in vitro and in vivo animal) studies [69]. *Ganoderma lucidum* is saprobic in nature and secondary metabolites have been extracted from the fruiting body using methanol extraction. The antistaphylococcal activity of extracts from fruiting body of *G. lucidum* strain was represented by inhibition zone diameter ranging from 2.2 to 2.5 cm recorded against *S. aureus* [15]. The isolated compound, 12β-acetoxy-3β, 7β-dihydroxy-11, 15, 23-trioxolanost-8-en-26-oic acid butyl ester (**1**) from *G. lucidum* displayed MIC value of 68.5 um against *S. aureus* [70]. *Ganoderma pfeifferi* is one of the phytochemically less investigated species of the family Ganodermataceae. *Ganoderma pfeifferi* produced farnesyl hydroquinones, namely ganomycins A (**2**) and B (**3**), which are potential inhibitors of *S. aureus* and MRSA [19,71]. The lack of data on mechanism of action of the antimicrobial compounds from *Ganoderma* spp. does not support their use as a sole antibiotic. Further studies are needed before these compounds are made available for human use [69].

The fruiting bodies of *Tapinella atrotomentosa* were extracted with methanol and the fractionated compounds were isolated using High Performance Liquid Chromatography analysis. Compounds from *T. atrotomentosa* include osmundalatone (**4**), 5-hydroxy-hex-2-en-4-olide (**5**) and spiromentin C (**6**), which showed positive growth inhibiting activity against MRSA (SZMC 6270) at higher MIC value of 250 μg/mL. Another compound from this fungus, spiromentin B, showed no activity against MRSA. Moreover, the interaction between cefuroxime and the isolated compounds was evaluated against MRSA. Results indicated that the compounds do not act as an adjuvant to increase the antibiotic activity. The antioxidant activity using 2,2-diphenyl-1-picryl-hydrazyl-hydrate(DPPH) and Oxygen Radical Absorbance Capacity (ORAC) assays showed that spiromentin C (**6**) and B exhibited high antioxidant effects (16.21 ± 0.38 and 11.23 ± 0.58 mmol TE/g, respectively) compared to the reference compound ascorbic acid (6.97 ± 0.01 mmol TE/g) [17].

Fruiting body extracts of *Pleurotus sajor-caju (Fr.) Singer* exhibited some extent of antistaphylococcal activity against MSSA and MRSA with a MIC value of 10 mg/mL with the presence of compounds such as *p*-hydroxybenzoic acid (**7**), *p*-Coumaric acid (**8**) and cinnamic acid (**9**) [72,73]. Cytotoxicity assay of the fungal fruiting body extract demonstrated strong inhibition against growth of non-small cell lung carcinoma (NCI-H460), breast (MCF-7) and cervical (HeLa) carcinoma cells. *Pleurotus sajor-caju* extract does not display cytotoxicity against non-tumour cells (primary PLP2) even at a test concentration of 400 μg/mL [72]. Ethanol extract of mycelia of the oyster mushroom, *P. ostreatus*, demonstrated a broad-spectrum antibacterial activity. Activity against the Gram-positive *S. aureus* showed a 24 mm diameter of inhibition zone in paper disc assay [74]. In another study, ethyl acetate mycelial extract of *P. ostreatus* exhibited potent antistaphylococcal activity with an inhibition zone of 8.5 mm diameter against *S. pyrogens*, comparable to the activity of control drug kanamycin (12.5 mg/mL) [75]. *Pleurotus aureovillosus* fruiting body extract showed antistaphyloccocal activity against *S. aureus* by displaying 8 mm diameter of inhibition zone [76]. This fungus has high potential to produce antistaphylococcal active compounds for drug development with further investigations.

The tropical basidiomycete fungus *Laxitextum incrustatum* from Kenya produced anticancer and antimicrobial compounds active against *S. aureus* strains. The isolated compound, laxitextine A (**10**), was discovered to show the highest activity against *S. aureus* (DSM 346) and MRSA with MIC value of 7.8 μg/mL (17.9 μM). Comparable cytotoxicity effect was also observed for this cyathane xyloside derived compound with IC_{50} values of 2.0–2.3 μM against breast cancer cell line (MCF-7) [77].

Caloboletus radicans possessed fruiting bodies that tasted bitter and are inedible due to the content of calopin compounds. The compound 8-deacetylcyclocalopin (**11**) showed inhibition against *S. aureus* ATCC 25923 and SA-1199B (MIC 16 μg/mL), and against *S. aureus* XU212 and EMRSA-15 at a MIC value of 32 μg/mL. The compound showed no activity against *S. aureus* RN4220. Different derivatives of calopins (100 μM) were inactive against the prostate cancer cell line PC3 and liver hepatoblastoma cell line HepG2 [78].

The compounds isolated from fruiting bodies of *Stereum hirsutum* mushroom were tested for different bioactivities. The isolated compounds, benzoate derivatives 1 (**12**) and 2 (**13**) revealed a strong antistaphylococcal activity with MIC value of 25 μg/mL and a significant cytotoxicity effect. However, derivative 1 (**12**) displayed stronger cytotoxicity against the growth of A549 cell line with an IC_{50} value of 13.14 ± 0.89 μM. Hence, this secondary metabolite was identified as a potential therapeutic agent [79].

Mycelia culture filtrate extract of *Pyrofomes demidoffi* exhibited strong inhibition against the in vitro growth of Gram-positive bacteria. *Staphylococcus* spp. were the most susceptible among the tested Gram-positive bacteria. Mycelial culture filtrate extract impregnated onto discs produced an inhibition zone with diameter of 46–47 mm against *S. aureus* strains. This indicated that culture filtrate extract of the fungus has a wide spectrum of activity against an array of bacteria strains [80].

The mushroom *Agaricus bisporus* identified from Njoro Kenya was tested for antimicrobial activity against an array of test bacteria and fungi. The mushroom metabolites from methanol extract revealed weak activity against Gram-positive bacteria. This was supported by diameters of inhibition zone of 16 ± 0.1 mm and 11.75 ± 0.96 mm against *S. aureus* exhibited by the mushroom extract from two different studies [81,82]. The metabolites responsible for the antistaphylococcal activities are 2,4-dihydroxybenzoic (**14**) and protocatechuic acid (**15**) [81]. Fruiting body extract of *A. brunnescens* prepared in chloroform and acetone solvents showed a higher antistaphylocccal activity with MIC value lower than 100 µg/mL [83].

Studies carried out by Reid et al. [84] presented that the fruiting body extracts of edible mushrooms from Zimbabwe, including *Amanita zambiana, Cantharellus miomboensis, Cantharellus symoensii, Cantharellus heinemannianus* and *Lactarius kabansus*, showed weak antistaphylococcal activity with inhibition zone range of 6.5–8.67 mm [84]. However, the non-edible mushroom *Trametes strumosa* had displayed slightly more potent activity than those edible mushrooms with a diameter of inhibition zone (DIZ) of 9.5 mm against *S. aureus*. Methanol extract of *Boletelus edulis* showed good activity against *S. aureus* with MIC value of 10.25 µg/mL [82].

The ethanol and aqueous extracts of *Auricularia sp.* showed activity against *S. aureus* with MIC value of 0.83 ± 0.29 µg/mL. High susceptibility of *S. aureus* towards the ethanol and aqueous extracts of *Termitomyces* sp. was observed with a MIC value of 0.67 ± 0.29 µg/mL. The extracts of this mushroom possessed potent antistaphyloccoccal activity [85]. On the other hand, the tested extracts of *Fomes fomentarius* showed lower antistaphyloccoccal activity with MIC value of 125 to 250 µg/mL when compared to the conventional antibiotic (ampicillin) with MIC value of 0.5 µg/mL [86].

Hot alkaline extracts obtained from fruiting bodies of the wild basidiomycete *Grifola frondosa* were examined for their antistaphyloccoccal activity. The antibacterial activity was determined through disc diffusion, Minimum Inhibitory Concentration (MIC) and Minimum Bactericidal Concentration (MBC) assays. The extract showed strong antistaphyloccoccal activity (Diameter of Inhibition Zone = 24 mm), almost similar to the control drug tetracycline (DIZ = 30 mm). This supported the strong MIC value observed for the fungal extract against *S. aureus* (39 µg/mL) [87].

Three wild mushrooms, *Trametes* sp. 1 (Arabuko-Sokoke forest), *Trametes* sp. 2 (Kakamega forest) and *Microsporus* sp. had shown promising antibacterial activity against *S. aureus* ATCC 25923). The chloroform, ethanol and hot water extracts of *Trametes* sp. 1 demonstrated good growth inhibitory activities against *Staphylococcus*. The MIC values (0.50–0.83 ± 0.29 mg/mL) against *S. aureus* indicated that the bacterium was highly susceptible to all of the tested extracts. Moreover, MRSA strain with MIC values of 0.83–1.00 mg/mL showed susceptibility towards the chloroform and ethanol extracts. MIC value of 0.50 mg/mL of *Trametes* sp. 2 displayed that *S. aureus* was the most susceptible to hot water extract. The chloroform and water extracts showed statistically significant antistaphylococcal activities compared to ethanol extract. The strong antistaphyloccoccal activity of *Microsporus* sp. was observed at MIC value of 0.67–1.00 mg/mL [88].

Agaricus brunnescens Peck and *Lactarius vellereus* (Fr.) Fr originated from Black Sea Region of Turkey possessed high antistaphylococcal activity in the chloroform and acetone extracts with a MIC value of 39 µg/mL. The author reported that a MIC value of less than 100 µg/mL was an indicator for high inhibitory activity [83].

Methanolic extracts of various mushrooms have shown a higher antibacterial activity against Gram-positive bacteria such as *S. aureus*. The basidiomycete *Lentinus* sp. presented potent antistaphyloccoccal activity with inhibition zone of 16 mm followed by antibacterial activity of *Schizophyllum commune* against *S. aureus* (9 mm) [76]. The extract and purified fraction of wild basidiomycete *Lentinus quercina* was tested against *S. aureus* and MRSA using agar well diffusion and disc diffusion assays. Extracts and the purified fraction of the macrofungus displayed a wide range of inhibition against *S. aureus* strains (8.7–22 mm) and MRSA strain (4–19 mm). The MIC results (3.125 to 6.25 μg/mL) showed that the wild basidiomycete has potent antistaphyloccoccal activity against both ATCC and clinical strains of *Staphylococcus* [89]. Oxalic acid from *L. edodes* (Berk.) Pegler was reported to be responsible for the growth inhibition of *S. aureus* [19].

Secondary metabolites from *Drechslera halodes* were investigated for potential antibacterial activity using disc diffusion and MIC assays. Fungal culture filtrate extracts displayed growth inhibition zone against *S. aureus* with a DIZ of 15 mm. The purified compound identified as 6-allyl-5,6-dihydro-5-hydroxypyran-2-one (**16**), which belongs to aromatic esterase group, revealed an MIC value of 25 μg/mL against *S. aureus*. The presence of other bioactive secondary metabolites in the culture filtrate extract of *D. halodes* attributed to a higher inhibitory activity against *S. aureus* [90].

Lewia infectoris derived bioactive compound namely pyrrocidine C (**17**) showed strong activity against *S. aureus* ATCC 29213 at MIC of 2 μg/mL. Previous reports mentioned that pyroccidine A and B are proven antibacterial against Gram-positive bacteria with similar MIC value of pyrrocidine C even though their relative stereochemistry is different to each other [91].

Fungus from genus *Cortinarius* produced compounds such as 6-methylxanthopurpurin-3-*O*-methyl ether (**18**), (1S,3S)-austrocortilutein (**20**), (1S,3R)-austrocortilutein (**21**), (1S,3S)-austrocortirubin (**22**) and erythroglaucin (**25**), which exhibited anti-*Staphylococcus* activity. The anthraquinones physcion (**19**) and emodin (**24**) exhibited the greatest potency against both Gram-positive and -negative bacteria. Ethanol fraction of most *Cortinus* sp., namely, *C. ardesiacus*, *C. archeri*, *C. austrosaginus*, *C. austrovenetus*, *C. austroviolaceus*, *C. coelopus*, *C. [Dermocybe canaria]*, *C. clelandii*, *C. [D. kula]*, *C. memoria-annae*, *C. persplendidus*, *C. sinapicolor* and *C. vinosipes* demonstrated IC_{50} values of ≤0.09 mg/mL against *S. aureus* [92].

Lactarius piperatus, *L. camphorates*, *L. volemus*, *Chanterellus cibarius*, *Ramaria flava*, *Macrolepiota procera*, *Leatiporus sulphureus*, *L. delicious*, *Boletus edulis*, *Hydnum repandum* and *Cortinarius* sp. are edible mushrooms commonly found in Black Sea Region of Turkey. The crude extracts of these wild mushrooms showed good activity against *S. aureus* with DIZ values ranging from 8.50 ± 1.00 to 11.75 ± 0.50 mm [82].

In another study carried out by Bala et al. [12], certain genera of macrofungi, namely, *Hohenbuehelia*, *Amanita* and *Agaricus* revealed strong antibacterial activity against *S. aureus* in their extracts. Within each sterile 96-well plate, the fungal fruiting body extract samples (ethanol and aqueous extracts) were prepared in serial dilutions. The initial absorbance reading was recorded as t_0, and subsequently the t_{22} absorbance reading was taken after incubation with fungal extracts for 22 h at 37 °C. The growth inhibition percentage was measured with formula: $\%\ \text{inhibition} = \{1 - (t_{22} - t_0)/(C_{22} - C_0)\} \times 100$, where C_0 is absorbance value of negative wells and C_{22} is the value corresponds to 22 h post incubation of negative wells. Both the water and ethanol extracts of *Hohenbuehelia* sp. exhibited strong inhibition against *S. aureus* at all four test concentrations (6.25%, 12.5%, 25% and 50%) as shown in Table S2.

4. Antibacterial Natural Products from Basidiomycetes against Other Drug-Resistant Pathogens

Staphylococcus aureus, *Enterococcus* spp., *Enterobacteriaceae*, *Pseudomonas aeruginosa* and *Acinetobacter* spp. have significant effects on contributing to the emergence of antimicrobial resistance that greatly impacts on healthcare system [93]. The presence of extended-spectrum beta-lactamase (ESBL) *Escherichia coli* strains causes high concern especially in healthcare setting due to their multidrug resistance properties. Some of the activities shown by secondary metabolites from Basidiomycetes against other drug-resistant pathogens (other than *S. aureus*) are given in Table S3.

Quambalaria cyanescens (Basidiomycota: Microstromatales) strains isolated from different insect species such as *Scolytus intricatus* (Bulgaria), *Phloetribus scarabeoides* (Croatia), *Scolytus amygdali* (Syria) and bark beetle feeding on *Arbutus unedo* (Tunnisia) were compared. Naphthoquinone derivatives, which are bright colored pigments identified from *Q. cyanescens*, exhibited potential antibacterial activity against *E. coli* with a DIZ value of 2–3mm. Quambalarine A (**26**) and mompain (**27**) are mainly active against bacterial strains. The antimicrobial activity described for both these compounds could be related to their selectivity towards mitochondria which centers the cellular energetic metabolism. Mompain (**27**) enables to transform the mitochondrial networks into vesicular formations while complete disappearance of mitochondria was observed when cells were treated with quambalarine B which leads to apoptosis phase. This indicated that alteration in mitochondrial function could be a potential mechanism of action for the compound in cells [11].

Ethanolic extract of *Pleurotus sajur-caju* displayed antimicrobial activity against *P. aeruginosa* and *Klebsiella pneumoniae* while the methanolic extract was active against *E. coli* [94,95]. *Pleurotus sajor-caju* ethanolic extract is composed of lipophilic compounds, *p*-Hydroxybenzoic (**7**), *p*-coumaric (**8**) and cinnamic acids (**9**) at respective concentration: 66 ± 4, 43.7 ± 0.9 and 37.5 ± 0.3 μg/g extract [72]. A previous report mentioned that the most suitable solvent to extract maximum antimicrobial bioactive compounds from the fungus is ethanol [94].

High anti-Gram-positive bacteria activity was observed in extracts of *Fistulina hepatica* and *Tapinella atrotomentosa* with MIC values ranging between 12.5–100 μg/mL. However, *F. hepatica* extract was inactive, whereas *T. atromentosa* was highly active against Gram-negative bacteria. The justification behind the observed difference could be that of type of solvent used for extraction process which fractioned out different compounds, and time and place of collection of the sample which may have affected the activity as well [17].

Tapinella atrotomentosa extracted by chloroform exhibited a broad spectrum of antibacterial activity against Gram-positive bacteria, ESBL *E. coli* and multidrug-resistant *P. aeruginosa* and *Acinetobacter baumanii*. Osmundalactone (**4**), 5-hydroxy-hex-2-en-4-olide (**5**), spiromentin C (**6**) and spiromentin B are metabolites with a terphenylquinone skeleton isolated from this fungus and assessed for their antibacterial activity using the broth microdilution assay. Although osmundalactone (**4**) and spiromentin C (**6**) were highly active against *A. baumanii* and *E. coli*, 5-hydroxy-hex-2-en-4-olide (**5**) was the most active antibacterial constituent against these multidrug-resistant bacteria [64]. Secondary metabolites from *Agaricus bisporus* and *Trametes gibbosa* inhibited growth of Gram-negative bacteria in previous studies (Table S3). Inhibition of the bacterial growth by *T. gibbosa* extracts was significantly higher when compared to inhibition exhibited by *A. bisporus* extracts [17].

Crude extracts from fruiting bodies and biomass of *Ganoderma* were reported to exhibit in vitro antibacterial effect against *Staphylococcus* spp., *E. coli* and *Enterobacter* spp. [96]. However, variations are present among *G. lucidum* strains regarding yield and production of bioactive compounds and their antimicrobial activity [69]. The methanolic extracts of biomass and fruiting body of *G. lucidum* strains were found to show low antibacterial activity against *K. pneumoniae* and *E. coli* while the aqueous extracts of biomass and fruiting body exhibited minimum inhibition of growth against *Pseudomonas* spp. Generally, methanolic extracts showed effective inhibition zone as compared to aqueous extracts [15].

Chloroform extract of *Auricularia* spp. exhibited activities with significant difference against resistant pathogens such as *E. coli*, *K. pneumoniae* (ATCC 13883) and *P. aeruginosa*. Ethanol and aqueous extracts displayed limited activity against *E. coli* and *P. aeruginosa*. From the results, extracts of *Auricularia* spp. were found to be less active against the Gram-negative bacteria *E. coli* [82]. Both chloroform and ethanol extracts of *Auricularia* and *Termitomyces* spp. exhibited a weaker antimicrobial activity. This can be explained by the absence of bioactive compounds in the extracts or the loss of functionality due to deviations associated with difference in the amount and type of bioactive constituents present in the extracts [82]. The dichloromethane, methanol, aqueous and cyclohexane extracts of *Fomes fomentarius* showed potential activity against *E. coli* at MIC 125 μg/mL. Generally, high total polyphenol content was found in methanol and aqueous extracts that possessed strong antimicrobial activity [93].

Drechslera halodes produced a potential bioactive compound that exhibited growth inhibition activity against *E. coli* with a DIZ value of 12 mm diameter and MIC 50 μg/mL. Secondary metabolites excreted into the culture medium during submerged fermentation attributed to this potential antibacterial activity. There are no cytotoxicity effects observed for the bioactive compound isolated from the fungal culture filtrate. The responsible alkaloid compound was identified as 6-allyl-5,6-dihydro-5-hydroxypyran-2-one (**16**). The mode of action of this compound was related to its inhibition effects on DNA synthesis [90].

Bioactive compounds isolated from fruiting bodies of *Cortinarius* sp. were investigated for anti-*Staphylococcus* and anti-*Pseudomonas* activity using microdilution assay. Compound (**19**) was more active against *P. aeruginosa* than *S. aureus*. Compounds (**20**) to (**24**), however, displayed lower activity against *P. aeruginosa*. *Cortinarius abnormis*, *C. austroalbidus*, *C. [D. kula]* and *C. persplendidus*, and eleven unknown *Cortinarius* spp. were reported to exhibit anti-*Pseudomonas* activity with $IC_{50} \leq 0.09$ mg/mL [92].

Potential antibacterial compounds have been isolated from organic extracts of the fungus *Pleurotus eous*. The petroleum ether extracts of *P. eous* possessed several fatty acids compounds such as stearic acid (**28**), heptadecanoic acid (**29**) and tartronic acid (**30**) that attributed to high growth inhibitory activity against *E. coli* (DIZ = 11 ± 0 mm) than *K. pneumoniae*. The antimicrobial activity of this fungus is related to the presence of fatty acids which traditionally could be used for pain, fever and inflammatory disorders [97].

5. Conclusions

The present review focuses on antistaphyloccocal activity of basidiomycetes globally and their isolated secondary metabolites. An exhaustive literature search was performed and only those basidiomycete extracts (and isolated compounds) with positive antistaphyloccoccal activity were included. For further extensive scientific studies, it will certainly prove useful to investigate the metabolites derived from this fungal group.

The emergence of several staphylococcal resistance strains linked to nosocomial infections requires a new antimicrobial solution. Literature reports indicated that extracts and/or compounds from various basidiomycete species exhibited potential antibacterial activity against *Staphylococcus* strains. Increasing reports in recent years have also shown the potential ability of some natural extracts in potentiating the activity of standard antibiotics [88,90]. Further pharmacological studies to look at synergistic interactions between antibiotics and these natural products can greatly improve the potential activities of moderately toxic antistaphylococcal compounds.

The numerous methodologies used for the assessment of antimicrobial activity of basidiomycete extracts or isolated compounds made it difficult to compare between the published data. The factors involved like type of solvent, conditions such as temperature and time, and the structure of compounds that causes variation in the extraction of bioactive compounds has to be optimised to attain a greater overall output and efficiency of the target compounds. The geographic location, different fungal cultivation method and types of growth media could possibly affect the content and amount of active compounds in the extracts [90]. Thus, the standardisation of methods and establishment of cut-off values for activity is important. The information on mechanisms of action of different fungal-derived compounds could assist in the development of new active pharmaceutical ingredients with interesting or novel activity. Moreover, the use of cytotoxicity assays is also important for the evaluation of effects of a compound in human at the in vitro efficacy concentrations studied.

The available literature studies are mostly focused on screening of antibacterial properties of basidiomycete extracts. The prospect of discovering new secondary metabolites from them is highly valued. The lack of data on mechanism of action of the antimicrobial compounds from Basidiomycete fungus does not support their use as a sole antibiotic. Further studies are needed before these compounds are made available for human use. Identifying individual compounds that exhibit antibacterial activity and explaining their mode of action is inevitable in drug discovery. This provides an avenue for the creation of pharmaceuticals that are effective against selected microorganisms resistant to traditional therapy in tandem with the growing multidrug resistance crisis in recent decades.

Supplementary Materials: The following are available online, Table S1: Antistaphylococcal activity reported from Basidiomycete fungi; Table S2: Antistaphylococcal activity showed by fruiting bodies of Basidiomycetes in water and ethanol extracts at various concentration; Table S3: Antibacterial activity of Basidiomycete compounds against other pathogens; Figure S1: Structures of antimicrobial compounds isolated from Basidiomycetes.

Funding: This research was funded by Ministry of Education Malaysia, FRGS/1/2018/SKK10/NRE/02/1.

Acknowledgments: Authors would like to thank Forest Research Institute Malaysia and Universiti Kebangsaan Malaysia for their support.

Conflicts of Interest: The authors declare no conflict of interest.

References

1. World Health Organization. WHO Report on Surveillance of Antibiotic Consumption: 2016–2018 Early Implementation. 2018. Available online: https://www.who.int/medicines/areas/rational_use/oms-amr-amc-report-2016-2018/en/ (accessed on 22 September 2020).
2. Magiorakos, A.P.; Srinivasan, A.; Carey, R.B.; Carmeli, Y.; Falagas, M.E.; Giske, C.G.; Harbarth, S.; Hindler, J.F.; Kahlmeter, G.; Olsson-Liljequist, B.; et al. Multidrug-resistant, extensively drug-resistant and pandrug-resistant bacteria: An international expert proposal for interim standard definitions for acquired resistance. *Clin. Microbiol. Infect.* **2012**, *18*, 268–281. [CrossRef]
3. Basak, S.; Singh, P.; Rajurkar, M. Multidrug resistant and extensively drug resistant bacteria: A study. *J. Pathog.* **2016**, *2016*, 1–5. [CrossRef]
4. Health Research & Educational Trust. Multi-Drug Resistant Organism Infection Change Package: 2017 Update. Chicago. Available online: www.hret-hiin.org (accessed on 20 September 2020).
5. World Bank. Drug-Resistant Infections: A Threat to Our Economic Future: Washington, DC, 2017. Available online: https://documents.worldbank.org/curated/en/323311493396993758/ (accessed on 22 September 2020).
6. Laxminarayan, R.; Matsoso, P.; Pant, S.; Brower, C.; Røttingen, J.A.; Klugman, K.; Davies, S. Access to effective antimicrobials: A worldwide challenge. *Lancet* **2016**, *387*, 168–175. [CrossRef]
7. Dhanalakshmi, T.A.; Umapathy, B.L.; Mohan, D.R. Prevalence of Methicillin, Vancomycin and Multidrug Resistance among *Staphylococcus aureus*. *J. Clin. Diagn. Res.* **2012**, *6*, 974–977.
8. Fleming, A. On the antibacterial action of cultures of a penicillium with, special reference to their use in the isolation of *B. influenzae*. *Br. J. Exp. Pathol.* **1929**, *10*, 226–236. [CrossRef]
9. Bills, G.F.; Gloer, J.B. Biologically active secondary metabolites from the fungi. *Fungal Kingd.* **2017**, 1087–1119. [CrossRef]
10. Betts, J.W.; Wareham, D.W. In vitro activity of curcumin in combination with epigallocatechin gallate (EGCG) versus multidrug-resistant *Acinetobacter baumannii*. *BMC Microbiol.* **2014**, *14*, 172. [CrossRef] [PubMed]
11. Stodulkova, E.; Cisarova, I.; Kolarik, M.; Chudickova, M.; Novak, P.; Man, P.; Flieger, M. Biologically active metabolites produced by the basidiomycete *Quambalaria cyanescens*. *PLoS ONE* **2015**, *10*. [CrossRef]
12. Bala, N.; Aitken, E.A.; Fechner, N.; Cusack, A.; Steadman, K.J. Evaluation of antibacterial activity of Australian basidiomycetous macrofungi using a high-throughput 96-well plate assay. *Pharm. Biol.* **2011**, *49*, 492–500. [CrossRef] [PubMed]
13. Alves, M.J.; Ferreira, I.C.F.R.; Martins, A.; Pintado, M. Antimicrobial activity of wild mushroom extracts against clinical isolates resistant to different antibiotics. *J. Appl. Microbiol.* **2012**, *113*, 466–475. [CrossRef]
14. De Silva, D.D.; Rapior, S.; Sudarman, E.; Stadler, M.; Xu, J.; Aisyah Alias, S.; Hyde, K.D. Bioactive metabolites from macrofungi: Ethnopharmacology, biological activities and chemistry. *Fungal Divers.* **2013**, *62*, 1–40. [CrossRef]
15. Kaur, H.; Sharma, S.; Khanna, P.K.; Kapoor, S. Evaluation of *Ganoderma lucidum* strains for the production of bioactive components and their potential use as antimicrobial agents. *J. Appl. Nat. Sci.* **2015**, *7*, 298–303. [CrossRef]
16. Wasser, S.P. Medicinal mushroom science: History, current status, future trends, and unsolved problems. *Int. J. Med. Mushrooms* **2010**, *12*, 1–16. [CrossRef]
17. Beni, Z.; Dekany, M.; Kovacs, B.; Csupor-Loffler, B.; Zomborszki, Z.P.; Kerekes, E.; Vanyolos, A. Bioactivity-guided isolation of antimicrobial and antioxidant metabolites from the mushroom *Tapinella atrotomentosa*. *Molecules* **2018**, *23*, 1082. [CrossRef]
18. Ling, L.L.; Schneider, T.; Peoples, A.J.; Spoering, A.L.; Engels, I.; Conlon, B.P.; Mueller, A.; Schaberle, T.F.; Hughes, D.E.; Epstein, S.; et al. A new antibiotic kills pathogens without detectable resistance. *Nature* **2015**, *517*, 455–459. [CrossRef] [PubMed]
19. Lindequist, U.; Niedermeyer, T.H.; Julich, W.D. The pharmacological potential of mushrooms. *Evid. Based Complement. Altern. Med.* **2005**, 2. [CrossRef]

20. Cecchini, M.; Langer, J.; Slawomirski, L. *Antimicrobial Resistance in G7 Countries and Beyond: Economic Issues, Policies and Options for Action*; Organization for Economic Co-operation and Development: Paris, France, 2015; pp. 1–75.
21. Mendelson, M.; Røttingen, J.A.; Gopinathan, U.; Hamer, D.H.; Wertheim, H.; Basnyat, B.; Balasegaram, M. Maximising access to achieve appropriate human antimicrobial use in low-income and middle-income countries. *Lancet* **2016**, *387*, 188–198. [CrossRef]
22. Jorgensen, J.H.; Ferraro, M.J. Antimicrobial susceptibility testing: General principles and contemporary practices. *Clin. Infect. Dis.* **1998**, *26*, 973–980. [CrossRef]
23. CLSI. *Performance Standards for Antimicrobial Susceptibility Testing*, 27th ed.; Clinical and Laboratory Standards Institute: Wayne, PA, USA, 2017; ISBN 1-56238-1-56238-805-3.
24. The European Committee on Antimicrobial Susceptibility Testing. *Breakpoint Tables for Interpretation of MICs and Zone Diameters*; Version 10; 2020; Available online: http://www.eucast.org (accessed on 5 December 2020).
25. Janes, D.; Kreft, S.; Jurc, M.; Seme, K.; Strukelj, B. Antibacterial activity in higher fungi (mushrooms) and endophytic fungi from Slovenia. *Pharm. Biol.* **2007**, *45*, 700–706. [CrossRef]
26. Lund, R.G.; Del Pino, F.A.B.; Serpa, R.; Nascimento, J.S.; DA Silva, V.M.; Ribeiro, G.A.; Rosalen, P.L. Antimicrobial activity of ethanol extracts of *Agaricus brasiliensis* against mutants streptococci. *Pharm. Biol.* **2009**, *47*, 910–915. [CrossRef]
27. Taufiq, M.M.J.; Darah, I. Anti-MRSA of the ethyl acetate crude extract from *Lasiodiplodia pseudotheobromae* IBRL OS-64, an endophytic fungus isolated from leaf of *Ocimum sanctum Linn. Int. J. Pharm. Pharm. Sci.* **2018**, *10*, 50–55.
28. Khameneh, B.; Diab, R.; Ghazvini, K.; Bazzaz, B.S.F. Breakthroughs in bacterial resistance mechanisms and the potential ways to combat them. *Microb. Pathog.* **2016**, *95*, 32–42. [CrossRef] [PubMed]
29. Liu, M.; Lu, J.; Muller, P.; Turnbull, L.; Burke, C.M.; Schlothauer, R.C.; Harry, E.J. Antibiotic-specific differences in the response of *Staphylococcus aureus* to treatment with antimicrobials combined with manuka honey. *Front. Microbiol.* **2015**, *5*, 779. [CrossRef]
30. Al-Talib, H.; Al-Khateeb, A.; Hassan, H. Antimicrobial resistance of *Staphylococcus aureus* isolates in Malaysian Tertiary Hospital. *Int. Med. J.* **2015**, *22*, 1–3.
31. Hyde, K.D.; Xu, J.; Rapior, S.; Jeewon, R.; Lumyong, S.; Niego, A.G.T.; Abeywickrama, P.D.; Aluthmuhandiram, J.V.S.; Brahamanage, R.S.; Brooks, S.; et al. The amazing potential of fungi: 50 ways we can exploit fungi industrially. *Fungal Divers.* **2019**, *97*, 1–136. [CrossRef]
32. Lazar, V.; Ditu, L.M.; Pircalabioru, G.G.; Gheorghe, I.; Curutiu, C.; Holban, A.M.; Picu, A.; Petcu, L.; Chifiriuc, M.C. Aspects of gut microbiota and immune system interactions in infectious diseases, immunopathology, and cancer. *Front. Immunol.* **2018**, *9*, 1830. [CrossRef]
33. Shaffer, R.K. The challenge of antibiotic-resistant *Staphylococcus*: Lessons from hospital nurseries in the mid-20th century. *Yale J. Biol. Med.* **2013**, *86*, 261.
34. Adhikari, R.; Pant, N.D.; Neupane, S.; Neupane, M.; Bhattarai, R.; Bhatta, S.; Lekhak, B. Detection of methicillin resistant *Staphylococcus aureus* and determination of minimum inhibitory concentration of vancomycin for *Staphylococcus aureus* isolated from pus/wound swab samples of the patients attending a tertiary care hospital in Kathmandu, Nepal. *Can. J. Infect. Dis. Med. Microbiol.* **2017**. [CrossRef]
35. Arunkumar, V.; Prabagaravarthanan, R.; Bhaskar, M. Prevalence of Methicillin-resistant *Staphylococcus aureus* (MRSA) infections among patients admitted in critical care units in a tertiary care hospital. *Int. J. Res. Med. Sci.* **2017**, *5*, 2362–2366. [CrossRef]
36. Onyeji, C.O.; Bui, K.Q.; Owens, R.C., Jr.; Nicolau, D.P.; Quintiliani, R.; Nightingale, C.H. Comparative efficacies of levofloxacin and ciprofloxacin against *Streptococcus pneumoniae* in a mouse model of experimental septicaemia. *Int. J. Antimicrob. Agents* **1999**, *12*, 107–114. [CrossRef]
37. Appelbaum, P.C. Reduced glycopeptide susceptibility in methicillin-resistant *Staphylococcus aureus* (MRSA). *Int. J. Antimicrob. Agents* **2007**, *30*, 398–408. [CrossRef] [PubMed]
38. Harkins, C.P.; Pichon, B.; Doumith, M.; Parkhill, J.; Westh, H.; Tomasz, A.; Holden, M.T. Methicillin-resistant *Staphylococcus aureus* emerged long before the introduction of methicillin into clinical practice. *Genome Biol.* **2017**, *18*, 130. [CrossRef] [PubMed]
39. Loomba, P.S.; Taneja, J.; Mishra, B. Methicillin and vancomycin resistant *S. aureus* in hospitalized patients. *J. Glob. Infect. Dis.* **2010**, *2*, 275. [CrossRef]

40. Okwu, M.U.; Olley, M.; Akpoka, A.O.; Izevbuwa, O.E. Methicillin-resistant *Staphylococcus aureus* (MRSA) and anti-MRSA activities of extracts of some medicinal plants: A brief review. *AIMS Microbiol.* **2019**, *5*, 117–137. [CrossRef]
41. Morales, E.; Cots, F.; Sala, M.; Comas, M.; Belvis, F.; Riu, M.; Salvado, M.; Grau, S.; Horcajada, J. Hospital costs of nosocomial multi-drug resistant *Pseudomonas aeruginosa* acquisition. *BMC Health Serv. Res.* **2012**, *12*, 122. [CrossRef]
42. Saaiq, M.; Ahmad, S.; Zaib, M.S. Burn wound infections and antibiotic susceptibility patterns at Pakistan Institute of Medical Sciences, Islamabad, Pakistan. *World J. Plastic Surg.* **2015**, *4*, 9.
43. Sit, P.S.; The, C.S.; Idris, N.; Sam, I.C.; Syed Omar, S.F.; Sulaiman, H.; Thong, K.L.; Kamarulzaman, A.; Ponnampalavana, S. Prevalence of methicillin-resistant *Staphylococcus aureus* (MRSA) infection and the molecular characteristics of MRSA bacteraemia over a two-year period in a tertiary teaching hospital in Malaysia. *BMC Infect. Dis.* **2017**, *17*, 1–14. [CrossRef]
44. Stefani, S.; Ryeon, D.; Lindsay, J.A.; Friedrich, A.W.; Kearns, A.M.; Westh, H.; Mackenzie, F.M. Methicillin-resistant *Staphylococcus aureus* (MRSA): Global epidemiology and harmonisation of typing methods. *Int. J. Antimicrob. Agents* **2012**, *39*, 273–282. [CrossRef]
45. Song, J.H.; Hsueh, P.R.; Chung, D.R.; Ko, K.S.; Kang, C.I.; Peck, K.R.; Yeom, J.S.; Kim, S.W.; Chang, H.H.; Kim, Y.S.; et al. Spread of methicillin-resistant *Staphylococcus aureus* between the community and the hospitals in Asian countries: An ANSORP study. *J. Antimicrob. Chemother.* **2011**, *66*, 1061–1069. [CrossRef]
46. Hiramatsu, K.; Hanaki, H.; Ino, T.; Yabuta, K.; Oguri, T.; Tenover, F.C. Methicillin-resistant *Staphylococcus aureus* clinical strain with reduced vancomycin susceptibility. *J. Antimicrob. Chemother.* **1997**, *40*, 135–136. [CrossRef]
47. Centers for Disease Control and Prevention. *Staphylococcus aureus* resistant to vancomycin—United States, 2002. *MMWR Morb. Mortal. Wkly. Rep* **2002**, *51*, 565–567.
48. Howden, B.P.; Davies, J.K.; Johnson, P.D.; Stinear, T.P.; Grayson, M.L. Reduced vancomycin susceptibility in *Staphylococcus aureus*, including vancomycin-intermediate and heterogeneous vancomycin-intermediate strains: Resistance mechanisms, laboratory detection, and clinical implications. *Clin. Microbiol. Rev.* **2010**, *23*, 99–139. [CrossRef]
49. Bandyopadhyay, D. In response to "Effectiveness of topical green tea against multidrug-resistant *Staphylococcus aureus* in cases of primary pyoderma: An open controlled trial". *Indian J. Dermatol. Venereol. Leprol.* **2018**, *84*, 309–310. [CrossRef]
50. Akanbi, B.O.; Mbe, J.U. Occurrence of methicillin and vancomycin resistant *Staphylococcus aureus* in University of Abuja Teaching Hospital, Abuja, Nigeria. *Afr. J. Clin. Exp. Microbiol.* **2013**, *14*, 10–13. [CrossRef]
51. Al-Talib, H.I.; Yean, C.Y.; Al-Jashamy, K.; Hasan, H. Methicillin-resistant *Staphylococcus aureus* nosocomial infection trends in Hospital Universiti Sains Malaysia during 2002-2007. *Ann. Saudi Med.* **2010**, *30*, 358–363. [CrossRef] [PubMed]
52. Rigby, K.M.; DeLeo, F.R. Neutrophils in innate host defense against *Staphylococcus aureus infections*. *Semin. Immunopathol.* **2012**, *34*, 237–259. [CrossRef] [PubMed]
53. Weinstein, R.A.; Fridkin, S.K. Vancomycin-intermediate and-resistant *Staphylococcus aureus: What* the infectious disease specialist needs to know. *Clin. Infect. Dis.* **2001**, *32*, 108–115. [CrossRef] [PubMed]
54. Gardete, S.; Tomasz, A. Mechanisms of vancomycin resistance in *Staphylococcus aureus*. *J. Clin. Investig.* **2014**, *124*, 2836–2840. [CrossRef] [PubMed]
55. Kali, A. Antibiotics and bioactive natural products in treatment of methicillin resistant *Staphylococcus aureus*: A brief review. *Pharmacogn. Rev.* **2015**, *9*, 29–34. [CrossRef]
56. Mc Nultty, C.; Rodgers, G.L.; Mortenson, J.E. An overview of the topical antimicrobial agents used in the treatment of burn wounds. *Contin. Edu. Top.* **2004**, 74–78.
57. Blunt, J.W.; Copp, B.R.; Munro, M.H.; Northcote, P.T.; Prinsep, M.R. Marine natural products. *Nat. Prod. Rep.* **2011**, *28*, 196–268. [CrossRef] [PubMed]
58. Paz, Z.; Komon-Zelazowska, M.; Druzhinina, I.S.; Aveskamp, M.M.; Shnaiderman, A.; Aluma, Y.; Yarden, O. Diversity and potential antifungal properties of fungi associated with a Mediterranean sponge. *Fungal Divers* **2010**, *42*, 17–26. [CrossRef]
59. Rateb, M.E.; Ebel, R. Secondary metabolites of fungi from marine habitats. *Nat. Prod. Rep.* **2011**, *28*, 290–344. [CrossRef] [PubMed]
60. Godtfredsen, W.O.; Vangedal, S. The structure of fusidic acid. *Tetrahedron* **1962**, *18*, 1029–1048. [CrossRef]

61. Crosbie, R.B. Treatment of staphylococcal infections with "fucidin". *Br. Med. J.* **1963**, *1*, 788. [CrossRef]
62. Kavanagh, F.; Hervey, A.; Robbins, W.J. Antibiotic substances from basidiomycetes: VIII. *Pleurotus multilus* (Fr.) Sacc. and *Pleurotus passeckerianus Pilat. Proc. Natl. Acad. Sci. USA* **1951**, *37*, 570. [CrossRef]
63. Rosa, L.H.; Machado, K.M.G.; Jacob, C.C.; Capelari, M.; Rosa, C.A.; Zani, C.L. Screening of Brazilian basidiomycetes for antimicrobial activity. *Mem. Inst. Oswaldo Cruz* **2003**, *98*, 967–974. [CrossRef]
64. Hemaiswarya, S.; Kruthiventi, A.K.; Doble, M. Synergism between natural products and antibiotics against infectious diseases. *Phytomedicine* **2008**, *15*, 639–652. [CrossRef]
65. Getha, K.; Hatsir, M.; Wong, H.J.; Lee, S.S. Submerged cultivation of basidiomycete fungi associated with root diseases for production of valuable bioactive metabolites. *J. Trop. For. Sci.* **2009**, *21*, 1–7.
66. Smania, A., Jr.; Monache, F.D.; Loguercio-Leite, C.; Smania, E.d.F.A.; Gerber, A.L. Antimicrobial Activity of Basidiomycetes. *Int. J. Med. Mushrooms* **2001**, *3*, 2–3. [CrossRef]
67. Quereshi, S.; Pandey, A.K.; Sandhu, S.S. Evaluation of antibacterial activity of different *Ganoderma lucidum* extracts. *J. Sci. Res.* **2010**, *3*, 9–13.
68. Wagner, R.; Mitchell, D.A.; Lanzi Sassaki, G.; Lopes de Almeida Amazonas, M.A.; Berovic, M. Current techniques for the cultivation of *Ganoderma lucidum* for the production of biomass, ganoderic acid and polysaccharides. *Food Technol. Biotechnol.* **2003**, *41*, 371–382.
69. Bhosle, S.; Ranadive, K.; Bapat, G.; Garad, S.; Deshpande, G.; Vaidya, J. Taxonomy and diversity of *Ganoderma* from the Western parts of Maharashtra (India). *Mycosphere* **2010**, *1*, 249–262.
70. Liu, D.Z.; Zhu, Y.Q.; Li, X.F.; Shan, W.G.; Gao, P.F. New triterpenoids from the fruiting bodies of *Ganoderma lucidum* and their bioactivities. *Chem. Biodivers.* **2014**, *11*, 982–986. [CrossRef] [PubMed]
71. Mothana, R.A.; Jansen, R.; Julich, W.D.; Lindequist, U. Ganomycins A and B, new antimicrobial farnesyl hydroquinones from the basidiomycete *Ganoderma pfeifferi*. *J. Nat. Prod.* **2000**, *63*, 416–418. [CrossRef]
72. Finimundy, T.C.; Barros, L.; Calhelha, R.C.; Jose, M.; Prieto, M.A.; Abreu, R.M.V.; Dillon, A.J.P.; Henriques, J.A.P.; Roesch-Ely, M.; Ferreira, C.F.R. Multifunctions of *Pleurotus sajor-caju* (Fr.) *Singer*: A highly nutritious food and a source for bioactive compounds. *Food Chem.* **2018**, *245*, 150–158. [CrossRef]
73. Vamanu, E. In vitro antimicrobial and antioxidant activities of ethanolic extract of lyophilized mycelium of *Pleurotus ostreatus* PQMZ91109. *Molecules* **2012**, *17*, 3653–3671. [CrossRef]
74. Kalyoncu, F.; Oskay, M.; Saglam, H.; Erdogan, T.F.; Tamer, A.U. Antimicrobial and antioxidant activities of mycelia of 10 wild mushroom species. *J. Med. Food* **2010**, *13*, 415–419. [CrossRef]
75. Roy, D.N.; Azad, A.K.; Sultana, F.; Anisuzzaman, A.S.M. In-vitro antimicrobial activity of ethyl acetate extract of two common edible mushrooms. *J. Pharmacol.* **2016**, *5*, 79–82. [CrossRef]
76. Karthiga, K.; Sivakumar, T. Antibacterial activity of methanolic extract of basidiomycetes. *Asian J. Chem.* **2010**, *22*, 1637–1638.
77. Mudalungu, C.M.; Richter, C.; Wittstein, K.; Abdalla, M.A.; Matasyoh, J.C.; Stadler, M.; Suussmuth, R.D. Laxitextines A and B, cyathane xylosides from the tropical fungus *Laxitextum incrustatum*. *J. Nat. Prod.* **2016**, *79*, 894–898. [CrossRef]
78. Tareq, F.S.; Hasan, C.M.; Rahman, M.M.; Hanafi, M.M.M.; Colombi Ciacchi, L.; Michaelis, M.; Spiteller, P. Anti-Staphylococcal Calopins from Fruiting Bodies of *Caloboletus radicans*. *J. Nat. Prod.* **2018**, *81*, 400–404. [CrossRef]
79. Ma, K.; Bao, L.; Han, J.; Jin, T.; Yang, X.; Zhao, F.; Li, S.; Song, F.; Liu, M.; Liu, H. New benzoate derivatives and hirsutane type sesquiterpenoids with antimicrobial activity and cytotoxicity from the solid-state fermented rice by the medicinal mushroom *Stereum hirsutum*. *Food Chem.* **2014**, *143*, 239–245. [CrossRef] [PubMed]
80. Bitew, A. Antibacterial and antifungal activities of culture filtrate extract of *Pyrofomes demidoffii* (Basidiomycete). *World Appl. Sci. J.* **2010**, *10*, 861–866.
81. Waithaka, P.N.; Gathuru, E.M.; Githaiga, B.M.; Onkoba, K.M. Antimicrobial activity of mushroom (*Agaricus bisporus*) and fungal (*Trametes gibbosa*) extracts from mushrooms and fungi of egerton main campus, Njoro Kenya. *J. Biomed. Sci.* **2017**, *6*, 1–6. [CrossRef]
82. Ozen, T.; Darcan, C.; Aktop, O.; Turkekul, I. Screening of antioxidant, antimicrobial activities and chemical contents of edible mushrooms wildly grown in the Black Sea region of Turkey. *Comb. Chem. High Throughput Screen.* **2011**, *14*, 72–84. [CrossRef]
83. Dogan, H.H.; Duman, R.; Ozkalp, B.; Aydin, S. Antimicrobial activities of some mushrooms in Turkey. *Pharm. Biol.* **2013**, *51*, 707–711. [CrossRef]

84. Reid, T.; Kashangura, C.; Chidewe, C.; Benhura, M.A.; Mduluza, T. Antibacterial properties of wild edible and non-edible mushrooms found in Zimbabwe. *Afr. J. Microbiol. Res.* **2016**, *10*, 977–984. [CrossRef]
85. Gebreyohannes, G.; Nyerere, A.; Bii, C.; Sbhatu, D.B. Investigation of Antioxidant and Antimicrobial Activities of Different Extracts of *Auricularia* and *Termitomyces* Species of Mushrooms. *Sci. World J.* **2019**. [CrossRef]
86. Kolundzic, M.; Grozdanic, N.D.; Dodevska, M.; Milenkovic, M.; Sisto, F.; Miani, A.; Farronato, G.; Kundakovic, T. Antibacterial and cytotoxic activities of wild mushroom *Fomes fomentarius* (L.) Fr., Polyporaceae. *Ind. Crops Prod.* **2016**, *79*, 110–115. [CrossRef]
87. Klaus, A.; Kozarski, M.; Vunduk, J.; Todorovic, N.; Jakovljevic, D.; Zizak, Z.; Pavlovic, V.; Niksic, M.; Van Griensven, L.J. Biological potential of extracts of the wild edible Basidiomycete mushroom *Grifola frondosa*. *Food Res. Int.* **2015**, *67*, 272–283. [CrossRef]
88. Gebreyohannes, G.; Nyerere, A.; Bii, C.; Sbhatu, D.B. Determination of antimicrobial activity of extracts of indigenous wild mushrooms against pathogenic organisms. *Evid. Based Complement. Altern. Med.* **2019**, *2019*. [CrossRef] [PubMed]
89. Ogidi, O.C.; Oyetayo, V.O.; Akinyele, B.J. In vitro evaluation of antimicrobial efficacy of extracts obtained from raw and fermented wild macrofungus, *Lenzites quercina*. *Int. J. Microbiol.* **2015**. [CrossRef]
90. Muhsin, T.M.; Mohammad, H.M. Antibacterial Bioactive Compound from The Fungus *Drechslera Halodes* (Drechsler) Subram. Jain Isolated from Soil of Basrah, Iraq. *Sci. J. Univ. Zakho* **2013**, *1*, 508–514.
91. Casella, T.M.; Eparvier, V.; Mandavid, H.; Bendelac, A.; Odonne, G.; Dayan, L.; Stien, D. Antimicrobial and cytotoxic secondary metabolites from tropical leaf endophytes: Isolation of antibacterial agent pyrrocidine C from *Lewia infectoria* SNB-GTC2402. *Phytochemistry* **2013**, *96*, 370–377. [CrossRef] [PubMed]
92. Beattie, K.D.; Rouf, R.; Gander, L.; May, T.W.; Ratkowsky, D.; Donner, C.D.; Gill, M.; Grice, I.D.; Tiralongo, E. Antibacterial metabolites from Australian macrofungi from the genus Cortinarius. *Phytochemistry* **2010**, *71*, 948–955. [CrossRef]
93. Richter, S.S.; Heilmann, K.P.; Dohrn, C.L.; Riahi, F.; Beekmann, S.E.; Doern, G.V. Changing epidemiology of antimicrobial-resistant *Streptococcus pneumoniae* in the United States, 2004–2005. *Clin. Infect. Dis.* **2009**, *48*, e23–e33. [CrossRef]
94. Ahmad, N.; Mahmood, F.; Khalil, S.A.; Zamir, R.; Fazal, H.; Abbasi, B.H. Antioxidant activity via DPPH, gram-positive and gram-negative antimicrobial potential in edible mushrooms. *Toxicol. Ind. Health* **2014**, *30*, 826–834. [CrossRef] [PubMed]
95. Gogavekar, S.S.; Rokade, S.A.; Ranveer, R.C.; Ghosh, J.S.; Kalyani, D.C.; Sahoo, A.K. Important nutritional constituents, flavour components, antioxidant and antibacterial properties of *Pleurotus sajor-caju*. *J. Food Sci. Technol.* **2014**, *51*, 1483–1491. [CrossRef]
96. Shikongo, L.T.; Chimwamurombe, P.M.; Lotfy, H.R. Antimicrobial screening of crude extracts from the indigenous *Ganoderma lucidum* mushrooms in Namibia. *Afr. J. Microbiol. Res.* **2013**, *7*, 4812–4816.
97. Suseem, S.R.; Saral, A.M. Analysis on essential fatty acid esters of mushroom *Pleurotus eous* and its antibacterial activity. *Asian J. Pharm. Clin. Res.* **2013**, *6*, 188–191.

Publisher's Note: MDPI stays neutral with regard to jurisdictional claims in published maps and institutional affiliations.

molecules

MDPI

Review

Cordycepin for Health and Wellbeing: A Potent Bioactive Metabolite of an Entomopathogenic Medicinal Fungus *Cordyceps* with Its Nutraceutical and Therapeutic Potential

Syed Amir Ashraf [1], Abd Elmoneim O. Elkhalifa [1], Arif Jamal Siddiqui [2], Mitesh Patel [3], Amir Mahgoub Awadelkareem [1], Mejdi Snoussi [2,4], Mohammad Saquib Ashraf [5], Mohd Adnan [2,*] and Sibte Hadi [6,*]

1 Department of Clinical Nutrition, College of Applied Medical Sciences, University of Hail, Hail PO Box 2440, Saudi Arabia; amirashrafy2007@gmail.com (S.A.A.); ao.abdalla@uoh.edu.sa (A.E.O.E.); mahgoubamir22@gmail.com (A.M.A.)
2 Department of Biology, College of Science, University of Hail, Hail PO Box 2440, Saudi Arabia; arifjamal13@gmail.com (A.J.S.); snmejdi@yahoo.fr (M.S.)
3 Bapalal Vaidya Botanical Research Centre, Department of Biosciences, Veer Narmad South Gujarat University, Surat 395007, Gujarat, India; patelmeet15@gmail.com
4 Laboratory of Bioresources: Integrative Biology and Valorization, (LR14-ES06), University of Monastir, Higher Institute of Biotechnology of Monastir, Avenue Tahar Haddad, BP 74, Monastir 5000, Tunisia
5 Department of Clinical Laboratory Sciences, College of Applied Medical Science, Shaqra University, Al Dawadimi PO Box 17431, Saudi Arabia; ashrafsaquib@gmail.com
6 School of Forensic and Applied Sciences, University of Central Lancashire, Preston PR1 2HE, UK
* Correspondence: drmohdadnan@gmail.com (M.A.); shadi@uclan.ac.uk (S.H.); Tel.: +966-533-642-004 (M.A.); +44-1772-894-395 (S.H.)

Academic Editors: Simona Fabroni, Krystian Marszałek and Aldo Todaro
Received: 25 May 2020; Accepted: 10 June 2020; Published: 12 June 2020

check for updates

Abstract: *Cordyceps* is a rare naturally occurring entomopathogenic fungus usually found at high altitudes on the Himalayan plateau and a well-known medicinal mushroom in traditional Chinese medicine. *Cordyceps* contains various bioactive components, out of which, cordycepin is considered most vital, due to its utmost therapeutic as well as nutraceutical potential. Moreover, the structure similarity of cordycepin with adenosine makes it an important bioactive component, with difference of only hydroxyl group, lacking in the 3′ position of its ribose moiety. Cordycepin is known for various nutraceutical and therapeutic potential, such as anti-diabetic, anti-hyperlipidemia, anti-fungal, anti-inflammatory, immunomodulatory, antioxidant, anti-aging, anticancer, antiviral, hepato-protective, hypo-sexuality, cardiovascular diseases, antimalarial, anti-osteoporotic, anti-arthritic, cosmeceutical etc. which makes it a most valuable medicinal mushroom for helping in maintaining good health. In this review, effort has been made to bring altogether the possible wide range of cordycepin's nutraceutical potential along with its pharmacological actions and possible mechanism. Additionally, it also summarizes the details of cordycepin based nutraceuticals predominantly available in the market with expected global value. Moreover, this review will attract the attention of food scientists, nutritionists, pharmaceutical and food industries to improve the use of bioactive molecule cordycepin for nutraceutical purposes with commercialization to aid and promote healthy lifestyle, wellness and wellbeing.

Keywords: *Cordyceps*; cordycepin; medicinal mushroom; nutraceutical; anti-diabetic; immunomodulator; anti-hyperlipidemia; Chinese medicine; DongChongXiaCao; bioactive compound

1. Introduction

Cordyceps, derived from two Latin words "cord" and "ceps" representing 'club' and 'head' respectively, describing it as club fungi. It is an entomopathogenic fungus where extensions of the stroma and fruiting body arise from insect larvae carcasses [1]. *Cordyceps* predominantly lives on the head of larvae of a particular moth species, *Hepialus armoricanus Oberthur* (Lepidoptera). It belongs to the Ascomycetes family and has been a very well-known fungus in Chinese traditional medicine for the last 300 years. *Cordyceps* is also known as 'Dong Chong Xia Cao', which means 'Worm in winter and grass in summer' in China [2–4]. According to the previous reports, around 1200 types of entomopathogenic fungi are known, out of which, *Cordyceps* is considered as one of the largest genus containing approximately 500 species. Several species of *Cordyceps* have been cultivated for their therapeutic properties such as *Cordyceps sinensis, Cordyceps sobolifera, Cordyceps cicadicola, Cordyceps liangshanesis, Cordyceps ophioglossoides and Cordyceps militaris* [5]. On the other hand, keeping in mind the therapeutic value, its major distribution location at approximately 14,000 ft altitude in the Himalayan regions of China, Nepal, Tibet and India makes it very expensive at around USD ($) 12,000 kg^{-1} [3,6,7]. Moreover, despite the harvesting difficulties and distribution, it is still considered a highly valued mushroom because of its abundant natural bioactive component resources with various potent biological activities and nutraceutical importance [2]. For hundreds of years, *Cordyceps* were used as a folk tonic food, but only in recent times, its potential pharmaceutical as well as nutraceutical application have been explored, which has attracted food scientists globally [8].

Currently, it has been observed that a majority of the population from developed as well as developing countries are suffering from chronic diseases, and the underlying causes are believed to be rapid urbanization and changes in eating and lifestyle behavior. Among the various underline causes, eating habits are considered one of the major risk factors for chronic diseases, such as obesity, diabetes, hypertension, hyperlipidemia and many more affecting both wellness and the wellbeing of mankind [9,10]. Therefore, the scientific community is working relentlessly to develop naturally occurring or naturally derived product, such as nutraceuticals, which could help in improving the human health status while not possessing harmful effects. *Cordyceps* are among the thousands of mushroom available containing various bioactive components with innumerable health benefits [11]. It has been used for its therapeutic values since long time and new promising features of cordycepin-based nutraceuticals are an advantage for the current population. There is a very well-known quote from Hippocrates stating that, "Let food be thy medicine and medicine be thy food", describing the importance of nutrition for the prevention, treatment and management of diseases. Therefore, *Cordyceps*, as an edible mushroom, could be an ideal nutraceutical containing both nutritionally bioactive components as well as a source of various physiological benefits [12,13]. Moreover, based on our literature search, we found that researchers have majorly discussed cordycepin for its anticancer potential, but other therapeutic applications and potential nutraceutical approaches have either not been discussed in detail or ignored. The main objective of this review is to focus on the nutraceutical potential of cordycepin (the major bioactive component of *Cordyceps*), using a mechanistic approach to study its pharmacological functions as well as to demonstrate the benefit of commercial availability of cordycepin-based nutraceuticals.

2. Nutritional Value of *Cordyceps*

Nutrition is considered as a fundamental pillar of human beings for maintaining health or in development across the entire life span. Moreover, it is very important to have a proper diet and enough nutrition for survival, physical growth, mental development, performance and productivity, health and well-being [14]. One often quoted phrase, "medicines and foods have a common origin" and based upon this idea, *Cordyceps* could be considered as one of the most significant mushrooms, enriched with various nutrients with possible nutraceutical value [12]. Abundant amounts of bioactive components are present in *Cordyceps* such as proteins, fats, essential amino acids, volatile oils, carotenoids, phenolic compounds, flavonoids, minerals (Fe, Ca, Mg, Ni, Sr, Na, Ti, Pi, Se, Mn, Zn, Al, Si, K, Cr, Ga, V and

Zr), vitamins (B1, B2, B12, E and K) as well as various types carbohydrates like monosaccharides, oligosaccharides, polysaccharides, sterols, nucleosides, etc. [15–19]. Proximate analysis of some of the *Cordyceps* species have reported that moisture, total ash, crude protein, fat, crude fibre and carbohydrate content are 7.18%, 7.48%, 21.46%, 1.80%, 6.40% and 55.68%, respectively [20]. Many authors have also reported proximate analyses of *Cordyceps* fruiting bodies and mycelial biomass. The protein, moisture, ash, fat, and carbohydrate compositions of *Cordyceps* fruiting bodies were reported as 59.8%, 5.7%, 5.1%, 8.8% and 29.1%, whereas, mycelial biomass contains 39.5%, 13.1%, 5.7%, 2.2% and 39.6% of protein, moisture, ash, fat and carbohydrates, respectively [21]. On the other hand, the amino acid contents of the corpus as well as fruiting bodies of *Cordyceps militaris (C. militaris)* were reported to be 14.03 mg/g and 69.32, respectively. Additional amino acid analysis indicates that the fruiting bodies contains abundant amounts of proline, lysine, threonine and glutamic acid. Moreover, fatty acid profiling indicates almost 70% of unsaturated fatty acids out of the total fat percentage. Importantly, the amount of cordycepin and adenosine in both corpus and fruit bodies was reported to be (0.97 and 0.36%) and (0.18 and 0.06%), respectively [17,22]. With the increasing interest in *Cordyceps* in recent times for pharmaceutical, nutraceutical or food biotechnological purposes, further research is necessary to obtain an overview of the potential of this medicinal mushroom [2].

3. Bioactive Components of *Cordyceps*

The significance of bioactive components present in foods has been gaining interest recently due to public health concerns. Moreover, with rising consumer awareness regarding promotion, prevention as well as maintenance of health, bioactive components present in food could play a very important role [23,24]. Bioactive compounds can be described as molecules present in food derived of plant or animal sources consumed to have a regular energy intake along with various therapeutic activities such as against metabolic disorders, anti-inflammatory, chronic diseases and many more [25]. The presence of these bioactive components, even in minute quantities, not only meets the basic nutritional needs, but also confers numerous health benefits. Epidemiological studies indicate that high consumption of foods rich in bioactive compounds with different phytochemicals such as antioxidants, vitamins, flavonoids and carotenoids has a positive effect on human health [26]. Mushrooms have been known and used for centuries for food and medicinal use. Among various mushrooms, very well-known medicinal mushrooms like *Cordyceps* producing bioactive metabolites are used or studied for the possible treatment of several diseases [15]. The presence of antioxidants acquired by modifying our diet with mushrooms could play an important role in the prevention of diseases. *Cordyceps* have a history of medicinal use spanning millennia in parts of Asia, but they can also be potentially used for their nutraceutical values [22,27]. *Cordyceps* have been reported to contain various bioactive components such as proteins, fat, carbohydrates, exopolysaccharides, cordycepin, phenolic compounds, polysaccharides, cordycepic acid, adenosine, proteoglucans, terpenoids, amphinol, steroids, ergosterol, lectins, etc. A detailed list of bioactive components present in *Cordyceps* is presented in Table 1 along with the chemical structures of some potent identified bioactive molecules (Figures 1 and 2). Out of these, cordycepin is the main active constituent which is most widely studied for its medicinal value along with its nutraceutical potential [2,8,22,28,29]. Moreover, various mechanisms have been reported for the pharmacological actions of cordycepin such as inhibition of DNA and RNA synthesis, post-transcriptional processing of hnRNA and activation of adenylate cyclase, inhibition of chemotaxis and particular protein synthesis of macrophage cell lines, anti-tumorigenic activity on some cell lines, enhancement of cell differentiation etc. [1].

Table 1. List of bioactive components present in *Cordyceps* with their biological activities and implication in therapeutics.

Bioactive Component	Biological Activity/Therapeutic Effect	References
	Nucleosides	
Cordycepin	Antitumor, anti-diabetic, anti-inflammatory, antimicrobial, inhibit platelet aggregation, hypolipidemic, analgesic, immunomodulatory	[15,30–32]
Adenosine	Anticonvulsant, Anti-inflammatory, Anti-tumor	[15,29,33]
	Polysaccharides	
Exopolysaccharide Fraction (EPSF)	Anti-tumor, antioxidant, anti-inflammatory, Immunomodulatory	[15,33,34]
Acid polysaccharides (APS)	Antioxidant, Immunomodulatory effect	[33]
CPS-1	Antioxidant	[15,33]
CPS-2	Cell proliferation inhibition	[33]
Mannoglucan	Cytotoxicity activity	[34]
CME-1	Antioxidant	[33]
PS-A	Inhibitory activity against cholesterol esterase	[34,35]
Cordyglucan	Anti-tumor	[33]
D-mannitol or Cordycepic acid	Diuretic, anti-tussive and anti-free radical activities	[2,29,31]
	Sterols	
Ergosterol	Antimicrobial, antiviral, anti-arrhythmic effects, Helps in bone development	[2,30,31,36]
β-Sitosterol	Protect from breast, colon and prostate cancer	[34,36]
H1-A	Immunoregulation	[33]
	Proteins, Amino acids and Polypeptides	
CSDNase	DNA hydrolysis, nucleolytic properties	[33]
CSP	Fibrinolytic activity	[33,34]
Cordymin	Anti-diabetic effect, antifungal	[22,33,37]
Cordycedipeptide A	Cytotoxic activity	[33,38]
Cordyceamides A and B	Cytotoxic activity	[34,38]
Tryptophan	Serves as precursor for the synthesis of the neurotransmitter's serotonin and tryptamine	[33,39]
	Others	
Xanthophylls	Anticancer	[30,31]
Fibrnolytic enzyme	Treatment of thrombosis	[30,31]
Proteoglucans	Enhanced anticancer effect on bladder cancer cells	[28,40]
Phenolic compounds	Antioxidant, antimicrobial, anti-arthritic, anti-carcinogenic, anti-hypertensive, cardio-protective, anti-inflammatory and anti-allergic	[28,41]
N-acetylgalactosamine	Necessary for intercellular communication	[22,42]
Exopolysaccharides	Nutraceutical, pharmaceutical	[22,43]
Chitinase, macrolides, cicadapeptins, myriocin, superoxide dismutase, protease, napthaquinone, cordyheptapeptide and dipicolinic acid	-	[22]
	Vitamins	
B1	Essential in neurologic activities	[22,44]
B2	Helps in energy production	[22,44]
B12	Help in cellular metabolism, DNA synthesis, methylation and mitochondrial metabolism	[16,45,46]
E	Antioxidant, Help in formation of blood cells, muscles, lung and nerve tissues, Increase immunity	[16,44,45]
K	Essential for blood clotting	[22,44]

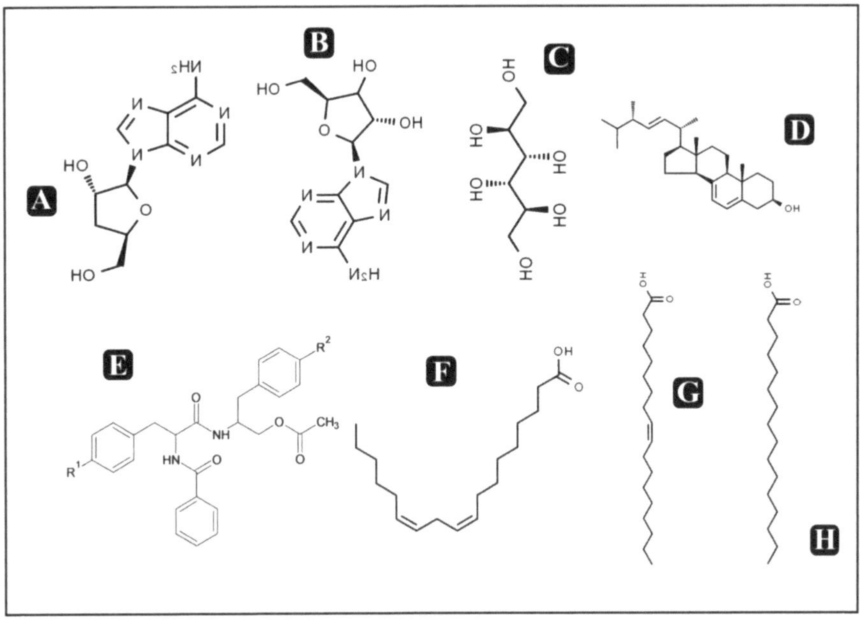

Figure 1. Chemical structures of some known and potent bioactive compounds in *Cordyceps*. (**A**) Cordycepin (**B**) Adenosine (**C**) Cordycepic acid (**D**) Ergosterol (**E**) R^1 = OH, R^2 = H (b) R^1 = OH, R^2 = OH (a) Cordyceamides A (b) Cordyceamides B (**F**) Linoleic acid (**G**) Oleic acid (**H**) Palmitic acid.

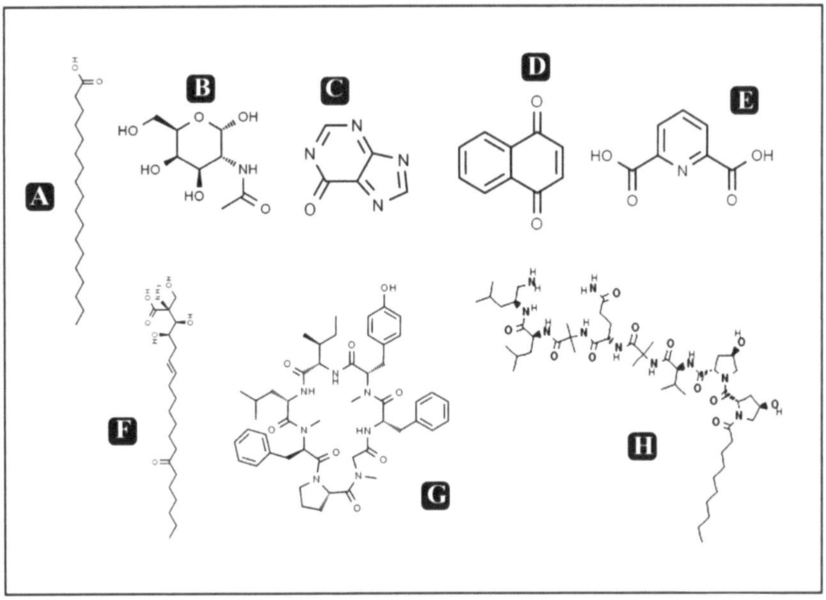

Figure 2. Chemical structures of some known and potent bioactive compounds in *Cordyceps*. (**A**) Stearic acid (**B**) *N*-acetyl muraminic acid (**C**) Hypoxanthine (**D**) Nephthaquinone (**E**) Dipliconic acid (**F**) Myriocin (**G**) Cordyheptapeptide (**H**) Cicadapeptins.

4. Cordycepin and Its Chemical Features

Cordycepin ($C_{10}H_{13}N_5O_3$, molecular weight 251.24 Da, melting point 228–231 °C) is chemically (9-(3-deoxy-β-D-ribofuranosyl) adenine), 3′–deoxyadenosine [5,47]. The structure of cordycepin consists of a purine molecule attached to one ribose sugar moiety [48]. The NMR spectrum of cordycepin shows a singlet at 3.4 ppm, attributed to the C-H proton and a -NH_2 peak reported to be present at 4.6 ppm, whereas, the peaks due to the different -OH groups are found to in the 8–8.5 ppm range [47]. The concentration of cordycepin in *C. militaris*, i.e., around 2–3 mg/kg, is very low compared to the needs of the commercial market, therefore various synthetic and semi-synthetic methods have been reported for the preparation of cordycepin [49]. Cordycepin can be synthesized chemically by replacing an OH group with an H group at the 30-position of the ribofuranosyl moiety to produce an adenosine analogue [50]. Cordycepin is found to be structurally very similar to adenosine except for the lack of a 30-hydroxyl group [49,51] Due to its very close structural similarity with adenosine, it is considered to be a very potent bioactive component with essential properties for its nutraceutical applications. There are various reports suggesting that cordycepin competitively inhibits the courses of synthesis and metabolism of DNA and RNA, as well as it affects the activity of adenosine deaminase and the mTOR signaling pathway [52]. Therefore, cordycepin has been reported to have a wide variety of pharmacological actions such as antioxidant, immunological, hypolipidemic, anti-inflammatory, anticancer, antimicrobial, antiviral, and hypoglycemic properties [49].

5. Cordycepin as a Nutraceutical and Its Role in Human Health

The concept of "nutraceutical" was first introduced in the survey study conducted in France, United Kingdom and Germany, where they found that rating of diet by the consumers was much higher compared to the other factors such as exercise or hereditary to achieve a good health [53]. DeFelice was the first to introduce the word "nutraceutical" by coalescing the two words "nutrition" and "pharmaceutical" and define it as "food or a part of food which not only impart health benefits but also contributes in preventing or treating various diseases". Moreover, in broad terms nutraceuticals can be summarized as bioactive components which play a vital role in human beings by maintaining their normal physiological functions and well-being [10,54]. Notably, nutraceuticals have been designed in such a way, that it could be useful for the betterment and maintenance of human health without causing any harm due to their natural occurrence [55,56]. Nutraceuticals of plant, animal origin or live microorganisms offer great opportunities for food scientists and food industries to produce novel foods or food components for the future needs of human beings to stay healthy [57,58]. Nutraceuticals have been classified in various ways based upon their understanding and applications. Naturally available food sources are considered for their nutraceutical values. They can be characterized as antioxidants, dietary fibre, prebiotics, polyunsaturated fatty acids, probiotics, vitamins, polyphenols and spices [9,53]. The sudden rise in demand of nutraceutical products is mainly due to their therapeutic properties in various ailments such as the common cold, hypertension, heart disease, arthritis, dyslipidemia, diabetes, cancer, depression, inflammatory bowel disease, and increased life span by postponing aging, integrity of the body and support of smooth normal functioning. Therefore, nutraceuticals could play an important role in the improvement of health as well in chronic disease prevention [58]. Nutritionally potent *Cordyceps* are considered a powerhouse of energy, because of their ability to revitalize several organ systems. Moreover, according to various scientific reports, the *Cordyceps* active component cordycepin is considered very useful due to its potential application in various ailments (Figure 3). Several pharmaceutical as well as nutraceutical preparations made from *Cordyceps* dry powder (Table 2) are marketed and reported to protect renal and hepatic functions, improve intracellular energy exchange, increases oxygenation and natural endurance, remove toxins from the body, control blood glucose level and lipid profile, delays the aging process, stimulates the natural metabolism of energy and nourishes the body's immune system [59].

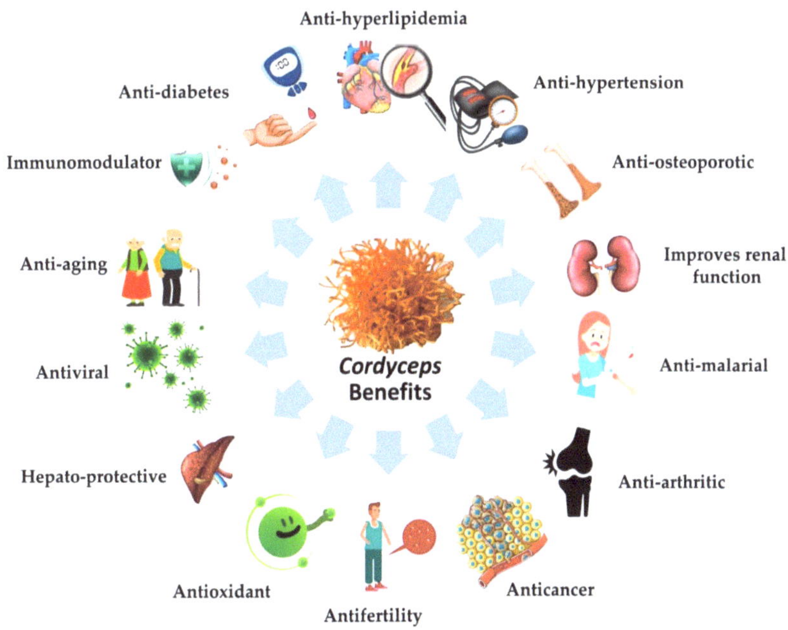

Figure 3. Pictorial representation of *Cordyceps* therapeutic potential in general.

Table 2. Nutraceutical products from *Cordyceps* available on the global market.

Product Name	Health Benefit Claim	Manufacturer
Mycoformulas Endurance™	Enhancement of intracellular energy exchange, increases oxygenation and natural endurance	Myco Formulas, USA
Nutricafe-organic	Increases physical endurance and helps to remove toxins from your body	Aloha Medicinals USA
Mushroom Plus	Supporting immune system, energy levels and cognition	Link Nutrition Ltd., UK
Dragon Herbs	Increases the primary stimulating force for life activities	Iherb Holdings LLC, USA
OM™ Maitake	Supports weight control and blood sugar balancing	Yukiguni Maitake CO., LTD, Japan
Host Defense Mushrooms	Energy support	Host Defense, USA
CaféCeps® Packets	Numbers of health-enhancing properties,	Madre Labs LLC, France
Bhutan *Cordyceps* Tea	Immunity booster, Anticancer, Anti-aging, Antioxidant, Improve kidney and gastrointestinal systems	Bhutan Natural, Singapore
MRM *Cordyceps* CS-4 Strain	It strengthens the immune system, respiratory system and cardiovascular system, Strengthens the natural metabolism of energy	All Star Health, USA
Ultra *Cordyceps* Plus	Help boost physical energy and stamina, improve vitality, support lung health, liver function, memory and mental ability	Doctor's Best, USA
Now Foods *Cordyceps*	Immune health support	Now Foods, USA
Exploding Buds *Cordyceps Sinensis*	Immune health support	Iherb Holdings LLC, USA
Fungiology from California Gold Nutrition	For healthy immunity and health promotion of the entire body	California Gold Nutrition, USA
Planetary Herbals *Cordyceps* POWER CS-4	Energy support	Michael Tierra, USA
Cordyceps Capsules, Extracts or Powders	Energy support and stamina	Host Defense Mushrooms, USA
Cordyceps active	Promotes mental health, Ensure perfect oxygenation of the heart and vascular system	Terezia, Czech
Collagen C ReLift Capsules	For less wrinkles and improved complexion	Zein Pharma, Germany
MycoNutri *Cordyceps* Organic	Immune system support	The Really Healthy, UK

6. Mechanism of Action and Pharmacological Implications of Cordycepin

6.1. Cordycepin and Diabetes

Diabetes has recently become one of the most prevalent epidemics worldwide, affecting almost 382 million people. According to the reports, it is believed that every year approximately 1.3 million people die from diabetes. According to International Diabetes Federation (IDF), it is estimated that around 629 million people will be diabetic around the globe by 2045. However, as per the recommendation from dieticians or physicians, a healthy life style and healthy food habits, could be one of the key answers to this problem [60,61]. It has also been observed that, in most diabetic patients, several other complications also arise, such as cardiovascular diseases, retinopathy, nephropathy, hyperlipidemia and neuropathy [62]. Of note, it is also suggested by various studies, that no single treatment programme can treat diabetes, although most of the treatments achieve normal blood glucose levels or improve microcirculation [63]. Currently, pharmaceutical products used for the management or therapeutic purposes against diabetes are sulfonylureas, biguanides, thiazolidinedione, α-glucosidase inhibitors or insulin injections. However, pharmaceutical products available for the treatment of diabetes have several adverse effects and their potency is sometimes controversial. Other non-medicinal strategies used in diabetes are exercise, weight loss plans and changes in food habits. Occasionally, complications of diabetes can cause morbidity because of pathophysiology flaws. Therefore, consumer interest has currently moved towards alternative medicinal approaches such as nutraceutical food products containing bioactive antidiabetic components. Some studies have reported that extract of *C. militaris* showed a significant decrease in blood glucose levels by virtue of increasing glucose metabolism as well as protection against diabetic nephropathy [64].

The mechanism of cordycepin's antidiabetic activity is not fully understood, but a few studies have explained a possible pathway. They found that cordycepin prevents the production of NO and pro-inflammatory cytokines like TNF-α, IL-1β, and IL-6 in LPS activated macrophages by inhibiting the protein expression of pro-inflammatory mediators. By virtue of this the expression of type 2 diabetes-regulating genes (11β-HSD1 and PPARλ) was reduced. Expression of co-stimulatory molecules such as ICAM-1 and B7-1/-2 was also decreased with the increase in cordycepin concentration as presented in Figure 4A [65]. Furthermore, cordycepin has been found to suppress the expression of diabetes-regulating genes through the inactivation of NF-κb-dependent inflammatory responses [66,67]. In another study, cordycepin's antidiabetic activity was reported in an alloxan-induced diabetic mouse model. The results suggested a significant improvement in glucose tolerance tests after administration of an effective dose of cordycepin [68]. Additionally, an effect of cordycepin on diabetic nephropathy by suppressing cell apoptosis, renal fibrosis and rescued cell autophagy in the diabetic nephropathy rat model was also reported [69]. There are several reports which suggest that cordycepin has very good potential for being a safe anti-diabetic pharmaceutical agent [64,65,70].

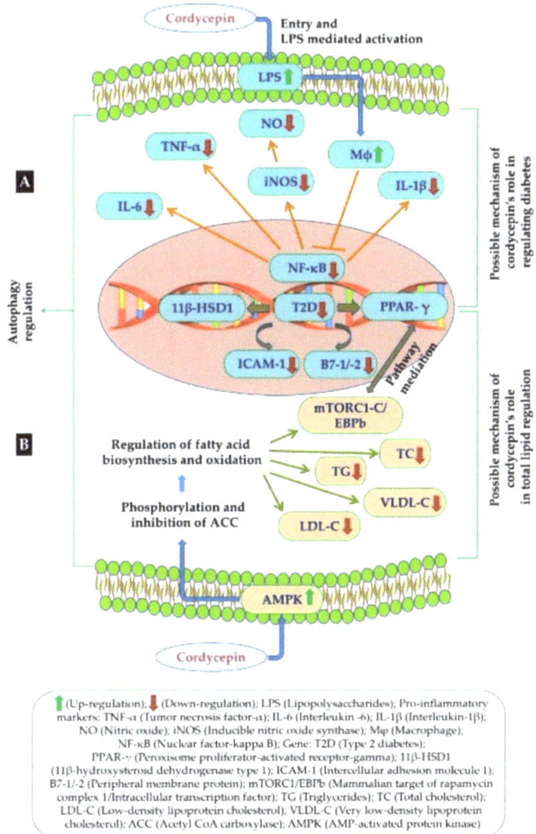

Figure 4. (**A**) Possible mechanism of cordycepin for its anti-diabetic activity (**B**) Possible mechanism of cordycepin in regulation of fat metabolism in hyperlipidemia [47].

6.2. Cordycepin and Cardiovascular Diseases (Hyperlipidemia)

Cardiovascular diseases have become one of the major causes of death around the world in developed as well as developing countries, and it has been assumed that there are various factors (excessive intake of tobacco, alcoholism, sedentary lifestyle, unhealthy eating habits etc.) associated with it. Among the various risk factors, rises in lipid levels are believed to be one of the main ones causing this chronic disease [71]. Hyperlipidemia is caused by or considered as the amount of fatty acids present in lipids, low density lipoprotein cholesterol, trans fats and triglycerides accumulated in human bodies causing cardiovascular diseases such as atherosclerosis and coronary heart disease [72–74]. Moreover, a mechanistic approach involved in the regulation of fat metabolism is due to AMP-activated protein kinase (AMPK), a main cell electricity sensor [75]. In addition, activation of AMPK causes a decline in levels of fatty acids through phosphorylation and inhibition of acetyl-CoA carboxylase (ACC), which helps in regulation of fatty acid biosynthesis and oxidation (Figure 4B). AMPK activation was also reported to decrease total cholesterol and triglycerides by inhibiting the activity of glycerol-3-phosphate acyltransferase (GPAT) and HMG CoA reductase, the two rate-limiting enzymes in TC and TG synthesis, respectively [76]. Wu et al. [77] proposed that cordycepin could additionally prevent intracellular lipid accumulation through activation of AMPK interplay with the ˠ1 subunit [77]. Thus, it was found that regulation of AMPK would provide a solution for overweight and obese people causing hyperlipidemia [78]. According to previous reports, cordycepin has been found very effective in lipid reduction, due to its chemical structural similarity with adenosine (an activator of AMPK). Similarly, administration of cordycepin was able to reduce the accumulation of low-density lipoprotein cholesterol, total cholesterol and triglycerides effectively, and could be a potent nutraceutical agent for reducing hyperlipidemia caused by high fat diets [79,80]. On the other hand, cordycepin was also evaluated for regulating autophagy as well as lipid metabolism. It has been found that cordycepin was effective against hepatic lipid accumulation induced by PA through autophagy induction and PKA/mTOR pathway could be the possible mechanism behind its efficacy. More importantly, it was observed that cordycepin was largely effective in reducing the intracellular levels of total lipids, total cholesterol, C and TG, LDL-C, VLDL-C as well as LDL-C/HDL-C and TC/HDL-C ratios. Therefore, cordycepin lipid-lowering activity could be of potential use in the treatment of hyperlipidemia [77,79,81].

6.3. Cordycepin and Anti-Inflammatory Effects

Inflammation, a natural response to injury, occurs in our body to eliminate harmful elements such as damaged cells, irritants and pathogens by initiating the healing system. Acute and chronic pulmonary inflammations have been reported in various respiratory diseases such as asthma, acute respiratory distress syndrome, cystic fibrosis (CF) and chronic obstructive pulmonary disease (COPD) [82]. Of note, cordycepin has been reported to suppress intestinal irritation in an acute colitis mouse model as well as in microglia through inhibiting pro-inflammatory mediators such as TNFα [83,84]. It has been found that cordycepin is effective in a mouse version of bronchial asthma as well as it improves both mucus clearance and airway surface hydration, as it is hyper-secreted in various respiratory problems, like COPD and asthma [85,86]. In addition, cordycepin has also been reported to attenuate airway remodeling in rats with COPD by preventing airway inflammation as well as TGF-β1/Smad signaling pathway. According to this report, cordycepin could be useful in the case of COPD [87,88]. Cordycepin extracts were reported to relieve fibrosis in the lung by inhibiting TGF-b1 expression [89,90] and the promotion of collagen degradation [91]. Therefore, based upon these collective reports and data, it could be summarized that cordycepin has all the potential to become a very potent bioactive anti-inflammatory component [83–85,92].

6.4. Cordycepin and Immunomodulatory Effects

Immunomodulation is usually defined as modulation of the immune system, and this can be done by any chemical agent that modifies the immune response or the functioning of the immune system via the stimulation of antibody formation or the inhibition of white blood cell activity [93]. Cordycepin has been reported to stimulate cytokine release of resting peripheral blood mononuclear cells (PBMCs) as well as influence PBMCs proliferation and transcription factors in a human acute monocytic leukemia cell line (THP-1). Moreover, cordycepin was found to regulate human immune cells functions in vitro [94,95]. It has also been observed that the antitumor activity of cordycepin is associated with its immunomodulatory effects [96]. Additionally, pure bioactive components extracted from *C. militaris* are reported to have good immunomodulatory effects by increasing the survival rate of lupus mice and reducing anti-ds-DNA production [34,65]. Zhang and Xia [97] observed that *C. sinensis* behaves like an immunosuppressant in a heterotopic heart allograft model in rats and increases the survival period. The same effect has been reported by other authors too, for example, Zhu and Hu [98], revealed that *C. sinensis* prolongs the mouse skin allograft survival time. Therefore, cordycepin has been proven a potentially effective immunomodulator, and it is even specifically used to control autoimmune disorders and transplant rejections after an organ transplant [99]. An increasing number of studies indicate that cordycepin is a bi-directional modulator with both suppressive as well as influencing effects on the immune system by regulating the adaptive and innate immunity [1,100–103].

6.5. Cordycepin and Anti-Osteoporosis Effect

Osteoporosis is a condition of low bone mineral density (BMD) and loss of the structural and bio-mechanical properties of bones. It increases the risk of fracture, as bones become more porous and fragile. Osteoporosis mainly occurs in aged people, specifically in post-menopausal women and patients who had long-term treatment of steroid therapy. Anti-osteoporotic effect of cordycepin was studied in ovariectomized osteopenic rats, it has been found that cordycepin was able to counteract the loss of bone in the experimental model. The mechanistic approach used in this study showed the decline in activity of tartrate-resistant acid phosphatase and alkaline phosphatase enzymes both in vitro and in vivo. Moreover, the results showed that oral intake of cordycepin could increase the level of osteocalcin (OC), a marker of bone formation, and decrease C-terminal cross-linked telopeptide of type I collagen (CTX) level, a marker of bone resorption, as well as restore oxidative stress levels in ovariectomized rats [104]. These results suggest that cordycepin can be a valuable bioactive compound for the treatment of osteoporosis and is able to prevent bone loss caused by estrogen deficiency. Cordycepin was also reported to inhibit RANKL-induced osteoclast differentiation (RANKL), receptor activator of nuclear factor kappa-B ligand, and down-regulate the mRNA expressions of osteoclastogenesis-related genes such as, matrix metalloproteinase (MMP)-9, cathepsin K, tartrate-resistant alkaline phosphatase (TRAP) and nuclear factor of activated T-cells, cytoplasmic 1 [105].

6.6. Cordycepin and Anti-Arthritic Effect

Arthritis, an autoimmune disease affecting bone joints, is mainly characterized by joint stiffness as well as joint pain, among other symptoms such as swelling, warmth, redness, and reduction in joint motility. There is no known specific effective treatment for arthritis, although many drugs such as glucocorticosteroids, non-steroidal anti-inflammatory drugs, and other biological agents are used to improve the symptoms, such as pain, fatigue, and disability. Long-term usage of these drugs decreases their effectiveness and increases side effects. Recently, studies were conducted looking for effective anti-arthritic drugs with increased therapeutic effects and fewer side effects. Traditional herbal medicine, which is shown to be more effective, safer, and economical, has attracted more attention in the area of arthritis treatment. Moreover, cordycepin has been found to modulate glycosaminoglycan (GAG) release by suppressing stimulation of IL-1β. In addition, levels of proteases that have been reported in cartilage matrix degradation, such as MMP-13, cathepsin K, MMP-1, cathepsin S, ADAMTS-5 and ADAMTS-4, were decreased by cordycepin in a dose-dependent manner. Chondroprotective effect of cordycepin by preventing cartilage denegation as well as interfering inflammatory response in osteoarthritis pathogenesis has also been reported [106]. Cordycepin has been reported to reduce excessive inflammatory cell infiltration via down-regulation of macrophages, interferon gamma-induced protein 10 (IP-10) and Mig expressions through terminating protein coding gene (STAT1) phosphorylation [107]. There are some reports suggesting that inflammation of T-cell infiltration could be inhibited by using a cordycepin concentration of 10 mg/kg. According to that report, cordycepin can regulate the T-cell receptor, a protein complex found on the surface of T-cells, that signals to suppress excessive T-cell activation in inflammation [108]. Therefore, based upon these reports, it can be concluded that cordycepin has therapeutic potential in both anti-catabolic and anti-inflammatory actions against arthritic diseases [107,108].

6.7. Cordycepin and Antioxidant Effect

Antioxidants are compounds that can prevent or slow down oxidation reactions producing free radicals, that ultimately cause cell damage in organisms. Oxidative stress, which is related with an increased formation of oxidizing species or a significant reduction of natural antioxidant levels, is involved in different human diseases (cellular necrosis, cardiovascular disease, cancer, neurological disorder, ageing) [109]. Non-toxic antioxidants from natural sources, particularly medicinal plants, are known to prevent oxidative damage due to their richness in polyphenolics and bioactive compounds [110–114]. The antioxidant activity of *Cordyceps* has been reported by various authors [115,116]. Cordycepin has been reported to significantly increase the levels of antioxidant enzymes such as superoxide dismutase and glutathione peroxidase activities in 6-OHDA-treated cells. Moreover, the results showed that cordycepin prevents 6-OHDA-induced neurotoxicity in adrenal pheochromocytoma cells (PC12 cells) via its potent antioxidant action [117]. In addition, cordycepin containing protein-bound polysaccharide causes a reduction in lipid peroxidation as well as an increase in the activity of antioxidant enzymes in the liver like catalase and superoxide dismutase [118]. Other authors have suggested the potential of cordycepin in reducing lipid peroxidation in mouse liver [119]. Therefore, per these reports, cordycepin could be considered as a potential antioxidant. A few studies have even suggested that the antioxidant potential of *Cordyceps* is close to that of ascorbic acid [120].

6.8. Cordycepin and Anti-Malarial Effect

Malaria, a very common disease globally with a high mortality rate, is caused by *Plasmodium*, a parasitic organism [121–124]. The parasite gets into the human body through the bites of infected mosquitoes [125–127]. It is one of the deadliest diseases in the world [128], as it is estimated that every 2 min, a child dies of malaria, and each year more than two hundred million new cases of the disease are reported [129]. Most of the people who die from the disease are young children in Africa. The effects of cordycepin on the malaria parasite in mice were first studied by Trigo et al. [130]; they suggested that the growth of the parasite was affected by cordycepin that affects the nucleic acid and protein synthesis of the parasite. This makes cordycepin a possible molecule which can be explored further as a probable anti-malarial agent.

6.9. Cordycepin and Other Diseases

Hyperuricemia is a long-time purine metabolic disorder recognized as a result of excessive serum uric acid status in blood and associated with gout, renal sicknesses, hypertension, hyperlipidemia, and atherosclerosis [96,131]. *C. militaris* has been reported for its anti-hyperuricemic effect in hyperuricemic mice at different doses, reaching the levels of normal mice [132]. In another study, Yong et al. [133] also reported cordycepin's potential as an anti-hyperuricemic in hyperuricemic mice. Infertility can be described as a disease condition where females are not able to become pregnant despite having frequent, unprotected sex for at least a year for most couples. According to the reports, cordycepin has been proved potent in increasing both the sperm quality and quantity. *C. militaris* supplementation has been stated to bring about an increase of serum cordycepin concentration, which concurrently enhances testosterone and estradiol-17 levels, ultimately increasing the percentage of motile sperm cells [134]. In addition to this cordycepin is also reported to increase semen production as well as sperm quality in boars [66]. The effect of cordycepin on testosterone levels in male rats was reported. It was found that the concentration of testosterone in the serum of the rats was significantly increased by *C. militaris*. Therefore, fruiting bodies of *C. militaris* grown on the drone bee medium could act as an integrative medicine for the treatment of reproductive problems caused by insufficient testosterone levels in human males [135]. On the other hand, chronic kidney disease (CKD) is a disease condition where the condition of the kidneys deteriorates steadily and has been related with both non-communicable diseases, e.g., diabetes and hypertension, and infectious diseases like hepatitis B, malaria, and HIV [136]. Clinical research exploring the possible cordycepin application has confirmed the beneficial effects in decreasing the progression of end-stage kidney disease in CKD patients [137]. Moreover, other pharmaceutical applications of cordycepin are also recommended, such as increasing creatinine clearance, serum albumin and hemoglobin, lowering serum creatinine levels as well as improving lipid metabolism [137–140].

7. Global Nutraceutical Market of Cordycepin: Current and Future Trend

Nutraceuticals are developed from foods or food derived components due to their health promoting properties to establish a clear link between 'nutrients' and 'pharmaceuticals' [9]. The health promoting effects of nutraceuticals have led to substantial increases in their market predominantly since the 1990s. Nutraceutical products have shown hope for various chronic diseases (diabetes, cancer, heart disease, hypertension, common cold, dyslipidemia, arthritis and many more) as well as help in delaying aging and eventually increasing life span [58]. Health awareness in consumers, caloric intake and weight management in countries like US, China and India, have already endorsed various applications of nutraceuticals, causing a significant impact on the industry growth. According to the data published online by the Grand View research team, the global nutraceutical market size could reach USD 722.49 billion by 2027, and is expected to expand at a CAGR of 8.3% over the forecast period [141]. Major key companies such as Abbott, Amway, Danone S.A., Nestle S.A., Glanbia Nutritionals, Herbalife International of America, Inc., The Archer Daniels Midland Company, BASF SE, PepsiCo, Inc., General Mills, Inc. are involved in the research & development of nutraceutical production [142]. On the other hand, the global cordycepin market is predicted to surpass US$ 1 billion by 2026. Asia Pacific (mainly China) is considered as the global leader in the production or extraction of cordycepin from *C. sinensis* and *C. militaris*, with an expected 47% of global nutraceutical market (Figure 5) [143].

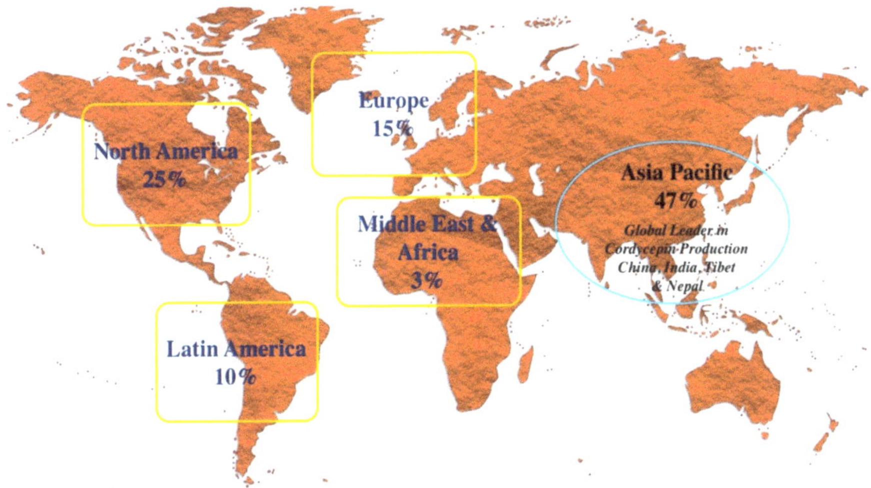

Figure 5. Expected global nutraceutical market by 2025 with China, India, Tibet and Nepal as global leaders for production and extraction of cordycepin [144].

8. Safety and Efficacy

Cordyceps, a well-known traditional Chinese medicine is also called a miracle mushroom because of its exceptional health benefits. Due to its natural occurrence, it is considered pharmacologically safe for human consumption. However, in some cases it has been reported to cause dry mouth, nausea, abdominal distension, throat discomfort, headache and diarrhea as well as allergic reactions [22,34,145]. It is recommended to avoid its consumption in the case of patients suffering from systemic lupus erythematosus, multiple sclerosis and rheumatoid arthritis. Moreover, there are a few reports indicating that *Cordyceps* intake can cause lead poisoning in some cases. Considering the potential for toxicity, resistance and its efficiency, potential drug derivatives designed from the natural products for the adenosine deaminase-resistant properties and low-toxicity causing effects of cordycepin, as well as organ-targeted nanoparticles for cordycepin delivery for in vivo therapy, are also needed [146].

Further clinical, experimental as well as epidemiological data are required for identification of other molecular targets to see the correlation between *Cordyceps* and other diseases like cancers and to estimate the validation of optimum dosing for its safety and efficacy [28]. The mechanism of action of cordycepin still needs further detailed scientific studies along with exploration of the various biochemical pathways involved. Thereby, its pharmacological and mechanism of actions will help in scientific scrutiny to answer every aspect of cordycepin [34]. Aside from some negative reports and clinical data, *Cordyceps* is still relatively considered safe and non-toxic for human consumption [22].

9. Conclusions and Future Perspectives

Over the past few years, people have shown confidence and trust in naturally occurring food-based products such as nutraceuticals for the treatment or management of various chronic diseases. Medicinal mushrooms have been proved to be very effective against numerous serious ailments. *Cordyceps*, some of the most important medicinal mushrooms, have been in use in the traditional Chinese medicine system since a long time ago. They is considered an excellent reservoir of various bioactive components, of which cordycepin is believed to have the most nutraceutical potential. Based on several research reports, cordycepin indicates its nutraceutical values by showing possible therapeutic activity against various diseases by modulating a number of cellular signaling pathways due to its redox behavior. In future, it would be important to uncover other unknown molecules present in *Cordyceps* and to understand their therapeutic potential. Likewise, it is equally important for the scientific community to explore the possibilities of nano-biotechnology-mediated targeted drug delivery systems for cordycepin and how to enhance its bioavailability. Further experimental and clinical studies are required to identify the exact mechanism behind the role of cordycepin, as well as its efficacy and safety alone and in combination. Since *Cordyceps* are edible, they will play a key role in the prevention and cure of various ailments caused by metabolic disorders or infections.

Author Contributions: Conceptualization, S.A.A., A.E.O.E., M.A., and S.H.; resources, A.E.O.E., A.M.A., M.S.A., and A.J.S.; data curation, A.J.S., M.P., A.E.O.E., A.M. and M.S.; visualization, M.S.A., M.A. and M.P.; writing—initial draft, S.A.A., M.P., M.S., A.M.A. and A.J.S.; writing—review and final editing, S.A.A., M.A., and M.S.; supervision, M.A. and S.H. All authors have read and agreed to the published version of the manuscript.

Funding: This research received no external funding.

Conflicts of Interest: The authors declare no conflict of interest.

References

1. Zhou, X.; Gong, Z.; Su, Y.; Lin, J.; Tang, K. Cordyceps fungi: Natural products, pharmacological functions and developmental products. *J. Pharm. Pharmacol.* **2009**, *61*, 279–291. [CrossRef]
2. Yue, K.; Ye, M.; Zhou, Z.; Sun, W.; Lin, X. The genus Cordyceps: A chemical and pharmacological review. *J. Pharm. Pharmacol.* **2013**, *65*, 474–493. [CrossRef] [PubMed]
3. Tuli, H.S.; Sharma, A.K.; Sandhu, S.S.; Kashyap, D. Cordycepin: A bioactive metabolite with therapeutic potential. *Life Sci.* **2013**, *93*, 863–869. [CrossRef] [PubMed]
4. Dong, C.; Guo, S.; Wang, W.; Liu, X. Cordyceps industry in China. *Mycology* **2015**, *6*, 121–129. [CrossRef] [PubMed]
5. Tuli, H.S.; Kashyap, D.; Sharma, A.K. Cordycepin: A Cordyceps Metabolite with Promising Therapeutic Potential. In *Fungal Metabolites*; Mérillon, J.-M., Ramawat, K.G., Eds.; Springer International Publishing: Cham, Switzerland, 2017; pp. 761–782.
6. Paterson, R.R. Cordyceps: A traditional Chinese medicine and another fungal therapeutic biofactory? *Phytochemistry* **2008**, *69*, 1469–1495. [CrossRef]
7. Meena, H.; Mohsin, M.; Pandey, H.K.; Negi, P.S.; Ahmed, Z. Estimation of cordycepin by improved HPLC method in the natural and cultured mycelia of high medicinal value Himalayan entomogenous fungus Cordyceps sinensis. *Electron. J. Environ. Agric. Food Chem.* **2010**, *9*, 1590–1603.
8. Yang, S.; Jin, L.; Ren, X.; Lu, J.; Meng, Q. Optimization of fermentation process of Cordyceps militaris and antitumor activities of polysaccharides in vitro. *J. Food Drug Anal.* **2014**, *22*, 468–476. [CrossRef]

9. Ahmad, M.F.; Ahmad, F.A.; Azad, Z.A.; Alam, M.S.; Ashraf, S.A. Nutraceutical is the need of hour. *World J. Pharm. Pharm. Sci.* **2013**, *2*, 2516–2525.
10. Ashraf, S.A.; Adnan, M.; Patel, M.; Siddiqui, A.J.; Sachidanandan, M.; Snoussi, M.; Hadi, S. Fish-based Bioactives as Potent Nutraceuticals: Exploring the Therapeutic Perspective of Sustainable Food from the Sea. *Mar. Drugs* **2020**, *18*, 265. [CrossRef]
11. Adnan, M.; Ashraf, S.A.; Khan, S.; Alshammari, E.; Awadelkareem, A.M. Effect of pH, temperature and incubation time on cordycepin production from Cordyceps militaris using solid-state fermentation on various substrates. *CyTA J. Food* **2017**, *15*, 617–621. [CrossRef]
12. Khan, M.; Tania, M.; Zhang, D.; Chen, H. Cordyceps Mushroom: A Potent Anticancer Nutraceutical. *Open Nutraceuticals J.* **2010**, *3*, 179–183. [CrossRef]
13. Smith, R. "Let food be thy medicine . . . ". *BMJ* **2004**, *328*. [CrossRef]
14. WHO, World Health Organization. Nutrition for Health and Development: A Global Agenda for Combating Malnutrition. 2000. Available online: https://apps.who.int/iris/handle/10665/66509 (accessed on 16 May 2020).
15. Elkhateeb, W.A.; Daba, G.M.; Thomas, P.W.; Wen, T.C. Medicinal mushrooms as a new source of natural therapeutic bioactive compounds. *Egypt. Pharm. J.* **2019**, *18*, 88–101.
16. Zhu, J.S.; Halpern, G.M.; Jones, K. The scientific rediscovery of a precious ancient Chinese herbal regimen: Cordyceps sinensis: Part II. *J. Altern. Complement. Med.* **1998**, *4*, 429–457. [CrossRef] [PubMed]
17. Hur, H. Chemical Ingredients of Cordyceps militaris. *Mycobiology* **2008**, *36*, 233–235. [CrossRef]
18. Yang, F.Q.; Feng, K.; Zhao, J.; Li, S.P. Analysis of sterols and fatty acids in natural and cultured Cordyceps by one-step derivatization followed with gas chromatography-mass spectrometry. *J. Pharm. Biomed. Anal.* **2009**, *49*, 1172–1178. [CrossRef]
19. Li, S.; Li, P.; Ji, H. RP-HPLC determination of ergosterol in natural and cultured Cordyceps. *Chin. J. Mod. Appl. Pharm.* **2011**, *18*, 297–299.
20. Sethy, N.K.; Singh, V.K.; Sharma, S.; Sharma, R.K.; Deswal, R.; Bhargava, K.; Mishra, K. Phytochemical and proteomic analysis of a high altitude medicinal mushroom cordyceps sinensis. *J. Proteins Proteom.* **2016**, *7*, 187–197.
21. Chan, J.S.; Barseghyan, G.S.; Asatiani, M.D.; Wasser, S.P. Chemical Composition and Medicinal Value of Fruiting Bodies and Submerged Cultured Mycelia of Caterpillar Medicinal Fungus Cordyceps militaris CBS-132098 (Ascomycetes). *Int. J. Med. Mushrooms* **2015**, *17*, 649–659. [CrossRef] [PubMed]
22. Tuli, H.S.; Sandhu, S.S.; Sharma, A.K. Pharmacological and therapeutic potential of Cordyceps with special reference to Cordycepin. *3 Biotech* **2014**, *4*, 1–12. [CrossRef]
23. Weaver, C.M. Bioactive foods and ingredients for health. *Adv. Nutr.* **2014**, *5*, 306S–311S. [CrossRef]
24. NIH, National Institute of Health Office of Dietary Supplements. Available online: https://ods.od.nih.gov/Research/Bioactive_Food_Components_Initiatives.aspx (accessed on 14 May 2020).
25. Siriwardhana, N.; Kalupahana, N.S.; Cekanova, M.; LeMieux, M.; Greer, B.; Moustaid-Moussa, N. Modulation of adipose tissue inflammation by bioactive food compounds. *J. Nutr. Biochem.* **2013**, *24*, 613–623. [CrossRef]
26. Santos, D.I.; Saraiva, J.M.A.; Vicente, A.A.; Moldão-Martins, M. 2-Methods for determining bioavailability and bioaccessibility of bioactive compounds and nutrients. In *Innovative Thermal and Non-Thermal Processing, Bioaccessibility and Bioavailability of Nutrients and Bioactive Compounds*; Barba, F.J., Saraiva, J.M.A., Cravotto, G., Lorenzo, J.M., Eds.; Woodhead Publishing: Cambridge, UK, 2019; pp. 23–54.
27. Gu, Y.-X.; Wang, Z.-S.; Li, S.-X.; Yuan, Q.-S. Effect of multiple factors on accumulation of nucleosides and bases in Cordyceps militaris. *Food Chem.* **2007**, *102*, 1304–1309. [CrossRef]
28. Mehra, A.; Zaidi, K.U.; Mani, A.; Thawani, V. The health benefits of Cordyceps militaris—A review. *KAVAKA* **2017**, *48*, 27–32.
29. Kuo, H.C.; Huang, I.C.; Chen, T.Y. Cordyceps s.l. (Ascomycetes) Species Used as Medicinal Mushrooms are Closely Related with Higher Ability to Produce Cordycepin. *Int. J. Med. Mushrooms* **2015**, *17*, 1077–1085. [CrossRef]
30. Bawadekji, A.; Al Ali, K.; Al Ali, M. A Review of the Bioactive Compound and Medicinal Value of Cordyceps militaris. *J. North Basic Appl. Sci. (JNBAS)* **2016**, *1*, 69–76. [CrossRef]
31. Cui, J.D. Biotechnological production and applications of Cordyceps militaris, a valued traditional Chinese medicine. *Crit. Rev. Biotechnol.* **2015**, *35*, 475–484. [CrossRef] [PubMed]

32. Wang, D.; Zhang, Y.; Lu, J.; Wang, Y.; Wang, J.; Meng, Q.; Lee, R.J.; Teng, L. Cordycepin, a Natural Antineoplastic Agent, Induces Apoptosis of Breast Cancer Cells via Caspase-dependent Pathways. *Nat. Prod. Commun.* **2016**, *11*, 63–68. [CrossRef] [PubMed]
33. Liu, Y.; Wang, J.; Wang, W.; Zhang, H.; Zhang, X.; Han, C. The Chemical Constituents and Pharmacological Actions of Cordyceps sinensis. *Evid. Based Complement. Altern. Med.* **2015**, *2015*, 575063.
34. Shashidhar, M.G.; Giridhar, P.; Udaya Sankar, K.; Manohar, B. Bioactive principles from Cordyceps sinensis: A potent food supplement—A review. *J. Funct. Foods* **2013**, *5*, 1013–1030. [CrossRef] [PubMed]
35. Kim, S.D. Isolation, structure and cholesterol esterase inhibitory activity of a polysaccharide, PS-A, from Cordyceps sinensis. *J. Korean Soc. Appl. Biol. Chem.* **2010**, *53*, 784–789. [CrossRef]
36. Lin, B.; Li, S. Cordyceps as an Herbal Drug. In *Herbal Medicine: Biomolecular and Clinical Aspects*, 2nd ed.; Benzie, I.F.F., Wachtel-Galor, S., Eds.; Taylor & Francis: Abingdon, UK; CRC Press: Boca Raton, FL, USA, 2011.
37. Wong, J.H.; Ng, T.B.; Wang, H.; Sze, S.C.; Zhang, K.Y.; Li, Q.; Lu, X. Cordymin, an antifungal peptide from the medicinal fungus Cordyceps militaris. *Phytomedicine* **2011**, *18*, 387–392. [CrossRef]
38. Jia, J.M.; Tao, H.H.; Feng, B.M. Cordyceamides A and B from the Culture Liquid of Cordyceps sinensis (BERK.) SACC. *Chem. Pharm. Bull. (Tokyo)* **2009**, *57*, 99–101. [CrossRef] [PubMed]
39. Heine, W.; Radke, M.; Wutzke, K.D. The significance of tryptophan in human nutrition. *Amino Acids* **1995**, *9*, 91–205. [CrossRef] [PubMed]
40. Zhang, M.; Zheng, K.; Choudhury, M.; Phillips, J.; Konno, S. Enhanced anticancer effect by combination of proteoglucan and Vitamin K_3 on bladder cancer cells. *Cancer Transl. Med.* **2018**, *4*, 117–122.
41. Bhuyan, D.J.; Basu, A. Phenolic compounds: Potential health benefits and toxicity. In *Utilisation of Bioactive Compounds from Agricultural and Food Waste*; Vuong, Q.V., Ed.; CRC Press: Boca Raton, FL, USA, 2017; pp. 27–59.
42. Wikipedia, N-Acetylgalactosamine. Available online: https://en.wikipedia.org/wiki/N-Acetylgalactosamine (accessed on 10 May 2020).
43. Nwodo, U.U.; Green, E.; Okoh, A.I. Bacterial exopolysaccharides: Functionality and prospects. *Int. J. Mol. Sci.* **2012**, *13*, 14002–14015. [CrossRef] [PubMed]
44. Chauhan, B.; Kumar, G.; Kalam, N.; Ansari, S.H. Current concepts and prospects of herbal nutraceutical: A review. *J. Adv. Pharm. Technol. Res.* **2013**, *4*, 4–8.
45. Pal, M.; Misra, K. Chapter 6—Cordyceps sp.: The Precious Mushroom for High-Altitude Maladies. In *Management of High Altitude Pathophysiology*; Misra, K., Sharma, P., Bhardwaj, A., Eds.; Academic Press: Cambridge, MA, USA, 2018; pp. 93–114.
46. Green, R.; Allen, L.H.; Bjørke-Monsen, A.-L.; Brito, A.; Guéant, J.-L.; Miller, J.W.; Molloy, A.M.; Nexo, E.; Stabler, S.; Toh, B.-H.; et al. Correction: Vitamin B12 deficiency. *Nat. Rev. Dis. Primers* **2017**, *3*, 17054. [CrossRef]
47. Soltani, M.; Malek, R.A.; Elmarzugi, N.A.; Mahoodally, M.F.; Uy, D.; Leng, O.M.; El-Enshasy, H.A. Cordycepin: A biotherapeutic molecule from medicinal mushroom. In *Biology of Macrofungi, Fungal Biology*; Springer Nature: Basel, Switzerland, 2018. [CrossRef]
48. Jin, C.Y.; Kim, G.Y.; Choi, Y.H. Induction of apoptosis by aqueous extract of Cordyceps militaris through activation of caspases and inactivation of Akt in human breast cancer MDA-MB-231 Cells. *J. Microbiol. Biotechnol.* **2008**, *18*, 1997–2003.
49. Huang, S.; Liu, H.; Sun, Y.; Chen, J.; Li, X.; Xu, J.; Hu, Y.; Li, Y.; Deng, Z.; Zhong, S. An effective and convenient synthesis of cordycepin from adenosine. *Chem. Pap.* **2018**, *72*, 149–160. [CrossRef]
50. Jin, Y.; Meng, X.; Qiu, Z.; Su, Y.; Yu, P.; Qu, P. Anti-tumor and anti-metastatic roles of cordycepin, one bioactive compound of Cordyceps militaris. *Saudi J. Biol. Sci.* **2018**, *25*, 991–995. [CrossRef] [PubMed]
51. Li, G.; Nakagome, I.; Hirono, S.; Itoh, T.; Fujiwara, R. Inhibition of adenosine deaminase (ADA)-mediated metabolism of cordycepin by natural substances. *Pharmacol. Res. Perspect.* **2015**, *3*, e00121. [CrossRef] [PubMed]
52. Wong, Y.Y.; Moon, A.; Duffin, R.; Barthet-Barateig, A.; Meijer, H.A.; Clemens, M.J.; de Moor, C.H. Cordycepin inhibits protein synthesis and cell adhesion through effects on signal transduction. *J. Biol. Chem.* **2010**, *285*, 2610–2621. [CrossRef] [PubMed]
53. Das, L.; Bhaumik, E.; Raychaudhuri, U.; Chakraborty, R. Role of nutraceuticals in human health. *J. Food Sci. Technol.* **2012**, *49*, 173–183. [CrossRef] [PubMed]

54. Ahmad, F.; Ashraf, S.A.; Ahmad, F.A.; Ansari, J.A.; Siddiquee, R.A. Nutraceutical Market and its Regulation. *Am. J. Food Technol.* **2011**, *6*, 342–347. [CrossRef]
55. Kumar, K.; Kumar, S. Role of Nutraceuticals in Health and diseaseprevention: A review. *South Asian J. Food Technol. Environ.* **2015**, *1*, 116–121.
56. Bashir, A.; Ashraf, S.A.; Khan, M.A.; Azad, Z.R.A.A. Development and Compositional Analysis of Protein Enriched Soybean-Pea-Wheat Flour Blended Cookies. *Asian J. Clin. Nutr.* **2015**, *7*, 76–83.
57. Dahiya, K. Nutraceuticals and Their Impact on Human Health. *J. Plant. Biochem. Physiol.* **2013**, *2013*, 1–4. [CrossRef]
58. Devi, A.; Chennakesavulu, S.; Suresh, C.; Reddy, A.B. Nutraceuticals and Their Role in Human Health and Disease. In *Functional Food and Human Health*; Rani, V., Yadav, U.C.S., Eds.; Springer: Singapore, 2018.
59. Kopalli, S.R.; Cha, K.M.; Lee, S.H.; Hwang, S.Y.; Lee, Y.J.; Koppula, S.; Kim, S.K. Cordycepin, an Active Constituent of Nutrient Powerhouse and Potential Medicinal Mushroom Cordyceps militaris Linn., Ameliorates Age-Related Testicular Dysfunction in Rats. *Nutrients* **2019**, *11*, 906. [CrossRef]
60. Alghamdi, A.A. Amir Mahgoub Awadelkarem, ABMS Hossain, Nasir A Ibrahim, Mohammad Fawzi, Syed Amir Ashraf, Nutritional assessment of different date fruits (*Phoenix dactylifera* L.) varieties cultivated in Hail province, Saudi Arabia. *Biosci. Biotechnol. Res. Commun.* **2018**, *11*, 263–269. [CrossRef]
61. IDF, International Diabetes Federation, The IDF Diabetes Atlas Eighth Edition. 2017. Available online: www.idf.org (accessed on 4 February 2020).
62. Winkler, G.; Hidvegi, T.; Vandorfi, G.; Jermendy, G. Risk-stratified screening for diabetes in adults: Results of the first investigation in Hungary. *Orv. Hetil.* **2010**, *151*, 691–696. [CrossRef] [PubMed]
63. Levterova, B.A.; Dimitrova, D.D.; Levterov, G.E.; Dragova, E.A. Instruments for disease-specific quality-of-life measurement in patients with type 2 diabetes mellitus—A systematic review. *Folia Med. (Plovdiv)* **2013**, *55*, 83–92. [CrossRef] [PubMed]
64. Dong, Y.; Jing, T.; Meng, Q.; Liu, C.; Hu, S.; Ma, Y.; Liu, Y.; Lu, J.; Cheng, Y.; Wang, D.; et al. Studies on the antidiabetic activities of Cordyceps militaris extract in diet-streptozotocin-induced diabetic Sprague-Dawley rats. *Biomed. Res. Int.* **2014**, *2014*, 160980. [CrossRef] [PubMed]
65. Shin, S.; Lee, S.; Kwon, J.; Moon, S.; Lee, C.K.; Cho, K.; Ha, N.J.; Kim, K. Cordycepin Suppresses Expression of Diabetes Regulating Genes by Inhibition of Lipopolysaccharide-induced Inflammation in Macrophages. *Immune Netw.* **2009**, *9*, 98–105. [CrossRef]
66. Patel, K.J.; Ingalhalli, R.S. Cordyceps militaris an important medicinal mushroom. *J. Pharmacogn. Phytochem.* **2013**, *2*, 315–319.
67. Ji, D.B.; Ye, J.; Li, C.L.; Wang, Y.H.; Zhao, J.; Cai, S.Q. Antiaging effect of Cordyceps sinensis extract. *Phytother. Res.* **2009**, *23*, 116–122. [CrossRef]
68. Ma, L.; Zhang, S.; Du, M. Cordycepin from Cordyceps militaris prevents hyperglycemia in alloxan-induced diabetic mice. *Nutr. Res.* **2015**, *35*, 431–439. [CrossRef]
69. Cao, T.; Xu, R.; Xu, Y.; Liu, Y.; Qi, D.; Wan, Q. The protective effect of Cordycepin on diabetic nephropathy through autophagy induction in vivo and in vitro. *Int. Urol. Nephrol.* **2019**, *51*, 1883–1892. [CrossRef]
70. Yun, Y.H.; Han, S.H.; Lee, S.J.; Ko, S.K.; Lee, C.K.; Ha, N.J.; Kim, K.J. Anti-diabetic Effects of CCCA, CMESS, and Cordycepin from Cordyceps militaris and the Immune Responses in Streptozotocin-induced Diabetic Mice. *Nat. Prod. Sci.* **2003**, *9*, 291–298.
71. Le, N.A. Hyperlipidemia and cardiovascular disease: Cardiovascular update. *Curr. Opin. Lipidol.* **2008**, *19*, 545–547. [CrossRef]
72. Das, P.P.; Malhotra, S.; Chakrabarti, S.; Sharma, S. Elevated total cholesterol in severely depressed patients: Role in cardiovascular risk? *World J. Biol. Psychiatry* **2010**, *11*, 321–328. [CrossRef] [PubMed]
73. Le, N.A.; Walter, M.F. The role of hypertriglyceridemia in atherosclerosis. *Curr. Atheroscler. Rep.* **2007**, *9*, 110–115. [CrossRef] [PubMed]
74. Rosenson, R.S. Low high-density lipoprotein cholesterol and cardiovascular disease: Risk reduction with statin therapy. *Am. Heart J.* **2006**, *151*, 556–563. [CrossRef] [PubMed]
75. Thomson, D.M.; Winder, W.W. AMP-activated protein kinase control of fat metabolism in skeletal muscle. *Acta Physiol.* **2009**, *196*, 147–154. [CrossRef] [PubMed]
76. Atkinson, L.L.; Kozak, R.; Kelly, S.E.; Onay Besikci, A.; Russell, J.C.; Lopaschuk, G.D. Potential mechanisms and consequences of cardiac triacylglycerol accumulation in insulin-resistant rats. *Am. J. Physiol. Endocrinol. Metab.* **2003**, *284*, E923–E930. [CrossRef]

77. Wu, C.; Guo, Y.; Su, Y.; Zhang, X.; Luan, H.; Zhu, H.; He, H.; Wang, X.; Sun, G.; Sun, X.; et al. Cordycepin activates AMP-activated protein kinase (AMPK) via interaction with the gamma1 subunit. *J. Cell. Mol. Med.* **2014**, *18*, 293–304. [CrossRef]
78. Hardie, D.G. AMP-activated protein kinase as a drug target. *Annu. Rev. Pharmacol. Toxicol.* **2007**, *47*, 185–210. [CrossRef]
79. Guo, P.; Kai, Q.; Gao, J.; Lian, Z.Q.; Wu, C.M.; Wu, C.A.; Zhu, H.B. Cordycepin prevents hyperlipidemia in hamsters fed a high-fat diet via activation of AMP-activated protein kinase. *J. Pharmacol. Sci.* **2010**, *113*, 395–403. [CrossRef]
80. Aymerich, I.; Foufelle, F.; Ferre, P.; Casado, F.J.; Pastor-Anglada, M. Extracellular adenosine activates AMP-dependent protein kinase (AMPK). *J. Cell Sci.* **2006**, *119*, 1612–1621. [CrossRef]
81. Gao, J.; Lian, Z.Q.; Zhu, P.; Zhu, H.B. Lipid-lowering effect of cordycepin (3′-deoxyadenosine) from Cordyceps militaris on hyperlipidemic hamsters and rats. *Yao Xue Xue Bao* **2011**, *46*, 669–676.
82. Aghasafari, P.; George, U.; Pidaparti, R. A review of inflammatory mechanism in airway diseases. *Inflamm. Res.* **2019**, *68*, 59–74. [CrossRef] [PubMed]
83. Han, E.S.; Oh, J.Y.; Park, H.J. Cordyceps militaris extract suppresses dextran sodium sulfate-induced acute colitis in mice and production of inflammatory mediators from macrophages and mast cells. *J. Ethnopharmacol.* **2011**, *134*, 703–710. [CrossRef] [PubMed]
84. Jeong, J.W.; Jin, C.Y.; Kim, G.Y.; Lee, J.D.; Park, C.; Kim, G.D.; Kim, W.J.; Jung, W.K.; Seo, S.K.; Choi, I.W.; et al. Anti-inflammatory effects of cordycepin via suppression of inflammatory mediators in BV2 microglial cells. *Int. Immunopharmacol.* **2010**, *10*, 1580–1586. [CrossRef] [PubMed]
85. Fung, C.K.; Ko, W.H. Cordyceps Extracts and the Major Ingredient, Cordycepin: Possible Cellular Mechanisms of Their Therapeutic Effects on Respiratory Disease. In *Respiratory Diseases*; Mostafa, G., Ed.; IntechOpen: London, UK, 2012; Available online: http://www.intechopen.com/books/respiratory-diseases/cordyceps-extracts-and-the-major-ingredientcordycepin-possible-cellular-mechanisms-of-their-therape (accessed on 11 March 2020).
86. Hsu, C.H.; Sun, H.L.; Sheu, J.N.; Ku, M.S.; Hu, C.M.; Chan, Y.; Lue, K.H. Effects of the immunomodulatory agent Cordyceps militaris on airway inflammation in a mouse asthma model. *Pediatr. Neonatol.* **2008**, *49*, 171–178. [CrossRef]
87. Antoniou, K.M.; Pataka, A.; Bouros, D.; Siafakas, N.M. Pathogenetic pathways and novel pharmacotherapeutic targets in idiopathic pulmonary fibrosis. *Pulm. Pharmacol. Ther.* **2007**, *20*, 453–461. [CrossRef]
88. Yang, L.; Jiao, X.; Wu, J.; Zhao, J.; Liu, T.; Xu, J.; Ma, X.; Cao, L.; Liu, L.; Liu, Y.; et al. Cordyceps sinensis inhibits airway remodeling in rats with chronic obstructive pulmonary disease. *Exp. Ther. Med.* **2018**, *15*, 2731–2738.
89. Wang, N.Q.; Jiang, L.D.; Zhang, X.M.; Li, Z.X. Effect of dongchong xiacao capsule on airway inflammation of asthmatic patients. *Zhongguo Zhong Yao Za Zhi* **2007**, *32*, 1566–1568.
90. Xu, H.; Li, S.; Lin, Y.; Liu, R.; Gu, Y.; Liao, D. Effectiveness of cultured Cordyceps sinensis combined with glucocorticosteroid on pulmonary fibrosis induced by bleomycin in rats. *Zhongguo Zhong Yao Za Zhi* **2011**, *36*, 2265–2270.
91. Li, F.H.; Liu, P.; Xiong, W.G.; Xu, G.F. Effects of corydyceps polysaccharide on liver fibrosis induced by DMN in rats. *Zhongguo Zhong Yao Za Zhi* **2006**, *31*, 1968–1971.
92. Cheng, Z.; He, W.; Zhou, X.; Lv, Q.; Xu, X.; Yang, S.; Zhao, C.; Guo, L. Cordycepin protects against cerebral ischemia/reperfusion injury in vivo and in vitro. *Eur. J. Pharmacol.* **2011**, *664*, 20–28. [CrossRef]
93. Zhang, W.; Ahmad, G.; Molehin, A.J.; Torben, W.; Le, L.; Kim, E.; Lazarus, S.; Siddiqui, A.J.; Carter, D.; Siddiqui, A.A. Schistosoma mansoni antigen Sm-p80: Prophylactic efficacy using TLR4 agonist vaccine adjuvant glucopyranosyl lipid A-Alum in murine and non-human primate models. *J. Investig. Med.* **2018**, *66*, 1124–1132. [CrossRef] [PubMed]
94. Cheng, Y.H.; Hsieh, Y.C.; Yu, Y.H. Effect of Cordyceps militaris Hot Water Extract on Immunomodulation-associated Gene Expression in Broilers, Gallus gallus. *J. Poult. Sci.* **2019**, *56*, 128–139. [CrossRef] [PubMed]
95. Zhou, X.; Luo, L.; Dressel, W.; Shadier, G.; Krumbiegel, D.; Schmidtke, P.; Zepp, F.; Meyer, C.U. Cordycepin is an immunoregulatory active ingredient of Cordyceps sinensis. *Am. J. Chin. Med.* **2008**, *36*, 967–980. [CrossRef]

96. Liu, R.; Han, C.; Wu, D.; Xia, X.; Gu, J.; Guan, H.; Shan, Z.; Teng, W. Prevalence of Hyperuricemia and Gout in Mainland China from 2000 to 2014: A Systematic Review and Meta-Analysis. *Biomed. Res. Int.* **2015**, *2015*, 762820. [CrossRef]
97. Zhang, Z.; Xia, S.S. Cordyceps Sinensis-I as an immunosuppressant in heterotopic heart allograft model in rats. *J. Tongji Med. Univ.* **1990**, *10*, 100–103. [CrossRef]
98. Zhu, X.Y.; Yu, H.Y. Immunosuppressive effect of cultured Cordyceps sinensis on cellular immune response. *Zhong Xi Yi Jie He Za Zhi* **1990**, *10*, 485–487. [PubMed]
99. Taylor, A.L.; Watson, C.J.; Bradley, J.A. Immunosuppressive agents in solid organ transplantation: Mechanisms of action and therapeutic efficacy. *Crit Rev Oncol Hematol* **2005**, *56*, 23–46. [CrossRef] [PubMed]
100. Wachtel-Galor, S. The biological and pharmacological properties of Cordyceps sinensis, a traditional Chinese medicine that has broad clinical applications. In *Herbal and Traditional Medicine*; Marcel Dekker: New York, NY, USA, 2004; pp. 657–682.
101. Ng, T.B.; Wang, H.X. Pharmacological actions of Cordyceps, a prized folk medicine. *J. Pharm. Pharmacol.* **2005**, *57*, 1509–1519. [CrossRef] [PubMed]
102. Wang, Y.; Li, S. Renggongchongcao. In *Pharmacological Activity-Based Quality Control of Chinese Herbs*; Li, S.P., Wang, Y.T., Eds.; Nova Science Publisher, Inc.: New York, NY, USA, 2008; pp. 155–178.
103. Kitamura, M.; Kato, H.; Saito, Y.; Nakajima, S.; Takahashi, S.; Johno, H.; Gu, L.; Katoh, R. Aberrant, differential and bidirectional regulation of the unfolded protein response towards cell survival by 3′-deoxyadenosine. *Cell Death Differ.* **2011**, *18*, 1876–1888. [CrossRef]
104. Zhang, D.W.; Deng, H.; Qi, W.; Zhao, G.Y.; Cao, X.R. Osteoprotective effect of cordycepin on estrogen deficiency-induced osteoporosis in vitro and in vivo. *Biomed. Res. Int.* **2015**, *2015*, 423869. [PubMed]
105. Kim, J.H.; Kim, E.Y.; Lee, B.; Min, J.H.; Song, D.U.; Lim, J.M.; Eom, J.W.; Yeom, M.; Jung, H.S.; Sohn, Y. The effects of Lycii Radicis Cortex on RANKL-induced osteoclast differentiation and activation in RAW 264.7 cells. *Int. J. Mol. Med.* **2016**, *37*, 649–658. [CrossRef] [PubMed]
106. Hu, P.; Chen, W.; Bao, J.; Jiang, L.; Wu, L. Cordycepin modulates inflammatory and catabolic gene expression in interleukin-1beta-induced human chondrocytes from advanced-stage osteoarthritis: An in vitro study. *Int. J. Clin. Exp. Pathol.* **2014**, *7*, 6575–6584. [PubMed]
107. Yang, R.; Wang, X.; Xi, D.; Mo, J.; Wang, K.; Luo, S.; Wei, J.; Ren, Z.; Pang, H.; Luo, Y. Cordycepin Attenuates IFN-gamma-Induced Macrophage IP-10 and Mig Expressions by Inhibiting STAT1 Activity in CFA-Induced Inflammation Mice Model. *Inflammation* **2020**, *43*, 752–764. [CrossRef] [PubMed]
108. Wang, X.; Xi, D.; Mo, J.; Wang, K.; Luo, Y.; Xia, E.; Huang, R.; Luo, S.; Wei, J.; Ren, Z.; et al. Cordycepin exhibits a suppressive effect on T cells through inhibiting TCR signaling cascade in CFA-induced inflammation mice model. *Immunopharmacol. Immunotoxicol.* **2020**, *42*, 119–127. [CrossRef]
109. Dalle-Donne, I.; Rossi, R.; Colombo, R.; Giustarini, D.; Milzani, A. Biomarkers of oxidative damage in human disease. *Clin. Chem.* **2006**, *52*, 601–623. [CrossRef]
110. Manach, C.; Scalbert, A.; Morand, C.; Remesy, C.; Jimenez, L. Polyphenols: Food sources and bioavailability. *Am. J. Clin. Nutr.* **2004**, *79*, 727–747. [CrossRef]
111. Bhatt, M.; Patel, M.; Adnan, M.; Reddy, M.N. Anti-Metastatic Effects of Lupeol via the Inhibition of MAPK/ERK Pathway in Lung Cancer. *Anticancer Agents Med. Chem.* **2020**. [CrossRef]
112. Reddy, M.; Adnan, M.; Alreshidi, M.; Saeed, M.; Patel, M. Evaluation of Anticancer, Antibacterial and Antioxidant Properties of a Medicinally Treasured Fern Tectaria coadunata with its Phytoconstituents Analysis by HR-LCMS. *Anticancer Agents Med Chem* **2020**. [CrossRef]
113. Adnan, M.; Patel, M.; Deshpande, S.; Alreshidi, M.; Siddiqui, A.J.; Reddy, M.N.; Emira, N.; De Feo, V. Effect of Adiantum philippense Extract on Biofilm Formation, Adhesion With Its Antibacterial Activities Against Foodborne Pathogens, and Characterization of Bioactive Metabolites: An in vitro-in silico Approach. *Front. Microbiol.* **2020**, *11*. [CrossRef]
114. Adnan, M. Bioactive potential of essential oil extracted from the leaves of Eucalyptus globulus (Myrtaceae). *J. Pharmacogn. Phytochem.* **2019**, *8*, 213–216.
115. Tsai, C.-H.; Stern, A.; Chiou, J.-F.; Chern, C.-L.; Liu, T.-Z. Rapid and Specific Detection of Hydroxyl Radical Using an Ultraweak Chemiluminescence Analyzer and a Low-Level Chemiluminescence Emitter: Application to Hydroxyl Radical-Scavenging Ability of Aqueous Extracts of Food Constituents. *J. Agric. Food Chem.* **2001**, *49*, 2137–2141. [CrossRef] [PubMed]

116. Li, S.P.; Li, P.; Dong, T.T.; Tsim, K.W. Anti-oxidation activity of different types of natural Cordyceps sinensis and cultured Cordyceps mycelia. *Phytomedicine* **2001**, *8*, 207–212. [CrossRef]
117. Olatunji, O.J.; Feng, Y.; Olatunji, O.O.; Tang, J.; Ouyang, Z.; Su, Z. Cordycepin protects PC12 cells against 6-hydroxydopamine induced neurotoxicity via its antioxidant properties. *Biomed. Pharmacother.* **2016**, *81*, 7–14. [CrossRef] [PubMed]
118. Shin, K.H.; Lim, S.S.; Lee, S.H.; Lee, Y.S.; Cho, S.Y. Antioxidant and immunostimulating activities of the fruiting bodies of Paecilomyces japonica, a new type of Cordyceps sp. *Ann. N. Y. Acad. Sci.* **2001**, *928*, 261–273. [CrossRef] [PubMed]
119. Liu, Y.; Wu, C.; Li, C. Anti-oxidation of Paecilomyces Sinensis (S. Pnov.). *Zhongguo Zhong Yao Za Zhi* **1991**, *16*, 240–242.
120. He, Y.T.; Zhang, X.L.; Xie, Y.M.; Xu, Y.X.; Li, J.R. Extraction and Antioxidant Property In Vitro of Cordycepin in Artificially Cultivated Cordyceps Militaris. *Adv. Mater. Res.* **2013**, *750–752*, 1593–1596. [CrossRef]
121. Bhardwaj, J.; Siddiqui, A.J.; Goyal, M.; Prakash, K.; Soni, A.; Puri, S.K. Repetitive live sporozoites inoculation under arteether chemoprophylaxis confers protection against subsequent sporozoite challenge in rodent malaria model. *Acta Trop.* **2016**, *158*, 130–138. [CrossRef]
122. Bhardwaj, J.; Siddiqui, A.J.; Goyal, M.; Prakash, K.; Soni, A.; Puri, S.K.; Srivastava, M. Host immune response is severely compromised during lethal Plasmodium vinckei infection. *Parasitol. Res.* **2015**, *114*, 3445–3457. [CrossRef]
123. Siddiqui, A.J.; Bhardwaj, J.; Goyal, M.; Prakash, K.; Adnan, M.; Alreshidi, M.M.; Patel, M.; Soni, A.; Redman, W. Immune responses in liver and spleen against Plasmodium yoelii pre-erythrocytic stages in Swiss mice model. *J. Adv. Res.* **2020**, *24*, 29–41. [CrossRef]
124. Siddiqui, A.J.; Bhardwaj, J.; Goyal, M.; Prakash, K.; Soni, A.; Tiwari, V.; Puri, S.K. Assessment of real-time method to detect liver parasite burden under different experimental conditions in mice infected with Plasmodium yoelii sporozoites. *Microb. Pathog.* **2015**, *89*, 35–42. [CrossRef]
125. Soni, A.; Goyal, M.; Prakash, K.; Bhardwaj, J.; Siddiqui, A.J.; Puri, S.K. Cloning, expression and functional characterization of heme detoxification protein (HDP) from the rodent malaria parasite Plasmodium vinckei. *Gene* **2015**, *566*, 109–119. [CrossRef] [PubMed]
126. Prakash, K.; Goyal, M.; Soni, A.; Siddiqui, A.J.; Bhardwaj, J.; Puri, S.K. Molecular cloning and biochemical characterization of iron superoxide dismutase from the rodent malaria parasite Plasmodium vinckei. *Parasitol. Int.* **2014**, *63*, 817–825. [CrossRef] [PubMed]
127. Azad, C.S.; Saxena, M.; Siddiqui, A.J.; Bhardwaj, J.; Puri, S.K.; Dutta, G.P.; Anand, N.; Saxena, A.K. Synthesis of primaquine glyco-conjugates as potential tissue schizontocidal antimalarial agents. *Chem. Biol. Drug Des.* **2017**, *90*, 254–261. [CrossRef] [PubMed]
128. Siddiqui, A.J.; Bhardwaj, J.; Puri, S.K. mRNA expression of cytokines and its impact on outcomes after infection with lethal and nonlethal Plasmodium vinckei parasites. *Parasitol. Res.* **2012**, *110*, 1517–1524. [CrossRef]
129. WHO. Malaria. Facts in Pictures, News Room, World Health Organization. Available online: https://www.who.int/news-room/facts-in-pictures/detail/malaria (accessed on 15 May 2020).
130. Trigg, P.I.; Gutteridge, W.E.; Williamson, J. The effects of cordycepin on malaria parasites. *Trans. R. Soc. Trop. Med. Hyg.* **1971**, *65*, 514–520. [CrossRef]
131. Choi, H.K.; Curhan, G. Independent impact of gout on mortality and risk for coronary heart disease. *Circulation* **2007**, *116*, 894–900. [CrossRef]
132. Yong, T.; Zhang, M.; Chen, D.; Shuai, O.; Chen, S.; Su, J.; Jiao, C.; Feng, D.; Xie, Y. Actions of water extract from Cordyceps militaris in hyperuricemic mice induced by potassium oxonate combined with hypoxanthine. *J. Ethnopharmacol.* **2016**, *194*, 403–411. [CrossRef]
133. Yong, T.; Chen, S.; Xie, Y.; Chen, D.; Su, J.; Shuai, O.; Jiao, C.; Zuo, D. Cordycepin, a Characteristic Bioactive Constituent in Cordyceps militaris, Ameliorates Hyperuricemia through URAT1 in Hyperuricemic Mice. *Front. Microbiol.* **2018**, *9*, 58. [CrossRef]
134. Chang, Y.; Jeng, K.C.; Huang, K.F.; Lee, Y.C.; Hou, C.W.; Chen, K.H.; Cheng, F.Y.; Liao, J.W.; Chen, Y.S. Effect of Cordyceps militaris supplementation on sperm production, sperm motility and hormones in Sprague-Dawley rats. *Am. J. Chin. Med.* **2008**, *36*, 849–859. [CrossRef]
135. Hong, I.P.; Choi, Y.S.; Woo, S.O.; Han, S.M.; Kim, H.K.; Lee, M.R.; Ha, N.G. Stimulatory Effect of Cordyceps militaris on Testosterone Production in Male Mouse. *Korean J. Mycol.* **2011**, *39*, 148–150. [CrossRef]

136. Nugent, R.A.; Fathima, S.F.; Feigl, A.B.; Chyung, D. The burden of chronic kidney disease on developing nations: A 21st century challenge in global health. *Nephron Clin. Pract.* **2011**, *118*, c269–c277. [CrossRef] [PubMed]
137. Jin, Z.H.; Chen, Y.P. Clinical study on the effect of Cordyceps sinensis mycelium on delaying progression of chronic kidney failure. *J. Nanjing Univ. Tradit. Chin. Med.* **2004**, *20*, 155–157.
138. Wu, W.C.; Meng, T. Clinical observation on the treatment effect of Jin Shui Bao Capsule for chronic kidney failure as a complementary approach. *J. Emerg. Tradit. Chin. Med.* **2007**, *16*, 1473.
139. Yu, X.M.; Tan, S. Clinical study on the treatment of chronic renal failure with Bailing. *J. Trop. Med.* **2003**, *3*, 203–204.
140. Quan, Y.H.; Xu, L. Clinical study on the Cordyceps preparation for the treatment of chronic kidney failure. *J. Tradit. Chin. Med.* **2004**, *26*, 11–12.
141. Research, G.V. Nutraceutical Market Size Worth $722.49 Billion By 2027 | CAGR: 8.3%. Available online: https://www.grandviewresearch.com/press-release/global-nutraceuticals-market (accessed on 5 May 2020).
142. Insights, F.B. Nutraceuticals Market Size to Reach USD 486.36 Billion by 2026; Stellar Demand for Functional Food to Aid Expansion, states Fortune Business Insights. Available online: https://www.globenewswire.com/news-release/2020/03/25/2006352/0/en/Nutraceuticals-Market-Size-to-Reach-USD-486-36-Billion-by-2026-Stellar-Demand-for-Functional-Food-to-Aid-Expansion-states-Fortune-Business-Insights.html (accessed on 14 May 2020).
143. Insight, C.M. Global Cordyceps Sinensis and Militaris Extract Market to Surpass US$ 1 Billion by 2026. Available online: https://www.globenewswire.com/news-release/2019/04/23/1807927/0/en/Global-Cordyceps-Sinensis-and-Militaris-Extract-Market-to-Surpass-US-1-Billion-by-2026.html (accessed on 14 May 2020).
144. Map, W. Credence Research, Nutraceutical Market by Ingredients. Available online: https://www.credenceresearch.com/report/nutraceutical-ingredients-market (accessed on 20 March 2020).
145. Yu, X.; Mao, Y.; Shergis, J.L.; Coyle, M.E.; Wu, L.; Chen, Y.; Zhang, A.L.; Lin, L.; Xue, C.C.; Xu, Y. Effectiveness and Safety of Oral Cordyceps sinensis on Stable COPD of GOLD Stages 2-3: Systematic Review and Meta-Analysis. *Evid. Based Complement. Altern. Med.* **2019**, *2019*, 4903671. [CrossRef]
146. Qin, P.; Li, X.; Yang, H.; Wang, Z.Y.; Lu, D. Therapeutic Potential and Biological Applications of Cordycepin and Metabolic Mechanisms in Cordycepin-Producing Fungi. *Molecules* **2019**, *24*, 2231. [CrossRef]

molecules

MDPI

Article

Detection of Fungi and Oomycetes by Volatiles Using E-Nose and SPME-GC/MS Platforms

Jérémie Loulier [1], François Lefort [1,*], Marcin Stocki [2], Monika Asztemborska [3], Rafał Szmigielski [3], Krzysztof Siwek [4], Tomasz Grzywacz [4], Tom Hsiang [5], Sławomir Ślusarski [6], Tomasz Oszako [2,6], Marcin Klisz [7], Rafał Tarakowski [8] and Justyna Anna Nowakowska [9,*]

1 InTNE (Plants & Pathogens Group), Hepia, University of Applied Sciences and Arts of Western Switzerland, 150 route de Presinge, 1254 Jussy, Switzerland; jeremie.loulier@bluewin.ch
2 Institute of Forest Sciences, Faculty of Civil Engineering and Environmental Sciences, Bialystok University of Technology, Wiejska 45E, 15-351 Bialystok, Poland; m.stocki@pb.edu.pl (M.S.); T.Oszako@ibles.waw.pl (T.O.)
3 Institute of Physical Chemistry, Polish Academy of Sciences, Kasprzaka 44/52, 01-224 Warsaw, Poland; monika@ichf.edu.pl (M.A.); ralf@ichf.edu.pl (R.S.)
4 Faculty of Electrical Engineering, Warsaw University of Technology, Koszykowa 75, 00-661 Warsaw, Poland; krzysztof.siwek@ee.pw.edu.pl (K.S.); tomasz.grzywacz@ee.pw.edu.pl (T.G.)
5 Environmental Sciences, University of Guelph, Guelph, ON N1G 2W1, Canada; thsiang@uoguelph.ca
6 Forest Protection Department, Forest Research Institute, Braci Leśnej 3, 05-090 Sękocin Stary, Poland; S.Slusarski@ibles.waw.pl
7 Department of Silviculture and Genetics, Forest Research Institute, Braci Leśnej 3, 05-090 Sękocin Stary, Poland; M.Klisz@ibles.waw.pl
8 Faculty of Physics, Warsaw University of Technology, Koszykowa 75, 00-662 Warsaw, Poland; Rafal.Tarakowski@pw.edu.pl
9 Institute of Biological Sciences, Cardinal Stefan Wyszynski University in Warsaw, Wóycickiego 1/3 Street, 01-938 Warsaw, Poland
* Correspondence: francois.lefort@hesge.ch (F.L.); j.nowakowska@uksw.edu.pl (J.A.N.); Tel.: +41-22-546-68-27 (F.L.); +48-22-569-6838 (J.A.N.)

Academic Editors: Simona Fabroni, Krystian Marszałek, Aldo Todaro and Wilfried Rozhon
Received: 13 October 2020; Accepted: 4 December 2020; Published: 5 December 2020

check for updates

Abstract: Fungi and oomycetes release volatiles into their environment which could be used for olfactory detection and identification of these organisms by electronic-nose (e-nose). The aim of this study was to survey volatile compound emission using an e-nose device and to identify released molecules through solid phase microextraction–gas chromatography/mass spectrometry (SPME–GC/MS) analysis to ultimately develop a detection system for fungi and fungi-like organisms. To this end, cultures of eight fungi (*Armillaria gallica, Armillaria ostoyae, Fusarium avenaceum, Fusarium culmorum, Fusarium oxysporum, Fusarium poae, Rhizoctonia solani, Trichoderma asperellum*) and four oomycetes (*Phytophthora cactorum, P. cinnamomi, P. plurivora, P. ramorum*) were tested with the e-nose system and investigated by means of SPME-GC/MS. Strains of *F. poae*, *R. solani* and *T. asperellum* appeared to be the most odoriferous. All investigated fungal species (except *R. solani*) produced sesquiterpenes in variable amounts, in contrast to the tested oomycetes strains. Other molecules such as aliphatic hydrocarbons, alcohols, aldehydes, esters and benzene derivatives were found in all samples. The results suggested that the major differences between respective VOC emission ranges of the tested species lie in sesquiterpene production, with fungi emitting some while oomycetes released none or smaller amounts of such molecules. Our e-nose system could discriminate between the odors emitted by *P. ramorum*, *F. poae*, *T. asperellum* and *R. solani*, which accounted for over 88% of the PCA variance. These preliminary results of fungal and oomycete detection make the e-nose device suitable for further sensor design as a potential tool for forest managers, other plant managers, as well as regulatory agencies such as quarantine services.

Keywords: fungi; oomycetes; VOCs; EI mass spectrometry; SPME-GC/MS; e-nose; sesquiterpenes

1. Introduction

Many phytopathogenic fungi and oomycetes species causing damping off and root rot diseases, are known to emit several secondary metabolites in the form of volatile organic compounds (VOCs), which can be genus- or species-specific. Other microbial secondary metabolites include, for instance, antibiotics, enzymes and toxins. Even though a great diversity of VOCs can be found from many different species, some volatiles are unique to particular species [1,2]. Volatile organic compounds are carbon molecules of molecular mass classically ranging from 100 to 500 Da and chain size up to C_{20} that occur in gaseous form above 20 °C and 10 Pa [3,4]. Generally lipophilic and thus poorly soluble in water, their boiling point is low and their vapor pressure at room temperature is high. Many of them have an associated characteristic odor [5].

Fungi produce VOC mixtures through primary and secondary metabolic pathways, and these diffuse through soil and into the atmosphere [6]. VOC production has been investigated in only ca. 100 of the over 100,000 described fungal species to date [7]. Fungal VOCs released through secondary metabolism seem to be at their highest during sporulation and mycotoxin production [8]. The secreted mixture profile varies according to many factors linked with the fungal strain itself and its growth environment, including physiological state, temperature, pH, oxygen, nutrients, incubation time, etc. [9,10].

Around one thousand different volatile compounds have to date been identified in a broad spectrum of microorganisms [11], produced mainly through glucose oxidation from diverse intermediates [6]. These molecules participate in various pathways of primary and secondary metabolism such as aerobic heterotrophic carbon metabolism, amino-acid catabolism, fatty acids degradation, fermentation, sulphur reduction and terpenoids biosynthesis [12]. Among these VOCs, more than one quarter are of unique fungal origin and include alcohols, aldehydes, alkenes, benzenoids, ketones, esters, terpenes and sulfuric compounds, although many other families are found as well, e.g., simple hydrocarbons, furans, heterocycles, phenols, thioalcohols, thioesters, etc. [6,13,14].

Sesquiterpenes are terpene hydrocarbons with an identical molecular formula of $C_{15}H_{24}$. Their skeletons are made of three isoprene units and can be acyclic or may include rings. Generally, sesquiterpenoids derive from sesquiterpenes through molecular rearrangements or oxidation. Sesquiterpene emission by fungi has been a field of particular interest over the past years [11,15], because this group of compounds may be key in discriminating between soil-borne fungi and *Phytophthora* spp. [16].

The exact reasons why microorganisms emit volatiles are not fully understood. It has been proposed that these molecules could be waste products resulting from microbial metabolism, and released for detoxification purposes [17]. Previous studies have shown that volatiles are involved in antimicrobial activity [8,18]. They have highlighted VOC involvement in interactions with distant microbial colonies in the form of infochemical compounds, influencing development, gene expression, and behavior of the recipient microorganisms [4]. During the last fifteen years, many trials were implemented in different research fields (for instance in environmental, food or medicinal sciences) to increase our understanding of microbial volatiles, facilitated by the development of efficient detection methods [19].

The first attempts of vapor characterization using sensing devices that can be considered as e-noses date back to 1960 [20], and they showed promise as tools for characterization of complex fungal VOC mixtures [21]. In general, an e-nose can refer to a wide range of devices including a set of gas sensors coupled with an information processing system and recognition software, founded on theoretical models as well as reference literature [22]. It was only in 1982 that intelligent e-noses able to classify odors really began to be developed [23], with the "electronic nose" appellation emerging for the first time [24]. A recent study managed to detect low concentrated matsutake alcohol by means of an e-nose using surface acoustic wave technology, which makes it a promising candidate for future developments in fungal e-nose detection [25].

In phytopathology, pathogen identification is usually achieved through study of symptoms, pathogen isolation, morphological description and possibly immunological testing or DNA sequence analyses. However, infection symptoms are sometimes invisible, hidden underground or non-specific for the pathogenic agent, which complicates proper diagnosis. Pathogen isolation, immunological tests, and DNA tests require laboratory work making the operation expensive, laborious and time-consuming, thus not well suited for large-scale monitoring [26]. In this regard, e-nose could serve as a preliminary screening tool allowing a quick diagnosis that could be achieved even without any symptoms being visible on the plant, as well as an efficient field monitoring applicable at the scale of single seeds, entire plants, or food storage installations.

To improve the performance of this technology for application to forest sciences or related fields, identification of pathogen-specific VOC profiles could be a key point for improved e-nose sensing efficiency. The aim of this work was to test an electronic nose developed by the Warsaw University of Technology, originally designed to identify bacterial and fungal rot in food, to the field of phytopathology. We focused on detection of eight well-known fungal pathogens i.e., *Armillaria* spp., *Fusarium* spp., and *Rhizoctonia solani* causing root rot damage e.g., in Scots pine, black alder and European oak seedlings [27]. The other tested fungus (*Trichoderma asperellum*) is not pathogenic and rather considered as a biocontrol agent in many forest and ornamental nurseries [27]. Furthermore, four *Phytophthora* species were investigated, since these are recognized as soil-borne pathogens of many forest tree species, including European oak, European beech, silver birch and black alder [28]. Two working hypotheses were tested. First, we assumed that the e-nose apparatus can be a suitable detection system for fungi and fungi-like organisms. The second hypothesis was that some VOCs are species-specific and because these VOCs emitted by microorganisms can be detected, this can be used to differentiate the investigated microbial species. To test these hypotheses we did the following: (i) selected and developed pure cultures of the studied microorganisms; (ii) investigated the gas composition in the air with e-nose technology; (iii) performed a precise detection of VOCs using two independent SPME-GC/MS analytical methods, and finally (iv) speculated on the extent the emitted compounds are useful for identification of a given microorganism.

2. Results

2.1. E-Nose Gas Detection

Distribution of the overall averaged values of gas detected by twelve differential sensor signals differed among all fourteen treatments (Figure 1). The two control treatments yielded somewhat divergent results because mean signal intensities obtained over four weeks of study showed clean empty glass flasks to be inducing a very weak response from e-nose sensors, whereas the potato dextrose agar (PDA) medium generated a stronger reaction from the sensing device on average (Figure 1a,b).

The e-nose detection showed that *A. gallica* and *A. ostoyae* cultures emitted low levels of volatiles similar to the empty flask control (Figure 1c,d). All *Fusarium* species provoked a clear reaction from e-nose sensors, with a slightly higher signal strength variation among replicates (Figure 1e–h), but still at the level comparable to the signal intensity for PDA medium (Figure 1e–g). The overall odor was thus moderate, except for *F. poae* which appeared to be one of the most odoriferous tested strains (Figure 1h), with an immediate and powerful rise in sensor signal.

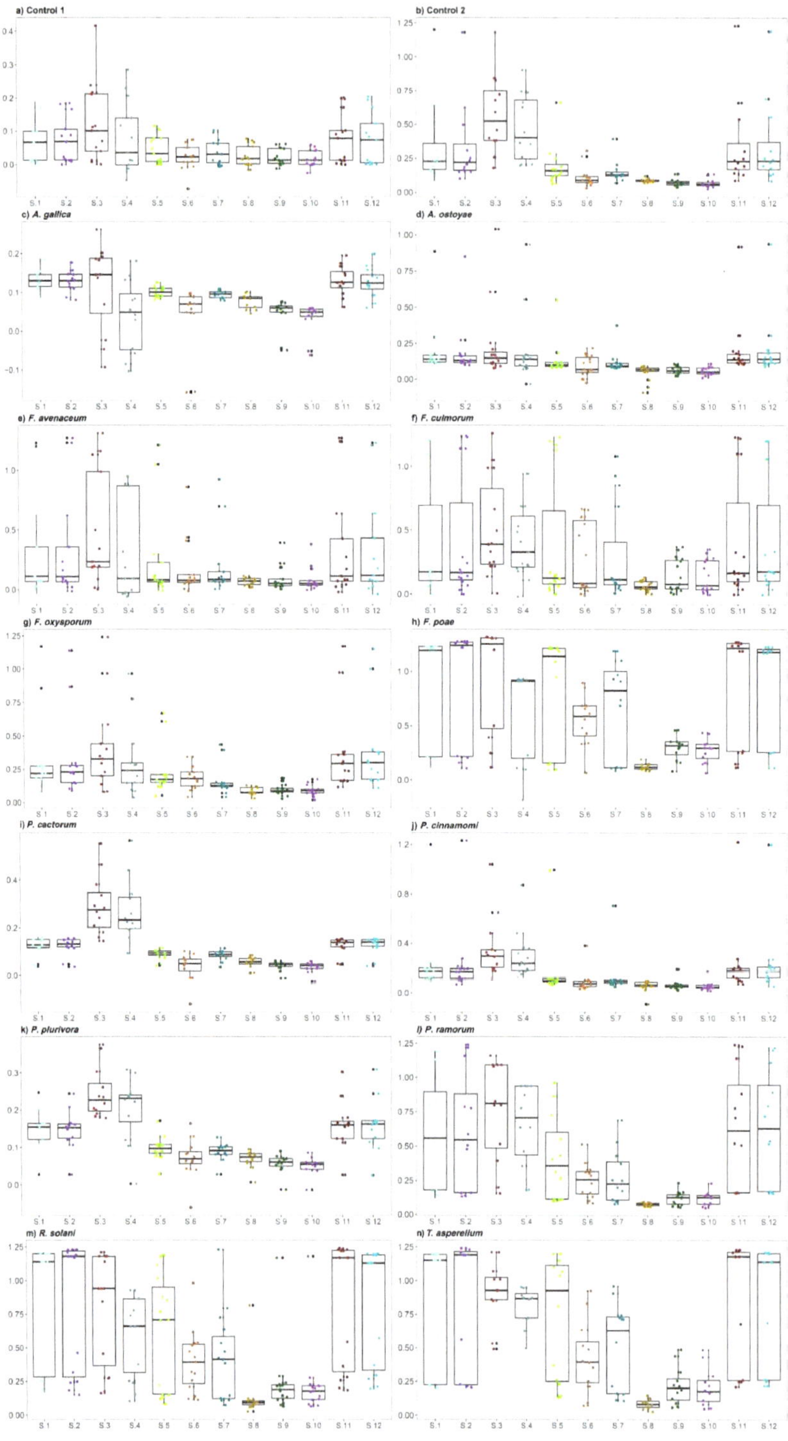
a) Control 1
b) Control 2
c) A. gallica
d) A. ostoyae
e) F. avenaceum
f) F. culmorum
g) F. oxysporum
h) F. poae
i) P. cactorum
j) P. cinnamomi
k) P. plurivora
l) P. ramorum
m) R. solani
n) T. asperellum

Figure 1. Distribution of the overall averaged values of twelve differential sensor original signals for the respective 14 treatments; e.g., (**a**) control 1 (empty flask); (**b**) control 2 (PDA medium); (**c**) *A. gallica;* (**d**) *A. ostoyae;* (**e**) *F. avenaceum;* (**f**) *F. culmorum;* (**g**) *F. oxysporum;* (**h**) *F. poae;* (**i**) *P. cactorum;* (**j**) *P. cinnamomi;* (**k**) *P. plurivora;* (**l**) *P. ramorum;* (**m**) *R. solani;* (**n**) *T. asperellum.*

Phytophthora species generated an overall feeble reaction from the e-nose (Figure 1i–l), except for *P. ramorum* which produced a strong odor (Figure 1l), only slightly weaker than *F. poae*, *R. solani* and *T. asperellum*. Signal strength usually increased quickly after the e-nose air inlet was introduced into the *P. ramorum* atmosphere. Like *F. poae*, *R. solani* and *T. asperellum* emitted a very strong odor highlighted during e-nose testing, with the introduction of the air inlet into a sample atmosphere which made the sensor signals instantly and strongly increase (Figure 1m,n). To sum up, the strains of *F. avenaceum*, *F. culmorum* and *F. oxysporum* along with *P. ramorum* generated less intense average reactions from the sensing device (Figure 1e–g,l), but these were however still stronger than for *Armillaria* and remaining *Phytophthora* strains whose odors appeared to be very weak (Figure 1c,d,i–k).

No clear pattern over time could be found among the tested species, probably because of the low level of repeated sampling (usually 3) per week of study which did not allow adequate assessment of temporal variation in emissions. The PDA medium was more constant between repeated observations, although the signal strength increased over time, possibly as the medium dried out. Sensors 1, 2, 3, 4, 11 and 12 overall systematically gave a stronger response to volatiles, while sensors 5, 6, 7 and even more 8, 9, 10 yielded weaker signals (Figure 1).

Based on correlation analyses between eigenvalues, we found that the variance between signals mostly depended on the first principal component (PC1) and only slightly on the second principal component (PC2) (93.87% and 03.26%, respectively; Figure 2, Table S2). This enabled us to deduce using the PCA plot that the odors generated by *P. ramorum*, *F. poae*, *T. asperellum* and *R. solani* were readily distinguishable due to the location of the microorganisms on the biplot and grouping pattern confirmed by hierarchical clustering analysis (Figure S1, Table S2).

2.2. Determination of VOC Emitted by Fungi and Oomycetes

In our study, two SPME fibers were compared (PDMS vs. PDMS/Carboxen). The results revealed that the PDMS fiber system was less applicable for volatiles released by fungi and oomycetes, because during GC/MS analyses a number of silicone-derived contaminants were observed. The latter could be explained by the chemical destruction of a DMS surface coating by biogenic volatiles released. Hence, only results obtained from the sampling of volatiles with a DMS/Carboxen fiber are presented.

The molecular identification was carried out using threefold approach: matching with GC/EI MS libraries, both commercial (NIST + Willey) and in-house built, comparison of Kovats indices (RI) (measured + reported elsewhere) and authentic standards. The last approach was applied only when standards were available, and thus was confined to common organics, such as hexanal, heptanal or benzaldehyde. However, a few unknowns from fungal emissions, e.g., 1,3,4,5,6,7-hexahydro-2,5,5-trimethyl-2*H*-2,4a-ethanonaphthalene, were tentatively elucidated using the two first approaches, because of the lack of authentic standards. The identification with standards (authentic not internal) is the gold standard for the thorough identification.

To identify components, both mass spectral data and the calculated retention indices were used. Mass spectrometric identification not confirmed by the retention index was considered as tentative. The background emissions from controls, empty containers and PDA medium were subtracted. Some of the volatiles found through GC/MS, such as aliphatic hydrocarbons, alcohols, aldehydes, esters and aromatic derivatives were present in samples from all treatments (including controls), but all the investigated fungi species (except *R. solani*) produced sesquiterpenes of variable quantities. However, the tested oomycete strains generally produced fewer detectable substances, even less than non-inoculated controls, including those compounds mentioned above (Table 1).

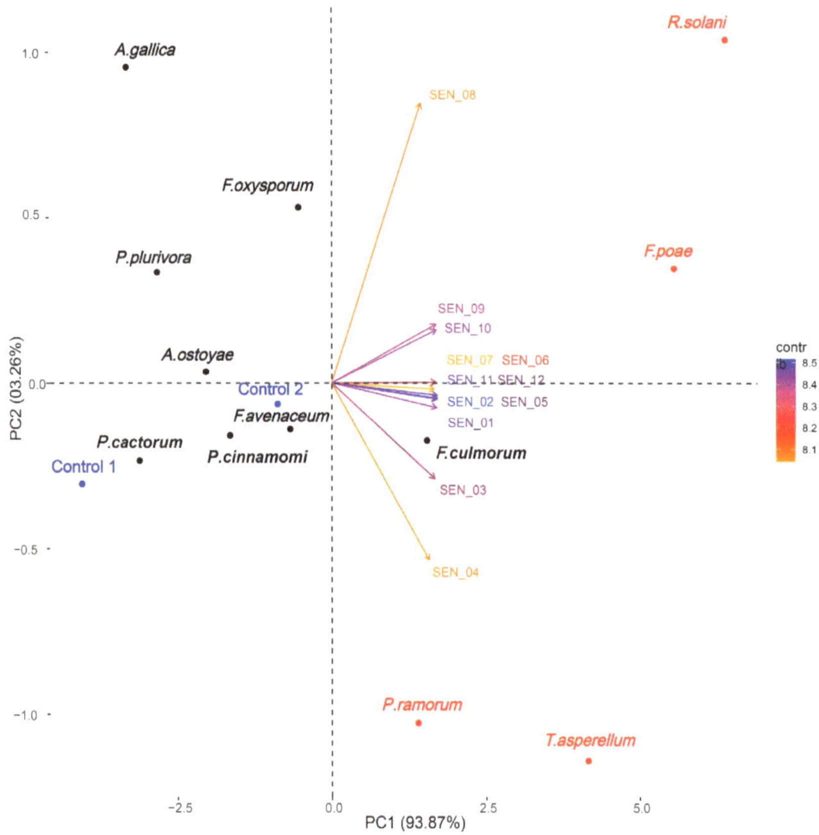

Figure 2. Variability of twelve differential sensor signals assigned to the twelve investigated microorganisms. Vectors were colored according to their contribution to total variance: orange—low, blue—high. Blue dots represent controls: 1—empty flask, and 2—PDA medium; red dots—microorganisms with the most pronounced eigenvalues, black dots—microorganisms with less pronounced eigenvalues (cf. Table S2, Figure S1).

Table 1. The VOC emission pattern detected using a SPME-GC/MS method in the headspace atmosphere of samples tested at BUT. Similarity indices (SI) were obtained by comparing a profile of the EI mass spectrum of a given VOC compound with the reference one in NIST Mass Spectral Database library; these values represent the confidence in the molecular identification.

t_R [1] [min]	SI [2] %	RI_{Exp} [3]	RI_{Lit} [4]	Ref.	Peak Area per Treatment	Name of Compound [5]
				(1) Empty Container		
7.19	65	902	901	[29]	17,674,234	Heptanal [++a]
10.05	80	1003	998	[29]	40,444,566	Octanal [++a]
10.82	28	1029	1027	[30]	21,189,032	2-Ethyl-hexanol [a]
13.05	79	1104	1100	[29]	76,538,207	Nonanal [++a]
15.98	53	1206	1201	[29]	14,513,924	Decanal [++a]
18.06	–	1281	–	–	14,724,741	Unknown compound
18.77	–	1307	–	–	13,630,063	Unknown compound
21.21	32	1400	1400	[29]	6,427,563	Tetradecane [++a]
21.44	58	1409	1408	[29]	17,444,416	Dodecanal [++a]

Table 1. *Cont.*

t_R [1] [min]	SI [2] %	RI_{Exp} [3]	RI_{Lit} [4]	Ref.	Peak Area per Treatment	Name of Compound [5]
24.07	78	1516	1514	[31]	47,388,559	Butylated Hydroxytoluene [++a]
					(2) PDA Medium	
4.77	50	804	801	[29]	19,939,473	Hexanal [+a]
5.48	57	834	828	[29]	11,223,384	Furfural [+a]
7.18	31	902	901	[29]	39,366,443	Heptanal [++a]
8.80	72	959	952	[29]	60,063,839	Benzaldehyde [a]
9.74	87	992	984	[29]	31,027,584	2-Pentyl-furan [b]
10.05	74	1003	998	[29]	11,751,703	Octanal [++a]
10.81	44	1029	1027	[30]	32,462,061	2-Ethyl-hexanol [a]
11.23	73	1043	1036	[29]	73,539,705	Benzeneacetaldehyde [++a]
12.96	21	1101	1100	[29]	12,884,326	Undecane [+a]
13.04	80	1104	1100	[29]	41,840,277	Nonanal [++a]
15.48	–	1189	–	–	4,687,953	Unknown compound
15.83	30	1201	1200	[29]	4,893,511	Dodecane [+a]
15.97	50	1206	1201	[29]	10,723,873	Decanal [++a]
18.13	–	1284	–	–	5,876,115	Unknown compound
18.77	–	1307	–	–	4,815,975	Unknown compound
21.43	44	1409	1408	[29]	11,027,672	Dodecanal [++a]
24.06	74	1515	1514	[31]	14,730,122	Butylated Hydroxytoluene [++a]
					(3) *A. gallica*	
8.80	47	959	n/a [6]	–	300,374,347	(2*E*)-4,4-Dimethyl-2-pentenal [c]
11.73	53	1060	1049	[29]	62,813,700	(*E*)-2-Octen-1-al [b]
20.87	10	1387	n/a	-	611,543,126	1,3,4,5,6,7-Hexahydro-2,5,5-trimethyl-2H-2,4a-ethanonaphthalene [c]
23.89	26	1506	1506	[29]	74,690,325	(*Z*)-α-Bisabolene [+a]
					(4) *A. ostoyae*	
8.80	49	959	n/a	-	87,776,838	(2*E*)-4,4-Dimethyl-2-pentenal [c]
11.74	35	1060	1049	[29]	20,166,139	(*E*)-2-Octen-1-al [b]
18.49	21	1297	1293	[29]	8,070,650	2-Undecanone [a]
20.83	43	1386	n/a	-	106,170,662	1,3,4,5,6,7-Hexahydro-2,5,5-trimethyl-2H-2,4a-ethanonaphthalene [c]
22.49	59	1451	1440	[29]	43,908,793	β-Barbatene [a]
24.42	34	1531	1522	[29]	24,134,794	δ-Cadinene [+a]
					(5) *F. avenaceum*	
3.28	76	712	735	[32]	2,653,753	2,4-Dimethylfuran [+b]
9.13	23	971	970	[33]	3,974,721	Mesitilene [+a]
10.91	76	1032	1029	[29]	9,540,610	β-Phellandrene [a]
21.59	35	1415	1407	[29]	11,961,247	Longifolene [a]
24.46	73	1532	1519	[29]	3,401,805	β-Bazzanene [+a]
					(6) *F. culmorum*	
1.62	77	<500	448	[34]	163,586,144	Ethanol [a]
2.25	96	573	606	[29]	45,234,113	Ethyl acetate [+a]
3.55	41	728	731	[29]	6,085,804	Isoamyl alcohol [+a]
3.61	56	732	724	[29]	3,371,660	2-Methyl-butanol [a]
10.91	18	1032	1029	[29]	177,256,946	β-Phellandrene [a]
21.80	43	1424	1408	[29]	20,993,355	Acora-3,7(14)-diene [+b]
22.23	man. [7]	1441	1432	[29]	138,760,275	β-Copaene [+a]
					(7) *F. oxysporum*	
6.02	78	854	864	[35]	845,609	2-Furanmethanol [+b]
7.96	59	930	927	[36]	2,621,568	Hexyl formate [+a]
8.86	79	961	952	[29]	49,523,292	Benzaldehyde [+a]
9.65	68	989	979	[29]	5,057,096	3-Octanone [a]
10.27	71	1010	960	[29]	2,393,859	Isoamyl propionate [+a]
10.90	54	1032	n/a	–	63,835,818	Heptyl formate [+a]
21.98	33	1431	1419	[29]	4,969,490	β-Cedrene [a]
22.49	90	1451	n/a	–	23,065,220	3-Chloro-4-methoxy-benzaldehyde [c]
22.77	20	1463	n/a	–	8,382,097	2,4-Dichloro-3-methoxy-1-benzene carbonyl chloride [+c]

Table 1. *Cont.*

t_R [1] [min]	SI [2] %	RI_{Exp} [3]	RI_{Lit} [4]	Ref.	Peak Area per Treatment	Name of Compound [5]
23.34	93	1486	864	[35]	2,740,362	3,4-Dimethoxy-benzaldehyde [+a]
					(8) *F. poae*	
9.43	81	981	974	[29]	93,704,125	1-Octen-3-ol [a]
9.79	36	994	988	[29]	166,758,547	Myrcene [+a]
13.39	59	1116	1122	[37]	100,643,564	1-Ethyl-4-methoxy-benzene [b]
16.68	41	1231	1223	[29]	36,663,640	Citronellol [+a]
20.10	28	1358	1350	[30]	106,101,000	α-Longipinene [+a]
20.83	16	1379	1371	[29]	108,987,407	Longicyclene [+a]
21.43	man.	1409	1400	[29]	248,606,376	β-Longipinene [+b]
21.65	37	1418	1407	[29]	1,390,824,920	Longifolene [a]
21.81	–	1427	–	–	482,683,621	Unknown sesquiterpene
21.91	22	1428	1419	[29]	195,664,192	β-Ylangene [+a]
22.31	–	1444	–	–	954,381,082	Unknown sesquiterpene
22.55	55	1454	1440	[29]	129,567,010	β-Barbatene [+a]
22.68	16	1459	1449	[29]	117,848,857	α-Himachalene [+a]
22.75	50	1462	1454	[29]	143,907,373	(*E*)-β-Farnesene [+a]
23.02	20	1473	1466	[29]	116,555,130	α-Acoradiene [+a]
23.38	16	1487	1481	[29]	134,783,119	γ-Himachalene [+a]
23.98	43	1512	n/a	–	44,779,678	8-Isopropenyl-1,5-dimethyl-1, 5-cyclodecadiene [+c]
24.05	17	1515	1505	[29]	51,262,732	β-Bisabolene [+a]
25.03	–	1557	–	–	42,987,721	Unknown sesquiterpene
26.31	60	1612	1599	[29]	1,576,920,976	Longiborneol [+a]
					(9) *P. cactorum*	
1.71	86	<500	500	[34]	5,827,788	Acetone [a]
3.75	97	740	744	[38]	4,459,515	Dimethyl disulfide [b]
4.12	–	763	–	–	4,321,348	Unknown compound
6.35	60	868	863	[29]	25,809,441	1-Hexanol [+a]
9.15	58	971	959	[29]	5,111,166	1-Heptanol [+a]
9.42	75	981	974	[29]	19,848,783	1-Octen-3-ol [a]
9.65	76	989	979	[29]	12,627,315	3-Octanone [a]
9.81	88	995	984	[29]	4,284,441	2-Pentyl-furan[a]
12.05	22	1071	1060	[29]	2,508,277	2-Octen-1-ol [+a]
12.13	19	1073	1063	[29]	4,044,266	1-Octanol [+a]
					(10) *P. cinnamomi*	
1.61	97	<500	448	[31]	9,483,339	Ethanol [a]
9.43	56	981	974	[29]	2,215,655	1-Octen-3-ol [a]
10.88	47	1031	1027	[30]	5,251,305	2-Ethyl-hexanol [a]
18.13	85	1284	1282	[39]	1,001,882	4-Ethyl-2-methoxy-phenol [+b]
					(11) *P. plurivora*	
1.71	84	<500	500	[34]	4,481,938	Acetone [a]
3.20	64	707	711	[40]	2,268,583	Acetoin [+a]
6.37	36	869	863	[29]	1,803,069	Hexanol [+a]
7.51	76	914	933	[41]	3,627,373	4-Hydroxy-butanoic acid [+b]
8.14	16	936	932	[29]	1,684,400	α-Pinene [+a]
10.35	14	1013	1008	[29]	1,103,050	Δ-3-Carene [+a]
					(12) *P. ramorum*	
1.16	98	<500	448	[34]	20,885,799	Ethanol [a]
3.55	74	728	730	[42]	80,424,948	3-Methyl-butanol [a]
3.61	74	732	724	[29]	26,123,097	2-Methyl-butanol [a]
9.43	72	981	974	[29]	8,370,575	1-Octen-3-ol [a]
9.65	70	989	979	[29]	11,144,929	3-Octanone [a]
13.37	87	1116	1106	[29]	48,377,677	2-Phenylethanol [+a]
					(13) *R. solani*	
1.71	87	<500	500	[34]	3,565,392	Acetone [a]
9.42	74	981	974	[29]	24,589,405	1-Octen-3-ol [a]
9.65	62	989	979	[29]	3,438,485	3-Octanone [a]
9.90	75	998	988	[29]	6,372,403	3-Octanol [+a]
15.56	–	1191	-	-	4,474,717	Unknown compound

Table 1. *Cont.*

t_R [1] [min]	SI [2] %	RI_{Exp} [3]	RI_{Lit} [4]	Ref.	Peak Area per Treatment	Name of Compound [5]
			(14) *T. asperellum*			
1.71	82	<500	500	[34]	3,920,210	Acetone[a]
3.56	62	729	730	[41]	7,870,509	3-Methyl-butanol[a]
3.62	45	733	724	[29]	4,424,960	2-Methyl-butanol[a]
3.76	98	741	744	[38]	11,829,339	Dimethyl disulfide[b]
7.26	–	905	–	–	153,721,663	Unknown compound
13.01	–	1103	–	–	20,319,451	Unknown compound
13.38	85	1116	1106	[29]	37,177,146	2-Phenyethanol [+a]
20.08	–	1357	–	–	70,272,034	Unknown sesquiterpene
20.91	44	1389	1380	[29]	229,806,310	Daucene [+b]
21.81	man.	1424	1412	[29]	25,748,186	2-epi-β-Funebrene [+b]
21.98	50	1431	1419	[29]	86,596,372	β-Cedrene[a]
23.93	28	1510	1500	[29]	118,470,969	Isodaucene [+b]
24.27	21	1524	1513	[29]	52,189,482	γ-Cadinene [+a]
24.63	40	1540	1530	[29]	194,592,295	Dauca-4(11)-8-diene [+a]
25.49	–	1576	–	–	8,894,971	Unknown sesquiterpene
27.81	–	1679	–	–	15,501,124	Unknown sesquiterpene
28.31	27	1701	n/a	–	15,437,838	1-Isopropyl-4,8-dimethylspiro [4.5]dec-8-en-7-one [+c]
30.26	–	1793	–	–	21,104,266	Unknown sesquiterpene

[1] retention time; [2] similarity index; [3] Kovats retention index calculated from experimental data; [4] Kovats retention index from literature; [5] compound exclusively present there (+) or present only in controls (++); [6] data not available; [7] manual matching; [a] compound identified using retention time of authentic standard, matching with MS library and comparison with reported KI; [b] compound identified by matching with MS library and comparison with reported KI; [c] compound identified by matching with MS library.

2.2.1. Control Treatments

The empty container control atmosphere appeared to be surprisingly rich in ten volatiles (Table 1, treatment 1), whose origin was probably linked to the silicone ring adjoined to the underside of the lid by the manufacturer to allow hermetic closure of the flask. This was consistent with the e-nose sensing a very weak, but still detectable, signal for this control treatment (Figure 1a). Even though the e-nose sensors showed more signals from PDA medium control samples than the empty container control ones (Figure 1b), respective molecule abundances of both chromatograms belonged to the same order of magnitude (10^7).

2.2.2. *Armillaria* Species

The SPME-GC/MS data indicated a similar emission pattern in two *Armillaria* treatments which is in accord with the results obtained from e-nose analysis (Figure 1c,d). Two *Armillaria* species shared three similar VOCs i.e., (2*E*)-4,4-dimethyl-2-pentenal, (*E*)-2-octen-1-al, and 1,3,4,5,6,7-hexahydro-2,5,5-trimethyl-2*H*-2,4a-ethanonaphthalene (Table 1, treatments 3 and 4). Some sesquiterpenes, i.e., δ-cadinene, (*Z*)-α-bisabolene, and 1,3,4,5,6,7-hexahydro-2,5,5,-trimethyl-2H-2,4a-ethanonaphthalene were present only in *Armillaria* atmosphere suggesting an inherent emission by the fungi (Table 1, treatments 3 and 4). Another sesquiterpene, β-barbatene, was detected in both *A. ostoyae* (a basidiomycete) and *F. poae* (an ascomycete) implying common VOC emission (Table 1, treatment 4 and 8) between two different fungal divisions.

2.2.3. *Fusarium* Species

The three *Fusarium* species showed an interesting pattern of VOC emission. Longifolene was detected among the most common volatiles in only *F. avenaceum* and *F. poae* (Table 1, treatments 5 and 8). Subsequently, production of β-phellandrene by *F. avenaceum* and *F. culmorum* was observed. α-pinene and Δ-3-Carene were detected in *P. plurivora* (Table 1, treatment 11). These compounds were not found in the non-inoculated media controls. Finally, *F. culmorum* contained ethanol, in common with *P. cinnamomi* and *P. ramorum* (Table 1, treatments 6, 10 and 12). Consistent with e-nose analysis results, the *F. poae* GC/MS data showed a higher molecular abundance magnitude (10^8) (and remarkable compound diversity) than other *Fusarium* species (usually at 10^7). The compound 1-octen-3-ol was found in *F. poae*, *P. cactorum*, *P. cinnamomi* and *R. solani* (Table 1, treatments 8, 9 and 13), supporting the idea of common molecules emitted across varied taxa.

2.2.4. *Phytophthora* Species

Acetone was shared between two oomycetes *P. cactorum* and *P. plurivora* while ethanol was shared between *P. cinnamomi* and *P. ramorum* (Table 1, treatments 9 and 11). Other volatiles were present in different species of oomycetes and fungi, e.g., 3-octanone detected in *P. cactorum*, *P. ramorum* and *R. solani* (Table 1, treatment 9 and 13), and 1-octen-3-ol found in the *P. cactorum*, P. ramorum, *F. poae* and *R. solani* (Table 1, treatments 8, 9 and 13). The α-pinene and Δ-3-Carene emitted by *P. plurivora* were the only terpene compounds found in a *Phytophthora* species during this research (Table 1, treatment 11). Abundance of *P. ramorum* compounds was higher than for other *Phytophthora* sp., in accordance with e-nose analysis results. Among the studied Oomycetes, only *P. ramorum* showed the highest level and similar order of overall VOC emission as *Armillaria* treatments (Table 1, treatments 3, 4 and 13). SPME-GC/MS analysis revealed presence of 2-phenylethanol in both *P. ramorum* and *T. asperellum* (Table 1, treatments 12 and 14).

2.2.5. *Rhizoctonia* and *Trichoderma* Species

The *R. solani* chromatogram differed from the signal indicated by e-nose sensors (Figure 1m). Higher amounts of 1-octen-3-ol were found in *R. solani* than in *F. poae* and *P. cactorum* (Table 1, treatments 8, 9 and 13). Consistent with e-nose results (Figure 1n), SPME-GC/MS analysis yielded for *T. asperellum* high richness of VOCs emitted with relatively high abundance magnitude (almost 10^8).

3. Discussion

The sustainability (durability) and biodiversity of forest stands in Poland depends not only sustainable forest management but also on integrated pest management. Foresters cannot afford to allow pests to escape from nurseries to forest plantations together with the sold seedlings. Visual inspection often fails (asymptomatic plants), so forest managers are looking for new tools to support their work. In particular, pesticides applied in nurseries mask diseases, which in suitable conditions after outplanting will start to develop e.g., *Phytophthora* in wet forest sites. New devices should allow a quick detection of potential pathogens, particularly for emerging diseases. Among them are the foreign, invasive oomycetes (*Phytophthora, Pythium*) and other soil borne pathogens in the genera *Fusarium, Rhizoctonia* or *Cylindrocarpon*. All those organisms growing in pure cultures have distinctive strong odors because of volatile secondary metabolite production, but no instrument has been used in practice for routine detection of pathogen by forest managers. In the present research, we focused on assessing the possibility of designing an electronic nose to recognize genera or species of some pathogenic organisms often found in forest nurseries. This e-nose apparatus was made to recognize the specific electronic footprints produced by a VOC mixture interacting with a set of sensors. The change in physicochemical properties of the sensors induced by interaction with VOCs is transduced in a characteristic electrical signal, which allows description of the compounds without having to isolate the different components of the mixture [43,44]. The instrument used in this study has been previously used for recognition of very diverse volatile emission sources such as gasoline, coffee, tobacco and even explosives [45–49]. Certainly, our laboratory studies on volatile emission by microorganisms were carried out under optimal growth conditions which may differ from natural environments [50]. Moreover, the volatile mixture produced by a mixed colony of microorganisms is very likely to differ from what a pure culture may release [2].

A comparison of substances found in our samples (and an examination of where else it has been found previously) allowed us to screen substances (that are listed in Table 1), and to point out those which were specific to particular organisms. A major goal was to elucidate compounds that were

21. Gardner, J.W.; Bartlett, P.N. (Eds.) *Sensors and Sensory Systems for An Electronic Nose*; Kluwer Academic: Boston, MA, USA; Dordrecht, The Netherlands; New York, NY, USA, 1992. [CrossRef]
22. Brudzewski, K.; Osowski, S., Ulaczyk, J. Differential electronic nose of two chemo sensor arrays for odor discrimination. *Sens. Actuators B Chem.* **2010**, *145*, 24–249. [CrossRef]
23. Persaud, K.; Dodd, G. Analysis of discrimination mechanisms in the mammalian olfactory system using a model nose. *Nature* **1982**, *299*, 352–355. [CrossRef] [PubMed]
24. Gardner, J.W. Pattern recognition in the Warwick Electronic Nose. In *8th Int Congress of European Chemoreception Research Organisation*; University of Warwick: Warwick, UK, 1987.
25. Di Pietrantonio, F.; Benetti, M.; Cannatà, D.; Verona, E.; Palla-Papavlu, A.; Fernández-Pradas, J.M.; Serra, P.; Staiano, M.; Varriale, A.; D'Auria, S. A surface acoustic wave bio-electronic nose for detection of volatile odorant molecules. *Biosens. Bioelectron.* **2015**, *67*, 516–523. [CrossRef] [PubMed]
26. Wilson, A.D. Diverse applications of electronic-nose technologies in agriculture and forestry. *Sensors* **2013**, *13*, 2295–2348. [CrossRef] [PubMed]
27. Okorski, A.; Pszczółkowska, A.; Gorzkowska, A.; Okorska, S.; Głuszek, P. Fungi associated with conifer seedlings grown in forest nurseries under different systems. *EEMJ* **2019**, *18*, 1509–1517. [CrossRef]
28. Jung, T.; Orlikowski, L.; Henricot, B.; Abad-Campos, P.; Aday, A.G.; Aguín Casal, O.; Bakonyi, J.; Cacciola, S.O.; Cech, T.; Chavarriaga, D.; et al. Widespread *Phytophthora* infestations in European nurseries put forest, semi-natural and horticultural ecosystems at high risk of *Phytophthora* diseases. *For. Pathol.* **2016**, *46*, 134–163. [CrossRef]
29. Adams, R. *Identification of Essential Oil Components by Gas Chromatography/Mass Spectrometry*, 4th ed.; Allured Publishing Corporation: Carol Stream, IL, USA, 2017.
30. Miyazawa, M.; Marumoto, S.; Kobayashi, T.; Yoshida, S.; Utsumi, Y. Determination of Characteristic Components in Essential Oils from Wisteria Braphybotrys Using Gas Chromatography—Olfactometry Incremental Dilution Technique, *Rec. Nat. Prod.* **2011**, *5*, 221–227.
31. Adams, R.P.; Morris, J.A.; Pandey, R.N.; Schwarzbach, A.E. Cryptic speciation between Juniperus deltoides and Juniperus oxycedrus (Cupressaceae) in the Mediterranean. *Biochem. Syst. Ecol.* **2005**, *33*, 771–787. [CrossRef]
32. Elmore, J.S.; Mottram, D.S.; Enser, M.; Wood, J.D. Effect of the polyunsaturated fatty acid composition of beef muscle on the profile of aroma volatiles. *J. Agric. Food Chem.* **1999**, *47*, 1619–1625. [CrossRef]
33. Wang, Z.; Fingas, M.; Li, K. Fractionation of a light crude oil and identification and quantitation of aliphatic, aromatic, and biomarker compounds by GC-FID and GC-MS, Part II. *J. Chromatogr. Sci.* **1994**, *32*, 367–382. [CrossRef]
34. Kotowska, U.; Zalikowski, M.; Isidorov, V.A. HS-SPME/GC-MS analysis of volatile and semi-volatile organic compounds emitted from municipal sewage sludge. *Environ. Monit. Asses.* **2012**, *184*, 2893–2907. [CrossRef]
35. Bonaiti, C.; Irlinger, F.; Spinnler, H.E.; Engel, E. An iterative sensory procedure to select odor-active associations in complex consortia of microorganisms: Application to the construction of a cheese model. *J. Dairy Sci.* **2005**, *88*, 1671–1684. [CrossRef]
36. Isidorov, V.A.; Zenkevich, I.G.; Krajewska, U.; Dubis, E.N.; Jaroszynska, J.; Bal, K. Gas chromatographic analysis of essential oils with preliminary partition of components. *Phytochem. Anal.* **2001**, *12*, 87–90. [CrossRef] [PubMed]
37. Larsen, T.O.; Frisvad, J.C. Characterization of volatile metabolites from 47 Penicillium taxa. *Mycol. Res.* **1995**, *99*, 1153–1166. [CrossRef]
38. Goodner, K.L. Practical retention index models of OV-101, DB-1, DB-5, and DB-Wax for flavor and fragrance compounds. *LWT* **2008**, *41*, 951–958. [CrossRef]
39. Steinhaus, P.; Schieberle, P. Characterization of the key aroma compounds in soy sauce using approaches of molecular sensory science. *J. Agric. Food Chem.* **2007**, *55*, 6262–6269. [CrossRef] [PubMed]
40. Jordan, M.J.; Margaria, C.A.; Shaw, P.E.; Goodner, K.L. Aroma active components in aqueous Kiwi fruit essence and Kiwi fruit puree by GC-MS and multidimensional GC/GC-O. *J. Agric. Food Chem.* **2002**, *50*, 5386–5390. [CrossRef]
41. Liu, Y.; Xu, X.-L.; Zhou, G.-H. Comparative study of volatile compounds in traditional Chinese Nanjing marinated duck by different extraction techniques. *Int. J. Food Sci. Technol.* **2007**, *42*, 543–550. [CrossRef]

42. Jalali-Heravi, M.; Zekavat, B.; Sereshti, H. Characterization of essential oil components of Iranian geranium oil using gas chromatography-mass spectrometry combined with chemometric resolution techniques. *J. Chromatogr. A* **2006**, *1114*, 154–163. [CrossRef]
43. Wilson, A.D.; Baietto, M. Applications and advances in electronic-nose technologies. *Sensors* **2009**, *9*, 5099–5148. [CrossRef]
44. Wilson, A.D.; Baietto, M. Advances in Electronic-Nose Technologies Developed for Biomedical Applications. *Sensors* **2011**, *11*, 1105–1176. [CrossRef]
45. Osowski, S.; Linh, T.H.; Brudzewski, K. Neuro-fuzzy TSK network for calibration of semiconductor sensor array for gas measurements. *IEEE Trans. Instrum. Meas.* **2004**, *53*, 630–637. [CrossRef]
46. Brudzewski, K.; Osowski, S.; Markiewicz, T.; Ulaczyk, J. Classification of gasoline with supplement of bio-products by means of an electronic nose and SVM neural network. *Sens. Actuators B Chem.* **2006**, *113*, 135–141. [CrossRef]
47. Brudzewski, K.; Osowski, S.; Pawlowski, W. Metal oxide sensor arrays for detection of explosives at sub-parts-per million concentration levels by the differential electronic nose. *Sens. Actuators B Chem.* **2012**, *161*, 528–533. [CrossRef]
48. Brudzewski, K.; Osowski, S.; Golembiecka, A. Differential electronic nose and support vector machine for fast recognition of tobacco. *Expert Syst. Appl.* **2012**, *39*, 9886–9891. [CrossRef]
49. Brudzewski, K.; Osowski, S.; Dwulit, A. Recognition of coffee using differential electronic nose. *IEEE Trans. Instrum. Meas.* **2012**, *61*, 1803–1810. [CrossRef]
50. Weise, T.; Kai, M.; Gummesson, A.; Troeger, A.; Von Reuß, S.; Piepenborn, S.; Kosterka, F.; Sklorz, M.; Zimmermann, R.; Francke, W.; et al. Volatile organic compounds produced by the phytopathogenic bacterium *Xanthomonas campestris* pv. *vesicatoria 85-10*. *Beilstein J. Org. Chem.* **2012**, *8*, 579–596. [CrossRef]
51. Pan, L.; Zhang, W.; Zhu, N.; Mao, S.; Tu, K. Early detection and classification of pathogenic fungal disease in post-harvest strawberry fruit by electronic nose and gas chromatography–mass spectrometry. *Food Res. Int.* **2014**, *62*, 162–168. [CrossRef]
52. Biondi, E.; Blasioli, S.; Galeone, A.; Spinelli, F.; Cellini, A.; Lucchese, C.; Braschi, I. Detection of potato brown rot and ring rot by electronic nose: From laboratory to real scale. *Talanta* **2014**, *129*, 422–430. [CrossRef]
53. Rettinger, K.; Burschka, C.; Scheeben, P.; Fuchs, H.; Mosandl, A. Chiral 2-alkylbranched acids, esters and alcohols. Preparation and stereospecific flavour evaluation. *Tetrahedron Asymmetry* **1991**, *2*, 965–968. [CrossRef]
54. Bahrmann, H.; Hahn, H.-D.; Mayer, D.; Frey, G.D. 2-Ethylhexanol. In *Ullmann's Encyclopedia of Industrial Chemistry*; Wiley-VCH Verlag GmbH & Co. KGaA: Weinheim, Germany, 2013; ISBN 978-3-527-30673-2.
55. Engels, H.W.; Weidenhaupt, H.J.; Abele, M.; Pieroth, M.; Hofmann, W. Rubber, 4. Chemicals and additives. In *Ullmann's Encyclopedia of Industrial Chemistry*; Wiley-VCH Verlag GmbH & Co. KGaA: Weinheim, Germany, 2000. [CrossRef]
56. Dickschat, J.S.; Brock, N.L.; Citron, C.A.; Tudzynski, B. Biosynthesis of sesquiterpenes by the fungus *Fusarium verticillioides*. *ChemBioChem* **2011**, *12*, 2088–2095. [CrossRef]
57. Dickschat, J.S. Fungal volatiles–a survey from edible mushrooms to moulds. *Nat. Prod. Rep.* **2017**, *34*, 310–328. [CrossRef] [PubMed]
58. Citron, C.A.; Gleitzmann, J.; Laurenzano, G.; Pukall, R.; Dickschat, J.S. Terpenoids are widespread in actinomycetes: A correlation of secondary metabolism and genome data. *ChemBioChem* **2012**, *13*, 202–214. [CrossRef] [PubMed]
59. Yamada, Y.; Kuzuyama, T.; Komatsu, M.; Shin-ya, K.; Omura, S.; Cane, D.E.; Ikeda, H. Terpene synthases are widely distributed in bacteria. *Proc. Nat. Acad. Sci. USA* **2015**, *112*, 857–862. [CrossRef] [PubMed]
60. Yamagiwa, Y.; Inagaki, Y.; Ichinose, Y.; Toyoda, K.; Hyakumachi, M.; Shiraishi, T. Talaromyces wortmannii FS2 emits β-caryphyllene, which promotes plant growth and induces resistance. *J. Gen. Plant Pathol.* **2011**, *77*, 336–341. [CrossRef]
61. Matsui, K.; Sasahara, S.; Akakabe, Y.; Kajiwara, T. Linoleic acid 10-hydroperoxide as an intermediate during formation of 1-octen-3-ol from linoleic acid in *Lentinus decadetes*. *Biosci. Biotechnol. Biochem.* **2003**, *67*, 2280–2282. [CrossRef]
62. Suda, M. Short-step syntheses of (±)-bazzanene and (±) trichodiene. *Tetrahedron Lett.* **1982**, *23*, 427–428. [CrossRef]

63. Zamir, L.O.; Gauthier, M.J.; Devor, K.A.; Nadeau, Y.; Sauriol, F. Trichodiene is a precursor to trichothecenes. *J. Chem. Soc. Chem. Commun.* **1989**, *9*, 598–600. [CrossRef]
64. Lin, H.; Phelan, P.L. Comparison of volatiles from beetle-transmitted *Ceratocystis fagacearum* and four non-insect-dependent fungi. *J. Chem. Ecol.* **1992**, *18*, 1623–1632. [CrossRef]
65. Wang, T.; Rabe, P.; Citron, C.A.; Dickschat, J.S. Halogenated volatiles from the fungus *Geniculosporium* and the actinomycete *Streptomyces chartreusis*. *Beilstein J. Org. Chem.* **2013**, *9*, 2767–2777. [CrossRef]
66. Schalchli, H.; Hormazábal, E.; Becerra, J.; Briceño, G.; Hernández, V.; Rubilar, O.; Diez, M.C. Volatiles from white-rot fungi for controlling plant pathogenic fungi. *Chem. Ecol.* **2015**, *31*, 754–763. [CrossRef]
67. Spinnler, H.E.; De Jong, E.; Mauvais, G.; Semon, E.; Le Quéré, J.L. Production of halogenated compounds by *Bjerkandera adusta*. *Appl. Microbiol. Biotechnol.* **1994**, *42*, 212–221. [CrossRef]
68. Zawirska-Wojtasiak, R. Optical purity of (R)-(−)-1-octen-3-ol in the aroma of various species of edible mushrooms. *Food Chem.* **2004**, *86*, 113–118. [CrossRef]
69. Thakeow, P.; Angeli, S.; Weißbecker, B.; Schütz, S. Antennal and behavioral responses of *Cis boleti* to fungal odor of *Trametes gibbosa*. *Chem. Senses* **2008**, *33*, 379–387. [CrossRef] [PubMed]
70. Mosandl, A.; Heusinger, G.; Gessner, M. Analytical and sensory differentiation of 1-octen-3-ol enantiomers. *J. Agric. Food Chem.* **1986**, *34*, 119–122. [CrossRef]
71. Spiteller, P. Chemical defence strategies of higher fungi. *Chem. A Eur. J.* **2008**, *14*, 9100–9110. [CrossRef]
72. Heddergott, C.; Calvo, A.M.; Latgé, J.P. The volatome of *Aspergillus fumigatus*. *Eukaryot. Cell* **2014**, *13*, 1014–1025. [CrossRef]
73. Berendsen, R.L.; Schrier, N.; Kalkhove, S.I.; Lugones, L.G.; Baars, J.J.; Zijlstra, C.; De Weerdt, M.; Wösten, H.A.B.; Bakker, P.A.H.M. Absence of induced resistance in *Agaricus bisporus* against *Lecanicillium fungicola*. *Antonie Van Leeuwenhoek* **2013**, *103*, 539–550. [CrossRef]
74. Chitarra, G.S.; Abee, T.; Rombouts, F.M.; Posthumus, M.A.; Dijksterhuis, J. Germination of *Penicillium paneum* conidia is regulated by 1-octen-3-ol, a volatile self-inhibitor. *Appl. Environ. Microbiol.* **2004**, *70*, 2823–2829. [CrossRef]
75. Nemčovič, M.; Jakubíková, L.; Ví-den, I.; Farkaš, V. Induction of conidiation by endogenous volatile compounds in *Trichoderma* spp. *FEMS Microbiol. Lett.* **2008**, *284*, 231–236. [CrossRef]
76. Müller, A.; Faubert, P.; Hagen, M.; Zu Castell, W.; Polle, A.; Schnitzler, J.P.; Rosenkranz, M. Volatile profiles of fungi–chemotyping of species and ecological functions. *Fungal Genet. Biol.* **2013**, *54*, 25–33. [CrossRef]
77. Zhang, M.; Zhao, J.L.; Liu, J.M.; Chen, R.D.; Xie, K.B.; Chen, D.W.; Feng, K.P.; Zhang, D.; Dai, J.G. Neural anti-inflammatory sesquiterpenoids from the endophytic fungus *Trichoderma* sp. Xy24. *J. Asian Nat. Prod. Res.* **2017**, *19*, 651–658. [CrossRef] [PubMed]
78. Takasu, K.; Mizutani, S.; Noguchi, M.; Makita, K.; Ihara, M. Total Synthesis of (±)-culmorin and (±)-longiborneol: An efficient construction of tricyclo [6.3. 0.03, 9] undecan-10-one by intramolecular double michael addition. *J. Org. Chem.* **2000**, *65*, 4112–4119. [CrossRef] [PubMed]
79. McCormick, S.P.; Alexander, N.J.; Harris, L.J. CLM1 of Fusarium graminearum encodes a longiborneol synthase required for culmorin production. *Appl. Environ. Microbiol.* **2010**, *76*, 136–141. [CrossRef] [PubMed]
80. Larsen, T.O. Volatile flavour production by *Penicillium caseifulvum*. *Int. Dairy J.* **1998**, *8*, 883–887. [CrossRef]
81. Brock, N.L.; Tudzynski, B.; Dickschat, J.S. Biosynthesis of sesqui-and diterpenes by the gibberellin producer *Fusarium fujikuroi*. *ChemBioChem* **2011**, *12*, 2667–2676. [CrossRef] [PubMed]
82. Studt, L.; Janevska, S.; Niehaus, E.M.; Burkhardt, I.; Arndt, B.; Sieber, C.M.; Humpf, H.U.; Dickschat, J.S.; Tudzynski, B. Two separate key enzymes and two pathway-specific transcription factors are involved in fusaric acid biosynthesis in *Fusarium fujikuroi*. *Environ. Microbiol.* **2016**, *18*, 936–956. [CrossRef] [PubMed]
83. Pollnitz, A.P.; Pardon, K.H.; Sefton, M.A. Quantitative analysis of 4-ethylphenol and 4-ethylguaiacol in red wine. *J. Chromatogr. A* **2000**, *874*, 101–109. [CrossRef]
84. Martorell, N.; Martı, M.P.; Mestres, M.; Busto, O.; Guasch, J. Determination of 4-ethylguaiacol and 4-ethylphenol in red wines using headspace-solid-phase microextraction-gas chromatography. *J. Chromatogr. A* **2002**, *975*, 349–354. [CrossRef]
85. Xiao, Z.; Xu, P. Acetoin metabolism in bacteria. *Crit. Rev. Microbiol.* **2007**, *33*, 127–140. [CrossRef]
86. Bach, B.; Meudec, E.; Lepoutre, J.P.; Rossignol, T.; Blondin, B.; Dequin, S.; Camarasa, C. New insights into γ-aminobutyric acid catabolism: Evidence for γ-hydroxybutyric acid and polyhydroxybutyrate synthesis in *Saccharomyces cerevisiae*. *Appl. Environ. Microbiol.* **2009**, *75*, 4231–4239. [CrossRef]

87. Singh, S.K.; Strobel, G.A.; Knighton, B.; Geary, B.; Sears, J.; Ezra, D. An endophytic Phomopsis sp. possessing bioactivity and fuel potential with its volatile organic compounds. *Microb. Ecol.* **2011**, *61*, 729–739. [CrossRef] [PubMed]
88. Strobel, G.; Singh, S.K.; Riyaz-Ul-Hassan, S.; Mitchell, A.M.; Geary, B.; Sears, J. An endophytic/pathogenic *Phoma* sp. from creosote bush producing biologically active volatile compounds having fuel potential. *FEMS Microbiol. Lett.* **2011**, *320*, 87–94. [CrossRef] [PubMed]
89. Fahlbusch, K.G.; Hammerschmidt, F.J.; Panten, J.; Pickenhagen, W.; Schatkowski, D.; Bauer, K.; Garbe, D.; Surburg, H. Flavors and fragrances. In *Ullmann's Encyclopedia of Industrial Chemistry*; Wiley-VCH Verlag GmbH & Co. KGaA: Weinheim, Germany, 2000. [CrossRef]
90. Lingappa, B.T.; Prasad, M.; Lingappa, Y.; Hunt, D.F.; Biemann, K. Phenethyl alcohol and tryptophol: Autoantibiotics produced by the fungus *Candida albicans*. *Science* **1969**, *163*, 192–194. [CrossRef] [PubMed]
91. Eshkol, N.; Sendovski, M.; Bahalul, M.; Katz-Ezov, T.; Kashi, Y.A.; Fishman, A. Production of 2-phenylethanol from by a stress tolerant *Saccharomyces cerevisiae* strain. *J. Appl. Microbiol.* **2009**, *106*, 534–542. [CrossRef]
92. Rapior, S.; Cavalié, S.; Croze, P.; Andary, C.; Pélissier, Y.; Bessière, J.M. Volatile components of ten frozen mushrooms (Basidiomycetes). *J. Essent. Oil Res.* **1996**, *8*, 63–66. [CrossRef]
93. Gao, F.; Daugulis, A.J. Bioproduction of the aroma compound 2-phenylethanol in a solid-liquid two-phase partitioning bioreactor system by *Kluyveromyces marxianus*. *Biotechnol. Bioeng.* **2009**, *104*, 332–339. [CrossRef]
94. Hua, D.; Xu, P. Recent advances in biotechnological production of 2-phenylethanol. *Biotechnol. Adv.* **2011**, *29*, 654–660. [CrossRef]
95. Drilling, K.; Dettner, K. Electrophysiological responses of four fungivorous coleoptera to volatiles of *Trametes versicolor*: Implications for host selection. *Chemoecology* **2009**, *19*, 109. [CrossRef]
96. Wawrzyn, G.T.; Quin, M.B.; Choudhary, S.; López-Gallego, F.; Schmidt-Dannert, C. Draft genome of *Omphalotus olearius* provides a predictive framework for sesquiterpenoid natural product biosynthesis in Basidiomycota. *Chem. Biol.* **2012**, *19*, 772–783. [CrossRef]
97. Ziegenbein, F.C.; Hanssen, H.P.; König, W.A. Chemical constituents of the essential oils of three wood-rotting fungi. *Flavour Fragr. J.* **2006**, *21*, 813–816. [CrossRef]
98. Ziegenbein, F.C.; König, W.A.; Hanssen, H.P. Volatile metabolites from the wood-inhabiting fungi *Bjerkandera adusta*, *Ganoderma applanatum*, and *Stereum hirsutum*. *J. Essent. Oil Res.* **2010**, *22*, 116–118. [CrossRef]
99. Costa, R.; De Grazia, S.; Grasso, E.; Trozzi, A. Headspace-solid-phase microextraction-gas chromatography as analytical methodology for the determination of volatiles in wild mushrooms and evaluation of modifications occurring during storage. *J. Anal. Methods Chem.* **2015**, 1–10. [CrossRef] [PubMed]
100. Dickschat, J.S.; Celik, E.; Brock, N.L. Volatiles from three genome sequenced fungi from the genus Aspergillus. *Beilstein J. Org. Chem.* **2018**, *14*, 900–910. [CrossRef] [PubMed]
101. Takigawa, H.; Kubota, H.; Sonohara, H.; Okuda, M.; Tanaka, S.; Fujikura, Y.; Ito, S. Novel Allylic Oxidation of α-Cedrene to sec-Cedrenol by a Rhodococcus Strain. *Appl. Environ. Microbiol.* **1993**, *59*, 1336–1341. [CrossRef] [PubMed]
102. Matysik, S.; Herbarth, O.; Mueller, A. Determination of volatile metabolites originating from mould growth on wall paper and synthetic media. *J. Microbiol. Methods* **2008**, *75*, 182–187. [CrossRef] [PubMed]
103. Hamad, Y.K.; Abobakr, Y.; Salem, M.Z.; Ali, H.M.; Al-Sarar, A.S.; Al-Zabib, A.A. Activity of plant extracts/essential oils against three plant pathogenic fungi and mosquito larvae: GC/MS analysis of bioactive compounds. *BioResources* **2019**, *14*, 4489–4511. [CrossRef]
104. Szmigielski, R.; Cieslak, M.; Rudziński, K.J.; Maciejewska, B. Identification of volatiles from Pinus silvestris attractive for Monochamus galloprovincialis using a SPME-GC/MS platform. *Environ. Sci. Pollut. Res.* **2012**, *19*, 2860–2869. [CrossRef] [PubMed]
105. Jeleń, H.H. Use of solid phase microextraction (SPME) for profiling fungal volatile metabolites. *Lett. Appl. Microbiol.* **2003**, *36*, 263–267. [CrossRef]
106. Gardner, J.W.; Pearce, T.C.; Friel, S.; Bartlett, P.N.; Blair, N. A multisensor system for beer flavour monitoring using an array of conducting polymers and predictive classifiers. *Sens. Actuators B Chem.* **1994**, *18*, 240–243. [CrossRef]
107. Stoppacher, N.; Kluger, B.; Zeilinger, S.; Krska, R.; Schuhmacher, R. Identification and profiling of volatile metabolites of the biocontrol fungus *Trichoderma atroviride* by HS-SPME-GC-MS. *J. Microbiol. Methods* **2010**, *81*, 187–193. [CrossRef]

108. Zhang, Z.; Li, G. A review of advances and new developments in the analysis of biological volatile organic compounds. *Microchem. J.* **2010**, *95*, 127–139. [CrossRef]
109. Nowakowska, J.A.; Stocki, M.; Stocka, N.; Ślusarski, S.; Tkaczyk, M.; Caetano, J.M.; Tulik, M.; Hsiang, M.; Oszako, T. Interactions between *Phytophthora cactorum*, *Armillaria gallica* and *Betula pendula* Roth. Seedlings Subjected to Defoliation. *Forests* **2020**, *11*, 1107. [CrossRef]
110. Senthilmohan, S.T.; McEwan, M.J.; Wilson, P.F.; Milligan, D.B.; Freeman, C.G. Real time analysis of breath volatiles using SIFT-MS in cigarette smoking. *Redox Rep.* **2001**, *6*, 185–187. [CrossRef] [PubMed]
111. Lindinger, W.; Hansel, A.; Jordan, A. On-line monitoring of volatile organic compounds at pptv levels by means of proton-transfer-reaction mass spectrometry (PTR-MS) medical applications, food control and environmental research. *Int. J. Mass Spectrom. Ion Process.* **1998**, *173*, 191–241. [CrossRef]
112. Ezra, D.; Jasper, J.; Rogers, T.; Knighton, B.; Grimsrud, E.; Strobel, G. Proton transfer reaction-mass spectrometry as a technique to measure volatile emissions of *Muscodor albus*. *Plant Sci.* **2004**, *166*, 1471–1477. [CrossRef]
113. Maenhaut, W.; Chi, X.; Wang, W.; Cafmeyer, J.; Yasmeen, F.; Vermeylen, R.; Szmigielska, K.; Janssens, I.; Claeys, M. Contribution from Selected Organic Species to PM2.5 Aerosol during a Summer Field Campaign at K.-Puszta, Hungary. *Atmosphere* **2017**, *8*, 221. [CrossRef]
114. Osowski, S.; Siwek, K.; Grzywacz, T.; Brudzewski, K. Differential electronic nose in on-line dynamic measurements. *Metrol. Meas. Syst.* **2014**, *21*, 649–662. [CrossRef]
115. R Core Team. *R: A Language and Environment for Statistical Computing*; R Foundation for Statistical Computing: Vienna, Austria. Available online: https://www.R-project.org/ (accessed on 17 November 2020).
116. Husson, F.; Josse, J.; Le, S.; Mazet, J. Factominer: Multivariate Exploratory Data Analysis and Data Mining. R Package Version 1.29. 2015. Available online: http://cran.r-project.org/package=FactoMineR (accessed on 17 November 2020).
117. Ward, J.H. Hierarchical grouping to optimize an objective function. *J. Am. Stat. Assoc.* **1963**, *58*, 236–244. [CrossRef]
118. Murtagh, F.; Legendre, P. Ward's hierarchical agglomerative clustering method: Which algorithms implement Ward's criterion? *J. Classif.* **2014**, *31*, 274–295. [CrossRef]
119. Kaufman, L.; Rousseeuw, P.J. *Finding Groups in Data: An Introduction to Cluster Analysis*; John Wiley & Sons, Inc.: Hoboken, NJ, USA, 2008.
120. Rousseeuw, P.; Struyf, A.; Hubert, M.; Studer, M.; Roudier, P.; Gonzalez, J. Package 'Cluster'. Available online: https://cran.r-project.org/web/packages/cluster/index.html (accessed on 17 November 2020).

Sample Availability: Samples of the compounds are not available from the authors.

MDPI
St. Alban-Anlage 66
4052 Basel
Switzerland
Tel. +41 61 683 77 34
Fax +41 61 302 89 18
www.mdpi.com

Molecules Editorial Office
E-mail: molecules@mdpi.com
www.mdpi.com/journal/molecules

revealed, and afterwards describe where they have been previously found and what significance they might have.

3.1. Accurateness of Analyses Performed by E-Nose

PCA revealed a clear distinction between the four fungal species detected by e-nose measurements, i.e., *P. ramorum*, *F. poae*, *T. asperellum* and *R. solani* compared to the controls. In strawberry for instance, a strong separation of pathogenic fungi i.e., *Botrytis* sp., *Penicillium* sp. and *Rhizopus* sp. was also based on the first two components of a PCA plot (accounting for 99.4% of variance) [51]. PCA of sensorial measurements under laboratory conditions also highlighted a strict relationship between the disease severity (potato brown rot) and the responses of the e-nose sensors [52].

A SPME-GC/MS revealed that sensors 1 and 2 were receptive to ethanol (for which GC/MS analysis detected it in *F. culmorum*, *P. cinnamomi* and *P. ramorum* with high confidence). Sensors 3, 4, 11 and 12 were receptive to VOCs in general (sesquiterpenes were frequently observed as well as other VOCs during GC/MS), and this may explain the high average corresponding signals observed. In contrast, the other sensors 5, 6, 9 and 10 may have been reacting to aliphatic hydrocarbons, as their sensitivity range included liquefied petroleum gases such as propane, butane, etc. On the other hand, sensors number 7, 8, 9 and 10 were sensitive to methane (7 and 8) or to methane and LP gases (9 and 10) and provided overall weak signals, which may have led to the conclusion that methane did not belong to the main compounds emitted by the tested samples. Furthermore, methane was not found in any sample during SPME-GC/MS.

3.2. Volatiles Identifying Fungal and Oomycete Species

3.2.1. Control Treatments

Empty Flask Control

SPME-GC/MS analysis yielded a low abundance for the empty flask control chromatogram (order of magnitude 10^7), compared to some other treatments. However, the empty flask control atmosphere contained some volatiles, which was consistent with the e-nose sensing a very weak, but still detectable, signal for this control treatment (Figure 1a). 2-ethyl-1-hexanol, tetradecane, and butylated hydroxytoluene are all compounds that may be found in adhesives, glues, gums, mastics, waxes, etc. For instance, 2-ethyl-1-hexanol is among the most highly produced worldwide synthetic alcohols and often serves as a solvent, while its ester derivatives have many industrial usages: adhesives, coatings, defoamers, emollients, inks, lubricants, plasticizers, etc. [53,54]. Paraffin wax and similar substances are frequently used as lubricants, insulators, water repellents or coating agents by industry. These include long-chain aliphatic alkanes whose oxidation by sample air could explain the presence of found aliphatic aldehydes (heptanal, octanal, nonanal, decanal and dodecanal). All these compounds were probably emitted by the silicone ring adjoined to the inner side of the lid by the manufacturer to allow hermetic closure of the flask, which thus appeared not to be chemically inert. Exposure to high temperatures during sterilization may have caused or facilitated volatile emission.

PDA Medium Control

The order of magnitude of the PDA control chromatogram was equivalent to what was obtained for the empty flask control (10^7), even though the e-nose analysis showed the former to release higher levels of detectable odors than the latter. Nonanal, decanal, dodecanal and butylated hydroxytoluene were found here again, suggesting these compounds may be released by the flask and not the nutritive medium itself. Furthermore, several of the molecules detected in PDA control that were not detected from the empty flask (benzaldehyde, dodecane, etc.) may also have arisen from the glue, rubber and other synthetic materials [55], making it uncertain the exact origin of these observed volatiles.

3.2.2. *Armillaria* Species

In accordance with e-nose analysis results, SPME-GC/MS indicated a similar chromatogram signal intensity between controls and *Armillaria* treatments (10^7). Detected sesquiterpenes in *Armillaria* treatment (1,3,4,5,6,7-hexahydro-2,5,5,-trimethyl-2*H*-2,4a-ethanonaphthalene, β-barbatene, (*Z*)-α-bisabolene and δ-cadinene) were likely to have been emitted by the fungi themselves. Possible presence of 1,3,4,5,6,7-hexahydro-2,5,5-trimethyl-2*H*-2,4a-ethanonaphthalene and 4,4-dimethylpent-2-enal, common to both *A. gallica* and *A. ostoyae*, suggest intrageneric similarities in VOC emission for the *Armillaria* genus. Furthermore, the detection of β-barbatene in both *A. ostoyae* (a basidiomycete) and *F. poae* (an ascomycete) suggested possible similarities in VOC emission ranges of two different fungal divisions. β-barbatene is known to have been highlighted among volatiles of several Ascomycota (*Fusarium verticillioides*) and Basidiomycota (*Fomitopsis pinicola, Piptoporus betulinus* and *Trametes suaveolens*) [56]. In addition to the widespread 1-octen-3-ol and its C8 derivatives (such as 2-octen-1-al), fungi frequently emit mixtures of non-ramified, saturated or unsaturated, alcohols, aldehydes, esters and ketones, as well as various ramified methylated molecules [57]. Even though fungi are known to produce large amounts of terpenes, these molecules are found in some bacteria as well, particularly actinomycetes [58,59]. Some of these compounds directly act on the emitting organism's close environment whereas others serve as intermediates in mycotoxins or other bioactive molecules biosynthetic pathways [60].

Detection of root rot by *Armillaria* species is currently very important in the green areas in contemporary cities. Harsh urban conditions such as soil compaction cause tree weakness and mortality of roots being covered by asphalt or bricks limiting access to oxygen and water. Most electronic devices such as resistographs or sound tomographs (PICUS) are designed for checking rot in tree trunks. There is no device for root examination without causing significant damage to plant tissues. In such circumstances the e-nose could address the issue of evaluating hazardous old trees and for the risk of dropping limbs or falling over, especially along streets where they might be poorly anchored.

3.2.3. *Fusarium* Species

The genus *Fusarium* is one of the most dangerous to germinating seeds, causing damping-off seedlings. There is also a significant menace of introducing the quarantine pathogen *F. circinatum* from southern Europe, so early warnings about their unintentional entrance and establishment are needed.

Fusarium avenaceum

The higher observed abundance magnitude (10^8) in the *F. poae* chromatogram compared to the other studied *Fusarium* species (usually at 10^7) was consistent with e-nose results indicating that this species was linked to a stronger sensor signal. The fact that longifolene was detected in *F. avenaceum* as well as in *F. poae* points towards possible intrageneric commonality in sesquiterpene emission. Furan and its derivatives (i.e., 2,4-dimethylfuran, but also 2-methylfuran, 3-methylfuran, 2,5-dimethylfuran, 2,3,5-trimethylfuran, 2-ethyl-5-methylfuran, etc.) are found in volatiles emitted by many fungal species. Moreover, the compound 1-octen-3-ol and its C8 co-metabolites can often be found together with 2-pentylfuran, which suggests a common biosynthetic pathway for these molecules. Other furan-derived compounds such as for instance 2-acetylfuran, 2-furanmethanol, 2-(methoxymethyl)furan and furfural are frequent as well [61]. β-bazzanene is, like trichodiene, an important precursor of various sesquiterpenoids in fungal metabolism [62]. Trichothecene mycotoxins for instance include powerful inhibiting compounds of eukaryotic protein biosynthesis [63].

Fusarium culmorum

Production of ethyl acetate is common in yeasts, but has been observed in filamentous fungi such as *Ceratocystis fagacearum* [64]. The *F. culmorum* flask atmosphere appeared to contain several sesquiterpenes, as did all fungi (except *R. solani*) that were investigated in this study. These molecules most likely did come from the fungi and were not emitted by inert bodies of the sample (flask or medium).

Fusarium oxysporum

Benzaldehyde, which was found in *F. oxysporum* and in the PDA control, may have been produced by the medium rather than the fungus itself. However, this molecule could theoretically have been emitted by the fungus, since benzene alkylated derivatives have been found to be produced by several fungal genera such as *Fusarium, Muscodor, Penicillium* and *Trichoderma* [57]. Whether these compounds fulfill any biological role or are just metabolic waste products still needs clarification. Benzaldehyde itself counts as one of the most widespread benzene derivatives among fungi (especially in the genera *Aspergillus, Botrytis, Fomes, Fusarium, Penicillium* and *Pleurotus*). Benzyl alcohol, methyl benzoate, ethyl benzoate and 4-methylbenzaldehyde are less common. Emission of benzothiazole was witnessed in *Aspergillus* and *Trichoderma* genera, even though pyrazines remain the most important group of nitrogenous fungal VOCs. Historically, the first aromatic fungal VOCs were identified in the middle of the 20th century in odoriferous decomposing wood. These molecules were thought to be synthetized from tyrosine and phenylalanine aromatic amino-acids or simply result from lignin degradation. The detected 3-chloro-4-methoxybenzaldehyde could have theoretically been released by the fungus as well, even though paraffine-derived chloroalkanes often serve in industry as ingredients for dyes and paste manufacturing. Indeed, chlorine aromatic compounds such as 4-chloro-1,2-dimethoxybenzene and 1,5-dichloro-2,3-dimethoxybenzene have been detected in the genus *Geniculosporium* [65]. Moreover, 3-chloro-4-methoxybenzaldehyde and 1,5-dichloro-2,3-dimethoxybenzene were identified in *Anthracophyllum discolor*, along with 3,5-dichloro-4-methoxybenzaldehyde [66]. The latter two molecules were furthermore observed in *Bjerkandera adusta*, together with their alcohol derivatives. Such compounds can be synthetized by fungi from the growth medium even if its chlorine concentration is very low. Analogous brominated compounds may be produced the same way, whereas iodine specks in growth medium can lead to diiodomethane or even chloroiodomethane biosynthesis according to Spinnler et al. [67].

Fusarium poae

We found Matsutake alcohol, chemically called 1-octen-3-ol, which is a fatty acid characteristic for fungi widespread in the form of its R-enantiomer [68,69]. The R-enantiomer releases a fruity odor reminiscent of mushrooms, whereas the L-enantiomer is associated with grassy smell [70]. Initially detected in other fungi which were not tested in our experiment, such as *Tricholoma matsutake*, the alcohol has since then been identified in a wide range of fungal genera such as *Agaricus, Aspergillus, Boletus, Fistulina, Fomes, Phomopsis, Lentinus, Penicillium, Pleurotus, Tuber* and *Verticillium* but also in cultivated *Fusarium* and *Trichoderma* [57]. The compound 1-octen-3-ol is often found together with some of its C8 co-metabolites: octan-3-ol, octan-3-one, oct-1-en-3-one, octan-1-ol, oct-2-en-1-ol, octanal, trans-oct-2-enal, oct-1-ene and 1,3-octadiene. Some 1-octen-3-ol, octan-1-ol or octan-3-ol ester derivatives can be present in the mixture as well, in variable amounts. Linoleic acid is considered to serve as primary substrate for matsutake alcohol production, even though the exact biosynthetic pathway remains unclear. Production intensifies when the fruiting body is wounded, which could correspond to a defensive strategy of the fungus [71]. This confirmation of its presence in fungi which can help distinguish them from oomycetes is a significant finding for this study. However, it is worth speculating why 1-octen-3-ol is synthesized. First of all, it probably stops growth of some competing fungi, and indeed it has been found to inhibit conidial germination of several fungal species, including *Aspergillus nidulans* [72], *Lecanicillium fungicola* [73] and *Penicillium paneum* [74]. On the other hand, it was able to induce *Trichoderma atroviride* conidia germination [75]. This suggests 1-octen-3-ol acts as a hormone influencing fungal development. Its exact role however remains to be clarified. In our experiment we did use a mixture of organisms, but this should be done in future tests since there are reports that interactions between fungi are very important. Very often the volatile compounds are produced to inhibit or stimulate growth. Diverse species belonging to Ascomycota and Basidiomycota phyla, such as *Trichoderma harzianum*, were shown to produce bisabolene [76,77]. We found it from tested *Armillaria* species. Similarly, several longiborneol sesquiterpene derivatives showing antifungal properties were observed in *Fusarium* spp. (for instance in *F. culmorum* and *F. graminearum*) [78,79]. It was also observed in our study.

3.2.4. *Phytophthora* Species

These organisms are considered to be serious primary pathogens of plants (including many forest tree species) and have often been unintentionally introduced to Europe from Asia. The observed increase of international trade of potted plants (e.g., bonsai) and seeds poses a new risk of establishment of alien, invasive species in forest stands. Also tourism, globalization, and quicker vessels with larger cargos accelerate this phenomenon. Therefore, an early detection of this group of organisms is of special importance.

Phytophthora cactorum

Acetone is produced by *Clostridium acetobutylicum* and is frequently used as solvent by industry and research. Another compound—dimethyl disulphide—identified in our study is one of the most common sulfur molecules produced by bacteria, together with dimethyl sulphide, dimethyl trisulphide and S-methyl methanethiosulfonate [58]. Generally, bacteria are known to be more abundant emitters of sulfur compounds than fungi, although these latter may produce such substances as well. Less frequent in fungi, these latter compounds have been observed in several *Fusarium* and *Penicillium* species [57], but in our case were found in *P. cactorum* and *T. asperellum*. Dimethyl disulphide and other compounds such as benzonitrile, trimethyl disulphide or S-methyl thioacetate are volatiles with important antifungal effects [2]. On the other hand, *Aspergillus* and *Trichoderma* genera were shown to emit benzothiazole, which contains both sulfur and nitrogen. Moreover, cheese *Penicillium* strains can produce sulfur dioxide [80]. According to our results, oomycetes seem to be able to release sulfur compounds as well. Emission of 1-hexanol was witnessed in several fungal species such as *Aspergillus flavus* or *Fusarium fujikuroi*, which probably synthesize it as well as other related compounds through fatty acid degradation [81,82].

Phytophthora cinnamomi

P. cinnamomi was rather poor in terms of VOC diversity. 4-ethyl-2-methoxyphenol is commonly found in beer and wine due to fermentation processes by yeasts from the genus *Brettanomyces* [83,84], which are fungi rather than oomycetes. 2-ethyl-1-hexanol was previously found in the PDA control, which makes it difficult to guess whether this compound was really emitted by the sample itself or originated from inert sample components. Some of the detected VOCs seemed to be common to *P. cinnamomi* and other species, including fungi.

Phytophthora plurivora

Acetoin is an important metabolite (mainly carbon storage, physiological acidification avoidance and NAD/NADH ratio regulation), widespread in nature, emitted by many micro-organisms as soon as these have access to a degradable carbon source [85]. Bach et al. [86] have shown production of 4-hydroxybutanoic acid by *Saccharomyces cerevisiae*. In our experiment, substances such as acetoin and 4-hydroxybutanoic acid, α-Pinene, and Δ-3-Carene were found only in *P. plurivora* samples.

Phytophthora ramorum

The peak abundances were higher in the *P. ramorum* chromatogram than for other *Phytophthora* species, in agreement with e-nose analysis results. 3-methyl-butanol and similar alcohols are often emitted by endophytic fungi belonging to *Phoma* and *Phomopsis* genera, which are able to break down cellulose [87,88]. More generally, 3-methyl-butanol is frequently observed in fungi, and is probably synthesized from leucine. *Ceratocystis paradoxa* was the first fungal strain to be witnessed producing 3-methyl-butanol in its volatile mixture [89]. 2-phenylethanol, widespread among microorganisms, is an intermediate of L-phenylalanine aromatic amino acid metabolism. Thus, numerous yeast species (including *Candida albicans* and *Saccharomyces cerevisiae*) synthesize it as an antibiotic [90,91]. Rapior et al. [92] observed the presence of 2-phenylethanol in VOC mixtures emitted by several Basidiomycetes species. It is common in many fungi genera: *Aspergillus, Chaetomium, Fusarium, Hypoxylon, Lasiodiploida, Penicillium, Polyporus, Trichoderma, Tuber*, etc. [57]. Production of 2-phenyethanol using microorganisms could serve the industry to avoid artificial synthesis with implied purification steps, which makes the whole process very expensive [93,94]. *P. ramorum* in Europe causes sudden larch decline in western Great Britain, so its further spread in Europe should be closely monitored.

3.2.5. *Rhizoctonia solani*

There appeared to be a big difference between the low abundance and diversity of the *R. solani* chromatogram and the strong induced reaction from e-nose sensors (Figure 1m). One possible hypothesis is that this species may emit many volatile inorganic compounds (VICs) detected by the e-nose sensors but not extracted by the SPME fiber. Alternatively, molecules released by *R. solani* may simply have intrinsically higher stimulation of the e-nose, meaning they "smell" stronger to the sensors. Drilling and Dettner [95] witnessed emission of 3-octanol by *Trametes versicolor*, independent of 1-octen-3-ol emission.

3.2.6. *Trichoderma asperellum*

Consistent with e-nose results (Figure 1), the *T. asperellum* chromatogram showed a high abundance magnitude (almost 10^8). Literature indicates that trans-dauca-4(11),8-diene was spotted in *Omphalotus olearius*, along with α-barbatene, β-barbatene and γ-cadinene [96]. It was also observed in *Schizophyllum commune* [97] and in *Bjerkandera adusta* [98]. Isodaucene was detected in several fungal species, such as *Tricholomopsis rutilans* [99] and *Aspergillus fischeri* [100]. Cedrene has been isolated in several fungi such as *Corynespora cassiicola* and *Beauveria sulfurescens*, but also in soil bacteria *Rhodococcus rhodochrous* [101]. VOCs emitted by the *Trichoderma* genus have been shown to play the role of signal compounds for communication between colonies as well as for growth regulation. For instance, molecules such as 3-octanol, 3-octanone and 1-octen-3-ol emitted by conidia of *Trichoderma* cultures induced conidiation of other colonies from the same genus [75].

3.2.7. Strategies for Improved Detection

The compounds mentioned above were often shared among organisms but some of them appeared only once suggesting specificity. However, our list of organisms was very limited so if we were to confirm specificity of some compounds, we would need to do so in the future broader screening of organisms. We could apply precise methods such as GC/MS is to identify chemical compounds or to use e-nose sensors. Both approaches have their advantages and disadvantages. Semiconductor sensors used in the e-noses usually show a wide sensitivity range and a non-linear response with respect to gas concentration. Such sensors provide a quick response, involve a simple (thus inexpensive) circuit design and have a long lifetime. The output signal of a given sensor arises from the superposition of individual effects of every component of the gaseous mixture. Because individual sensors (e.g., TGS—Figaro Gas Sensors) show a lack of selectivity, combining several of them with different cross sensitivities allowed improved performance. On the other hand, too many sensors taken together will increase measurement noise.

In order to know which sensor is suitable for a particular chemical compound or their group, other techniques should be applied e.g., multidimensional data analysis tries to highlight odor patterns that can be used to characterize gaseous mixtures based on identification, comparison and classification principles. First studies investigating microbial volatiles involved steam distillation and liquid-liquid extraction followed by compounds concentration and identification [7]. Since the gas chromatography (GC) is becoming an affordable and reliable detection method, scientists have merged the separation, identification and quantification steps into one single analytic process, GC/MS. We also used this technique in our experiments because of its high sensitivity and strong discriminatory ability. GC/MS is today the most used analytical tool for fungal VOC identification [102–104] and is easily coupled to solid-phase microextraction (SPME) techniques for the VOC extraction and concentration, which allows easy progression to environmental sampling and subsequent laboratory analysis [100]. However, it should be noted that results observed depend on both the nature of the fiber used and the extraction method [105]. Artifacts can sometimes appear due to solid sorbents serving for headspace analysis [7]. The atmospheric water content of the sample can also bias GC/MS data since fungal VOC formation is easier in humid ambient air. Furthermore, such physical analyses require some time to be implemented [106], even though GC/MS remains a relatively quick way of analyzing VOC mixtures, which can be furthermore automated for real time profiling of compounds emitted from living fungi [107–109].

Alternatively, other analytical methods may be used for VOC characterization. For example, the simultaneous distillation extraction (SDE) method relies on simultaneously occurring vapor distillation and solvent extraction [105]. It is especially applied for the extraction of high boiling volatiles with the flaw of potentially making false features appear because of the longstanding effect of heat. The selected-ion flow-tube mass spectrometry (SIFT-MS) is a technique allowing for a real time measurements of VOC concentration in a sample's atmosphere, with a high degree of sensitivity (up to a few ppb) [110]. It allows quick characterization of a gaseous mixture composed of a wide range of molecules. The proton-transfer-reaction mass spectrometry (PTR-MS) involves the VOC ionization using H_3O^+ primary ions, which results in the MH^+ ions production (where M stands for a neutral organic molecule). These are in turn detected by the means of a quadrupole mass spectrometer [111]. This method provides a sensitivity which is comparable to GC/MS, however in contrast to GC/MS affords the robust analysis of a sample without any pre-processing or pre-concentration [112,113].

Based on what was found in our samples and the above literature review, we hypothesize that some substances (listed in Table 1) are also specific to particular organisms. Even if we do not know their putative functions, we believe that they can be used to discriminate between individual fungal and oomycetes species or at least between genera. The only problem could be obtaining sensors for reasonable prices for the most specific compounds for each organism in order to construct new models of e-nose.

Alternatively, further developments may go in the other direction of determining not just single or a few specific compounds, but to train neural networks (artificial intelligence) how to recognize differences among samples e.g., like pictures. In such a case we are not concerned about the chemistry of VOCs (its content), but the focus is to point out differences between organisms. For this purpose, much empirical data will be needed, as well as technical details that need to be worked out such as timing of each measurement, its temperature, and the required humidity. Furthermore, specific cultural conditions for microbial growth, may also affect the quality and quantity of emitted VOCs.

4. Materials and Methods

4.1. Sample Material Preparation

In total, 144 jars were tested by e-nose containing pure cultures of strains (one to four-week-old fungi and oomycetes prepared in three repetitions). They consisted of 12 different species important in forestry, and were obtained from stocks kept in the laboratory of the Forest Research Institute (IBL) (Table 2). In addition, three copies of aforementioned organisms were grown for VOC analysis by the GC-SM method.

Table 2. General characteristics of the fungal and oomycete species tested in the experiment.

Treatment	Species	Reference from GenBank
Control	–	–
1	*Armillaria gallica* (Marxm. & Romagn.) 1987	DQ115578 [1]
2	*Armillaria ostoyae* (Romagn.) Herink 1973	DQ115574
3	*Fusarium avenaceum* (Fr.) Sacc. 1886	MK560761
4	*Fusarium culmorum* (Wm.G. Sm.) Sacc. 1892	KP008988
5	*Fusarium oxysporum* (Schltdl.) 1824	MF162321
6	*Fusarium poae* (Peck) Wollenw. 1913	MF162318
7	*Phytophthora cactorum* (Lebert & Cohn) J. Schröt. 1886	KX242303
8	*Phytophthora cinnamomi* Rands 1922	KF682434
9	*Phytophthora plurivora* T. Jung & T.I. Burgess 2009	JX276032
10	*Phytophthora ramorum* Werres, De Cock & Man in 't Veld 2001	JF771575
11	*Rhizoctonia solani* J.G. Kühn 1858	KU901561
12	*Trichoderma asperellum* Samuels, Lieckf. & Nirenberg 1999	MT197117

[1] number available from www.ncbi.nlm.nih.gov.

Transfers of mycelia were made from stored pure colonies onto PDA (20 g glucose + 15 g agar + 4 g potato extract dissolved in 1 L distilled water) using powder purchased at BTL Ltd. (Łódź, Poland), and cultured at room temperature. The investigated microorganisms were raised in specially constructed 300 mL glass flasks or 40 mL glass vials. Each flask was fitted with a 66-mm-diameter lid made of galvanized steel with a silicone ring on the underside to prevent air exchange. Each lid was made a 9-mm-diameter hole covered with sterilization tape to allow subsequent collection of gaseous samples by introducing the e-nose inlet or a SPME syringe inside the flask through the hole, thus removing or piercing the closing tape. Similarly, all vials were sealed by a polypropylene cork provided with a polyisobutylene-polytetrafluoroethylene (PIB-PTFE) septum contiguous to its bottom side. As previously, the cork was perforated so that a SPME syringe could later on be introduced in the vial atmosphere by only piercing the underlying septum. All the dishes were distributed in the prepared PDA medium, sealed and autoclaved at 117 °C and 0.08 MPa for 20 min, prior to microorganisms' inoculation under sterile conditions. Subsequently, in every week (up to four weeks), each of the three jars for every organism was measured with the e-nose devices.

4.2. E-nose Device and Measurements

The e-nose was developed by the Warsaw University of Technology (WUT) based on the e-nose sensing device, which has been used in previous research [48,49] and was made of two homologous sets of gas sensors produced by Figaro Engineering Inc. (Osaka, Japan), which include a semiconductor tin oxide layer and arrays made of six different heated metal oxide gas sensors types (Figaro Engineering Inc. Osaka, Japan). Four sensor types were duplicated inside each array (Table S1). Duplicated sensors of the same array had a slightly different loading setting, so that they don't show exactly the same sensitivity spectrum: their output signals were therefore somewhat different, and for instance the 26xx sensor series family was chosen for its small size and high stability of operation. The sensors were not wired in series but were connected to different analog inputs of the data acquisition unit, one by one, and their loading was tuned in a clean air environment using a potentiometer. Moreover, two sensors measuring, respectively, relative humidity (HIH-3610-002 from Honeywell, Morristowne, NJ, USA), and temperature (LM35DH from Texas Instruments, Dallas, TX, USA) were included in both arrays, since those parameters impact sensor sensitivity (Table S1).

Each of the two independent sensor arrays were seated on a custom designed printed circuit board, and installed in a specially designed optimized test chamber consisting of an aluminum black cylinder (approximately 500 cm^3 volume) with both extremities connected to a socket outlet consisting of an 8.5 mm external diameter rubber tube (Figure 3a,b). Both cylindrical channels were standing nearby, in the same ambient environment. The sensor electrical conductivity (resistance) changed according to the concentration of molecules belonging to their respective sensitivity spectra in the test chamber and was converted by a simple electrical circuit into an output tension voltage signal. One set formed the measuring device and was put in contact with the sample atmosphere to be analyzed, whereas the other kept sampling ambient air thus providing a reference signal (Figure 3c). Both sets of sensors delivered their own real time signal when the measurement was launched. However, only the difference was continuously saved and transferred to the computer through a serial communication interface consisting of two 8-Channel Analog Input Modules Rev. D1 type ADAM-4017 built by Advantech (Taipei, Taiwan).

(**a**)

(**b**)

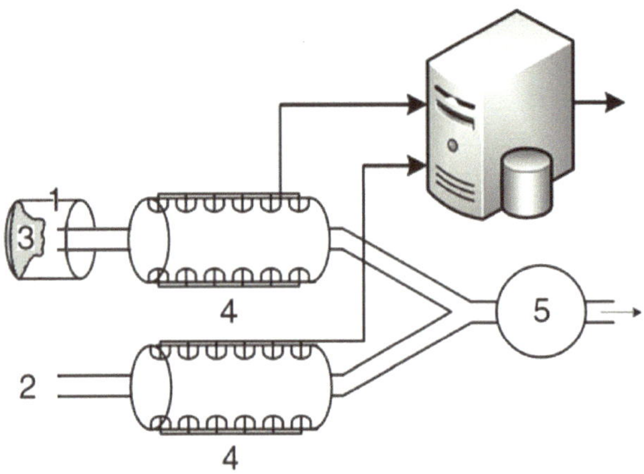

(**c**)

Figure 3. Representative view of the differential electronic nose apparatus. (**a**) Introduction of the air intake tube inside the sample; (**b**) Twelve chemical sensors in an array chamber; (**c**) General scheme of the e-nose principle of measurement: (1) Sample flask; (2) Ambient air; (3) Tested material; (4) Sensor chambers; (5) Inducting pump and flow-meter. Adapted from Brudzewski et al. [22], Osowski et al. [114].

The deduction between analog signals from both sensor arrays was done by a differential amplifier, whereupon an A/D converter turned the signal into digital format. Differential profiles obtained this way were made of the weighted sum of sensor reaction to the gaseous mixture respective components. This differential functioning removed the need to perform a systematic calibration of the reference signal prior to each measurement at changing conditions of ambient air. It also greatly reduced the influence of natural drift due to ageing of the metal oxide sensors as well as the common distortion effect linked with variations of pressure or temperature in the test chamber. Furthermore, the impact of potential intrinsic differences between two homologous sensors or operation errors was in this way minimized. Finally, it increased the sensitivity towards low concentration compounds.

An inducting pump was used to set up the air intake with a laminar and almost one-dimensional flow entering the channel through the socket outlet. It was kept constant by a flowmeter set to 1 L·min^{-1}. Twelve samples of each treatment were tested by introducing the rubber tube inlet in the sample's atmosphere through the hole on the flask's lid, from which the sterilization tape had been previously removed. The samples were tested at a rate of three per week, up to four weeks after inoculation. Each sample was analyzed only once. The volumes of the test chamber as well as the volume of the sample atmosphere were kept constant during the measurement process. The measurement window duration was arbitrarily set to 300 sec with a resistance sampling rate of 60 times per min. During measurements, which were performed in a dynamic on-line mode, the sensor temperature would fluctuate in a range of 27 to 35 °C, with an almost constant sensor chamber RH of 14–15%.

A washing interval was implemented between every two successive measurements during which the system was let running on itself for typically 10 min so that the measuring array could desaturate from the preceding sample's atmosphere molecules. However, implementing a more accurate procedure would have meant injecting synthetic air into both test chambers for 15 min. Furthermore, a blank analysis (without submitting any sample to the sensing device) was carried out every time before each new measurement series. Due to imperfect coordination in the setting of homologous sensors of the two sensing arrays, a very low intensity baseline signal would appear which was in turn removed from the pattern yielded from each subsequent analysis.

The experimental data set for each sample consisted of a matrix including 300 vectors in a twelve-dimensional space. Only measured signal values from the final period of the measuring window were used to obtain diagnostic features, so that the signal could settle down and reach a steady state before being considered. Thus, the initial rapidly increasing stage of the sensor signals derived from the transient effect was not taken into account for descriptive features generation, which was done by averaging the second half (last 150 measurements) of every temporal series of sensor resistances $R_{(j)}$, referring to each j-th sensor of the array:

$$\mathrm{r}_{(j)} = \mathrm{R}_{(j)} - \mathrm{R}_{0(j)} \tag{1}$$

with $R_{(j)}$ representing the averaged measured resistance of the j-th sensor of the array and $R_{0(j)}$ standing for the averaged baseline value of resistance measured during the blank analysis, both calculated for the j-th sensor of the array based on the last 150 measurements of the corresponding temporal series. Thus, each analyzed sample would yield a final diagnostic feature consisting of a 12-dimensional vector (r1, ... , r12) quantifying the reaction of e-nose sensors exposed to the tested material. Overall, twelve such vectors were obtained per treatment, corresponding to the 12 repetitions performed. Calculating the mean of these 12 diagnostic features allowed us to obtain a global averaged signal vector for each treatment, which could be pictured as a diagram. This final 12-dimensional vector was considered as representing the mean e-nose sensor reaction to the treatment, thus providing an assessment of the tested material odor intensity [22,110].

4.3. SPME-GC/MS Analysis

To validate the previously described method of detection with e-nose, VOCs emitted by tested microorganisms were investigated using headspace SPME-GC/MS in two different laboratories at the BUT and at the Institute of Physical Chemistry of the Polish Academy of Sciences in Warsaw (IPC-PAS). A two way approach was performed in order to better detect the VOCs emitted by tested microorganisms, i.e., the first one based on the 85 µM PDMS/Carboxen fiber (in the BUT laboratory), and the second one based on 100 µM diameter PDMS fiber in the IPC-PAN laboratory. All SPME-GC/MS measurements were for the first time applied to the fungal species, and hence the detailed explanation.

In the first approach, the samples were analyzed with a SPME syringe including a 85 μM diameter PDMS/Carboxen fiber (Supelco, Bellefonte, PA, USA). The fiber was heated to 250 °C for 1–2 h after the purchase to remove any potential contaminant adsorbed on the coating. Similarly, this process was repeated for 5–10 min before testing each new set of samples. The fiber was introduced in the flask through the lid hole and placed in an incubator under 40 °C where a 30 min extraction took place. The fiber was then introduced for 10 min in the chromatograph injection device, where a temperature of 250 °C would allow the compounds to be desorbed. The injection was done in a splitless mode. An Agilent 7890A gas chromatograph including a 30 m × 0.25 mm × 0.25 μM HP-5MS semipolar capillary column connected to an Agilent 5975C mass spectrometer (Agilent Technologies, Santa Clara, CA, USA) was used for GC/MS analysis, which lasted 43 min overall. From an initial temperature of 35 °C, the oven was heated at a rate of 5 °C min^{-1} up to 250 °C. The electron ionization potential was set at 70 eV and the electron ionization (EI) source worked at the temperature of 230 °C. The temperature of the quadrupole analyzer was 150 °C. Helium circulating through the column at a steady flow of 1 mL·min^{-1} served as a carrying gas. The spectrometer was working in a full scan mode over a 29–600 mass range. Recorded EI mass spectra were compared against the NIST Mass Spectral Database and Willey libraries for analyte identification. For some unknowns, the identification was supported by an in-house constructed library comprising the EI mass spectra for available standards. In addition to the MS spectra, also RI and authentic standards were used for identification of the compounds.

The IPC-PAS laboratory pursued analyses of the samples with a SPME syringe equipped with a 100 μM diameter PDMS fiber (Supelco, Bellefonte, PA, USA). The fiber was preconditioned before each analysis in a GC/MS injector where it was heated at 250 °C during 30 min. Then, it was introduced in a vial by piercing a cork's underlying septum to perform the extraction, which took place at room temperature and lasted 60 min. Subsequently, a 3 min desorption was performed at 250 °C in a GC injector. The injection was done in a splitless mode. The chromatographic analysis lasted overall 28 min and was performed in a Thermo Trace 1300 gas chromatograph equipped in a 30 m × 0.25 mm × 0.25 μM Rtx-5MS semipolar capillary column coupled with a Thermo ITQ 700 mass spectrometer (Thermo Scientific, Waltham, MN, USA). The initial oven temperature was kept isocratic at 100 °C for 5 min, and then increased at a rate of 10 °C min^{-1} to reach a peak temperature of 280 °C at which it was kept for 5 min. The MS detector was equipped with a 70 eV electron ionization (EI) source. The temperature of the EI ion source was 250 °C, while the quadrupole ion trap analyzer was 250 °C. The carrying gas was helium circulating through a column at a steady flow of 1 mL·min^{-1}. The spectrometer was working in a full scan mode for a 50–650 mass range. The EI mass spectra obtained were matched with these from the NIST and Willey Mass Spectral Databases for the molecular elucidation.

In both analyses, a ranking showing a list of candidates with the best matches and corresponding estimated confidence rates was generated by the analyzing software. For GC/MS analyses, we used the following software: MSD ChemStation E.02.02.1431, Agilent Technologies and NIST MS Search 2.0. The default analytical software obtained with each instrument was used for calculations for each spotted compound [22,114].

4.4. Statistical Analysis

The variability among signals obtained from twelve sensors of the e-nose device was computed by principal component analysis (PCA) in "R" software [115]. PCA analyses as well as the biplot were created with fviz_pca_biplot functions from the "FactoMineR" 1.41 package [116]. The variables with the strongest impact on the distribution of the microorganisms along the principal components were identified on the basis of Pearson correlation coefficients.

To group microorganisms according to their signal similarity among twelve sensors, a hierarchical clustering using Euclidean distance (root sum-of-squares of differences) as the similarity measure and Ward [117] clustering method with the criterion proposed by Murtagh and Legendre [118] were applied. Four different clustering methods, single and complete linkage, the unweighted pair group method with arithmetic mean (UPGMA), and Ward's method were tested according to the clustering structure of the dataset [119]. Ward's method was chosen because it expressed the highest value of the agglomerative coefficient. Hierarchical clustering was performed using the Cluster 2.0.7-1 package [120]. Final grouping of microorganisms according to their locations on the PCA biplot were determined visually in Figure 2.

5. Conclusions

For the first time, we demonstrated that the e-nose apparatus was able to distinguish between VOCs emitted by the investigated fungi or oomycetes, but further technical developments are still needed for its practical use in the forestry sector in the field or in practice, i.e., nurseries, plantations, stands, or in quarantine laboratories.

Results of testing with the e-nose prototype showed that certain fungal species, such as *F. poae*, *R. solani* and *T. asperellum*, were the most odoriferous among the studied organisms, and gave rise to the strongest signals. In our in vitro study, *fusaria* (*F. avenaceum*, *F. culmorum* and *F. oxysporum*) and an oomycete (*P. ramorum*) generated moderately intense signals as detected by the e-nose. Two tested *Armillaria* species, *A. gallica*, *A. ostoyae*, and three oomycetes, *P. cactorum*, *P. cinnamomi* and *P. plurivora*, generated specific odors detected by the e-nose sensors, e.g., α-Pinene and Δ-3-Carene for *P. plurivora*. The Principal Component Analysis plot revealed that our system of e-nose detection could discriminate between the odors emitted by *P. ramorum*, *F. poae*, *R. solani* and *T. asperellum*, making this device suitable for practical use in laboratory situations, at least for the species tested.

Identification of VOCs detected by e-nose was revealed when using two carbon fibers SPMEs. The tested PDMS/Carboxen fiber was more efficient for fungal detection compared to PDMS fiber. The majority of VOCs detected were specific compounds to the genus or species level, due to a complex mixture of (un)saturated (non-)ramified hydrocarbons and their oxygenated derivatives (aldehydes, alcohols, esters). We also noticed that all tested fungal species released sesquiterpenes in variable amounts, apart from *R. solani*. All tested *Phytophthora* strains emitted none of these compounds. Therefore, in the future, development of future stage e-nose apparatus will rely on unique volatile compounds identified and specific to investigated microorganisms.

This research should be of special interest for quarantine organizations (e.g., National Plant Protection Organizations or international ones like EPPO) dealing with alien invasive species such as *P. ramorum*. Further studies need to be carried out with protocol designs allowing control of intraspecific and temporal variability of VOC mixture profiles dependent on environmental conditions, especially in forest nurseries.

Supplementary Materials: The following are available online. Table S1: Composition of the sensor array used for e-nose sample detection tests. 1–8: Tagoshima gas sensors were purchased from Figaro Engineering Inc. (Osaka, Japan); 9–10: Relative humidity and temperature sensors were manufactured by Honeywell (Morristowne, NJ, USA) and Texas Instruments (Dallas, TX, USA), respectively; Table S2: Pearson's correlation coefficients between variables and the first two major components, and contribution of variables in principal components and variation explained; Figure S1: Hierarchical clustering of the microorganism-induced sensor signals based on Euclidean distance and Ward's minimum variance clustering method. The groups of similarities were clustered. Control 1—Empty flask; Control 2—PDA medium.

Author Contributions: Conceptualization, F.L. and T.O.; methodology, M.S., M.A., R.S.; chemical analyses and critical interpretation, M.S., M.A., R.S.; investigation, K.S., T.G.; statistical computation, J.A.N., M.K.; writing—original draft preparation, J.L.; writing—review and editing, J.A.N., T.O., M.A., T.H., R.S.; supervision, F.L.; project administration, S.Ś., R.T.; funding acquisition, T.O., R.T. All authors have read and agreed to the published version of the manuscript.

Funding: This paper was developed with a Scholarship Fund of the Forest Research Institute (Poland). This work was financially supported by the National Centre for Research and Development by the grant "PROZEL" agreement BIOSTRATEG 3/347105/9/NCBR/2017.

Acknowledgments: The authors are grateful to Małgorzata Lissy for technical assistance during sample preparation, to Anna Żółciak and Miłosz Tkaczyk for their advice concerning culture monitoring, and to Leszek Adamowicz, Bronisław Pura and Ryszard Siegoczyński for help in work coordination.

Conflicts of Interest: The authors declare no conflict of interest. The funders had no role in the design of the study; in the collection, analyses, or interpretation of data; in the writing of the manuscript, or in the decision to publish the results.

References

1. Schulz, S.; Dickschat, J.S. Bacterial volatiles: The smell of small organisms. *Nat. Prod. Rep.* **2007**, *24*, 814. [CrossRef]
2. Garbeva, P.; Hordijk, C.; Gerards, S.; De Boer, W. Volatile-mediated interactions between phylogenetically different soil bacteria. *Front. Microbiol.* **2014**, *5*. [CrossRef] [PubMed]
3. Pagans, E.; Font, X.; Sanchez, A. Emission of volatile organic compounds from composting of different solid wastes: Abatement by biofiltration. *J. Hazard. Mater.* **2006**, *131*, 179–186. [CrossRef] [PubMed]
4. Schmidt, R.; Cordovez, V.; De Boer, W.; Raaijmakers, J.; Garbeva, P. Volatile affairs in microbial interactions. *ISME J.* **2015**, *9*, 2329–2335. [CrossRef] [PubMed]
5. Herrmann, A. *The Chemistry and Biology of Volatiles*; Wiley: Chichester, UK; Hoboken, NJ, USA, 2010; ISBN 978-0-470-77778-7.
6. Korpi, A.; Järnberg, J.; Pasanen, A.-L. Microbial Volatile Organic Compounds. *Crit. Rev. Toxicol.* **2009**, *39*, 139–193. [CrossRef] [PubMed]
7. Hung, R.; Lee, S.; Bennett, J.W. The effects of low concentrations of the enantiomers of mushroom alcohol (1-Octen-3-Ol) on *Arabidopsis thaliana*. *Mycology* **2014**, *5*, 73–80. [CrossRef]
8. Zeringue, H.J.; Bhatnagar, D.; Cleveland, T.E. $C_{(15)}H(24)$ volatile compounds unique to aflatoxigenic strains of *Aspergillus flavus*. *Appl. Environ. Microbiol.* **1993**, *59*, 2264–2270. [CrossRef]
9. Insam, H.; Seewald, M.S.A. Volatile organic compounds (VOCs) in soils. *Biol. Fertil. Soils* **2010**, *46*, 199–213. [CrossRef]
10. Romoli, R.; Papaleo, M.C.; De Pascale, D.; Tutino, M.L.; Michaud, L.; LoGiudice, A.; Fani, R.; Bartolucci, G. GC–MS volatolomic approach to study the antimicrobial activity of the antarctic bacterium *Pseudoalteromonas* sp. TB41. *Metabolomics* **2014**, *10*, 42–51. [CrossRef]
11. Lemfack, M.C.; Nickel, J.; Dunkel, M.; Preissner, R.; Piechulla, B. mVOC: A database of microbial volatiles. *Nucleic Acids Res.* **2014**, *42*, D744–D748. [CrossRef]
12. Peñuelas, J.; Asensio, D.; Tholl, D.; Wenke, K.; Rosenkranz, M.; Piechulla, B.; Schnitzler, J.P. Biogenic volatile emissions from the soil: Biogenic volatile emissions from the soil. *Plant Cell Environ.* **2014**, *37*, 1866–1891. [CrossRef] [PubMed]
13. Ortíz-Castro, R.; Contreras-Cornejo, H.A.; Macías-Rodríguez, L.; López-Bucio, J. The role of microbial signals in plant growth and development. *Plant Signal. Behav.* **2009**, *4*, 701–712. [CrossRef] [PubMed]
14. Piechulla, B.; Degenhardt, J. The emerging importance of microbial volatile organic compounds: The emerging importance of microbial volatile organic compounds. *Plant Cell Environ.* **2014**, *37*, 811–812. [CrossRef] [PubMed]
15. Kramer, R.; Abraham, W.-R. Volatile sesquiterpenes from fungi: What are they good for? *Phytochem. Rev.* **2012**, *11*, 15–37. [CrossRef]
16. Peghaire, E.; Hamdache, S.; Galien, A.; Sleiman, M.; Ter Halle, A.; El Alaoui, H.; Kocer, A.; Richard, C.; Goupil, P. Inducing plant defense reactions in tobacco plants with phenolic-rich extracts from red maple leaves: A characterization of main active ingredients. *Forests* **2020**, *11*, 705. [CrossRef]
17. Claeson, A.-S.; Sandström, M.; Sunesson, A.-L. Volatile organic compounds (VOCs) emitted from materials collected from buildings affected by microorganisms. *J. Environ. Monit.* **2007**, *9*, 240–245. [CrossRef] [PubMed]
18. Dalilla, C.R.; Mauricio, B.F.; Simone, C.B.; Silvia, B.; Sergio, F.P. Antimicrobial activity of volatile organic compounds and their effect on lipid peroxidation and electrolyte loss in *Colletotrichum gloeosporioides* and *Colletotrichum acutatum* mycelia. *Afr. J. Microbiol. Res.* **2015**, *9*, 1527–1535. [CrossRef]
19. Šimpraga, M.; Ghimire, R.P.; Van Der Straeten, D.; Blande, J.D.; Kasurinen, A.; Sorvari, J.; Holopainen, T.; Adriaenssens, S.; Holopainen, J.K.; Kivimäenpää, M. Unravelling the functions of biogenic volatiles in boreal and temperate forest ecosystems. *Eur. J. For. Res.* **2019**, *138*, 763–787. [CrossRef]
20. Wilkens, W.F.; Hartman, J.D. An Electronic Analog for the Olfactory Processes. *J. Food Sci.* **1964**, *29*, 372–378. [CrossRef]